P R I N C I P L E S   O F

EXTRACTIVE
METALLURGY

PRINCIPLES OF

EXTRACTIVE METALLURGY

SECOND EDITION

Terkel Rosenqvist

Professor Emeritus of Extractive Metallurgy
The Norwegian University of Technology and Science
(NTNU)

꿰 tapir academic press

Printed by Tapir Uttrykk
Binding: Grafisk Produksjonsservice AS

Tapir Academic Press
N–7005 TRONDHEIM

Tel.:   + 47 73 59 32 10
Fax: + 47 73 59 32 04
E-mail: forlag@tapir.no
www.tapir.no/forlag

CONTENTS

Appendixes 453

PREFACE TO THE SECOND EDITION

This second edition of *Principles of Extractive Metallurgy* has the same objective as the first: to give a broad review of the general principles which apply in the extraction of metals from their ores, and the same chapter sequence is used. As it is assumed that the reader is familiar with elementary chemical thermodynamics Chaps. 2 to 4 have been somewhat condensed, and greater emphasis is put on the *application* of thermodynamics to metallurgical problems. The one major change has been in Chap. 6, which has been given a new "profile," one of reactor design. The remaining chapters have been somewhat expanded in order to include some more recent processes and developments, and some obsolete material is discarded. Finally the SI system has been introduced consistently, with revised thermodynamic data in joules rather than in calories, and some minor changes have been made in the choice of symbols.

Like in the first edition the general principles which apply to all metallurgical processes are presented in Chaps. 1 to 6, and are followed by the more applied and specialized chapters. This is done in order to systematize the material. It does not necessarily mean that the same sequence should be followed in classroom teaching or self-study. Thus an instructor may choose to start with some application, for example, the iron blast furnace process (Chap. 9) and, as the need appears, to proceed to the necessary sections on heat balances, chemical equilibria, kinetics, fluid dynamics, and heat and mass transfer, as these are treated in Chaps. 2 to 6.

The author is grateful to his colleagues, Dr. Knut L. Sandvik and Dr. Gunnar Thorsen for valuable advice on Chaps. 7 and 15 respectively. As for the first edition, the typing as well as the necessary encouragement and patience during the work was expertly taken care of by Linda Tidosaar. Thanks are due to her.

In this reprinted edition of *Principles of Exctractive Metallurgy* misprints have been corrected and minor corrections are made in the text.

Trondheim, January 2004.
Terkel Rosenqvist

PREFACE TO THE FIRST EDITION

In the past most textbooks on extractive metallurgy have been of a rather descriptive nature. Various processes for the production of various metals have been listed and described, and the emphasis has been on the technology rather than on the basic principles involved. The chemistry of the processes has often been limited to a list of chemical reactions which are believed to have taken place. As the amount of industrial experience increases it becomes more and more difficult to give a comprehensive review of all possible and impossible metallurgical processes. Also, by limiting the teaching to the technology of yesterday the students will be less prepared to develop the technology of tomorrow.

By concentrating on the fundamental principles of metal extraction the author hopes to overcome some of these obstacles. The emphasis of the present text is not on *how* the various processes are performed, but rather on *what* is actually happening and *why* the processes are carried out in a certain way. Such an understanding may show what possibilities exist with respect to future development.

The teaching and learning of the principles of extractive metallurgy is connected with certain inherent difficulties. A metallurgical process is first of all governed by chemical reactions. The extractive metallurgist, therefore, should be well schooled in chemistry, in particular in chemical thermodynamics and reaction kinetics. Second, the design of a metallurgical reactor is based on the application of engineering principles of heat and mass balance, and of heat and mass flow. Finally, the extractive metallurgist should know something about existing techniques, and he should be trained to use his imagination to improve these techniques.

Present university courses give a certain, but often inadequate, background in chemistry and chemical thermodynamics. Thermodynamics courses are often very formal, and have a tendency to become sterile. In the present text it is, therefore, felt necessary to give a review of thermodynamics, based on first principles and with emphasis on its application to metallurgy. Also, present courses in general engineering are not always geared to the need of the extractive metallurgist. Subjects such as heat transfer and fluid flow are therefore discussed in the present text. The first six chapters, therefore, represent a review of those fundamental principles: Thermodynamics, kinetics, and engineering principles, which are of importance to extractive metallurgy. These chapters may be used as a separate text, or they may be omitted entirely by those readers who already have an adequate background in these fields.

The major part of the text is devoted to the various metallurgical unit processes: roasting, reduction, smelting, electrolysis, etc., and is illustrated by existing techniques for the extraction of the most common metals. The emphasis is mainly on the chemistry and dynamics of the processes, and with only brief reference to reactor design. In the description of metallurgical reactors the principal concern has been to show how these function and not how they actually look. A more detailed discussion of reactor design is considered outside the scope of the text.

Metal extraction is in the end always decided by economic considerations. The most elegant use of thermodynamics and reactor design is of little value if the process is uneconomical or if there is no market for the product. A discussion on plant economics is considered outside the scope of the present text, and only incidental reference is made to the economics of the processes discussed. Both the operating metallurgist and the person engaged in industrial research are well advised, however, always to keep their eyes open to the economic consequences of their activities.

With the exception of certain key publications it has not been possible to include references to all information given. A great deal is part of the common heritage of the metallurgical profession, and the author has drawn information also from other textbooks in the field. As a tribute to these books and as a guide for the reader who wants further information, each chapter includes a bibliography of recommended reading. Also each chapter includes a list of problems ranging from simple calculations to problems which require imagination and ability for creative synthesis.

The appendixes include tables and graphs of thermodynamic quantities for most substances of metallurgical importance, and may be used to calculate heat (enthalpy) balances and chemical equilibrium constants.

The text is intended to give a broad review of metal extraction based on first principles, and is primarily aimed at the junior or senior undergraduate student. It may be supplemented by more specialized texts on subjects of special interest to the course or to the individual student. Some parts, as for example Chap. 14 and parts of other chapters may also be used at the graduate student level. Furthermore, it is hoped that the text may give some viewpoints of value to the person engaged in metallurgical research and development.

The philosophy of the text has been greatly influenced by the author's work at American universities and by his discussions with American colleagues. Most of the writing was done during the years 1967–68 when the author served as a

UNESCO expert at the Middle East Technical University in Ankara, Turkey. The author is indebted to a number of friends and colleagues for encouraging advice and constructive criticism: to Professor Olaf G. Paasche of the Oregon State University and Dr. Jomar Thonstad who both have read the entire manuscript: to Professors Marcus Digre, Håkon Flood, and M. Brostrup Müller who have given valuable advice on Chaps. 7, 11, and 9 and 13, respectively. The appendixes have been compiled with the help of former students: Torgeir Alvs-åker, Per-T. Torgersen, and Georg B. Jensen. These people are not responsible for the content of the text, however. That responsibility rests entirely with the author, who welcomes further comments and criticism from readers. Finally the author wishes to thank his secretary, Miss Linda Tidosaar, who, in addition to typing most of the manuscript, also animated him to complete it.

Terkel Rosenqvist

Throughout the world various sets of symbols are used in various fields of science and engineering. Unfortunately, no consistent and unambiguous set has been agreed on. In this text an attempt has been made to use symbols that are most generally accepted. This means that one and the same symbol may be used to denote different quantities. As this is done in definitely different connections and the symbols are explained in the text, this should not represent any problem. In a few cases it is found necessary to use different symbols for the same quantity (in different chapters), this being done in order to avoid confusion.

A = first component
A = Helmholtz energy, area
a = activity
B = second component
C = third component
C = concentration, molar heat capacity $[J/(K \cdot mol)]$, number of components
c = specific heat $[J/(K \cdot g)]$
D = diffusion coefficient
d = diameter
E = age distribution function, electromotive force
F = number of degrees of freedom, radiation efficiency factor, Faraday's number
f = fugacity, henrian activity coefficient
G = Gibbs energy, mass velocity
g = gravitational acceleration
H = enthalpy, height

H_0 = integration constant

h = heat transfer coefficient, Planck's constant

I = integration constant, intensity, electric current

J = flow per unit area

K = equilibrium constant, proportionality constant

k = rate constant, heat conductivity, mass transfer coefficient, Boltzmann's constant, size modulus

L = length

M = molecular weight, mass

m = mass, mass per unit time, molality

N = mole fraction, Avogadro's number = 6.02×10^{23}

n = number of moles, charge of ions

P = pressure, pressure in SI units

p = relative pressure, pressure in atmospheres

q = heat flow

R = gas constant = 0.082 liter-atm/$(K \cdot mol)$
$\quad$ = 8.3144 J/$(K \cdot mol)$

r = radius, reflectivity

S = entropy, surface area

s = distance

T = absolute temperature in K

t = time, temperature in °C

U = internal energy, overall heat transfer coefficient

V = volume, voltage

v = linear velocity, bond energy

W = probability, width

w = work done on surroundings, exchange bond energy

X = any state function, fraction reacted

Z = height above base, coordination number

Greek

α = absorptivity

β = thermal expansion coefficient, transmissivity

γ = activity coefficient (raoultian and for aqueous electrolytes)

δ = thickness of surface layer

ε = emissivity, void fraction

η = overvoltage

θ = relative time (dimensionless)

κ = electrical conductivity

μ = viscosity

ρ = density
σ = surface or interface energy
ϕ = shape factor
ω = angular velocity

Subscripts

M = molar quantity (often omitted)
C = by concentration
m = mean value, melting
P = constant pressure
p = particle
s = sublimation, at surface, solid
T = constant temperature
tot = total
trf = transformation
V = constant volume
v = vaporization, boiling
vol = by volume
__ (underscore) = dissolved

Superscripts

$°$ = standard state
$*$ = at equilibrium, activated compound
— (overbar) = partial quantity, mean value

Some less used symbols are explained in the text.

ONE

INTRODUCTION

The art of extracting metals from their ores dates back to the dawn of human civilization. The first metals used by man were gold and copper, which were found in nature in metallic or *native* form. Around 4000 B.C. man learned to produce copper and bronze by the smelting of copper and tin ores in a charcoal fire. Throughout the history of mankind the processes of extractive metallurgy were developed further by trial and error. The knowledge of the smelter or the blacksmith passed on from father to son. New developments were sometimes the result of an ingenious imagination, but perhaps more frequently a result of accidents. A visitor to a modern metallurgical plant will be struck by the large number of complex operations. Particularly in the field of nonferrous metallurgy the operations vary considerably from one metal to another and even between different plants producing the same metal. In this text we shall see how the many different metallurgical processes may be understood as the result of a relatively small number of fundamental principles. But first a few words about the ores.

1-1 ORES

According to traditional definition an ore is a rock which can be mined economically to serve as raw material for the production of metals. The economic aspect is an important point, which makes the limit between what is ore and what is worthless rock or gravel depend on the existing state of technology and by the market price for the metal in question. A hundred years ago a copper ore had to contain at least 5% of copper in order to be accepted as an ore. Today, due to improved technology, rocks with as little as 0.5% copper may be mined and processed economically in spite of the market price for copper, relative to other commodities, in the meantime having decreased. In the future we may expect other rocks which today are considered worthless, such as the aluminum silicates or the deep-sea nodules, to becomes ores for aluminum, copper, nickel, and cobalt.

The question of what is called ore and what is rock depends also on alternative uses of the rock. Traditionally rocks like limestone, dolomite, quartz, and rock salt are not called ores even though, in addition to other uses, they may also serve as raw materials for the production of metals like calcium, magnesium, silicon, and sodium. On the other hand bauxite should definitely be called an ore as it is mined mainly to serve as a raw material for the production of aluminum. Seawater, which contains 0.13% Mg and may be used as a raw material for the production of magnesium, can hardly be called an ore, other uses being much more dominating.

Ore Types

A great many metals are present in nature as *oxides*. Most important are the *iron* ores, which usually contain the mineral hematite, Fe_2O_3, but which also may contain magnetite, Fe_3O_4. Iron hydroxides and carbonates, such as limonite and siderite, may also be classified as oxide ores as the bound water and carbon dioxide are easily expelled during the iron-making processes. Other heavy metals which are mainly produced from oxide ores are *manganese*, *chromium, titanium, tungsten, uranium,* and *tin.* Aluminum is, as already mentioned, almost exclusively derived from bauxite, which is a hydroxide, whereas some magnesium is derived from dolomite, which is a carbonate. Silicon is produced from quartz. Some copper and nickel is derived from oxide ores, which have been formed by weathering of sulfide deposits, but in general these metals are derived mainly from sulfide ores.

The *sulfide ores* represent a large and important group. Most important are the *copper* ores, which often consist of mixed sulfides of copper and iron. Other metals which are mainly derived from sulfides are *nickel, zinc, lead, mercury,* and *molybdenum.* Even though sulfide ores often contain iron, e.g., as pyrite, FeS_2, they serve only in exceptional cases as sources for iron. Sulfide ores often contain metalloids like *arsenic, antimony, selenium,* and *tellurium.* Of greater economic importance is the fact that sulfide ores often contain smaller or larger amounts of *silver, gold,* and the *platinum* metals. These noble metals may in some cases also occur in nature in the metallic state, i.e., as *native* ores.

The last imported group of ores are the *halides,* such as rock salt for *sodium,* and magnesium chloride brine which, together with seawater, serve as sources for *magnesium.*

Most crystalline ores contain a number of different *minerals*, i.e., crystals of given composition. Thus a typical sulfide ore may contain *chalcopyrite*, $CuFeS_2$, together with *sphalerite*, (Zn,Fe)S, and *galena*, PbS, as well as pyrite and silicates or other gangue minerals. Such ores are usually first treated to separate the different minerals, to give separate concentrates of copper, zinc, and lead, as well as to discharge the worthless gangue minerals. This treatment, which is called ore-dressing, will be discussed further in Chap. 7, whereas in the present chapter a brief and general discussion on metal-producing processes will be given.

1-2 FLOW SHEETS

The process or combination of processes which is used in a given plant is illustrated conveniently by means of a flow sheet. An example of such a flow sheet is shown in Fig. 1-1 which illustrates the production of copper from a low-grade copper sulfide ore. If flow sheets for different copper smelters are examined they usually show some variations. This may be the result of differences in the composition of the ore, differences in the local supply of energy such as fuel and electric power, differences in the demand for final products, and differences in the size of the plant. Flow sheets for the production of steel and zinc are shown in Figs. 1-2 and 1-3. These differ again considerably from the copper flow sheet, and it would seem that every metal and every locality required its own extraction process.

Unit Processes and Unit Operations

If the flow sheets are examined more closely it is seen that they consist of combinations of steps or operations, and that some of the same steps or operations are found in the production of different metals and in different locations. Thus for the copper flow sheet (Fig. 1-1) we have first some steps which involve crushing, grinding, and flotation. The purpose of these steps, commonly called ore-dressing, is to separate the ore mineral from the gangue or from other ore minerals. The next step may be a roasting operation where some of the sulfur in the ore is removed. This is followed by a smelting operation where copper is enriched in a sulfide melt (matte), whereas iron and gangue minerals are discarded in a slag. The matte is then blown in a converter where the remaining iron is slagged and copper sulfide is oxidized to a crude metal (blister copper). This is refined further in a refining furnace and finally refined by electrolysis to produce the commercial product. Some deviations from this flow sheet are found in different plants. In some plants the roasting step is omitted. In other cases the electrolytic refining step is omitted, whereas sometimes the crude blister copper is cast directly into anodes for electrolytic refining.

If we look at the flow sheets for the production of steel, zinc, or lead we find some of the same steps as in copper production: roasting is used also in the production of zinc and lead, a smelting operation is used for the production of most metals, the blowing in a converter is used in the steel industry, etc. Thus a large variety of flow sheets is possible by combinations of a relatively small number of different single steps. In analogy with the usage in chemical engineering we call these steps "Unit Processes" or "Unit Operations." Unfortunately there is no generally accepted definition of the two terms. By "Unit Operation" is usually understood a step which is characterized by certain physical features. Thus crushing, screening, and transportation are typical unit operations. "Unit

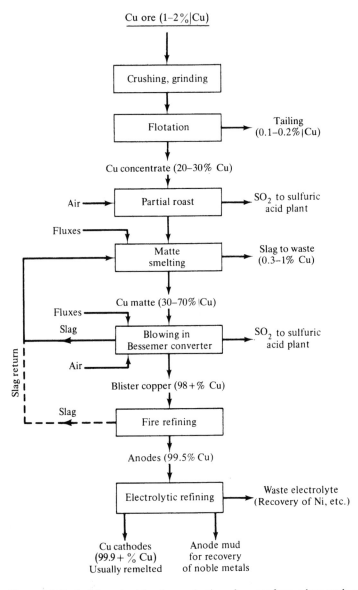

Figure 1-1 Typical flow sheet for the production of copper from a low-grade copper sulfide ore.

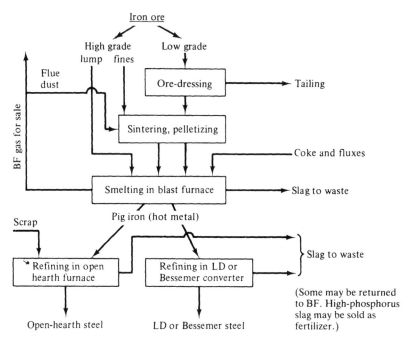

Figure 1-2 Flow sheet for iron- and steel-making. Left: open hearth; right: pneumatic steel-making.

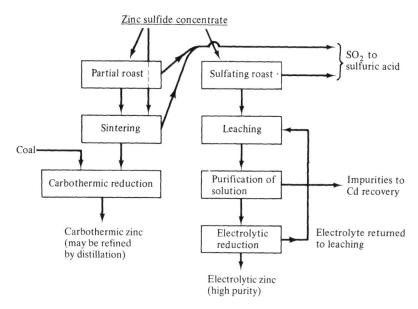

Figure 1-3 Flow sheet for production of zinc from zinc sulfide concentrate. Left: carbothermic reduction; right: leaching and electrolysis.

Processes" on the other hand means steps characterized by a certain chemical reaction. Roasting of a sulfide ore or gas reduction of an oxide ore could be called unit processes. It is evident that there will be a number of borderline cases, and also cases where different steps have characteristic physical features in common but differ with respect to their chemistry and vice versa. Thus a blast furnace may be used for reduction of oxides, but may also be used for oxidation of sulfides. A retort may be used for the coking of coal as well as for the reduction of zinc oxide, and an electrolytic cell may be used for electrolytic reduction as well as for electrolytic refining.

Even with the vague definition of unit operations and unit processes given above, classification is more difficult in extractive metallurgy than in chemical engineering. Three major types of classification could be considered:

1. According to phases involved
   a. Gas–solid. Examples: Roasting, gas reduction.
   b. Gas–liquid. Examples: Bessemer blowing, distillation.
   c. Liquid–liquid. Examples: Slag-metal reactions.
   d. Solid–liquid. Examples: Liquation, leaching, and precipitation.
2. According to equipment
   a. Fixed bed. Examples: Sintering, percolation leaching.
   b. Fluidized bed. Examples: Fluidized roasting and reduction.
   c. Shaft furnace. Examples: Iron blast furnace, lime kiln.
   d. Rotary kiln. Examples: Drying and calcination.
   e. Retort. Examples: Coke-oven, carbothermic zinc production.
   f. Reverberatory furnace. Examples: Matte smelting, open-hearth steel-making.
   g. Electric furnace. Examples: Electric matte smelting and steel-making.
   h. Cell for fused salt electrolysis. Examples: Production and refining of aluminum.
   i. Cell for aqueous electrolysis. Examples: Electrolytic reduction and refining.
3. According to chemical reactions
   a. Oxidation. Examples: Roasting, sintering, Bessemer blowing.
   b. Reduction. Examples: Iron-making, carbothermic zinc production.
   c. Slag-metal reactions. Examples: Steel-making, matte smelting.
   d. Chlorination. Examples: Production of titanium tetrachloride.
   e. Electrolytic reduction. Examples: Electrolytic zinc and aluminum production.
   f. Electrolytic refining. Examples: Refining of copper and nickel.

Further processes could be added to the above list.

It appears that whatever system of classification is used, it will tend to emphasize certain similarities, whereas it will separate others. In the present text the classification will mainly be according to the type of chemical reactions. It is almost impossible to follow this classification consistently, however, and certain digressions will be made into other systems of classification.

Before the different unit processes are discussed, it will be necessary for the reader to be familiar with some fundamental chemical and physical principles. A unit process has as its main purpose to carry out a given chemical reaction. It is therefore necessary to know how to make a reaction run in the desired direction, and how to avoid unwanted side reactions. This is first of all a question of chemical thermodynamics and chemical kinetics. Furthermore, a certain reaction temperature is needed. This requires a knowledge of heats of reactions and of the principles of heat transfer. From the heats of reactions also the theoretical energy requirements may be calculated. Finally, the process has as its purpose to produce the wanted product preferably as fast as possible. An evaluation of the production capacity and of the factors which determine the size and design of the reactors is obtained partly from chemical kinetics and partly from heat transfer and fluid flow considerations. The first six chapters in this text will, therefore, be devoted to the fundamental chemical and physical principles which are common to all unit processes.

In the discussion of the individual unit processes, which will follow in later chapters, emphasis will be put on the prevailing chemical equilibria and heat balances. Furthermore, examples will be given of industrial applications, illustrated by the production of important metals: iron and steel, copper, lead, zinc, light metals, and others.

1-3 FEATURES OF METAL EXTRACTION

The final combination of unit processes into a complete flow sheet is to a large extent dependent on economic considerations: the cost of raw materials and market conditions. Very often the same metal can be produced by several different methods. This was shown in Figs. 1-2 and 1-3 for iron (steel) and for zinc. Zinc may be produced from zinc sulfide ores either by a carbothermic process (roasting, sintering, and reduction) or by hydrometallurgy (roasting, leaching, and electrolysis). The choice will depend first of all on the cost of fuel relative to electricity, second on the market conditions. Carbothermic zinc contains impurities such as lead, arsenic, etc., and is used for making brasses and also for galvanizing purposes. In these cases the impurities are not particularly harmful and may be tolerated. Electrolytic zinc is preferably used for the production of special zinc alloys, where a low impurity content is essential. Carbothermic zinc could meet the requirements of the alloy industry if it were purified by redistillation, but that would add to the fuel bill.

The above two alternatives illustrate another feature of extractive metallurgy: most metals are produced from *impure* ores. Most commonly these are reduced to give an impure metal: pig iron, blister copper, lead bullion, or crude zinc, etc. The impure metal may then be refined further to remove impurities and give a product of the required purity. Alternatively the ores could be treated to give an essentially pure chemical compound of the metal, and the compound may be reduced to give pure metal directly. The latter procedure, which is illustrated in hydrometallurgical zinc production, is more commonly used for the more reactive metals: titanium, niobium, and the light metals. These metals are more reactive than the impurities, which means that in most refining processes the metal will be attacked before the impurities. The only way to obtain a pure metal would, in that case, be to do the purification before the reduction.

If we examine various flow sheets, we will observe certain other common features. One of these is the circulation or return of intermediate products, slags, etc., to previous steps. This is done in order to recover, as completely as possible, the valuable constituents. Thus in the flow sheet for copper (Fig. 1-1), the slags from the Bessemer converter and from the fire-refining furnace are returned to the matte smelting stage. Here the copper content of the slags is recovered, and the only slag which is discarded is that from the matte smelting stage. This slag contains almost all the impurities and as little as possible of the valuable copper.

Recirculation cannot be used in all cases, however. In iron- and steel-making (Fig. 1-2) one could consider the possibility of returning the steel-making slag to the blast furnace in order to recover the iron and manganese contents. This would be all right if the ore had been free of phosphorus. If phosphorus is present in the ore, however, it will be reduced in the blast furnace and enter the pig iron. In the steel-making processes phosphorus is oxidized and transferred to the slag. If this slag were returned to the blast furnace, phosphorus would accumulate in a closed circuit. Therefore, for every element introduced into the flow sheet, there has to be an outlet.

The return of refining slags to previous steps is an example of the so-called *counter-current principle*, which is used extensively in extractive metallurgy. Thus if the wanted metal preferentially dissolves in one phase, and the impurities preferentially in another phase, the best separation is obtained by letting the two phases flow counter-current to each other. This gives the highest possible concentration of the wanted product in the product phase and the lowest possible concentration in the unwanted or discarded phase, i.e., the highest recovery. Thus in the hydrometallurgical leaching of certain ores the raw ore is first brought into contact with a solution of relatively high content of the valuable element. The partly leached ore then passes through stages where it is treated with solutions with successively lower contents of the valuable element, these solutions being enriched, and the ore being washed out before it is discarded. Another example of the counter-current system is in heat exchange, which will be discussed in more detail in Sec. 6-7. When heat is transferred from one phase to another, as in the heating of a solid burden by means of hot combustion gases, the solids and the gases are made to flow counter-current to each other, the cold solids first being brought into contact with the partly cooled gases and then gradually into contact with gases of increasing temperature.

In many flow sheets we observe the recovery of *by-products*. Thus, in Fig. 1-1 sulfur is recovered in the roasting and Bessemer steps, gold and silver from the anode mud, and nickel from the spent electrolyte. Most other ores also contain, in addition to the major elements, smaller or larger quantities of by-product elements. These are enriched in certain phases at certain stages in the treatment: volatile metals in the flue dust, metals with high affinity for oxygen in the refining slags, noble metals in anode mud, etc. Also, special processes may be developed to recover valuable by-products which do not separate out during the normal refining processes. Thus small amounts of silver will usually do no harm in lead, but because of its value special methods have been developed for desilverization of lead.

If the plant produces more than one main product, we may talk about *integrated processes*. In an integrated plant the by-product from one part may serve as raw material for another. Thus, in a plant which treats both zinc and lead concentrates, the ashes from the zinc plant usually contain some lead, and are passed on to the lead smelter, whereas the slag from the lead smelter may contain enough zinc to justify its recovery and utilization in the zinc plant. The efficient utilization of by-products in an integrated plant contributes greatly to the economy of such a plant.

The recovery, both of the main product and of the by-products, will always depend on economic considerations. Even though it may be technically possible to get close to 100 percent recovery this is not necessarily the most economic solution. Sometimes it is better to let some of the valuable elements go to the slag dumps. When extraction techniques have improved or when the market conditions have changed, these slag dumps may once again be "mined" to recover with profit those elements which earlier generations could only recover with loss.

1-4 UNITS OF MEASURE

The remaining part of this chapter will be devoted to some general considerations of importance for the quantitative evaluation of metallurgical processes. One difference between the art and the science of metallurgy is that the latter tries to express the phenomena in a form which may be subject to quantitative evaluation: measurement and calculation. This requires a well-established system of measuring units. It is true that the choice of units in numerical calculations is immaterial as long as they are consistent. Certain unit systems are more convenient than others, however, and in recent years the SI system (Système International d'Unités) has been adopted almost exclusively in the industrialized countries and will be used in the present text.

Units for Length and Mass

The basic units in the SI system are the meter (m) and the kilogram (kg). For larger masses the metric ton (1000 kg), commonly written tonne, for smaller masses the gram (10^{-3} kg) are used.

In chemistry the mass is sometimes expressed by the number of moles of the various constituents. One mole is equal to as many grams of the substance as is given by its atomic or molecular weight. Thus for iron, one mole is equal to 55.84 grams, and one kilogram of iron is equal to 17.9 moles. Sometimes a substance may be described by different formulas. Thus oxygen may be described by the symbol O, with an atomic weight of 16, or by O_2 with a molecular weight of 32. One kilogram of oxygen is then equal to 62.5 moles of O or 31.25 moles of O_2.

Units for Force, Pressure, and Energy

The unit for force is the newton (N), which is the force which gives a mass of one kilogram an acceleration of one meter per second per second. As the acceleration of gravity on Earth is about 9.81 m/s^2 this means that the *weight* of a one kilogram mass on Earth is about 9.81 N, whereas it is considerably less on the Moon and considerably more on Jupiter.

The consistent unit for pressure would be the pascal ($= 1 \ N/m^2$). Compared to atmospheric pressure this is a very small unit, however. More common is, therefore, the kilopascal ($= 10^3$ Pa) or the bar ($= 10^5$ Pa). Atmospheric pressure (1 atm) corresponds to 1.013 bar. For chemical thermodynamic calculations a special problem arises as all available thermodynamic data are based on atmospheric pressure as the standard state. All derived pressures will, therefore, also appear in atmospheres. Actually, however, the derived pressures may be considered dimensionless, as they are given *relative* to the chosen standard state which is 1.013×10^5 Pa, and will in this text be denoted by the symbol p.

For energy, heat, and work, the consistent unit is the joule (J) which is equal to one newton-meter, and also equal to one watt-second. This unit will be used exclusively in this text, even though most available thermodynamic data are published elsewhere in the more traditional unit, the calorie, which is equal to 4.184 joules.[1] Other units for energy which are still in use are the kilowatt-hour and the liter-atmosphere, as well as the traditional British unit (Btu) and foot-pound, and in Table 1-1 conversion factors for the various units for energy are summarized.

Units for Temperature

We distinguish between *temperature scale* and *temperature units*. The temperature scale expresses how a temperature difference at one temperature level relates to temperature differences at other temperature levels. Thus we need to establish a temperature scale in order to compare the temperature difference

---

[1] This number applies for the thermochemical calorie relative to the absolute joule. See footnote to Table 1-1.

Table 1-1 Conversion factors for units of energy (heat and work)

Units	joule (abs) = W·s = N·m	calorie (thermochem.)	kWh	liter-atm	kg (force)·m	ft·lb (force)	Btu
1 joule (abs) = 1 W·s = 1 N·m =	1	0.2390	2.778×10^{-7}	9.869×10^{-3}	0·1020	0.7376	9.478×10^{-4}
1 calorie (thermochem.) =	4.184	1	1.162×10^{-6}	4.131×10^{-2}	0.4269	3.086	3.966×10^{-3}
1 kWh =	3.600×10^{6}	8.604×10^{5}	1	3.553×10^{4}	3.671×10^{5}	2.655×10^{6}	3412
1 liter-atm =	101.33	24.22	2.815×10^{-5}	1	10.33	74.74	9.607×10^{-2}
1 kg (force)·m =	9.807	2.344	2.724×10^{-6}	9.678×10^{-2}	1	7.233	9.295×10^{-3}
1 ft·lb (force) =	1.356	0.3240	3.766×10^{-7}	1.338×10^{-2}	0.1383	1	1.285×10^{-3}
1 Btu =	1055	252.2	2.931×10^{-4}	10.41	107.6	778.2	1

Note: Unfortunately two different joules (the absolute and the international joule) as well as two different calories (the thermochemical and the International Table calorie) have been defined. The differences between these two sets are less than 0.1 percent and do not significantly affect the above table, which is based on the former set of units.

between, say, the melting and boiling points of water with the temperature difference between the melting and boiling points of, say, zinc. The *thermodynamic temperature scale*, which we use, is derived from the equation of state for a perfect gas, that is, the thermodynamic (absolute) temperature is defined by the relation $T = PV_M/R$. Here R is a constant, whereas PV_M represents the limiting value which the product of pressure and molar volume of any gas approaches when the pressure approaches zero. The *temperature unit* depends on the numerical value for the constant R. For the *Kelvin unit* this constant is chosen to make the difference between the normal melting and boiling points of water equal to 100 units (degrees). This gives R the numerical value of 0.082 liter-atm/(K·mol) = 8.314 J/(K·mol) and the temperature for the ice point equal to 273.16 K.

In practical work the temperature is often expressed in the *Celsius scale*. This has the same unit as the Kelvin scale, but the zero point is chosen at the melting point of water. This fixes the boiling point of water to 100°C and, in general, °C = K − 273.16.

Gases and Gas Compositions

Whereas for solids and liquids the quantities and the concentrations are usually given by masses and weight percentages, a different procedure is used for gases. Most gases will under ordinary conditions obey the perfect gas law, that is, for a given temperature and pressure the gas volume is proportional to the number of moles. The quantity of a gas may then be expressed by its volume at some standardized temperature and pressure. Most common is 0°C and one atmosphere pressure, where one cubic meter contains 1000/22.4 = 44.6 moles of gas, and is denoted one normal cubic meter (Nm³) or m³ (STP). By means of the perfect gas law the volume at any other temperature and pressure may be calculated.

The concentration of a component in a gas mixture is usually expressed by its volume percentage which is equal to its mole percentage. Thus (vol % x) = $100·n_x/n_{tot}$ where n_x is the number of moles of component x and n_{tot} is the total number of moles in the gas mixture. Also, the relative partial pressure of a component in a gas mixture is given by the relation $p_x = (n_x/n_{tot})p_{tot}$, where p_{tot} is the relative total gas pressure. The two last relations are independent of the temperature of the gas.

It should be pointed out that the above relations are strictly valid only for gases which obey the perfect gas law, but they are usually rather closely obeyed at and above room temperature and at pressures up to a few atmospheres. For higher pressures and extremely low temperatures special equations will have to be applied.

Combustion gases often contain water vapor. The composition may then be expressed either on *wet* or on *dry* basis. The wet composition includes the water vapor, whereas in the dry composition the water vapor is disregarded, the amounts of the remaining components adding up to 100 percent.

1-5 STOICHIOMETRY

In chemical and metallurgical calculations we often need to calculate the relative amounts of reactants that are needed for a given reaction, as well as the amounts of products. Such calculations, which are based on the chemical reaction equations and the molar weights of the various compounds, are called *stoichiometry*. The questions that we want to answer by such calculations are:

1. What are the relative quantities of reactants to go into a certain process?
2. What are the quantities of products obtained?
3. What is the composition of these products?

Stoichiometric calculations may be carried out in many different ways and each person will, by experience, find his own method. The following procedure is therefore not the only way to do such calculations, but is a way which has proved to be most useful and gives few possibilities for mistakes.

As a typical case we shall consider the combustion of methane with air to give carbon dioxide and water vapor according to the reaction:

$$CH_4(g) + 2O_2 = CO_2 + 2H_2O(g)$$

We may then proceed as follows:

1. Choose as a *basis* a given amount of reactant. This could be one mole, one kilogram, one tonne, or any other quantity.
2. Calculate the number of moles of reactants in the chosen basis.
3. From the reaction equation calculate the number of moles of other reactants and the number of moles of products.
4. Convert the number of moles of reactants and products into suitable units, weights, or volumes.
5. Calculate the composition of the product in suitable units.

As an example we may choose as our basis 1 Nm³ of methane, which we consider to be an ideal gas giving 44.6 mol of CH_4. This reacts with 89.2 mol of O_2 to give 44.6 mol of CO_2 and 89.2 mol of H_2O. However, since air contains 21 mol percent of O_2 and 79 mol percent of N_2, 335.3 moles of nitrogen will enter with the air and appear in the product. The air volume will be $(89.2 + 335.3) 22.4 = 9.5 \times 10^3 Nl = 9.5$ Nm³ per unit. The corresponding combustion gas volume will be $(44.6 + 89.2 + 335.3)22.4 = 10.5 \times 10^3 Nl = 10.5$ Nm³, and the wet composition will be 9.5 mole percent of CO_2, 19 mole percent of $H_2O(g)$ and 71.5 mole percent of N_2. The corresponding *dry* volume and composition will be 8.5 Nm³ with 11.75 mole percent of CO_2 and 88.25 mole percent of N_2. Notice that it is perfectly legitimate to give the wet gas volume in normal cubic meters, even though most of the water would condense at 0°C.

In most combustion processes a surplus of air is used. If for example a 10 percent surplus was used in the above case, an additional 8.9 mol of O_2 and 33.5 mol of N_2 would have to be added to the air as well as to the combustion gas. If, on the other hand, a deficiency of air is used the matter becomes more complicated. For a moderate deficiency the combustion gas will consist of a mixture of CO_2, CO, H_2O, and H_2. If the percentage of each product in the gas were known the corresponding amount of air can be easily calculated. If, on the other hand, only the amount of air per unit of methane were known, only the number of moles of $CO_2 + CO$ and of $H_2O + H_2$ as well as the *ratio* between the number of moles of $CO + H_2$ and of $CO_2 + H_2O$ can be calculated. In order to know the percentage of each gaseous component it will be necessary to know the relevant chemical equilibria, as will be discussed in Chap. 3.

Example 1-1 Limestone with 56 percent CaO and 44 percent CO_2 is calcined in a rotary kiln giving a calcine of pure CaO. For each kilogram of limestone, 0.15 kg of fuel oil, with 85 percent carbon and 15 percent hydrogen is used, and the volume of the combustion air is 2.10 Nm^3. The fuel burns completely to CO_2 and H_2O, which mix with CO_2 expelled from the limestone. *Calculate* the volume (in Nm^3) of the furnace gas, as well as its wet and dry composition. Air is regarded as containing 21 volume percent O_2 and 79 volume percent N_2.

SOLUTION For a basis of 1 kg of limestone the amount of CO_2 in the limestone is 0.44 kg = 10 mol, the amount of carbon in the fuel is 0.1275 kg = 10.63 mol, and the amount of hydrogen in the fuel is 0.0225 kg = 11.17 mol of H_2. The combustion air corresponds to a total of 93.8 moles with 19.7 mol of O_2 and 74.1 mol of N_2. For the burning of the fuel $10.63 + 5.59 = 16.22$ mol of O_2 are needed, leaving 3.48 mol of O_2 unused. The furnace gas, therefore, contains $10 + 10.63 = 20.63$ mol of CO_2, 11.17 mol of H_2O, 3.48 mol of O_2, and 74.1 mol of N_2, or a total of 109.4 moles, corresponding to 2.45 Nm^3 with the wet composition: 18.8 percent CO_2, 10.2 percent $H_2O(g)$, 3.2 percent O_2, and 67.8 percent N_2. The corresponding dry composition will be 21.0 percent CO_2, 3.5 percent O_2, and 75.5 percent N_2.

1-6 MATERIAL BALANCES

We distinguish between material balances and charge calculations. In charge calculations the *necessary* amounts of raw materials are calculated. This requires a knowledge of chemical equilibria and kinetics, heats of reaction and heat losses, as they will be discussed in the following chapters. *Material balances*, on the other hand, is a form of bookkeeping for a process which is already operating. The basic equation in material balances is the law of conservation of matter. For each element x we have:

$$\sum_{i=1}^{i=n} m_i(\% \ x) = \sum_{j=1}^{j=m} m_j(\% \ x) + \Delta m_x$$

input = output + accumulation

Here m_i represents the mass of each type of raw material which enters, and m_j the mass of each product which leaves the process over a given period of time. $(\% \ x)$ is the weight percentage of element x in that raw material or product, and Δm_x is the mass of x which accumulates in the process. Notice that under certain conditions, for example when a furnace lining is being eroded, Δm_x may be negative. If Δm_x is zero, we say that *steady state conditions* exist. In Table 1-2 is illustrated steady state conditions for components x, y, and z in a process with the raw materials A, B, and C and the products L, M, and N.

Very often we do not know all the masses or all the concentrations. During a slag refining process, for example, we may know the weights and composition of all raw materials, as well as the weight and composition of the refined metal, but we do not know the weight and composition of the slag. Assuming that there has been no exchange of material with the atmosphere, or assuming that this exchange is known, the missing data may in that case be calculated from the material balance.

Similarly, if the composition but not the mass of the slag is known, the latter may be calculated. In that case we will have one equation for each element, and we may calculate several values for the slag weight. If all weights and analyses were exact, these slag weights would agree. Usually this will not be the case. We must then choose the values which give the highest accuracy. This means choosing the elements for which the chemical analyses are most reliable, and also the elements which occur in the larger quantities in the slag. Having assessed the slag weight on this basis, the amount and concentration of the remaining elements in the slag may be calculated and compared with the analytical values. If the two sets of values differ by much more than the expected accuracy of the procedure, this may indicate that some element has been accumulated in the furnace, alternatively dissolved from the furnace lining, or lost in some yet unknown reaction product, for example, the flue dust.

The efficiency of an extraction process may conveniently be expressed by two figures: *the recovery* and the *ratio of concentration*. The recovery is the

Table 1-2 Material balance under steady state conditions

Component	Raw materials			Products		
	A	B	C	L	M	N
x	$m_A(\% \ x) + m_B(\% \ x) + m_C(\% \ x) = m_L(\% \ x) + m_M(\% \ x) + m_N(\% \ x)$					
y	$m_A(\% \ y) + m_B(\% \ y) + m_C(\% \ y) = m_L(\% \ y) + m_M(\% \ y) + m_N(\% \ y)$					
z	$m_A(\% \ z) + m_B(\% \ z) + m_C(\% \ z) = m_L(\% \ z) + m_M(\% \ z) + m_N(\% \ z)$					

amount of a given element in a given product relative to the total amount of the same element in the raw materials. In principle we may speak about the recovery in all products (metal, slag, and gases), which necessarily must add up to 100 percent, but usually we are only interested in the recovery of the valuable element in the valuable product. The remaining amounts then represent the loss.

The ratio of concentration is of particular importance in ore-dressing, and is equal to the weight of the raw material (the ore) relative to the weight of the valuable product (the concentrate). For a given process there is a kind of inverse relationship between the recovery and the ratio of concentration, a high recovery giving a low ratio of concentration and vice versa. Of several processes the one is the most efficient which combines a high recovery with a high ratio of concentration.

Example 1-2 A copper ore contains 1.5 percent Cu. After ore-dressing, 4.5 kg of concentrate with 30 percent Cu is obtained from 100 kg of ore. *Calculate* (a) the ratio of concentration, (b) the recovery, and (c) the weight and Cu-percentage of the discarded gangue (tailing).

SOLUTION (a) The ratio of concentration is given by $100/4.5 = 22.2$.
(b) The ore contains a total of $0.015 \times 100 = 1.5$ kg Cu. The concentrate contains $0.30 \times 4.5 = 1.35$ kg Cu. This gives the recovery $= 1.35/1.5 = 0.90$ or 90 percent.
(c) The weight of the tailing is $100 - 4.5 = 95.5$ kg, and its copper content 0.15 kg. This gives its Cu-percentage equal to 0.157.

Example 1-3 A zinc sulfide concentrate contains 50 percent Zn, 13 percent Fe, 32 percent S, and 5 percent SiO_2. When 100 kg is roasted with air, 85 kg of calcine with 54.5 percent Zn, 14.1 percent Fe, 2.7 percent S, and 5.4 percent SiO_2 is obtained. The remaining percentages are assumed to be oxygen. The flue dust is assumed to have the same composition as the calcine. The roast gas contains 8 volume percent SO_2. *Calculate* (a) the weight of the flue dust and its content of the various components, as well as (b) the volume (in Nm^3) and (c) the composition of the roast gas, and (d) the volume of air used. Air may be regarded as containing 21 volume percent of O_2 and 79 volume percent of N_2.

SOLUTION (a) For the concentrate and solid products we get the following material distribution (basis 100 kg of concentrate).

	Concentrate	Calcine	Flue dust
Zn:	50 kg	$0.544 \times 85 = 46.2$ kg	3.8 kg = 54.5 percent
Fe:	13 kg	$0.141 \times 85 = 12.0$ kg	0.99 kg = 14.1 percent
S:	32 kg	$0.027 \times 85 = 2.3$ kg	0.19 kg = 2.7 percent
SiO_2:	5 kg	$0.054 \times 85 = 4.6$ kg	0.38 kg = 5.4 percent
O_2:		balance $= 19.9$ kg	1.64 kg = 23.4 percent
Total	100 kg	85.0 kg	7.00 kg = 100 percent

Here the weight of the flue dust (7.00 kg) is estimated from the difference between the zinc content of the concentrate (50 kg) and that of the calcine (46.2 kg) and from its assumed zinc percentage (54.4), zinc being the major component. From the weight of dust and its assumed composition the contents of the remaining components are calculated.

(b) The sulfur content of the gas is now obtained as the difference between that of the concentrate and that of the calcine plus flue dust, and amounts to $32.0 - 2.5 = 29.5$ kg $= 0.92$ kmol of S, which in the roast gas is present as 0.92 kmol of SO_2. This gives the total amount of roast gas equal to $0.92/0.08 = 11.50$ kmol or 258 Nm^3.

(c) The composition of the roast gas is obtained from the following reasoning: the oxygen in the calcine and flue dust amounts to $19.9 + 1.64 = 21.54$ kg $= 0.67$ kmol of O_2, which together with 0.92 kmol in SO_2 gives a total of 1.59 kmol of O_2, all derived from air. This amount is associated with $(1.59 \times 79)/21 = 6.0$ kmol of N_2, which must also be present in the roast gas. The remaining part of the roast gas, $11.50 - (0.92 + 6.0) = 4.58$ kmol represents a surplus of air. This surplus contains $4.58 \times 0.21 = 0.96$ kmol of O_2 and 3.62 kmol of N_2. Thus we have the total composition of the roast gas:

$$
\begin{array}{lll}
SO_2 & = \ 0.92 \text{ kmol} = & 8.0 \text{ volume percent} \\
N_2 = 6.0 + 3.62 = & 9.62 \text{ kmol} = & 83.7 \text{ volume percent} \\
O_2 & = \ 0.96 \text{ kmol} = & 8.3 \text{ volume percent} \\
\text{Total} & \ 11.50 \text{ kmol} = & 100 \text{ volume percent}
\end{array}
$$

(d) The amount of air used is then:

$$
\begin{array}{lll}
N_2 & = & 9.62 \text{ kmol} \\
O_2 & = & 2.55 \text{ kmol} \\
\text{Total} & & 12.17 \text{ kmol} = 273 \ Nm^3
\end{array}
$$

PROBLEMS

1-1 Express in mole (atomic) percents the composition of a stainless steel with 18 wt % chromium, 8 wt % nickel, and the balance iron.

1-2 Air contains 21 volume percent O_2, 78 volume percent N_2, and 1 volume percent Ar (argon). Express its composition in weight percents.

1-3 Pyrite (FeS_2) is roasted with an excess of air to give Fe_2O_3 and SO_2. The roast gas contains 6.3 percent SO_2, the balance being O_2 and N_2. Calculate per metric ton of pyrite: (a) theoretical air requirement (in Nm^3), (b) actual air requirement, (c) composition of roast gas in volume percents, (d) volume of roast gas at 500°C.

1-4 A blast furnace burden contains 160 kg Fe_2O_3, 54 kg SiO_2, 20 kg Al_2O_3. 100 kg $CaCO_3$, and 78 kg C. For the smelting of the above burden 266 Nm^3 of air is used. The hot metal produced contains 4 percent C and 1 percent Si, the balance being iron, and it is assumed that all the iron in the burden enters the hot metal. The remaining oxides form a slag, whereas CO_2 from the limestone is expelled and mixes with the furnace gas. Carbon in the gas is present partly as CO_2 and partly as CO and there is no free oxygen. (a) Calculate the weight of the hot metal, as well as the weight and composition of the slag; (b) calculate the volume and composition of the furnace gas, all for the above basis.

BIBLIOGRAPHY

Bray, J. L.: "Non Ferrous Production Metallurgy" and "Ferrous Process Metallurgy," John Wiley and Sons Inc., New York, 1947 and 1954.

Dennis, W. H.: "Metallurgy of the Non-Ferrous Metals" and "Metallurgy of the Ferrous Metals," Sir Isaac Pitman and Sons Ltd, London, 1961 and 1963.

Gilchrist, J. D.: "Extraction Metallurgy," 2d ed. Pergamon Press, London, 1980.

Gill, C. B.: "Non-Ferrous Extractive Metallurgy," John Wiley and Sons Inc., New York, 1980.

Habashi, F.: "Principles of Extractive Metallurgy," Vol. I, Gordon and Breach, New York, 1969.

Liddell, D. M.: "Handbook of Non-Ferrous Metallurgy," McGraw-Hill Book Co. Inc., New York, 1945.

Newton, J.: "Extractive Metallurgy," John Wiley and Sons Inc., New York, 1959

Schuhmann, R., Jr.: "Metallurgical Engineering," vol. I, Addison-Wesley Publ. Co. Inc., Reading, Mass., 1952.

TWO

THERMOCHEMISTRY

It is not the purpose of this text to give a comprehensive treatment of thermo-dynamics. This subject is dealt with in a number of excellent texts, some of which are listed in the bibliography. Sometimes it is useful, however, to sum-marize the main thoughts of a science and show how this science is used to solve practical problems. No attempt has been made to derive fully all thermo-dynamic equations. The main emphasis has been put on showing which basic assumptions are implied, what are the main thoughts behind the derivation, and as a consequence what are the limitations in their use. For readers who are already acquainted with thermodynamics the first chapters of this text will have the nature of a refresher course. For others this presentation of thermodynamics in a nutshell should be a first introduction only, and should be followed by a more comprehensive study.

Historically thermodynamics was developed in order to describe the rela-tions between heat and work, for example in a steam engine. The traditional introduction to thermodynamics shows this historical derivation. In chemistry and in metallurgy the application has been to relations between heat, temper-ature, pressure, and chemical compositions. This field is often called chemical thermodynamics. It may be divided roughly into two parts. The first part is concerned primarily with heat effects associated with changes in temperature, pressure, and chemical composition. These relations can be derived from the first law of thermodynamics and are the subject of this chapter. The second part is concerned with the tendency for a reaction to take place, with the equilibrium composition of a system, and with the effect of temperature and pressure upon this equilibrium. These relations are derived from the second law of thermodynamics and are discussed in Chap. 3. Finally, in Chap. 4, thermodynamic properties of melts and solutions, such as molten metals and slags, are discussed.

2-1 THE FIRST LAW OF THERMODYNAMICS

If we consider a *system*, such as shown in Fig. 2-1, this contains a certain quantity of material, as given by n_1, n_2, ..., n_n moles of the various components. It will have a certain temperature T, pressure P, and volume V. Here T, P, and V are called *state properties* or *state functions*. In thermodynamics other functions will be defined, and it will be shown that some of these are also state functions.

For a given composition and quantity of material, i.e., for a *closed system*, there will be a relation between T, P, and V in such a way that if two of these functions are fixed the third function is also fixed. Usually T and P are chosen as the independent variables, in which case V becomes the dependent variable. This relation is called the *equation of state* of the system. For an *ideal* or *perfect* gas the equation of state is given by the well-known equation $PV = nRT$, where n is the number of moles of gas, or $PV_M = RT$, where V_M is the molar volume of the gas. For other systems the equation of state has to be determined experimentally.

The system may exchange *energy* with the surroundings. This may be potential energy, mechanical energy, magnetic energy, electric energy, and heat. Of these the four first types may be classified as work energies and the last type as heat energy. By experiments and experience it has been found that work may be converted into heat and vice versa. Experiments and experience have also shown that from a given amount of one type of energy there is always obtained the same amount of another kind regardless of the process by which the conversion is carried out. This led to the formulation of the first law of thermo-

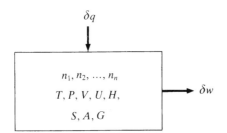

Figure 2.1 A thermodynamic system.

dynamics, which may be expressed in statements like: "Energy cannot be created, nor can it be destroyed," or "The energy of the universe is constant."†

The first law may be expressed mathematically by the following two statements:

1. There exists a property, U, the *internal energy* of a system, which is a state function

$$U = f(T, P, n_1, n_2, \ldots, n_n)$$

Some authors use for the internal energy the symbol E.

If the internal energy of a system is divided by the number of moles, n_{tot}, we get the molar energy content, U_M, which is a function only of T and P (or of T and V_M or P and V_M). In chemical thermodynamics we are usually dealing with molar properties. Therefore the subscript M is often omitted. Except when otherwise stated the thermodynamic functions discussed in the following will be molar functions.

2. The energy content of a system may change as a result of exchange of heat and work with the surroundings. As heat and work are different forms of energy, we may write for an infinitesimal change

$$dU = \delta q - \delta w \qquad (2\text{-}1)$$

By convention q here represents heat absorbed *from* the surroundings, and w work done *on* the surroundings. Other texts define w as work done *on* the system, in which case the right-hand side of Eq. (2-1) would read $\delta q + \delta w$.

Since the energy content is a state function, the symbol dU is used to express an *exact differential*, which can be integrated independent of the path. Heat and work are not state functions and the symbols δq and δw are used to express inexact differentials.

In going from one state of the system, 1, to another state 2, the internal energy changes by an amount

$$\Delta U \equiv U_2 - U_1 = \int_1^2 dU = q - w$$

The identity $\Delta U \equiv U_2 - U_1$ makes it arbitrary which of these two expressions is used. It is customary to limit the use of the Δ sign to isothermal changes and to use the minus sign for processes which involve changes in temperature. Notice that when the Δ sign is used, it always means the property of the final state minus that of the initial state. Because the energy is a state

---

† In nuclear processes matter may be converted into energy and vice versa. This relation is given by the *Einstein formula* $\Delta E = \Delta mc^2$ where ΔE and Δm represent changes in energy and mass, and c is the velocity of light. With this conversion factor the first law of thermodynamics may be restated: "The sum of energy and matter is constant." In metallurgical processes we are normally not concerned with nuclear reactions and the traditional formulation is adequate.

property the change in going from one state to the other is independent of the path. This is in contrast to the total amounts of absorbed heat, q, and work done, w.

In trying to integrate Eq. (2-1) we fall short of an integration constant. This means that we cannot assign an absolute value to the energy content of a system. We can only determine changes in energy associated with changes in temperature, pressure, or volume or changes associated with chemical reactions. These changes can only be determined for closed systems, because any addition or removal of material would involve addition or removal of energy contents for which we have no absolute measure.

Work of Expansion

In Eq. (2-1) the work done on the surroundings may be of different kinds

$$\delta w = \delta w_{vol} + \delta w_{mech} + \delta w_{electr} + \delta w_{magn}, \text{ etc.}$$

Of these the first term, the volume work, requires special attention. When a system expands against an external pressure P, the volume work $\delta w_{vol} = P \, dV$, or integrated,

$$w_{vol} = \int_1^2 P \, dV$$

Since δw_{vol} is an inexact differential it is necessary to specify the path of expansion, for example at constant pressure, temperature, or energy content. If, for example, the outer pressure is that of the atmosphere the work takes place at constant pressure and

$$w_{vol} = P \int_1^2 dV$$

If the expansion is unaccelerated and without friction the difference between inner and outer pressures is infinitesimal. The expansion is then called *reversible*. This is the case for thermal expansion of a solid or liquid or of a gas in a cylinder with a frictionless piston. For such systems we may rewrite Eq. (2-1)

$$dU = \delta q - P \, dV - \delta w' \tag{2-2}$$

Here $\delta w'$ represents all other work with the exception of volume work. If the system is kept at constant volume and also if no other work is done on the surroundings $dU = \delta q$, and $(\Delta U)_V = U_2 - U_1 = q$, that is, equal to the heat absorbed.

2-2 THE ENTHALPY OR HEAT FUNCTION

Very few processes are carried out at constant volume, and the energy function U is therefore of more theoretical than practical interest. Since most chemical and metallurgical processes are carried out at constant pressure it has been found convenient to define a new function, the *enthalpy*

$$H \equiv U + PV \tag{2-3}$$

Since U, P, and V are all state functions the enthalpy must also be a state function. Like the energy U, we do not know its absolute value; it can only be given in terms of differences.

In order to give the relation between changes in enthalpy and absorbed heat we differentiate Eq. (2-3) and combine it with Eq. (2-2)

$$dH = dU + P\,dV + V\,dP = \delta q - P\,dV - \delta w' + P\,dV + V\,dP$$

which gives the important relation

$$dH = \delta q + V\,dP - \delta w' \tag{2-4}$$

At constant pressure and if no work other than reversible volume work is done, $dH = \delta q$, and the change in enthalpy $(\Delta H)_P = H_2 - H_1 = q$, where q is the heat absorbed. Because of this correspondence, the enthalpy is often called the *heat function*.

Variation of U and H with Pressure and Volume

The molar energy and enthalpy of a system are state properties, i.e., they are functions of any two of the three parameters, pressure, volume, and temperature. This dependence can, in a general form, be expressed by total differentials, for example

$$dU = \left(\frac{\partial U}{\partial V}\right)_T dV + \left(\frac{\partial U}{\partial T}\right)_V dT$$

$$dH = \left(\frac{\partial H}{\partial P}\right)_T dP + \left(\frac{\partial H}{\partial T}\right)_P dT$$

The first law of thermodynamics gives no information about the partial differentials $(\partial U/\partial V)_T$ and $(\partial H/\partial P)_T$. On the basis of the second law of thermodynamics, however, it can be shown that

$$\left(\frac{\partial U}{\partial V}\right)_T = T\left(\frac{\partial P}{\partial T}\right)_V - P \tag{2-5}$$

$$\left(\frac{\partial H}{\partial P}\right)_T = V - T\left(\frac{\partial V}{\partial T}\right)_P \tag{2-6}$$

These expressions correlate the variation in energy and enthalpy with the relation between P, V, and T of the system, i.e., with its equation of state. One very special case is that of an ideal gas. Inserting the relation $PV_M = RT$ into Eqs. (2-5) and (2-6) we find that at constant temperature, U is independent of V and H is independent of P. Consequently at constant temperature U and H are also independent of P and V respectively. For real gases the dependence of U

and H on V and P is small, and as the pressure approaches zero the energy and enthalpy approach constant values. Also for condensed substances the effect of pressure is small. This will be illustrated by an example:

For water at room temperature the molar volume $V_M = 0.018$ liter and its thermal expansion coefficient $(\partial V/\partial T)_P(1/V) = 2 \times 10^{-4}$. This gives the variation of the enthalpy with pressure at room temperature (≈ 300 K)

$$\left(\frac{\partial H}{\partial P}\right)_T = 0.018 - 0.001 = 0.017 \frac{\text{liter-atm}}{\text{atm}}$$

As one liter-atmosphere is equal to 101.33 J the increase in molar enthalpy with pressure is 1.72 J/atm. In comparison, an increase in temperature of one degree Celsius results in an increase in molar enthalpy of 75.5 J. Thus a one-atmosphere pressure change has the same effect on the enthalpy as a temperature change of only 0.023°C. In metallurgical processes we are concerned with rather large variations in temperature but small variations in pressure, and it is evident that under these conditions the effect of pressure on the enthalpy is negligible. It becomes significant only for substances with large expansion coefficients caused, for example, by structural changes.

Standard States

Because, strictly speaking, the energy and enthalpy are functions of pressure and volume it is convenient to define standard molar energies and enthalpies corresponding to a chosen standard pressure or *standard state*. These standard functions are denoted $U°$ and $H°$ respectively and are functions only of temperature. As a general rule the standard energy and enthalpy are those at 1 atm pressure ($= 1.013 \times 10^5$ Pa). One exception is for nonideal gases where the standard energy and enthalpy are the limiting values which the functions approach when the pressure approaches zero. For most gases this energy and enthalpy differ insignificantly from those which apply at amospheric pressure. Furthermore, as the effect of pressure is small for all substances the superscript o is often omitted in metallurgical calculations.

Variation of U and H with Temperature

From Eqs. (2-2) and (2-4) it follows that, in the absence of other work, $(U_2 - U_1)_V = q$ and $(H_2 - H_1)_P = q$. If the addition of a heat quantity δq results in an increase in temperature dT the ratio $\delta q/dT$ is called the *heat capacity* of the substance. Therefore, for one mole of material, $(\partial U/\partial T)_V = C_V$ and $(\partial H/\partial T)_P = C_P$, where C_V and C_P are the molar heat capacities at constant volume and pressure respectively.

Thermodynamics gives no information about the numerical value of the heat capacities or on their variation with temperature. This is primarily a subject for experimental determination, even though some basis for calculation exists in statistical thermodynamics and quantum mechanics. In general the

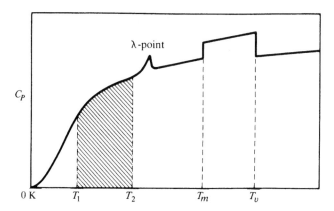

Figure 2-2 Molar heat capacity as function of temperature. The enthalpy increment between T_1 and T_2 is given by the hatched area.

heat capacity will vary with the temperature as shown in Fig. 2-2. In going from one temperature T_1, to another, T_2, and at constant pressure, the enthalpy increases by an amount

$$H_{T_2} - H_{T_1} = \int_{T_1}^{T_2} C_P \, dT$$

as illustrated by the hatched area on Fig. 2-2.

When the substance is heated through a transformation point the enthalpy increases by an amount given by the *enthalpy of transformation* ΔH_{trf}.† If the heat of transformation is absorbed at a constant temperature this means that the specific heat is infinite at that temperature. We call this a transformation of the *first order*. This is the case for transformations like melting and boiling at constant pressure and for most crystallographic transformations. In a few other cases, such as magnetic transformations and order-disorder reactions, the heat of transformation may be absorbed over a finite temperature range. This results in an abnormally high, but finite, heat capacity. Such transformations are called transformations of *second* and *higher* order. Because of the shape of the heat capacity curve (Fig. 2-2) they are also called λ transformations. In this case the increase in enthalpy could be obtained by integration of the heat-capacity curve, but usually it is done by assuming a smooth heat-capacity curve and adding the remaining heat as the heat of transformation. For the total difference in enthalpy between two temperatures and at standard pressure we therefore write

$$H_{T_2}^{\circ} - H_{T_1}^{\circ} = \int_{T_1}^{T_2} C_P \, dT + \sum \Delta H_{trf}$$

† In this text the enthalpy of fusion will be denoted ΔH_m and that of vaporization ΔH_v.

This is called the *enthalpy increment,* and since we shall use it considerably in metallurgical calculations we will save space by using the shorthand expression $H_{T_1}^{T_2}$. The superscript which indicates standard state is omitted since changes in pressure only have an insignificant effect.

In chemical and metallurgical calculations we are particularly interested in the heat required to heat a substance from room temperature to some higher temperature. Therefore we define as our *reference temperature* 25°C = 298 K. Referred to this temperature the enthalpy increment at any temperature can be assigned an absolute value, and is sometimes called the *heat content* of the substance.

Within a certain temperature range, molar heat capacities are sometimes expressed arithmetically by expressions like

$$C_P = a + bT + cT^2 + \cdots + eT^{-2}$$

Such expressions have the advantage that they can be integrated arithmetically to give the enthalpy increment

$$H_{T_1}^{T_2} = a(T_2 - T_1) + \frac{b}{2}(T_2^2 - T_1^2) + \frac{c}{3}(T_2^3 - T_1^3) - e(T_2^{-1} - T_1^{-1})$$

We should always remember that such expressions are approximately valid only within a certain temperature range and may not be used beyond this range without serious loss in accuracy.

Very often the heat capacities are not easily expressed arithmetically. In these cases the enthalpy increment may either be tabulated as function of temperature or it may be shown graphically. Tabulated values are the most accurate and have been published by several authors.[1,2,3] For many purposes a graphical representation is adequate. A collection of such curves is given in App. B.

In Fig. 2-3 are shown the molar (atomic) enthalpy increment curves for sulfur and iron. At the Curie temperature, 760°C, the enthalpy of iron increases continuously, this being a second-order transformation. The α–γ transformation at 910°C and the γ–δ transformation at 1400°C as well as the melting at 1535°C are first-order transformations, and the enthalpy increases discontinuously. The curve also shows that the enthalpies of transformation are much smaller than the enthalpy of fusion.

The curve for sulfur is not accurate enough to show the transformation at 96°C from orthorhombic to monoclinic sulfur but shows the melting point at

[1] K. K. Kelley: "Contributions to the Data on Theoretical Metallurgy XIII. High-Temperature Heat-Content, Heat-Capacity, and Entropy Data for the Elements and Inorganic Compounds," *U.S. Bur. Mines Bull.,* 584, 1960.

[2] J. F. Elliott and M. Gleiser: "Thermochemistry for Steelmaking," vol. I, Addison-Wesley Publ. Co. Inc., Reading, Mass., 1960.

[3] "JANAF Thermochemical Tables," *U.S. Bur. Standards,* Washington D.C., 1965–1971.

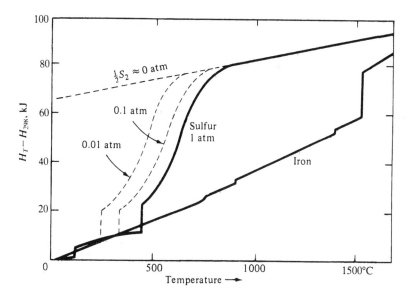

Figure 2-3 Molar enthalpy increment for iron and sulfur as function of temperature.

119°C. The convex shape of the curve for liquid sulfur is a result of structural changes in the liquid. At 445°C sulfur boils at atmospheric pressure and the heat of vaporization causes an enthalpy increase of about 8 kJ. Of particular interest is the very high heat capacity of the vapor between 500 and 800°C. This is caused by the gradual dissociation of S_8 molecules into S_6 and eventually into S_2 molecules. At 1000°C the vapor is essentially S_2, and the heat capacity has a value which is normal for diatomic gases.

Sulfur vapor is one of the few substances where the enthalpy changes considerably with pressure, as illustrated by the enthalpy curves for 1, 0.1, and 0.01 atm. At a given temperature, for example 600°C, the enthalpy increases with decreasing pressure, this being a result of increasing dissociation of the molecules. Expressed in terms of Eq. (2-6) it means that sulfur vapor has a very large thermal expansion coefficient, i.e., deviates considerably from ideal gas behavior. At decreasing pressure the enthalpy approaches a limit which is shown by the upper curve on Fig. 2-3. This is the *standard* enthalpy for one half mole of S_2 gas. Above 1000°C this curve coincides with the curve for the vapor of one atmosphere because at that temperature the vapor is essentially in ideal S_2 gas. At very low pressure and high temperatures further dissociation into monatomic vapor takes place.

At 25°C the standard enthalpy curve for $\frac{1}{2}S_2$ has a value of $+64.9$ kJ relative to orthorhombic sulfur. This is the *standard enthalpy of sublimation* S(rh) $= \frac{1}{2}S_2$(g) at 25°C. In a similar way enthalpies of sublimation to form S_8, S_6, and monatomic sulfur at 25°C can be derived.

2-3 ENTHALPY CHANGES IN CHEMICAL REACTIONS

For a chemical reaction $A \rightarrow B$ the *isothermal* enthalpy of reaction at a temperature T is defined as $\Delta H_T = H_{T(B)} - H_{T(A)}$. Here A represents the sum of all the reactants and B the sum of all the reaction products. If the reaction is carried out at constant pressure and in the absence of other work, the isothermal enthalpy of reaction is equal to the amount of heat absorbed from the surroundings. Thus for endothermic reactions ΔH_T is positive, whereas for exothermic reactions ΔH_T is negative. If both reactants and products are present in their standard states the enthalpy of reaction is denoted ΔH_T°. For most substances the standard enthalpy of reaction may be used also at pressures slightly different from one atmosphere. In metallurgical calculations this is usually done.

For reactions which involve substances which are partly dissociated, as for example sulfur vapor, the standard enthalpy of reaction may be calculated for each molecular species. The enthalpy of reaction for the gas mixture is then the sum of the enthalpies of reaction for all constituents in the proportions in which they are present at the temperature and pressure in question.

The enthalpy of reaction is a function of temperature. If the enthalpy of reaction at $25°C = 298$ K is denoted ΔH_{298} and at some other temperature ΔH_T, then

$$\Delta H_T = H_{T(B)} - H_{T(A)} = \Delta H_{298} + (H_{298(B)}^T - H_{298(A)}^T) \qquad (2\text{-}7)$$

The term in the parentheses may also be expressed by

$$\int_{298}^T \Delta C_P \, dT + \sum \Delta H_{\text{trf}}$$

where $\Delta C_P = C_{P(B)} - C_{P(A)}$ and $\sum \Delta H_{\text{trf}}$ is the sum of all transformation enthalpies for the products minus the same sum for the reactants. This expression is known as *Kirchhoff's law*. If ΔC_P is expressed arithmetically by

$$\Delta C_P = \Delta a + \Delta bT + \Delta cT^2 + \Delta eT^{-2}$$

and in the absence of phase transformations, the enthalpy of reaction becomes

$$\Delta H_T = \Delta H_{298} + \int_{298}^T \Delta C_P \, dT = \Delta H_0 + \Delta aT + \frac{\Delta b}{2} T^2 + \frac{\Delta c}{3} T^3 - \Delta eT^{-1}$$

where ΔH_0 is an integration constant composed of

$$\Delta H_{298} \qquad \text{and} \qquad \Delta a(298) + \frac{\Delta b}{2} (298)^2, \qquad \text{etc.}$$

Kirchhoff's law is illustrated in Fig. 2-4 which shows the enthalpies associated with the reaction Fe + S = FeS. The upper curve in Fig. 2-4 gives the sum of the standard enthalpies for iron and sulfur, and the lower curve the standard enthalpy for iron sulfide. At 25°C the latter curve is 96 kJ more negative than the former curve, i.e., for the reaction Fe + S(rh) = FeS: $\Delta H_{298}^\circ = -96$ kJ. At

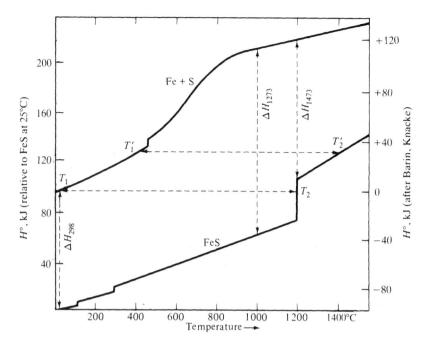

Figure 2-4 Molar enthalpy of iron plus sulfur and iron sulfide, relative to each other, as function of temperature.

1000°C the enthalpy for the reaction $Fe + \frac{1}{2}S_2 = FeS(s)$ is -151 kJ and at 1200°C the enthalpy of formation of FeS(l) from Fe and $\frac{1}{2}S_2$ is -113 kJ.

Standard enthalpies of formation at 25°C are listed in App. A for a number of compounds of metallurgical interest. When used in connection with the enthalpy increment curves the enthalpy of formation at any other temperature may be calculated. Some authors[1] *assign* the value of *zero* enthalpy to all *elements* in their *stable* state at 25°C = 298 K. According to this convention the enthalpy of all elements will be positive above 25°C and negative below that temperature. The enthalpy of *compounds* formed by exothermic reactions will be negative, and equal to ΔH_{298}°, at 25°C, zero at the *adiabatic* reaction temperature and positive at temperatures above the adiabatic one (see Fig. 2-4). Although this type of listing undoubtedly has certain merits the more traditional type with separate listings of ΔH_{298}° and $H_T^\circ - H_{298}^\circ$ will be used in the present text.

The enthalpy of a compound is a state property and is therefore independent of the method by which the compound is prepared. As a consequence, for a reaction between chemical compounds the enthalpy change may be obtained

[1] I. Barin and O. Knacke: "Thermochemical properties of inorganic compounds," Springer Verlag, Berlin, 1973, Supplement 1977.

as the difference between the enthalpies of formation of the products and the enthalpies of formation of the reactants, all at the same temperature. This is known as *Hess' law*, as illustrated below:

$$\Delta H_{298}^{\circ}$$

$$
\begin{array}{ll}
2Cr + \tfrac{3}{2}O_2 = Cr_2O_3 & -1130 \text{ kJ} \\
2Al + \tfrac{3}{2}O_2 = Al_2O_3 & -1677 \text{ kJ} \\
\hline
Cr_2O_3 + 2Al = Al_2O_3 + 2Cr & -547 \text{ kJ}
\end{array}
$$

Adiabatic Processes

In adiabatic processes there is no exchange of heat with the surroundings. If the pressure is constant the enthalpy of the system is constant. For a chemical reaction $A \rightarrow B$ the enthalpy of reaction will remain in the system and cause a change in temperature. If the initial temperature is denoted T_1 and the final temperature T_2, then $H_{T2(B)} = H_{T1(A)}$ or, as

$$H_{T2(B)} = H_{298(B)} + H_{298(B)}^{T_2} \quad \text{and} \quad H_{T1(A)} = H_{298(A)} + H_{298(A)}^{T_1}$$

then

$$H_{298(B)}^{T_2} = H_{298(A)}^{T_1} - \Delta H_{298} \tag{2-8}$$

where $\Delta H_{298} = H_{298(B)} - H_{298(A)}$. This *enthalpy balance* corresponds to the picture that we assume the reactant cooled from T_1 to 298 K before reaction, and the total enthalpy liberated is used to heat the product to T_2. The fact that the reaction actually takes place at some intermediate temperature is immaterial, the change in enthalpy being independent of the reaction path.

The enthalpy balance for an adiabatic process is illustrated in Fig. 2-4. If iron and sulfur, both at 25°C, are ignited the *adiabatic* temperature will be $T_2 = 1190°C$, and the product will be a mixture of $\tfrac{5}{8}$ liquid and $\tfrac{3}{8}$ solid iron sulfide. If the initial temperature had been $T_1' = 400°C$ a final temperature T_2' of about 1400°C would be obtained, and the product would be completely molten.

If the different reactants have different initial temperatures the enthalpy increment for each reactant is used in the enthalpy balance. This is the case for the combustion of fuel with preheated air. If, on the other hand, the reactants contain inert material which is heated with the reaction products its enthalpy increment must be added to those of the reaction products. A typical example is the combustion of fuel with air. In that case the nitrogen in the air has to be heated to the temperature of the flame. If the combustion is carried out with pure oxygen a much higher flame temperature is obtained, not because the heat of reaction is different but because the heat capacity of the products is less than for combustion with air. Also, if the reaction is incomplete or if an excess of some reactant is used the unused amounts must be included in the product.

Available Heat

If an exothermic reaction, for example, fuel combustion, is used to supply heat to some other process, its enthalpy balance is written

$$-H^{T_1}_{298(A)} + \Delta H_{298} + H^{T_3}_{298(B)} = q \qquad (2\text{-}9)$$

Here q represents the heat absorbed from the surroundings, or $-q$ is the *gross available heat* or the heat left over after the reaction products are heated to T_3, which is the temperature at which the combustion products leave the reaction zone.

For many metallurgical processes heat has to be supplied above a given temperature. Thus, in open-hearth steel-making the heat needed to melt the steel must be supplied at or above the melting temperature of the steel. Since heat can only flow from high to low temperature the temperature of the combustion products leaving the open-hearth furnace must be at least as high as the melting temperature of the steel. The lowest temperature at which the combustion products may leave the reaction zone is called the *critical temperature* of the process, and is for steel-making about 1600°C. If, in Eq. (2-9), T_3 is equal to the critical temperature, $-q$ is called the gross available heat above the critical temperature. By comparison with Eq. (2-8) it is easily seen that this is also equal to $H^{T_2}_{T_3(B)}$, where T_3 represents the critical temperature and T_2 the *adiabatic* temperature of the combustion process.

The fraction of the total heat of combustion which is available above a certain critical temperature increases with increasing adiabatic temperature. The adiabatic temperature for combustion of most fuels with air of 25°C is about 2100°C. If a surplus of air is used or if the reaction is incomplete the adiabatic temperature is less. For open-hearth steel-making the critical temperature is 1600°C and only a very small fraction of the heat of combustion is available heat. In practice steel-melting with cold air is impossible. If the air is preheated, the available heat increases by an amount equal to the heat content of the air, and open-hearth steel-making was only possible after preheated air had been introduced. A similar result is obtained if the combustion air is enriched with oxygen. Oxygen enrichment is of little value if the critical temperature is low, however.

The available heat is used partly to supply heat to the metallurgical process and partly to cover heat losses. The heat losses may be calculated from knowledge of the temperature in the reaction furnace and the heat conductivity and heat-transfer coefficient of the furnace materials, as will be discussed in Chap. 6. The heat available after heat losses are covered is called the *net available heat*.

2-4 CALORIMETRY

The purpose of calorimetry is to measure heat quantities. These quantities are usually measured by measuring the temperature increase in a substance with known heat capacity, called a calorimeter. Most common is the water calorimeter. In order to determine the heat capacity of the water bath, including its container, stirrer, etc., a known heat quantity is evolved by means of an electric heating element. In this way the calorimeter is calibrated. In order to prevent heat losses, the calorimeter has to be well insulated, for example by the use of a Dewar flask.

Calorimetric measurements may have two different objectives:

1. To measure enthalpy increments and heat capacities
2. To measure heats of chemical reactions

The first type of measurement is usually carried out by heating the substance to a given temperature, T_2. It is then dropped into the calorimeter which is kept at T_1. After the temperature in the calorimeter is stabilized a temperature increase ΔT is measured. This should be small compared to the difference $T_2 - T_1$. From the temperature increase and the heat capacity of the calorimeter the amount of heat which is given off by the substance is calculated. If the substance is dropped at constant pressure the heat is equal to $H_{T_2} - H_{T_1}$; if it is dropped at constant volume it is equal to $U_{T_2} - U_{T_1}$. From a number of measurements the enthalpy increment is obtained as function of temperature, and by differentiation of this curve the heat capacity may be obtained.

One weakness of the dropping technique is that high-temperature phases may freeze in because of low transformation rates, and the heat of phase transformation is not liberated. In such cases an alternative method is to measure the heat capacity directly. This can be done by supplying a known quantity of heat, e.g., electrically, to the substance which is kept in an insulating container. From the temperature increase of the substance its heat capacity can be calculated. This method is very useful at low temperatures, but becomes increasingly difficult with increasing temperature because of the increased rate of heat loss from the substance.

The second objective of calorimetry is to measure heats of reaction. If the reaction is carried out at constant volume the heat is equal to $-\Delta U$. If it is carried out at constant pressure it is equal to $-\Delta H$. One important example is combustion calorimetry, for example for the combustion of fuels. The reaction is carried out in a steel bomb which contains the fuel and oxygen at high pressure. The entire bomb is immersed in a water calorimeter. By means of an electric current the fuel is ignited and the reaction goes to completion. After the reaction products have cooled to the temperature of the calorimeter, its temperature increase is measured, and the heat of reaction is obtained. A correction must be made for the electrical energy used for the ignition.

Since the combustion was carried out at constant volume the heat corresponds to ΔU. In order to calculate the enthalpy of reaction use is made of the relation $\Delta H = \Delta U + \Delta(PV)$. Since the product PV is much larger for gases than for solids or liquids, only gaseous constituents need to be considered. For ideal gases $PV = nRT$, where n is the number of moles. For a reaction at constant temperature $\Delta(PV)$ is therefore equal to ΔnRT when Δn is the increase in the number of moles of gas during the reaction. For the reaction $C + O_2 = CO_2$, $\Delta n = 0$ and $\Delta U = \Delta H$. For fuels which contain hydrogen, as for example gas, oil, or coal, liquid water is formed after cooling of the combustion products, Δn will be negative, and $\Delta H < \Delta U$. Thus knowing the chemical composition of the fuel, the value of Δn and the $\Delta(PV)$-correction may be calculated.

By the above procedure the *gross calorific power* of the fuel at constant pressure, i.e., the heat of combustion to form CO_2 and liquid water, is obtained. In most industrial processes the waste gases leave the combustion zone above the dew point of the water, and the heat of condensation of the water is not utilized. By subtracting the heat of vaporization of the water from the gross heat of combustion the *net calorific power* is obtained. The heat of vaporization should be taken not only for the water formed as a combustion product but also for moisture present in the original fuel.

Example 2-1 One gram of coal, containing 80 percent carbon, 5 percent hydrogen, 5 percent moisture, and 10 percent ash, was burned completely with an excess of oxygen in a bomb calorimeter. The coal was ignited by means of an electric current and for ignition a total of 1500 W · s were used. After combustion the temperature of the calorimeter had increased from 24.0 to 26.0°C. The calorimeter is calibrated by means of an electric energy input, and after an input of 18 000 W · s a temperature increase of 1.0°C was observed. *Calculate* (a) the heat capacity of the calorimeter, (b) the gross calorific power at constant volume as well as (c) at constant pressure, and calculate (d) the net calorific power at constant pressure, all values (b)–(d) for 1 kg of coal.

SOLUTION. (a) The input for calibration, 18 000 W · s, is equal to 18 000 J. This gives the heat capacity of the calorimeter equal to 18 000 J/°C.

(b) A temperature increase of 2°C corresponds to 36 000 J. From this is subtracted the input for ignition, 1500 W · s = 1500 J, giving a heat of reaction of 34 500 J/g or 34 500 kJ/kg coal. This is the gross heat of combustion at constant volume.

(c) One kilogram of coal contains 800 g = 66.7 mol of C, 50 g = 49.6 mol of H (or 24.8 mol of H_2) and 50 g = 2.8 mol of H_2O. For the reaction $C(s) + O_2(g) = CO_2(g)$, Δn is zero, whereas for the combustion of 49.6 mole of H *in solid coal*, corresponding to the reaction $H(s) + \frac{1}{4}O_2(g) = \frac{1}{2}H_2O(l)$, Δn is equal to +12.4. This gives the gross calorific power at constant pressure equal to $34\,500 - (12.4 \times 8.3144 \times 298 \times 10^{-3}) = 34\,530$ kJ/kg. Notice that the $\Delta(PV)$ correction amounts to only 30 kJ/kg, and that for this correction a mean temperature of 25°C = 298 K is taken.

(d) The heat of vaporization of water at 25°C is 44 kJ/mol. This gives the net calorific power equal to $34\,530 - 24.8 + 2.8)44 = 33\,316$ kJ/kg coal. Notice that the heat of vaporization is taken not only for the water formed during combustion (24.8 mol) but also for the moisture in the coal (2.8 mol).

Other types of reaction calorimeters are solution calorimeters and high-temperature reaction calorimeters. In solution calorimetry a compound AB is dissolved in a suitable solvent, for example an acid. Separately a mixture of the constituents A and B is dissolved in the same acid. Since in both cases the same final solution is obtained, the heat of formation of the compound AB is the difference between the heat evolved in dissolving the constituents and in dissolving the compound. Since the heat of formation usually is small compared to the heat of solution, this method involves subtracting two large values, and the accuracy of the differences may be small. For intermetallic compounds the acid may be replaced by a low-melting metal which dissolves the compound. Since the heat of solution in a metal is smaller than in acid the accuracy of the heat of formation may be improved. Similarly for compounds between oxides, such as silicates, a low-melting slag may be used as solvent.

2-5 INDUSTRIAL ENTHALPY BALANCES

We have in the preceding sections seen various examples of enthalpy balances. In general any enthalpy balance is based on the integrated form of Eq. (2-4) where, for processes at constant pressure, the term $V\,dP$ is zero. In order to facilitate the calculations for industrial processes it is customary to divide the total enthalpy change, as well as the heat and work terms, into suitable groups. Since, for a given process, the heat supplied usually represents the unknown quantity, such balances are often called *heat balances.*

For a process which in a general form may be written as $A_{T_1} \rightarrow B_{T_2}$, we may choose a suitable reference temperature, which we shall call T_0. The total enthalpy change is then given by:

$$H_{T_2(B)} - H_{T_1(A)} = H_{T_0(B)}^{T_2} + \Delta H_{T_0} - H_{T_0(A)}^{T_1} = q - w'$$

This means that we consider the reactant A cooled from T_1 to T_0, the reaction to take place at T_0 and the product B heated from T_0 to T_2. The fact that the reaction actually takes place at some intermediate temperature is immaterial. For a process which involves a number of reactants and products, and various reactant and product temperatures, the enthalpy increments above the reference temperature are taken for each reactant and product compound. Similarly the enthalpy of reaction ΔH_{T_0} may be taken equal to the enthalpy of formation, from the elements, of all product compounds minus the same enthalpy of formation for all reactant compounds.

In practice all negative terms are collected on one side of the balance, the "input" side, and all positive terms on the other side, the "output" side. The input side contains the enthalpy increments, above the reference temperature, for all reactants, as well as the negative enthalpies of formation of all products. The output side contains the enthalpy increments, above the reference temperature, of all products, as well as the negative enthalpies of formation, i.e., the enthalpy of dissociation, of all reactants. If the enthalpy of formation of some products or reactants are positive they are listed on the opposite side. The same is the case with enthalpy increments if some reactant or product temperature is below the reference temperature.

Similar grouping is made with the heat and work terms. Positive values of q, that is, heat supplied from the surroundings, are listed on the input side, whereas heat losses to the surroundings enter the output side. Work supplied from the surroundings, e.g., electrical energy, is listed on the input side, whereas possible work on the surroundings is listed on the output side.

Some authors prefer, instead of counting the enthalpies of formation of the reactants and products, to count the enthalpy of the reactions that actually take place. This has the advantage that the enthalpies of reaction appear with their proper value and not as a difference between two large figures. On the other hand, as the actual reactions may be rather complex, the procedure recommended here is safer and gives fewer chances of error. The overall balance remains the same, of course.

The reference temperature usually chosen is 298 K = 25°C. This has the advantage that the enthalpy increment of reactants supplied at room temperature may be ignored. Also most enthalpies of formation are reported for 25°C. The choice of reference temperature is immaterial, however, since the enthalpy balance is determined by the initial and final states only. This is illustrated by Table 2-1 which gives enthalpy balances for the oxidation of

Table 2-1 Effect of reference temperature on heat balance

Reaction: $Cu_2S(1300°C) + (O_2 + 3.76N_2)(25°C) = (2Cu + SO_2 + 3.76N_2)(1250°C)$

(a) Reference temperature = 25°C = 298 K

Input		kJ	Output		kJ
$Cu_2S(l)$	H_{298}^{1573}	125.6	2Cu		95.4
$O_2 + 3.76N_2$	H_{298}^{298}	0.0	SO_2 $\Big\}$ H_{298}^{1523}		63.2
			3.76N_2		147.7
$S(rh) + O_2 = SO_2$	$-\Delta H_{298}^{\circ}$	297.0	$Cu_2S = 2Cu + S(rh)$	ΔH_{298}°	82.0
			Heat surplus		34.3
Total input		422.6	Total output		422.6

(b) Reference temperature = 1300°C = 1573 K

Input		kJ	Output		kJ
2Cu		3.3	O_2		43.1
SO_2 $\Big\}$ H_{1523}^{1573}		3.0	3.76N_2 $\Big\}$ H_{298}^{1573}		154.0
3.76N_2		6.3	$Cu_2S = 2Cu + \frac{1}{2}S_2(g)$	ΔH_{1573}°	141.7
$\frac{1}{2}S_2 + O_2 = SO_2$	$-\Delta H_{1573}^{\circ}$	360.5	Heat surplus		34.3
Total input		373.1	Total output		373.1

molten cuprous sulfide (white metal) with air in a Bessemer converter for two different reference temperatures. The overall reaction is

$$Cu_2S + O_2 + 3.76N_2 = 2Cu + SO_2 + 3.76N_2$$

The cuprous sulfide is assumed to have an initial temperature of 1300°C and the air to enter at 25°C. The temperature of all reaction products is assumed to be 1250°C. The basis for the enthalpy balance is the number of moles which appear in the above equation.

Notice that in case (a) the enthalpy increments of all reactants and products are positive, and occur on the input and output sides respectively, in the balance. In case (b) the temperatures of the air and of the products are below the reference temperature, and their enthalpy increments occur on the output and input sides respectively. Notice also that the heats of formation of Cu_2S and SO_2 are based on orthorhombic sulfur in case (a), whereas they are based on diatomic sulfur gas in case (b).

It follows that for the chosen assumptions the heat surplus is the same, 34.3 kJ, in both cases, even though the individual terms differ considerably from one case to the other. This, of course, is not unexpected, since the enthalpies of formation at 1300°C are derived from those at 25°C by means of Kirchhoff's law, Eq. (2-7) which in itself is an enthalpy balance.

In the above example the heat surplus represents the unknown, and may be assumed to cover heat losses. If, on the other hand, the heat losses were known or estimated to be 34.3 kJ, the temperature of the reaction products would represent the unknown term. From the heat balance it may be concluded that the enthalpy increment H_{298}^T of the reaction products $2Cu + SO_2 + 3.76N_2$ amounts to a total of 306.3 kJ. This is compared with the enthalpy curves of the products, and by trial and error a common temperature of 1250°C = 1523 K is found.

It is important, in the enthalpy balance, to include the enthalpy changes of all steps between the reactant and the product states. Each component or element should be traced through the balance to make sure that some steps have not been overlooked or counted twice.

For reactions which involve the combustion of fuels containing hydrogen a particular point should be emphasized. If the *net heat* of combustion at 25°C is used, the enthalpy increment of water in the combustion gases should be relative to water vapor at 25°C. If the *gross heat* of combustion is used the enthalpy increment should be relative to liquid water at 25°C. As the difference between the gross and the net heats of combustion is the heat of vaporization of the water the final result will be the same in the two cases.

Industrial processes are often carried out in reactors (furnaces) where the reactions occur at different temperature levels. A typical example is the iron blast furnace, which will be discussed in Sec. 9-3. An enthalpy balance for the entire reactor would, in such a case, give little information about the temperature distribution inside the reactor. Instead the enthalpy balance should be worked out for the different temperature levels or zones in the furnace. Examples of this will be shown in Secs. 8-3 and 9-3.

PROBLEMS

2-1 For iodine, I_2, the following data exist:
Solid: $C_P = 54.3 + 13.4 \times 10^{-4}T$ J/(K mol)
Melting point: 114°C. $\Delta H_m = 15\,774$ J/mol
Liquid: $C_P = 80.5$ J/(K mol)
Boiling point: 183°C. $\Delta H_v = 41\,700$ J/mol
Gas: $C_P = 37.2$ J/(K mol)
Calculate the enthalpy increment between 298 and 473 K.

2-2 For iron, Fe, the following data exist:
Solid α: $C_P = 17.5 + 24.8 \times 10^{-3}T$ J/(K mol)
α–β transformation: 760°C. $\Delta H_{trf} = 2760$ J/mol
Solid β: $C_P = 37.7$ J/(K mol)
β–γ transformation: 910°C. $\Delta H_{trf} = 920$ J/mol
Solid γ: $C_P = 7.7 + 19.5 \times 10^{-3}T$ J/(K mol)
γ–δ transformation: 1400°C. $\Delta H_{trf} = 1180$ J/mol
Solid δ: $C_P = 44$ J/(K mol)
Melting point: 1535°C. $\Delta H_m = 15\,680$ J/mol
Liquid: $C_P = 42$ J/(K mol)
Calculate the enthalpy increment between 298 and 1873 K.

2-3 (a) From the data listed in App. A, calculate the enthalpy of combustion of methane according to the reaction $CH_4 + 2O_2 = CO_2 + 2H_2O$ and with the water formed as $H_2O(l)$ as well as $H_2O(g)$.

(b) On the basis of the enthalpy increment curves (App. B) calculate the adiabatic temperature for combustion of methane with stoichiometric amounts of air (21 vol % O_2) as well as for combustion with 20 percent air excess. The methane and the air are both initially at 25°C.

(c) Calculate the gross available heat above 1600°C for the two cases given under (b).

(d) The air is now preheated to 1000°C. Calculate the adiabatic temperature as well as the gross available heat above 1600°C for stoichiometric as well as for 20 percent air excess combustion.

(e) The air is enriched with oxygen to contain 30 vol % O_2. Calculate the same values as under (d) for the case that the reactants all are at 25°C.

2-4 2.7 g of aluminum powder is reacted with oxygen under pressure to give Al_2O_3 in a constant volume combustion calorimeter at 25°C. The evolved heat is measured as 83.7 kJ.

(a) Calculate ΔU and ΔH for the reaction $2Al(s) + \frac{3}{2}O_2(g) = Al_2O_3(s)$ under these conditions. Oxygen is regarded as an ideal gas.

(b) The reaction had been carried out at a mean pressure of 50 atm. On the basis of Eq. (2-6) calculate the value of ΔH_{298}°. The volume expansion coefficient $\alpha = (\partial V/\partial T)_P(1/V)$ is for aluminum 7×10^{-5} and for aluminum oxide 2×10^{-5}. The molar volume V_M is, for Al, 10.0 ml/mol and for Al_2O_3, 25.6 ml/mol. The results should not be given with more significant figures than the calorimetric measurement.

2-5 A high-temperature solution calorimeter consists of a well-insulated bath of about 200 g of molten tin, kept at 250°C. The heat capacity of the calorimeter is determined electrically: after an input of 67 W · s the temperature of the bath has increased by 1.00°C.

(a) Calculate the heat capacity in J/°C for the calorimeter.

(b) Into this bath are dropped, in successive order, 1.972 g of gold. 1.079 g of silver and 1.634 g of a gold–silver alloy with 75 wt % gold, all with a temperature of 0°C. Each addition is allowed to dissolve completely in the molten tin, and the temperature change is recorded, before the next addition. The first addition results in a temperature increase of 2.77°C, the second and third additions in temperature drops of 3.33°C and 0.10°C respectively. Calculate the enthalpy of formation of the given amount of alloy from the pure metals. To which temperature does this value apply?

(c) In the temperature range 0 to 250°C the heat capacity of the metals and the alloy are: Au, 25.8 J/(K mol); Ag, 25.6 J/(K mol); alloy, 24.9 J/(K mol). (One mole of alloy is defined as a total of one mole of Au + Ag.) Calculate the molar enthalpies of solution of the two metals and the alloy in molten tin at 250°C, and calculate the molar enthalpy of formation of the alloy at 25°C and at 250°C. The enthalpies of solution in molten tin are assumed to be independent of composition in this low concentration range.

2-6 Zinc sulfide is roasted with air according to the reaction: $ZnS + \frac{3}{2}O_2 = ZnO + SO_2$.

(a) On the basis of the data given in Apps. A and B, calculate for this reaction ΔH°_{298} and ΔH°_{1173}.

(b) In practice a 50 percent air surplus is used. The zinc sulfide and the air are introduced at 25°C and the reaction products are withdrawn at 900°C. Make an enthalpy balance for the process and calculate possible heat surplus or deficiency.

2-7 Chromium oxide and aluminum powder are mixed in stoichiometric amounts for the reaction $Cr_2O_3 + 2Al = Al_2O_3 + 2Cr$. The mixture is ignited at 25°C and the reaction goes to completion. On the basis of the data given in Apps. A and B calculate the adiabatic reaction temperature. The heat introduced for ignition is ignored.

2-8 The enthalpy of solution of various minerals in a large volume of a low-melting lead–silicate melt is measured calorimetrically at 695°C = 968 K, and the following values are obtained.

Al_2SiO_5 (andalusite)	24.9 kJ/mol
$Al_6Si_2O_{13}$ (mullite)	42.2 kJ/mol
SiO_2 (quartz)	−15.3 kJ/mol
Al_2O_3 (α-corundum)	31.8 kJ/mol

Calculate ΔH_{968} for the formation of andalusite and mullite from SiO_2 and Al_2O_3, and calculate ΔH_{968} for the reaction

$$2Al_2SiO_5 \text{ (and.)} + Al_2O_3 = Al_6Si_2O_{13} \text{ (mull.)}$$

BIBLIOGRAPHY

Darken, L. S., and R. W. Gurry: "Physical Chemistry of Metals," McGraw-Hill Book Co. Inc., New York, 1953.

Gaskell, D. R.: "Introduction to Metallurgical Thermodynamics," 2d ed., McGraw-Hill Book Co. Inc., New York, 1981.

Kubaschewski, O., and C. B. Alcock: "Metallurgical Thermochemistry," 5th ed., Pergamon Press, London, 1979.

Lewis, G. N., and M. Randall: "Thermodynamics," revised edition by K. S. Pitzer and L. Brewer, McGraw-Hill Book Co. Inc., New York, 1961.

THREE

CHEMICAL EQUILIBRIUM

Around the middle of the nineteenth century Berthelot (1867) and Thomson (1853) suggested that the heat evolved by a chemical reaction is a measure of its tendency to take place, its *affinity*. It had been observed that spontaneous reactions usually took place under evolution of heat, and the larger the heat of reaction the more stable the product seemed to be. It can easily be shown that this postulate is incorrect. Thus when a salt is dissolved in water this is usually accompanied by an absorption of heat even though the process is spontaneous. Another example is calcium carbonate which at about 900°C dissociates spontaneously into calcium oxide and carbon dioxide, a strongly endothermic process. It is, therefore, evident that the heat or enthalpy of reaction is not a measure of the affinity. A satisfactory criterion for the affinity of reaction, and for the chemical equilibrium which is obtained, can de derived from the second law of thermodynamics.

3-1 THE SECOND LAW OF THERMODYNAMICS

This law can be expressed in many different ways. The following formulation was essentially formulated by Guggenheim,[1] and refers to the system which was shown in Fig. 2-1 (notice the similarity with the formulation of the first law):

1. There exists a function S, *the entropy*, which is a property only of the state of the system, and which is the sum of the entropy of the parts of the system.

---

[1] E. A. Guggenheim: "Thermodynamics," Interscience Publishers, New York, 1949.

2. If heat flows into a system, the change in entropy is given by $dS = \delta q/T$, where δq is an infinitesimal heat quantity and T the absolute temperature of the system.
3. In a thermally insulated system $dS = 0$ for reversible changes within the system, whereas for irreversible changes $dS > 0$. The entropy of a thermally insulated system can never decrease.

From this definition of entropy it follows that for a reversible process $T\,dS = \delta q$, whereas for an irreversible process $T\,dS > \delta q$, where δq is the heat supplied from the surroundings.

In the definition above, the word *reversible* is used. This in turn requires a definition: By a reversible process we understand a process where the direction may be reversed by an infinitesimal change in the system. Reversible processes are never attained in practice, but may be visualized as limiting cases. Examples of reversible processes are heat-flow in the absence of a temperature gradient and chemical reactions occurring at chemical equilibrium. Irreversible processes on the other hand occur in one direction only and may also be called *spontaneous*. Thus for all spontaneous processes within a thermally insulated system the entropy increases.

Point 2 above says that if heat is supplied from the surroundings to one mole of a substance, the increase in molar entropy is $dS_M = \delta q/T$, where T is the absolute temperature of the substance. If the heating is carried out at constant pressure and in the absence of any other work, $\delta q = C_P\,dT$ and $dS_M = (C_P/T)\,dT$, or integrated

$$S_M = \int_0^T (C_P/T)\,dT + S_0$$

Here the integration constant S_0 represents the entropy at the absolute zero. According to the so-called *Third Law of Thermodynamics* this is zero for all crystalline substances in internal equilibrium.

If the substance undergoes phase transformations in the temperature range, the entropy increases further by the entropy of transformation $\Delta S_{trf} = \Delta H_{trf}/T_{trf}$. Thus, in general, the molar entropy is given by

$$S_M = \int_0^T (C_P/T)\,dT + \sum \Delta H_{trf}/T_{trf}$$

Here $\sum \Delta H_{trf}/T_{trf}$ denotes the sum of all entropies of transformation between 0 and T K. Notice that, contrary to the cases of energy and enthalpy, an absolute value can be assigned to the entropy function.

The molar entropy of a substance will vary not only with the temperature but also with the pressure. It can be shown[1] that this variation is expressed by

[1] See, e.g., L. S. Darken and R. W. Gurry: "Physical Chemistry of Metals," McGraw-Hill Book Co. Inc., New York, 1953, p. 186.

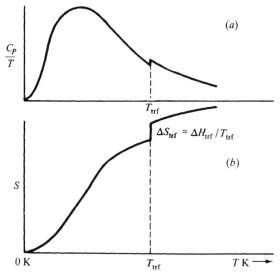

Figure 3-1 (a) The ratio C_P/T, and (b) the entropy $S = \int_0^T C_P/T \, dT + \Sigma \, \Delta H_{trf}/T_{trf}$ as functions of temperature.

the relation $(\partial S/\partial P)_T = -(\partial V/\partial T)_P$ and is, therefore, determined by the equation of state of the substance. In the same way as we previously have defined standard molar enthalpies, we define standard molar entropies as the entropy of the substance in its standard state, which is usually chosen as the pure substance at atmospheric pressure. Again for nonideal gases a slightly different standard state is used. The standard entropy is denoted $S°$.

As the unit of heat is one joule it follows that the unit of molar entropy is one joule per kelvin-mole. In Fig. 3-1 the ratio C_P/T and the molar entropy of a substance at constant pressure are shown as function of temperature. Notice the similarity with the enthalpy function shown in Fig. 2-3, and notice how at the temperature of transformation the entropy increases by an amount $\Delta H_{trf}/T_{trf}$.

Entropy as Criterion for Equilibrium

The great value of the entropy function in chemistry and metallurgy is its ability to show the direction in which a process will proceed and the final equilibrium state of the process. If no heat is supplied from the surroundings, any irreversible change within the system would result in an entropy increase, $dS > 0$, whereas for reversible changes $dS = 0$. If all changes within the system are reversible, the system is said to be in *internal equilibrium*. Thus it can be said that a chemical reaction within a system will have a positive *affinity* to take place if it leads to an increase in the entropy, without there being any exchange of heat with the surroundings. For noninsulated isothermal systems the criterion for equilibrium is expressed by the equation $\delta q/T = dS$, whereas for spontaneous reactions $\delta q/T < dS$. Most chemical reactions may be carried out either reversibly or irreversibly, but more heat is absorbed from the surroundings if the reaction is carried out in a reversible manner than if the reaction is spontaneous.

The *work* done by a process may be derived from a combination of the first and second laws. For any process or reaction is, according to the first law, $dU = \delta q - \delta w$. Since $\delta q \leq T \, dS$ it follows that $dU \leq T \, dS - \delta w$. The work δw may again be divided into two terms, $P \, dV$ which is the reversible work against pressure, and $\delta w'$ which is any other work done on the surroundings. This gives

$$\delta w' \leq T \, dS - P \, dV - dU = T \, dS + V \, dP - dH \qquad (3\text{-}1)$$

The equality sign applies for a reversible, the inequality for a spontaneous process. Thus the work which a process does on the surroundings gives a measure of the reversibility of the process. For a given process the right-hand side of Eq. (3-1) is fixed. Hence if a given process is carried out in a reversible manner, the work done on the surroundings will have its maximum value, $\delta w'_{max}$. This is also called the *available work*. For any spontaneous process the work will be less than the available work.

3-2 THE FREE ENERGY FUNCTIONS

Usually we are not working with thermally insulated systems and we would like to have a way of expressing chemical affinity for reactions which take place at constant temperature. For this purpose we define two new functions (free energy functions):

$$A = U - TS \qquad \text{(Helmholtz energy)}$$

$$G = U + PV - TS = H - TS \qquad \text{(Gibbs energy)}$$

Since U, H, and S are all state functions the same is the case with A and G. They may therefore be subjected to total differentiation

$$dA = dU - T \, dS - S \, dT$$

$$dG = dU + P \, dV + V \, dP - T \, dS - S \, dT = dH - T \, dS - S \, dT$$

If Eq. (3-1) is inserted in these equations we obtain

$$dA \leq -P \, dV - S \, dT - \delta w' \qquad \text{and} \qquad dG \leq V \, dP - S \, dT - \delta w'$$

Hence for a process which takes place at constant volume and temperature $dA \leq -\delta w'$, and for a process at constant temperature and pressure $dG \leq -\delta w'$. If the process is carried out in the absence of work beyond volume work, and at constant temperature and pressure $dG < 0$ for a spontaneous process, whereas for a reversible process $dG = 0$. Thus at constant temperature and pressure the Gibbs energy gives a measure of the affinity of the process in the same way as the entropy function did for thermally insulated systems. It is easily seen that $dA = -\delta w'_{max}$ at constant temperature and volume, whereas $dG = -\delta w'_{max}$ at constant temperature and pressure.

Variation of G with Pressure and Temperature

For a closed system or substance which does no work beyond volume work $dG = V\,dP - S\,dT$. As G is a state function, dG is a perfect differential

$$dG = \left(\frac{\partial G}{\partial T}\right)_P dT + \left(\frac{\partial G}{\partial P}\right)_T dP \tag{3-2}$$

Here $(\partial G/\partial T)_P = -S$ and $(\partial G/\partial P)_T = V$. In the same way as shown for the energy and enthalpy functions, the free energy functions have no absolute value. They can only be given in terms of differences. Since both the volume and the entropy are positive values, the Gibbs energy decreases with increasing temperature and increases with increasing pressure. In the same way as shown for the energy and entropy functions, we are at liberty to define a *standard state* for each temperature. Usually this is chosen to be the pure substances at one atmosphere pressure.

For *ideal gases* the variation of the molar Gibbs energy with pressure has a particularly simple form. Since for such gases $V_M = RT/P$ it follows that $(\partial G/\partial \ln P)_T = RT$. If integrated at constant temperature

$$G_M = G_M^\circ + RT \ln (P/P^\circ) = G_M^\circ + RT \ln p \tag{3-3}$$

Here the symbol $^\circ$ denotes the chosen standard state, and $p = P/P^\circ$ is the (dimensionless) pressure, relative to the standard pressure P° which usually is 1.013×10^5 Pa. The same relation holds for each component in an *ideal gas mixture*. Here p is the *relative* partial pressure defined as $p_A = (n_A/n_{tot}) P_{tot}/P^\circ$, where n denotes the number of moles in the gas mixture.

For *nonideal gases* it has been found convenient to define a standard state different from one atmosphere. Since all gases approach ideal gas behavior at low pressure, the relation $G = G^\circ + RT \ln p$ will also be approached at low pressure. If G represents the actual Gibbs energy of the gas at low pressure, G° will no longer represent the Gibbs energy at one atmosphere pressure, but at some slightly different pressure, which is chosen as the standard state. For any actual pressure the relation between the Gibbs energy of the nonideal gas and its standard Gibbs energy may be expressed by an equation $G = G^\circ + RT \ln f$, where f is called the *fugacity* of the gas, and is a function of the gas pressure and approaches this at low pressure. For nonideal gases, therefore, the standard state is one of unit fugacity. The relation between pressure and fugacity, and the pressure at unit fugacity, may be derived from the equation of state of the gas. At high temperatures and moderate pressures the difference between the pressure and the fugacity is usually small for most gases, and for most metallurgical calculations ideal gas behavior may be assumed.

For *condensed systems*, solids, or liquids, the molar volume is usually small, and the variation of Gibbs energy with pressure for pressure changes of a few atmospheres is small and may be ignored.

Variation of G with temperature The variation of the Gibbs energy with temperature for one mole of an arbitrary substance at constant pressure is shown in Fig. 3-2, which also shows the corresponding variation in the enthalpy. We see that at any given temperature the slope of the G curve is $(\partial G/\partial T)_P = -S$, and the difference between the enthalpy and the Gibbs energy is TS. Whereas the enthalpy increases discontinuously at a first-order transformation or melting point by $\Delta H_{trf} = T_{trf}\,\Delta S_{trf}$ the Gibbs energy curve is characterized by a change in slope $\Delta(\partial G/\partial T)_P = -\Delta S_{trf}$. At the transformation temperature the high- and the low-temperature forms have the same Gibbs energy in accordance with the two forms being in mutual equilibrium.

Although no absolute value can be given for the Gibbs energy of individual elements or compounds some other characteristic expressions can be listed. Most common is the so-called *Gibbs energy function (Gef)* defined as:

$$Gef = \frac{G_T^\circ - H_{298}^\circ}{T}$$

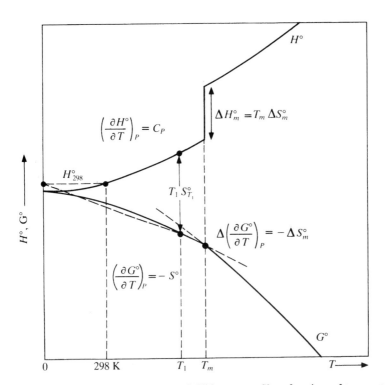

Figure 3-2 Standard enthalpy, H°, and Gibbs energy, G°, as functions of temperature.

As can be seen from Fig. 3-2 the Gibbs energy function at temperature T_1 is equal to the *slope* of the tie-line between H°_{298} as read on the *ordinate* axis and the G° value read at T_1. It thus bears a close relation to the entropy. Like the entropy it changes relatively little with changing temperature, and a tabulation of *Gef* values at different temperatures allows easy interpolation.

In other texts another Gibbs energy function is defined equal to $(G^\circ_T - H^\circ_0)/T$ where H°_0 is the enthalpy at the absolute zero. This function will not be used here.

Another type of listing may be derived from the *Barin–Knacke convention* mentioned in Sec. 2-3. According to this the *enthalpy* of all *elements* in their stable form at 298 K is assigned the value of zero, and conventional absolute enthalpy values may be derived for all elements and compounds as function of temperature. If these are combined with absolute *entropy* values as derived from the third law, corresponding conventional Gibbs energy values may be calculated. These show greater variation with temperature than the *Gef* values and do not as easily allow interpolation.

The Gibbs–Helmholtz Equation

In the absence of work beyond volume work the Gibbs energy of a substance is

$$G = H - TS = H + T(\partial G/\partial T)_P$$

This is known as the *Gibbs–Helmholtz equation*. By mathematical manipulation it may be converted into a more useful form

$$\left(\frac{\partial(G/T)}{\partial(1/T)}\right)_P = H \tag{3-4}$$

As this equation gives a direct relation between the G and H functions, it may be integrated at constant pressure and is very useful in chemical thermodynamics. Thus if H is expressed arithmetically

$$H = H_0 + aT + \frac{b}{2}T^2 + \cdots - eT^{-1}$$

integration of the Gibbs–Helmholtz equation gives

$$G = H_0 - aT \ln T - \frac{b}{2}T^2 + \cdots - \frac{e}{2}T^{-1} + IT$$

where I is an integration constant. This expression may also be written

$$G = H_0 - \iint C_P \, d \ln T \, dT + IT$$

The Clapeyron Equation

As was shown in Fig. 3-2 a phase transformation in a substance, e.g., melting, occurs when the Gibbs energy of the low- and high-temperature forms are equal. The Gibbs energy of both forms are functions of pressure as well as temperature, however. This makes the transformation temperature a function of pressure. If the Gibbs energy of transformation is denoted ΔG_{trf} then $(\partial \Delta G_{trf}/\partial P)_T = \Delta V_{trf}$, and $(\partial \Delta G_{trf}/\partial T)_P = -\Delta S_{trf}$, where ΔV_{trf} and ΔS_{trf} denote, respectively, the volume change and the entropy change on transformation. As ΔG_{trf} at the transformation temperature is zero, it is easily seen that

$$\frac{dT_{trf}}{dP} = \frac{\Delta V_{trf}}{\Delta S_{trf}} = \frac{T\Delta V_{trf}}{\Delta H_{trf}}$$

This is known as the *Clapeyron equation*. Since ΔV_{trf} is small for reactions among all condensed phases, the effect of pressure on such reactions is small and of minor interest to the extraction metallurgist. It becomes large for vaporization processes, however. For vaporization ΔV_v is mainly given by the volume of the gas, which for one mole of an ideal gas is RT/P. This gives the effect of temperature on the vapor pressure

$$\frac{d \ln P}{dT} = \frac{\Delta H_v}{RT^2} \quad \text{or} \quad \frac{d \ln P}{d(1/T)} = \frac{-\Delta H_v}{R}$$

where ΔH_v is the enthalpy of vaporization. This is known as the *Clausius–Clapeyron equation*.

3-3 GIBBS ENERGY CHANGES IN CHEMICAL REACTIONS

For an isothermal chemical reaction $A \rightarrow B$ the change in molar Gibbs energy $\Delta G = G_B - G_A$. Whereas we have no absolute value for G, a definite value can always be assigned to ΔG. Also as $G = H - TS$, $\Delta G = \Delta H - T\Delta S$ for reactions at constant temperature. We remember that at constant temperature and pressure and in the absence of work beyond volume work dG is negative for spontaneous processes and zero for reversible processes. Under these conditions a reaction $A \rightarrow B$ will have a positive affinity to take place if ΔG is negative, and it is able to do an amount of work $w'_{max} = -\Delta G$ on the surroundings. Conversely if ΔG is positive, an amount of work equal to ΔG is required to make the reaction take place. If ΔG is zero, A and B will be in chemical equilibrium. If both A and B are present in their standard states, the change in Gibbs energy is denoted ΔG°.

The Equilibrium Constant

For gas reactions the gases may occur with relative partial pressures different from unity and with free energies much different from the standard ones. In this case there will be some reaction, and a *chemical equilibrium* will be established.

The same is the case for reactions which involve substances in solid or liquid solutions. For a reaction between ideal gases $kA + lB \rightarrow mC + nD$, where $G_A = G_A^\circ + RT \ln p_A$, etc., the change in Gibbs energy becomes

$$\Delta G = \Delta G^\circ + RT \ln \frac{(p_C)^m (p_D)^n}{(p_A)^k (p_B)^l} \tag{3-5}$$

Here p_A, p_B, etc., are the relative partial pressures of the gases in the reacting mixture. If ΔG is negative and in the absence of other work, the reaction will proceed from left to right; if it is positive it will proceed in the reverse direction. At chemical equilibrium $\Delta G = 0$ and

$$\Delta G^\circ = -RT \ln \frac{(p_C^*)^m (p_D^*)^n}{(p_A^*)^k (p_B^*)^l} = -RT \ln K \tag{3-6}$$

Here p_A^*, p_B^*, etc., are the relative partial pressures of the gases at chemical equilibrium, and K is called the *equilibrium constant*. Frequently the asterisk, which denotes equilibrium, is omitted. Equation (3-6) is one of the most important equations in chemical metallurgy. Its main limitation is that it is confined to systems which do no work, except volume work, on the surroundings. This is usually the case for chemical reactions. The one major exception is electrolytic processes which will be discussed in Chap. 16.

Equations (3-5) and (3-6) also apply for reactions which involve nonideal gases if the relative partial pressure p is replaced by the fugacity f. For reactions which involve solid or liquid solutions an analogous quantity, *the activity* is defined by the equation $\bar{G}_A = G_A^\circ + RT \ln a_A$. Here $\bar{G}_A$ is the partial molar Gibbs energy of the substance A in the solution, and a_A its activity.* For the substance in its standard state the activity is unity. For chemical equilibria involving substances in solution the equilibrium constant may, therefore, be expressed in terms of activities. Further discussion of the relation between the partial Gibbs energy, the activity, and the composition of a solution will be given in Chap. 4.

It should be noted that Eq. (3-6) is derived independently of whether the system is homogeneous or heterogeneous. For heterogeneous reactions which involve gases as well as solutions the equilibrium constant is expressed in terms of relative partial pressures or fugacities for the gases, and activities for the solutions. It is very important to remember that partial pressures, fugacities, and activities always should be relative to the standard states which are used in evaluating the standard Gibbs energy change and the equilibrium constant. In writing chemical equilibria or reactions the standard states should, in case of doubt, be indicated: (s) = pure solid, (l) = pure liquid, (g) = gas, unit fugacity. Also other standard states may be used.

On the writing of chemical reactions Chemical reaction equations may have different meanings. Using the reaction between carbon dioxide and solid carbon as an example we may distinguish between the following types:

* See Sec. 4.3.

1. *Standard state reactions*

$$CO_2(g, p = 1) + C(s) = 2CO(g, p = 1)$$

In this case CO_2 of unit pressure or fugacity reacts with solid carbon (graphite) to give CO of unit pressure or fugacity. The change in Gibbs energy is in this case denoted $\Delta G°$. Standard state reactions occur rarely in practice, but may serve as theoretical references.

2. *Reactions in general*

$$CO_2(g, p_{CO_2}) + C(s) = 2CO(g, p_{CO})$$

Here CO_2 of pressure p_{CO_2} reacts with carbon to give CO of pressure p_{CO}. The change in Gibbs energy is in this case denoted ΔG. If the reaction is in chemical equilibrium $\Delta G = 0$.

3. *Stoichiometric process reactions*

$$CO_2 + xC = 2xCO + (1 - x)CO_2$$

This equation describes what actually occurs in a process where the reaction product is a mixture of CO and unreacted CO_2. If chemical equilibrium is established between solid carbon and the effluent gas mixture the value of x may be calculated from the corresponding equilibrium constant and the total gas pressure. It will usually have a noninteger value. If equilibrium is not established the value of x will depend also on the kinetics of the reaction.

4. *Kinetic reactions*

$$CO_2(g) + C_f \rightarrow CO + C_O$$

This equation describes one of the kinetic steps which occurs between CO_2 and solid carbon. Here C_f denotes a free active site on the carbon surface, and C_O denotes a carbon site on which an oxygen atom is adsorbed. See also Sec. 5-3.

Both for standard state reactions and for reactions in general the parentheses which describe the standard states or actual gas pressures are often omitted if these are evident from the context in which the equations are used.

Variation of K with Pressure and Temperature

Since the standard Gibbs energy refers to a chosen pressure, $\Delta G°$ and consequently also the equilibrium constant K are independent of pressure at constant temperature. This does not mean that the equilibrium *composition* of a mixture may not be affected by the changes in the total pressure. The relative partial pressure of a component A in an ideal gas mixture is defined by the expression $p_A = (n_A/n_{tot})p_{tot} = N_A p_{tot}$, where the n's are the number of moles, N the molar fraction in the gas mixture, and p_{tot} is the *relative total* pressure of the gas $= P_{tot}/P°$. For a reaction $mA = nB$ the equilibrium constant

$$K = \frac{(p_B^*)^n}{(p_A^*)^m} = \frac{(N_B^*)^n (p_{tot})^{n-m}}{(N_A^*)^m} \qquad (3\text{-}7)$$

From this follows that the *mole fraction ratio* at equilibrium, $(N_B^*)^n/(N_A^*)^m$, increases with increasing pressure if $m > n$, that is, if the number of gas molecules decreases during the reaction, whereas it decreases with increasing pressure if $m < n$.

Some authors use an equilibrium constant by molar fraction: $K_N = K/(p_{tot})^{n-m}$ or an equilibrium constant by molar concentration: $K_C = (C_B^*)^n/(C_A^*)^m = K/(RT)^{n-m}$, where $C_B = n_B/V$. The latter expression will be employed in Chap. 5.

For solutions, the relation between activity and composition is affected little by changes in pressure, and the equilibrium composition is more or less independent of pressure even for reactions which involve changes in the number of dissolved molecular species.

Whereas the equilibrium constant is independent of pressure it changes with changing temperature. For $\Delta G°$ Eq. (3-4) gives the following relation:

$$\frac{d(\Delta G°/T)}{d(1/T)} = \Delta H°$$

The subscript which denotes constant pressure is here unnecessary since $\Delta G°$ is independent of pressure. If $\Delta G°$ is replaced by $-RT \ln K$, the following relation is obtained:

$$\frac{d \ln K}{d(1/T)} = \frac{-\Delta H°}{R} \qquad (3\text{-}8)$$

This very important equation, which is known as the *van't Hoff equation*, makes it possible to calculate the variation of K with temperature if $\Delta H°$ is known. The equation shows that if $\Delta H°$ is positive, that is for an endothermic reaction, the equilibrium constant increases with increasing temperature, whereas the opposite is the case for exothermic reactions. The equation may be integrated to give the equilibrium constant at any temperature T

$$\ln K_T = -\frac{1}{R} \int_{T_1}^{T} \Delta H° \, d(1/T) + \ln K_{T_1}$$

In order to evaluate K_T it is necessary to know the equilibrium constant at some temperature T_1 and $\Delta H°$ as function of temperature. Within a limited temperature range, and in the absence of phase transformations, $\Delta H°$ may be taken as constant and

$$\ln K_T = \ln K_{T_1} - (\Delta H°/R)(1/T - 1/T_1)$$

If $\ln K_T$ is plotted versus the inverse absolute temperature a straight line is obtained, the slope of which is proportional to the enthalpy of the reaction. Examples of such plots are given in Figs. 8-4 and 9-1.

3-4 GIBBS ENERGY DATA

In order to calculate the equilibrium constant for a reaction it is necessary to know its change in standard Gibbs energy. Gibbs energy data are either derived from calorimetric measurements or from known chemical equilibria. Electrochemical and spectroscopic data are also used. The best values are usually obtained by a combination of the results of various measurements. In thermodynamic tabulations the standard Gibbs energy is usually given as function of temperature for the formation of different compounds from their elements. Values for a complex reaction may then be obtained by addition or subtraction of the Gibbs energies of formation of the components.

Standard Gibbs energies of formation are sometimes listed in the following form:

$$\Delta G^\circ = \Delta H^\circ + AT \ln T + BT^2 + \cdots + ET^{-1} + IT$$

This rather cumbersome form is the result of integration of Eq. (3-4) and the use of arithmetic expressions for the heat capacities. Very often the measurements on which the equation is based are not sufficiently accurate to justify such an elaborate expression, however.

Other methods of tabulation are based on the *Gibbs energy function* and the *Barin–Knacke convention*, mentioned in Sec. 3-2. For a chemical reaction the *Gef of reaction* may be obtained by subtracting the *Gef* for the reactants from that of the products

$$\Delta Gef = \frac{\Delta G^\circ_T - \Delta H^\circ_{298}}{T}$$

When combined with listed values for ΔH°_{298}, ΔG°_T may be calculated. Similarly the Barin–Knacke Gibbs energies may be subtracted to give ΔG°_T directly.

A simpler but less accurate method is to plot graphically ΔG° as function of temperature. Such graphs have been worked out by several authors[1] and are extremely useful. In App. C graphs of ΔG° versus $T^\circ C$ are given for the formation of a number of compounds of metallurgical interest: oxides, sulfides, chlorides, fluorides, and carbides, as well as complex compounds: carbonates, sulfates, silicates, etc. In the latter cases the Gibbs energy is given for the formation of the compounds from metal oxides and CO_2, SO_3, SiO_2, etc. If needed, the Gibbs energy of formation from the elements may then be derived.

---

[1] H. J. T. Ellingham: *J. Soc. Chem Ind.*, **63**: 125 (1944). F. D. Richardson and J. H. E. Jeffes: *J. Iron. & Steel Inst.*, **160**: 261 (1948); **171**: 165 (1952). C. J. Osborn: *Trans. AIME*, **188**: 600 (1950). H. H. Kellogg: *Trans. AIME*, **191**: 137 (1951).

CHEMICAL EQUILIBRIUM **51**

The curves are all based on one mole of the common component. Thus for the oxides the curves refer to one mole of O_2

$$4Na + O_2 = 2Na_2O$$

$$2Mg + O_2 = 2MgO$$

$$\tfrac{4}{3}Al + O_2 = \tfrac{2}{3}Al_2O_3$$

$$Si + O_2 = SiO_2$$

This makes possible a simple comparison between the oxygen affinity of the different elements and also makes it easy to obtain the Gibbs energy for a reaction of the type $A + BO \rightarrow AO + B$ by simple subtraction of the $\Delta G°$ values for the two compounds.

The standard states chosen are the pure elements and compounds in their stable form at atmospheric pressure. For the gases the standard state is, strictly speaking, unit fugacity. For moderate pressures and high temperatures this is closely equal to one atmosphere. For the sulfides the standard state for sulfur is S_2 gas even though this is not stable at unit fugacity at lower temperatures. This need not worry us, however, since for any exchange reaction the element itself is canceled out. For use with reactions which involve solid or liquid sulfur as well as polymerized sulfur gas below 900°C a special curve gives the Gibbs energy of the stable form at one atmosphere relative to S_2 gas.

On examining the curves we see that in regions where there is no phase transformation they are close to straight lines. This is in agreement with the relation $\Delta G° = \Delta H° - T\,\Delta S°$, where $\Delta H°$ and $\Delta S°$ both vary little with temperature. The slope of the curve $d\Delta G°/dT$ is equal to $-\Delta S°$, and if the tangent to the curve is extrapolated to the absolute zero the intercept gives $\Delta H°$. At temperatures of transformation the curve shows a break. Increased slope corresponds to a phase transformation in the reactant, whereas decreased slope shows a phase transformation in the product. The change in slope corresponds to the entropy of transformation. It is small for melting and large for boiling.

We see further that the curves for the formation of a compound from an element and one mole of gas all have almost the same slope. This means that $\Delta S°$ is almost the same, about -180 J/K, as long as both metal and compound are condensed phases. The reason is that one gas molecule is consumed during the reaction, and this is mainly responsible for the entropy change. The same is the case for the reaction $2CO + O_2 = 2CO_2$, where there is a net consumption of one gas molecule. For the reaction $C + O_2 = CO_2$, on the other hand, the number of gas molecules remains unchanged, and the change in entropy is almost zero. Finally the reaction $2C + O_2 = 2CO$ shows a net increase of one gas molecule and $\Delta S°$ is positive, about 180 J/K. For reactions among all condensed phases, such as the formation of carbides and silicates, $\Delta S°$ is close to zero.

In most cases the curves refer to the formation of the compounds directly from the elements. This is, for example, the case for titanium where curves are drawn for the reactions $2Ti + O_2 = 2TiO$, $\frac{4}{3}Ti + O_2 = \frac{2}{3}Ti_2O_3$, $\frac{6}{5}Ti + O_2 = \frac{2}{5}Ti_3O_5$ and $Ti + O_2 = TiO_2$. From these curves the Gibbs energy of intermediate reaction steps, for example $2Ti_3O_5 + O_2 = 6TiO_2$, may be calculated. In a few other cases, for example for iron, the Gibbs energy is given for each reaction step: $2Fe + O_2 = 2FeO$, $6FeO + O_2 = 2Fe_3O_4$, and $4Fe_3O_4 + O_2 = 6Fe_2O_3$. By arithmetic addition of these values the Gibbs energy of formation of Fe_3O_4 and Fe_2O_3 from the elements may be obtained.[1]

The use of the Gibbs energy curves is shown best by a few examples. In order to obtain $\log K$ instead of $\ln K$, use is made of the conversion factor $\ln K/\log K = 2.303$ which together with $R = 8.3144$ J/(K mol) gives $\Delta G^\circ = -19.144T \log K$.

Calculation of Metallurgical Equilibria

Decomposition pressure and decomposition temperature For a reaction $2Me + O_2 \rightarrow 2MeO$

$$\Delta G^\circ = -RT \ln \left[(a_{MeO}/a_{Me})^2/p_{O_2} \right]$$

If the metal and the oxide are both present in their standard states during the reaction, their activities are unity, and $p_{O_2} = \exp \Delta G^\circ/(RT)$. This is the *decomposition pressure* of the oxide. For $2Ag_2O$, ΔG° at $25°C$ is -21 kJ. This gives $\log p_{O_2} = -3.68$, or $p_{O_2} = 2.1 \times 10^{-4}$ atm $= 21$ Pa. The oxygen pressure becomes one atmosphere when ΔG° is zero, i.e., at $200°C$. This is the *decomposition temperature* for silver oxide. In the same way we find the decomposition temperature for PdO at about $880°C$. For the partial dissociation of pyrite: $2FeS_2 = 2FeS + S_2$, p_{S_2} becomes one atmosphere at about $700°C$. Similarly $CaCO_3$ dissociates into CaO and CO_2 at about $900°C$, etc.

Heterogeneous gas equilibria. Example: $FeO + CO = Fe + CO_2$ The free energy is obtained by subtraction of ΔG° for the components:

		600°C	800°C	1000°C
$2Fe + O_2 = 2FeO$	$\Delta G^\circ =$	-410	-385	-360 kJ
$2CO + O_2 = 2CO_2$	$\Delta G^\circ =$	-414	-381	-343 kJ
$2Fe + 2CO_2 = 2FeO + 2CO$	$\Delta G^\circ =$	$+4$	-4	-17 kJ
$FeO + CO = Fe + CO_2$	$\Delta G^\circ =$	-2	$+2$	$+8.5$ kJ
$\dfrac{a_{Fe} \cdot p_{CO_2}}{a_{FeO} \cdot p_{CO}} = \exp \dfrac{-\Delta G^\circ}{RT}$	$=$	1.3	0.8	0.45

[1] A special complication arises in this case because FeO is actually nonstoichiometric and has the formula $Fe_{1-x}O$. For this compound ΔG° applies to one mole of O_2 and not to two moles of Fe. Arithmetic addition is, in this case, not strictly correct. A way to handle this problem will be discussed in Sec. 4-5.

As Fe and FeO can coexist with activities of unity or close to unity, the equilibrium constant is equal to the gas ratio CO_2/CO. Similarly the gas ratio can be calculated for the reduction $Fe_3O_4 + CO = 3FeO + CO_2$ and for the reduction of other metal oxides with carbon monoxide or hydrogen. See Sec. 9-1.

Equilibrium among metals and their oxides The Gibbs energy of reaction is obtained by subtraction of $\Delta G°$ for the formation of the different oxides:

		500°C	600°C
$2Fe + O_2 = 2FeO$	$\Delta G° = -423$	-410 kJ
$6FeO + O_2 = 2Fe_3O_4$	$\Delta G° = -431$	-406 kJ
$2Fe + 2Fe_3O_4 = 8FeO$	$\Delta G° = \quad +8$	-4 kJ

We see that FeO is unstable under atmospheric pressure at 500°C and will decompose to form Fe and Fe_3O_4 whereas at 600°C Fe and Fe_3O_4 will react to form FeO. The three phases can only coexist at one temperature, 570°C.

In the above examples the metals and the oxides were assumed to coexist as pure, or practically pure, phases and their activities were taken equal to unity. In actual equilibria this is not necessarily the case. Thus the reduction of TiO_2 takes place through a number of intermediate oxides, and Ti and TiO_2 cannot coexist. In this case the equilibrium constant has to be calculated for each reduction step. Even then the phases are not pure. Metallic titanium, for example, dissolves considerable amounts of oxygen, and for coexistence with the TiO phase its activity is less than one. Detailed knowledge of the activities as function of composition is, in that case, needed in order to calculate the oxygen pressure or gas ratio for the reduction process. The result of combined calculations and measurements for the titanium–oxygen system is shown in Fig. 14-5.

For equilibria which involve several metals, these may form solid and liquid solutions with each other. Similarly, their oxides may react to form complex compounds, and, above the melting point, molten slags. Also in these cases calculations are dependent on activity data for the metallic and the oxide solutions. Finally, new phases may be formed which were not anticipated in the calculation. Thus if titanium oxides are reduced with carbon, the product will be a carbide rather than metallic titanium. A calculation of the equilibrium conditions for reduction to metal will, in such a case, be of little value.

3-5 MULTICOMPONENT EQUILIBRIA

Metallurgical reactions usually involve a number of different reactants, and a number of different reaction products may be formed. The types, composition, and amounts of the different reaction products will be determined partly by the relative amounts of the various reactants and partly by the chemical equilibria which can be established between the reactants and reaction products. Furthermore, the reactants and reaction products may occur in a number of different *phases*, such as solid or liquid metals, slags, and gases. We shall first discuss the effect of the number of phases on the chemical equilibria which can be established.

The Gibbs Phase Rule

A *phase* is defined as a part of a system, and is in itself homogeneous and separated from the rest of the system by an interface. The interface need not be continuous; many parts of the same homogeneous material are counted as one phase. The phases may be solid, liquid, and gaseous. In a given system we can have only one gaseous phase, as all gases are mutually miscible, whereas we may have several liquid and solid phases. Typical examples of mutually immiscible liquids are molten metals, mattes, and slags, or at room temperature, oil, water, and mercury. Typical examples of solid phases are metal oxides and carbides as well as various metallic phases. A given phase need not correspond to a definite chemical composition. In the same way as gases are mutually miscible, liquid and solid compounds may be more or less miscible to give liquid and solid solutions. It is the solution, and not the compound, which is the phase. The *number* of phases which may exist in mutual equilibrium is given by the *Gibbs phases rule:*

We consider a system which contains a number, C, of *components*, which in the following will be taken equal to the participating chemical elements. These are distributed over a number, P, of phases. This will be the case even though the concentration of some of the components in some of the phases may be too low to be determined by ordinary methods. At chemical equilibrium the *partial Gibbs energy* of each component i must be the same in all phases

$$\bar{G}_i^{(1)} = \bar{G}_i^{(2)} = \bar{G}_i^{(3)} = \cdots = \bar{G}_i^{(P)}$$

where the parentheses denote the different phases. Instead of partial Gibbs energy, the activity or the partial vapor pressure of each component could have been used. In any case, for each component this gives $(P-1)$ equations, i.e., in total $C(P-1)$ equations. This number is to be compared with the number of *unknowns* or *independent variables*. These are first of all the variables in phase compositions. In order to specify the composition of one phase it is necessary to specify $(C-1)$ composition variables, e.g., concentrations, in that phase. For a number, P, of phases this gives a total of $P(C-1)$ composition variables. In addition the temperature and the pressure of the system must be specified, which makes the total number of variables equal to $P(C-1) + 2$. If the number of variables is equal to the number of equations, the system is fixed or *nonvariant*, i.e., the composition of each phase as well as the temperature and pressure are fixed. If the number of variables is larger than the number of equations, the difference represents the number, F, of *degrees of freedom:*

$$F = P(C-1) + 2 - C(P-1) = C + 2 - P \tag{3-9}$$

Thus the number of degrees of freedom gives the number of variables which can be changed without this affecting the number of phases. The variables which can be changed are temperature, pressure, and one or more composition variables in one of the phases. As all phases are in mutual equilibrium this will again lead to corresponding changes in composition variables of all other phases, even though these changes are not always large enough to be detected.

Restrictions to the phase rule As already mentioned we have taken the chemical elements in the system to represent the components. This may occasionally lead to results which apparently disagree with experience. Thus if one considers the dissociation of limestone according to the equation

$$CaCO_3(s) = CaO(s) + CO_2(g)$$

and chooses as components the three elements calcium, carbon, and oxygen, one gets with three phases (limestone, burned lime, and gas) the number of degrees of freedom

$$F = 3 + 2 - 3 = 2$$

This is against all experience, which tells that the system is monovariant: the decomposition pressure is a single function of temperature. We have, however, here forgotten that in defining the system we have also fixed the composition of the three phases to obey the *stoichiometric relation* $n_O = n_{Ca} + 2n_C$. Thus we have put a *restriction* on the system by fixing one of the composition variables, and we cannot operate freely within the ternary system Ca–C–O. Every time we put a restriction on the system we reduce the number of degrees of freedom, and we may rewrite the phase rule:

$$F = C + 2 - P - R \tag{3-10}$$

where R is the number of independent restrictions. Thus if the decomposition of limestone is considered in terms of the ternary Ca–C–O system we have one composition restriction, and with three phases we have one degree of freedom which agrees with experience. If, on the other hand, the decomposition of limestone took place in an atmosphere which also contained CO, that restriction is removed and we will have two degrees of freedom, e.g., temperature and gas composition, which together determine the total gas pressure.

In addition to fixing one or more composition variables we may put further restrictions on the system by specifying a given constant temperature or constant pressure. As an example, if we consider the slag system $CaSiO_3$–TiO_2 we have a total of four elements. On the other hand we have two composition restrictions: $n_O = n_{Ca} + 2n_{Si} + 2n_{Ti}$, and $n_{Ca} = n_{Si}$. This gives

$$F = 4 + 2 - P - 2 = 4 - P$$

as for a simple two-component system. If also we specify constant temperature and pressure we get $F = 2 - P$, that is, the system can at the most contain only two phases in mutual equilibrium.

It should finally be pointed out that the phase rule applies only to systems which are in complete chemical equilibrium. When a system is away from equilibrium, as when a chemical reaction is progressing, the number of phases may be larger than what can be calculated from the phase rule. Thus in the reduction of Fe_2O_3 with CO we may, in addition to the gas phase, have simultaneous occurrence of the phases Fe_2O_3, Fe_3O_4, FeO, and Fe as well as solid carbon, even though these phases cannot all coexist under equilibrium conditions. Furthermore, and particularly at lower temperatures, we may have compounds and phases which are *metastable* but which have a considerable life-time, like Fe_3C. This again will lead to an apparent increase in the number of degrees of freedom of the system.

The Phase Rule Applied to Metallurgical Reactions

In metallurgical processes various raw materials are mixed and brought to reaction at some temperature. The equilibrium state of the system may then be described by (1) the number and type of phases formed, and (2) the composition and pressure of these phases. These may be calculated from thermodynamic data for the compounds and phases involved. As an example we shall consider a system Me–C–O which may contain the phases Me, MeO, MeC, C, and gas. For simplicity we have assumed the four condensed phases to be rather pure stoichiometric, but this need not be the case as long as they appear as separate phases. At a fixed temperature but unspecified pressure we have

$$F = 3 + 2 - P - 1 = 4 - P$$

Thus in addition to the gas phase we may have one, two, or three condensed phases coexisting, but not all four. In order to see which condensed phases may coexist we look at the reaction Me + C = MeC. If $\Delta G°$ for this is negative the metal and carbon phases cannot coexist, and we may have the two following *four-phase* combinations:

(*A*)   MeO, MeC, C, gas

(*B*)   Me, MeO, MeC, gas

For each of these combinations the *pressure* and *composition* of the gas phase is fixed at a given temperature. In many cases the gas phase will consist mainly of carbon monoxide, the pressure of which may be calculated from the two reaction equilibria:

(*A*)   MeO + 2C   = MeC + CO

(*B*)   MeO + MeC = 2Me + CO

As the condensed phases are either pure or of some fixed equilibrium composition the partial pressure of CO is fixed and may be calculated as function of temperature from known $\Delta G°$ values for the reaction. If in addition the gas should also contain some carbon dioxide, its partial pressure can be calculated from the reaction equilibria:

$$(A) \quad 2\text{MeO} + 3\text{C} \quad = 2\text{MeC} + \text{CO}_2$$

$$(B) \quad 2\text{MeO} + \text{MeC} = 3\text{Me} + \text{CO}_2$$

At high temperature the gas phase may also contain some metal vapor or some volatile suboxide like $\text{Me}_2\text{O}(g)$. For example, for phase combination (A) their partial pressures may be calculated from the equilibria:

$$\text{MeC} = \text{Me}(g) + \text{C}$$

$$\text{MeO} + \text{MeC} = \text{Me}_2\text{O}(g) + \text{C}$$

and similarly for phase combination (B).

Finally the total pressure is calculated as the sum of all significant partial pressures.

Notice that in all cases the reaction equations are balanced to give only one gaseous product. Reactions like

$$3\text{MeO} + 5\text{C} = 3\text{MeC} + \text{CO} + \text{CO}_2$$

or

$$4\text{MeO} + 7\text{C} = 4\text{MeC} + 2\text{CO} + \text{CO}_2$$

although thermodynamically permissible, are unable to give the relative amounts of CO and CO_2 in the gas phase.

So far we have considered combinations of three condensed phases and a gas. In addition we may have combinations with only two or even only one condensed phase. This leads to additional degrees of freedom, and in order to calculate the pressure and composition of the gas it is necessary to specify some additional parameter. This could for example be the CO pressure. Thus for the two condensed phases MeO and MeC we may calculate the equilibrium

$$\text{MeO} + 3\text{CO} = \text{MeC} + 2\text{CO}_2$$

If we specify the CO pressure we may calculate the CO_2 pressure, which together give p_{tot}. Alternatively we may calculate the gas ratio $(p_{CO_2})^2/(p_{CO})^3$, but we cannot calculate the CO and CO_2 pressures independent of each other.

The relation between the CO and CO_2 pressures in the Me–C–O system at some constant temperature is illustrated in Fig. 3-3. Here the lines which describe equilibrium with two condensed phases or with solid carbon are defined by the equilibria

$$(1) \quad \text{MeO} + \text{CO} = \text{Me} + \text{CO}_2$$
$$(2) \quad \text{Me} + 2\text{CO} = \text{MeC} + \text{CO}_2$$
$$(3) \quad \text{MeO} + 3\text{CO} = \text{MeC} + 2\text{CO}_2$$
$$(4) \quad 2\text{CO} = \text{C} + \text{CO}_2$$

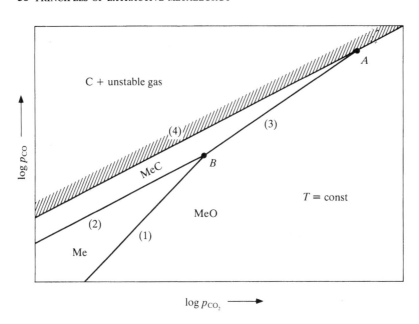

Figure 3-3 Stability regions for condensed phases in the Me–C–O system.

It will be noticed that the *slope* of each line is given by the *stoichiometry* of the corresponding reaction. Thus for reaction 1 the slope $d \log p_{CO}/d \log p_{CO_2} = 1/1$, and for reaction 3 it is $2/3$. The intersection of the lines 1, 2, and 3 corresponds to phase combination B, and the intersection of the lines 3 and 4 to phase combination A. Reaction 4 is the so-called *Boudouard reaction*. It will be noticed that any gas mixture with a CO pressure above line 4 is unstable and will decompose with precipitation of soot. Finally for gas compositions within the areas between the lines only one condensed phase is stable, and the CO_2 and CO pressures may be varied independently of each other.

Diagrams like the one shown in Fig. 3-3 may be adapted to most ternary systems and some further examples will be shown in Sec. 8-4 (roasting of sulfides). Similar diagrams may be constructed which also show the effect of temperature at some constant pressure on the various phase equilibria. One example will be shown in Sec. 14-1 (carbothermic reduction of silica). The same general principle may also be expanded to cover systems with more than three components.

Simultaneous Equilibria

If complete chemical equilibrium is to be established within a multicomponent system, all chemical reactions that can be written between the various components and compounds must be in equilibrium. This is called *simultaneous equilibrium*. Although the number of chemical reactions which can be written is very large the number of *independent* equilibria is given by the number of chemical *species* minus the number of *components*. Usually the number of components is equal to the number of elements. However, if there are one or more composition restrictions imposed on the system the number of components is equal to the number of elements minus the number of such restrictions.

For a gas mixture which contains the elements carbon, hydrogen, and oxygen we may for a certain temperature and composition range consider the five gaseous species CO_2, CO, H_2O, H_2, and CH_4. As there are no composition restrictions, the system may be described by the two equilibria:

$$(1) \quad CO + H_2O = CO_2 + H_2 \qquad K_1 = \frac{p_{CO_2} \cdot p_{H_2}}{p_{CO} \cdot p_{H_2O}} = \frac{n_{CO_2} \cdot n_{H_2}}{n_{CO} \cdot n_{H_2O}}$$

$$(2) \quad CO + 3H_2 = CH_4 + H_2O \qquad K_2 = \frac{p_{CH_4} \cdot p_{H_2O}}{p_{CO} \cdot p_{H_2}^3} = \frac{n_{CH_4} \cdot n_{H_2O} \cdot n_{tot}^2}{n_{CO} \cdot n_{H_2}^3 \cdot p_{tot}^2}$$

Any other reaction such as $CO_2 + 4H_2 = CH_4 + 2H_2O$ may be obtained by arithmetic addition or subtraction of the first two, and are therefore not independent equilibria. On the other hand, if we also want to include gaseous oxygen among the species, a third equilibrium is needed, e.g.,

$$(3) \quad 2CO + O_2 = 2CO_2 \qquad K_3 = \frac{p_{CO_2}^2}{p_{CO}^2 \cdot p_{O_2}} = \frac{n_{CO_2}^2 \cdot n_{tot}}{n_{CO}^2 \cdot n_{O_2} \cdot p_{tot}}$$

Similarly additional equilibria have to be added if we want to consider other hydrocarbons, such as C_2H_6, or monatomic hydrogen or oxygen. Furthermore, if solid carbon is present in the system this gives rise to the additional equilibrium

$$(4) \quad CO_2 + C(s) = 2CO \qquad K_4 = \frac{p_{CO}^2}{p_{CO_2}} = \frac{n_{CO}^2 \cdot p_{tot}}{n_{CO_2} \cdot n_{tot}}$$

the activity of solid carbon being unity.

We see that in these cases the number of unknowns, i.e., the partial pressures or number of moles of the gaseous species, is larger than the number of independent equilibria. In order to calculate the composition of the equilibrium gas mixture additional equations are therefore needed. These may be obtained from the *stoichiometry* of the reactions. Thus if an initial gas mixture had consisted of $n_{H_2O}^\circ$ moles of H_2O and n_{CO}° moles of CO, and if, after equilibrium is established, it contains n_{H_2O}, n_{H_2}, n_{CO_2}, n_{CO} and n_{CH_4} moles of the five species, then a *mass balance* gives

$$n_{CO} = n_{CO}^\circ - x - y$$
$$n_{CO_2} = x$$
$$n_{H_2O} = n_{H_2O}^\circ - x + y$$
$$n_{H_2} = x - 3y$$
$$n_{CH_4} = y$$

$$\overline{n_{tot} = n_{CO}^\circ + n_{H_2O}^\circ - 2y}$$

Here x and y are the number of moles of CO_2 and CH_4 formed during the reaction. By inserting the number of moles into the known equilibrium constants K_1 and K_2 the values for x and y may in principle be calculated, and the equilibrium gas composition derived. This will, however, involve solving equations of higher order, and in most cases exact arithmetic solutions are impossible. The problem may then be solved in three different ways: (1) successive approximations (iteration), (2) graphical methods, (3) Gibbs energy minimization.

Successive approximation If only a few calculations are needed iteration may be the easiest method and will be illustrated by the following case: We shall try to calculate the equilibrium composition at 800°C for a gas which originally contained 1 mol of CO, and 2 mol of H_2O at 1 atm total pressure. For this temperature $K_1 = 1.2$ and $K_2 = 0.01$. As $K_2 \ll K_1$, we assume in the first approximation that n_{CH_4}, that is, $y = 0$. This gives, by inserting in K_1

$$K_1 = \frac{x^2}{(1-x)(2-x)} = 1.2 \qquad \text{that is, } x = 0.70$$

This value for x is inserted in K_2 taking p_{tot} equal to unity, and disregarding y in all terms where it is small compared to the rest of the term. This gives

$$K_2 = \frac{11.7y}{0.103} = 0.01, \qquad \text{that is, } y = 9 \times 10^{-5}$$

We see that our original assumption is verified, and we can calculate the equilibrium gas composition

$n_{CO_2} = 0.70$	$p_{CO_2} = 0.233$
$n_{CO} = 0.30$	$p_{CO} = 0.100$
$n_{H_2O} = 1.30$	$p_{H_2O} = 0.433$
$n_{H_2} = 0.70$	$p_{H_2} = 0.233$
$n_{CH_4} = 9 \times 10^{-5}$	$p_{CH_4} = 3 \times 10^{-5}$
$n_{tot} = 3.00$	$p_{tot} = 1.00$

We will now try to do the same for 500°C. We will assume that CO and H_2 occur in small amounts, but we see that if one or both of these are put equal to zero in the equilibrium expressions, everything else becomes zero. Instead we put n_{CO} and n_{H_2} equal to zero in the mass balance. This gives $x = n_{CO_2} = 0.75$, $y = n_{CH_4} = 0.25$ and $n_{H_2O} = 1.5$. If these values are inserted in the equilibrium constants, we find as a first approximation $n_{CO} = 0.062$ and $n_{H_2} = 0.72$. These values are inserted back into the mass balance to give $x = 0.89$ and $y = 0.055$. These values are once more inserted into the equilibrium constants to give

$n_{CO} = 0.059$ and $n_{H_2} = 0.45$. In this way we may continue, and after the next round we get

$$x = n_{CO_2} = 0.85 \qquad p_{CO_2} = 0.301$$
$$n_{CO} = 0.06 \qquad p_{CO} = 0.021$$
$$n_{H_2O} = 1.24 \qquad p_{H_2O} = 0.440$$
$$n_{H_2} = 0.58 \qquad p_{H_2} = 0.206$$
$$\underline{y = n_{CH_4} = 0.09 \qquad p_{CH_4} = 0.032}$$
$$n_{tot} = 2.82 \qquad p_{tot} = 1.000$$

With this result we declare ourselves satisfied. We see that our initial assumption that n_{H_2} is close to zero was rather far from the truth, but nevertheless the method proved successful. It would have been better for the above case if we had chosen $n_{CH_4} = 0$. We see that under *equilibrium conditions* only 58 percent of the initial CO is converted to hydrogen and 9 percent to methane.

In these calculations we have disregarded oxygen gas or higher hydrocarbons as possible species. This is permissible because their partial pressures will be too low to affect the mass balance significantly. Nevertheless we may now calculate, for example, the relative oxygen pressure for the above case at 800°C. At this temperature K_3 has a value of 4×10^{18}. Inserting the calculated values for p_{CO} and p_{CO_2} the relative oxygen pressure is calculated to be 1.36×10^{-18} and n_{O_2} to 4.08×10^{-18}. We see that these values would not have affected the mass balance.

If solid carbon is present in the system we have to consider reaction 4, the *Boudouard reaction*. This gives one additional equation. But on the other hand we get one additional unknown, the number of moles of solid carbon *consumed* during the reaction. If this is denoted z we obtain the revised mass balance

$$n_{CO} = n_{CO}^{\circ} - x - y + 2z$$
$$n_{CO_2} = x - z$$
$$n_{H_2O} = n_{H_2O}^{\circ} - x + y$$
$$n_{H_2} = x - 3y$$
$$\underline{n_{CH_4} = y}$$
$$n_{tot} = n_{CO}^{\circ} + n_{H_2O}^{\circ} - 2y + z$$

If calculations are carried out for the given initial gas mixture in the presence of solid carbon we will find that z has a positive value and that the concentration of CO and CH_4 as well as the total number of moles had increased as compared to the case where carbon had been absent.

We may now summarize the method of iteration:

1. The values for the unknowns, x, y, z, etc., are estimated by means of those compounds which seem to occur in the smallest amounts. These amounts may either be taken equal to zero, or, by experience, to some small value.

2. The amounts of the remaining compounds are calculated from the mass balance. Since these compounds occur in the larger amounts, the relative precision of these values will be higher than of the estimated values for the first set of compounds.
3. The values thus obtained are inserted in the equilibrium expressions and new values are obtained for the originally estimated ones. If possible, these values should be obtained from those equilibrium expressions where they are raised to some high power. Thus n_{CO} should preferably be calculated from K_3 and n_{H_2} from K_2. In that way the highest precision is obtained.
4. The values thus obtained are inserted in the mass balance for the calculation of the amounts of the remaining compounds, and the whole operation is repeated until a satisfactory precision is obtained for all compounds.

The above procedure refers to systems where complete chemical equilibrium is established. Often this is not the case, and in such cases the composition of the effluent flow has to be evaluated from combined kinetic and stoichiometric calculations, as will be discussed in Chap. 5.

Graphical methods If more than a few calculations are needed graphical methods may be useful. For this purpose a number of arbitrary equilibrium gas compositions are first calculated from the equilibrium constants. By means of the mass balance corresponding compositions of the initial reaction mixture are calculated. A set of curves are then prepared where different equilibrium compositions are plotted against different initial reactant mixtures. Thus, as compared to the previous treatment, the calculations are performed in the reverse order, starting with the results and calculating the initial mixtures. From this set of curves the equilibrium gas composition for a given reaction mixture may now be read, if necessary by interpolation. The types of graphs may be varied according to the type of reaction studied, and may include gaseous as well as solid phases. Examples of such graphs which describe the equilibria between CO_2–CO–H_2O–H_2–CH_4 mixtures and solid iron, iron oxide, and iron carbide are given by Darken and Gurry.[1]

Gibbs energy minimization This method requires a computer, and is particularly useful if a large number of calculations are required. The principle of the method is that the Gibbs energy of a system at constant pressure and temperature has its minimum value when the system is in internal equilibrium. The total Gibbs energy is the sum of the molar Gibbs energy—for solutions the partial molar Gibbs energy—for each species multiplied by the number of moles of that species. For a multiphase system the summation is done for each phase

$$G_{tot} = \sum n_{ij} G_i$$

[1] L. S. Darken and R. W. Gurry: "Physical Chemistry of Metals," McGraw-Hill Book Co. Inc., New York, 1953, pp. 218–222.

where i and j represent the different species and phases respectively. The molar and partial molar Gibbs energies may be referred to the pure elements as standard state.

Whereas for gaseous mixtures the variation of the molar Gibbs energy is given by Eq. (3-3) the partial molar Gibbs energies in solid or liquid solutions require a knowledge of the thermodynamics of those solutions as will be discussed in Chap. 4. For given total amounts of the reacting elements the computer calculates the total Gibbs energy of the system for all possible combinations of phases and phase compositions which satisfy the mass balance, and that one combination is selected which gives the lowest total Gibbs energy. Various computer programs have been developed as, e.g., Solgasmix[1] and will not be discussed further in this text.

PROBLEMS

3-1 By means of the data given in Prob. 2-1 calculate the increase in entropy when one mole of I_2 is heated from 298 to 473 K at atmospheric pressure, and calculate the entropy at 473 K when $S^\circ_{298} = 116.8$ J/K.

3-2 One mole of supercooled liquid zinc solidifies at 400°C in a container with a large heat capacity. Calculate the change in entropy, (a) for the zinc, and (b) for the container, and calculate the overall change in entropy. The melting point of zinc is 420°C and $\Delta H_m = 7280$ J/mol. The heat capacity of solid and liquid zinc are assumed to be equal.

3-3 For tin the densities of the solid and the liquid are: 7.184 and 6.99 g/cm^3. The melting point is 231.9°C, and the heat of fusion at the melting point is 58.6 J/g. Calculate the change in melting temperature when the pressure is increased from 1 to 2 atm.

3-4 For the reaction H_2 (g, 1 atm) $+ \frac{1}{2}O_2$(g, 1 atm) $= H_2O$(l), $\Delta H^\circ_{298} = -285.9$ kJ and $\Delta G^\circ_{298} = -238.7$ kJ. (a) Calculate ΔS°_{298}. (b) The reaction may be carried out at 298 K and 1 atm either reversibly, e.g., in a fuel cell, or completely irreversibly, i.e., by combustion without any work beyond volume work. Calculate for the two cases the heat, q, absorbed from the surroundings, and for the first case the work, w', done on the surroundings. Show that these findings are in agreement with the second law of thermodynamics.

3-5 The α-γ transformation in Fe at $P = 1$ atm occurs at 1183 K with $\Delta H = 900$ J/mol. C_P is for Fe(α): 37.7 and for Fe(γ): 30.6 J/(K·mol). The molar volumes are for Fe(α): 7.39 and for Fe(γ): 7.33 cm^3, and are regarded as independent of temperature and pressure. Calculate ΔC_P, ΔS, and ΔV, and derive an expression for ΔG for the reaction Fe(α)$\rightarrow$ Fe(γ) as function of temperature and pressure.

3-6 The standard molar free energy change for the conversion C(graph) = C(diam) is at 1000 K: $+6.7$ kJ and at 2000 K: $+11.3$ kJ. The densities are 2.25 g/cm^3 for graphite and 3.51 g/cm^3 for diamond. The volume change ΔV is assumed to be independent of temperature and pressure. (a) Calculate the minimum pressure required to convert graphite into diamond at the two temperatures. (b) In practice synthetic diamonds are made at temperatures in excess of 2000 K. Give the reason for this.

3-7 The vapor pressure of liquid zinc is given by the approximate equation: log p_{Zn} (atm) $= -6411/T + 5.56$. (a) Calculate the heat of vaporization and calculate the normal boiling point as well as the vapor pressure at the melting point, 693 K. (b) The heat of fusion for zinc is 7300 J/mol.

[1] G. Eriksson: *Chemica Scripta* **8**:100 (1975).

Derive an expression for the vapor pressure of solid zinc as function of temperature. (c) In this problem, which assumptions have been made with respect to the heat capacity of the solid, liquid, and gas?

3-8 The mean heat capacity of liquid magnesium is 32.7, and of magnesium vapor 20.8 J/(K · mol). The heat of vaporization and the vapor pressure are, at the melting point of 650°C: 133 500 J/mol and 3.6×10^{-3} atm respectively. Derive an expression for the vapor pressure as function of temperature, and calculate the boiling point as well as the heat and entropy of vaporization at the boiling point.

3-9 (a) Show qualitatively by means of a sketch how to expect the following functions for nickel to vary in the temperature range 0 to 1800 K: C_P, $S°$, $H°$, and $G°$. Nickel has a magnetic transformation point (λ point) about 630 K and melts at 1728 K.

(b) Show qualitatively by means of a sketch how you expect $\Delta H°$ and $\Delta G°$ to vary with temperature for the reaction $Ni + Cl_2(g) = NiCl_2$. $NiCl_2$ sublimes without melting at 1243 K. State any assumptions you have made.

3-10 Nitrogen is purified for oxygen by passage over finely divided and heated copper. (a) By using the data given in App. C calculate the oxygen pressure for the reaction $4Cu + O_2 = 2Cu_2O$ at 300, 500, and 800°C. (b) Calculate the residual percentage of oxygen in the gas if equilibrium is established at these temperatures and for pressures of one and ten atmospheres. (c) In practice a temperature of about 500°C is used. What is the reason for this choice?

3-11 (a) From the data given in App. C calculate the equilibrium constant for the reaction $Cr_2O_3(s) + 3H_2(g) = 2Cr(s) + 3H_2O(g)$ at 1200°C and calculate the equilibrium percentage of water vapor in hydrogen at total pressures of one and ten atmospheres. Cr_2O_3 and Cr are assumed to be present in their standard states. (b) From the same data calculate $\Delta H°$ and $\Delta S°$ for the above reaction and temperature.

3-12 From the data given in App. C calculate the equilibrium constant for the reaction $C + CO_2 = 2CO$ at 700°C and calculate the composition of the gas mixture for $p_{CO} + p_{CO_2} = 0.2, 1,$ and 10 atm.

3-13 The free energy change for the following reactions are known (each compound represents a separate phase):

$$Al_2O_3(s) + SiO_2(s) = Al_2SiO_5(s) \qquad \Delta G° = -8320 - 0.4T \text{ J}$$

$$3Al_2O_3(s) + 2SiO_2(s) = Al_6Si_2O_{13}(s) \qquad \Delta G° = +22\,770 - 31.8T \text{ J}$$

(a) Calculate $\Delta G°$ as function of temperature for the two reactions $3Al_2SiO_5 = Al_6Si_2O_{13} + SiO_2$ and $Al_6Si_2O_{13} = 2Al_2SiO_5 + Al_2O_3$, and calculate the temperatures where $\Delta G°$ is zero.

(b) Construct a phase diagram (all solid and stoichiometric phases) for the system SiO_2–Al_2O_3 between 45 and 65 mole percent Al_2O_3 for atmospheric pressure, and covering the temperatures calculated under (a).

(c) For the reaction $3Al_2SiO_5 = Al_6Si_2O_{13} + SiO_2$, $\Delta V > 0$. How will the temperature for coexistence between these three phases be affected by an increase in pressure?

3-14 For BN(s) and $B_4C(s)$ the following standard enthalpies and entropies for formation from the elements are known:

	$\Delta H°_{298}$, kJ	$\Delta S°_{298}$, J/K
$B(s) + \frac{1}{2}N_2 = BN(s)$	-251.0	-86.2
$4B(s) + C(s) = B_4C(s)$	-58.6	-1.7

For both reactions ΔC_P is assumed equal to zero. Calculate the temperature at which $p_{N_2} = 10^{-3}$ atm for the following two phase combinations: (a) BN + B, and (b) BN + C + B_4C.

BIBLIOGRAPHY

Bodsworth, C., and A. S. Appleton: "Problems in Applied Thermodynamics," Longmans, London, 1965.
Darken, L. S., and R. W. Gurry: "Physical Chemistry of Metals," McGraw-Hill Book Co. Inc., New York, 1953.
Gaskell, D. R.: "Introduction to Metallurgical Thermodynamics," 2d ed., McGraw-Hill Book Co. Inc., New York, 1981.
Kubaschewski, O., and C. B. Alcock: "Metallurgical Thermochemistry," 5th ed., Pergamon Press, London, 1979.
Lewis, G. N., and M. Randall: "Thermodynamics," revised edition by K. S. Pitzer and L. Brewer, McGraw-Hill Book Co. Inc., New York, 1961.
Reed, Th. B.: "Free Energy of Formation of Binary Compounds: An Atlas of Charts for High-Temperature Chemical Calculations," The MIT Press, Cambridge, Mass., 1971.

CHAPTER
FOUR

MELTS AND SOLUTIONS

Molten phases play an important role in extractive metallurgy. Thus in the production of iron or copper one of the first steps is a smelting operation in which the valuable elements are enriched in one molten phase, whereas the gangue and other impurities are discarded in another (the slag). In later steps the metallic melt is refined further and other impurities are removed.

The importance of smelting processes lies in the tendency for different elements to separate into different phases. There are usually two or more liquid phases, for example metal and slag, but two metallic melts may also occur. In other cases the separation is between a liquid and a solid phase. Finally the separation may be between a liquid and a gas as for example in distillation or degassing.

The tendency for the different elements to become enriched in the different phases depends on their chemical affinity for these phases, i.e., on the thermodynamics of the phases. An understanding of the thermodynamics of melts and solutions is therefore of great importance.

4-1 PURE MELTS

The thermodynamic functions for a pure element, Me, around its melting point at constant pressure was shown in Fig. 3-2. For the solid as well as for the liquid phase the Gibbs energy curve is described by the equation

$(\partial G/\partial T)_P = -S$. At the melting point the Gibbs energy of the two phases is equal, and the entropy of fusion $\Delta S_m = \Delta H_m/T_m$. For most metals the molar entropy of fusion is between 8 and 12 J/K. This gives an approximate relation between the enthalpy of fusion and the melting point, known as *Richard's rule:* $\Delta H_m \approx 9T_m$. The entropy of fusion is a result of the increase in disorder associated with the melting process. Most metals have a close-packed structure with a coordination number of 12 in the solid state. On melting the long-range order of the structure breaks down and the average coordination number may be about 11. For all such metals the entropy of fusion is approximately the same.

Certain metalloids such as bismuth and silicon as well as water have a complex structure in the solid state. On melting, the complex structure breaks down. This gives rise to a somewhat larger entropy of fusion of about 17 to 25 J/K per gram atom (for water: per mole). For ionic compounds (salts) the entropy of fusion is about 13 J/K per gram ion in the salt. Thus for a di-ionic salt such as the alkali halides the entropy of fusion is about 26 J/K per mole. This shows that for ionic compounds the change in order on melting is also small. Exceptionally small entropies of fusion are found for the silicates, particularly for silica, where the entropy of fusion is less than 4 J/K. This shows that the large silicate network in the solid is only slightly distorted on melting. Other compounds with small entropies of fusion are Ag_2S and Ag_2SO_4. These compounds are already strongly disordered in the solid state.

Around the melting point the entropy of fusion varies with the temperature according to the relation $d\Delta S_m = \Delta C_P \, d \ln T$ where ΔC_P is the change in heat capacity on melting. By integration we get

$$\Delta S_m = \int_{T_m}^{T} \Delta C_P \, d \ln T + \Delta H_m/T_m$$

This again gives ΔG_m at constant pressure as a function of temperature

$$\Delta G_m = (\Delta H_m/T_m)(T_m - T) - \int\int_{T_m}^{T} \Delta C_P \, d \ln T \, dT \qquad (4\text{-}1)$$

The heat capacity of the melt is usually larger than that of the solid, but for temperatures not very far from the melting point the second term in Eq. (4-1) is usually small and may be disregarded.

4-2 THERMODYNAMICS OF SOLUTIONS

The following thermodynamic relations for multicomponent systems have their major application for solutions. It should be emphasized, however, that the relations are equally valid for multiphase systems.

A solution or a mixture may be described by *partial* and *integral* thermodynamic functions. These functions may be for example the enthalpy, the entropy, the Gibbs energy, or the volume.

We consider a system which contains n_1 moles of component 1, n_2 moles of component 2, etc. Any extensive function X of the system is a function of temperature and pressure, and of the number of moles of components

$$X = f(T, P, n_1, n_2, \ldots, n_n)$$

At constant temperature and pressure we have

$$dX = \left(\frac{\partial X}{\partial n_1}\right)_{T, P, n_i} dn_1 + \left(\frac{\partial X}{\partial n_2}\right)_{T, P, n_j} dn_2 \cdots \text{etc.} \qquad (4\text{-}2)$$

Here n_i denotes the number of moles of all components except 1, and n_j denotes the number of moles of all components except 2. The partial derivative $(\partial X/\partial n_1)_{T, P, n_i}$ is called the *partial molar* function of component 1 in the solution. We will express this function by the symbol $\bar{X}_1$. Thus the partial molar volume is denoted $\bar{V}$, the partial molar enthalpy $\bar{H}$, the partial molar entropy $\bar{S}$, and the partial molar Gibbs energy $\bar{G}$.[1]

It follows from Eq. (4-2) that the partial molar functions for component 1 represent the increase in that function for the solution when one mole of component 1 is added to a large quantity of solution at constant temperature and pressure, keeping the quantity of all other components constant.

In the same way as we can assign absolute values to the total volume and entropy of a system, we can assign absolute and measurable values to the partial molar volume and entropy. In contrast, we cannot assign absolute values to $\bar{H}$ or $\bar{G}$. These may only be expressed in terms of differences, as for example relative to a chosen standard state. These differences are called *relative partial molar functions:* $\Delta\bar{X} = \bar{X} - X^\circ$. Here X° is the molar function for the component in a chosen standard state, which usually is the pure component at the same temperature and pressure as the solution. Other standard states may also be chosen, however, and it is obvious that the magnitude of $\Delta\bar{X}$ depends on the choice of standard state. We shall later see some examples of standard states different from the pure component.

Whereas the total thermodynamic function X of the system is an extensive property, i.e., proportional to the total number of moles, the partial molar functions $\bar{X}$ and $\Delta\bar{X}$ are intensive properties and are only functions of the composition of the solution.

Integral Quantities

The total function X for the solution is called the *integral function* or *integral quantity*. The relation between the integral and partial quantities can be visualized if we consider the system being built up by infinitesimal additions of the

---

[1] The partial molar Gibbs energy is sometimes denoted μ and called the "chemical potential." There is no need for this distinction from other partial quantities, and in this text the symbol $\bar{G}$ will be used exclusively.

various components in such a way as to keep the ratio between the different elements constant, i.e., at constant composition. After a total of n_1 moles of component 1, n_2 of component 2, etc., have been added the integral function for the system is obviously equal to

$$X = n_1 \bar{X}_1 + n_2 \bar{X}_2 + \cdots + n_n \bar{X}_n \qquad (4\text{-}3)$$

If the integral function is divided by the total number of moles $n_1 + n_2 + \cdots + n_n$, we get the *integral molar function* X_M

$$X_M = \frac{X}{n_{\text{tot}}} = N_1 \bar{X}_1 + N_2 \bar{X}_2 + \cdots + N_n \bar{X}_n \qquad (4\text{-}4)$$

Here $N_1 = n_1/n_{\text{tot}}$ is the molar fraction of component 1 in the solution, etc.

The *relative integral molar function* is equal to the difference between the integral molar function and the sum of the function for each component in their standard states: $\Delta X_M = X_M - (N_1 X_1^\circ + N_2 X_2^\circ + \cdots + N_n X_n^\circ)$. We see that this may also be written

$$\Delta X_M = N_1 \Delta \bar{X}_1 + N_2 \Delta \bar{X}_2 + \cdots + N_n \Delta \bar{X}_n \qquad (4\text{-}5)$$

This means that if all relative partial molar quantities and the composition of the system are known we may calculate the relative integral molar function.

The Gibbs–Duhem Equation

Since the integral function X is a state function, Eq. (4-3) may be subjected to total differentiation

$$dX = \bar{X}_1 \, dn_1 + n_1 \, d\bar{X}_1 + \bar{X}_2 \, dn_2 + n_2 \, d\bar{X}_2 + \cdots + \bar{X}_n \, dn_n + n_n \, d\bar{X}_n$$

If this expression is subtracted from Eq. (4-2), and remembering that $(\partial X/\partial n_1)_{T, P, n_i} = \bar{X}_1$ the following relation is obtained:

$$n_1 \, d\bar{X}_1 + n_2 \, d\bar{X}_2 + \cdots + n_n \, d\bar{X}_n = 0 \qquad (4\text{-}6)$$

If this equation is divided through by n_{tot}, the more useful form is obtained:

$$N_1 \, d\bar{X}_1 + N_2 \, d\bar{X}_2 + \cdots + N_n \, d\bar{X}_n = 0 \qquad (4\text{-}7)$$

It can easily be shown that the same relation holds for relative partial quantities

$$N_1 \, d\Delta \bar{X}_1 + N_2 \, d\Delta \bar{X}_2 + \cdots + N_n \, d\Delta \bar{X}_n = 0 \qquad (4\text{-}8)$$

The above equations are different forms of the *Gibbs–Duhem equation*, which is one of the most important relations in the thermodynamics of solutions. Its great value is that it makes possible a calculation of partial molar functions for one component if partial molar functions for the other components are known. It is equally valid for any thermodynamic function which obeys Eq. (4-3), that is for all extensive functions such as enthalpy, entropy, or volume. But it is not valid for intensive functions such as pressure and temperature. A procedure for integration of the Gibbs–Duhem equation will be given in Sec. 4.3.

The Tangent Intercept Method

By further combination of Eqs. (4-4) and (4-6) one obtains the relation

$$\bar{X}_1 = X_M + (1 - N_1)\left(\frac{\partial X_M}{\partial N_1}\right)_{N_2:N_3:\cdots:N_n} \tag{4-9}$$

Here $N_2 : N_3 : \cdots : N_n$ denotes that the ratio between the mole fractions of all components, except 1, is kept constant. For a binary system, Eq. (4-9) becomes

$$\bar{X}_1 = X_M + (1 - N_1)\frac{dX_M}{dN_1} \tag{4-10}$$

These relations are equally valid for relative partial molar quantities in which case $\bar{X}_1$ and X_M are replaced by $\Delta\bar{X}_1$ and ΔX_M. Also they are equally valid for the other components by exchange of the subscripts 1 and 2, etc. Equations (4-9) and (4-10) make it possible to calculate partial functions if the integral function is known as function of composition. This is shown in Fig. 4-1, which illustrates the molar volume of a binary solution as function of composition. The partial molar volumes are obtained by the tangent intercept method: At a certain composition N_1 the partial molar volumes of components 1 and 2 are obtained by the intercept of the tangent to the integral curve with the

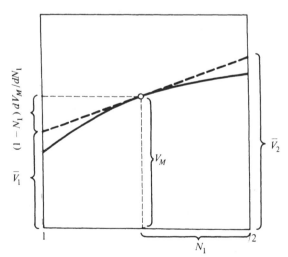

Figure 4-1 Integral molar volume as function of composition. The partial volumes $\bar{V}_1$ and $\bar{V}_2$ are obtained by the tangent intercept method.

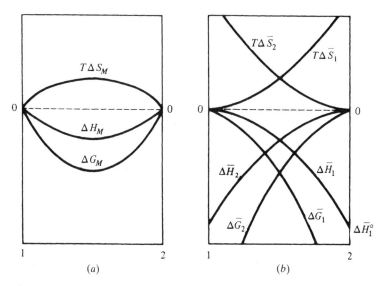

Figure 4-2 (*a*) Relative integral, and (*b*) relative partial functions for a binary system.

ordinate axes 1 and 2. Similarly for a ternary system the partial molar values may be obtained from the intercept of the tangent plane to the integral function surface with the three ordinate axes, and so on.

Although the partial and integral thermodynamic functions are interrelated through Eqs. (4-3) to (4-10), their values and variations with composition of the solution must be determined experimentally. For metallurgical calculations the functions H, S, and G are of particular importance. Examples of the variation of these functions relative to the pure components as standard states in a binary system are shown in Fig. 4-2(*a*) and (*b*).

For mixtures which are stable with respect to their components the relative Gibbs energy function is always negative. The relative enthalpy function may be negative or positive, and the relative entropy function is *usually* positive. On the basis of statistical mechanics, Sec. 4-6, it may be shown that $\Delta \bar{S}_1$ approaches infinity at infinite dilution of component 1. Inserted into Eq. (4-10) this means that $d\Delta S_M/dN_1$ is infinite for $N_1 = 0$. Since $G = H - TS$, it may be concluded that $\Delta \bar{G}_1$ and $d\Delta G_M/dN_1$ approach minus infinity at infinite dilution of component 1. The relative partial enthalpy and $d\Delta H_M/dN_1$ on the other hand approach limiting values at infinite dilution.

4-3 THE ACTIVITY

For calculations of chemical equilibria we are particularly interested in the relative partial Gibbs energy. For equilibrium calculations it is more convenient to use the *chemical activity*, a, which is defined by the relation $\Delta \bar{G} = RT \ln a$.

Like $\Delta \bar{G}$ the value of a depends on the chosen standard state and it has a value of unity for the standard state. For infinite dilution where $\Delta \bar{G}$ is minus infinity, the activity has a value of zero. The activity has the advantage that it can be inserted directly into the equilibrium constant, Eq. (3-6), where it has the same significance as the partial pressure of an ideal gas. Another significance is its relation to the partial pressure of the component in the solution. For equilibrium between a solution and its vapor the partial molar Gibbs energy of each component is the same in the solution and in the vapor. Since in the vapor $\bar{G} = G°(g) + RT \ln f$, where $G°(g)$ is the standard Gibbs energy and f the fugacity of the component in the vapor, it follows that $RT \ln a = RT(\ln f - \ln f°)$, where $f°$ is the fugacity in the vapor above the standard state chosen for the component in the solution. Hence $a = f/f°$. For most metals the vapor pressure is sufficiently low to make the gas behave like an ideal gas, and as a consequence the fugacities may be replaced by the partial pressures, i.e., $a = p/p°$, where $p°$ is the vapor pressure above the standard state.

If the activity is inserted into Eq. (4-8), this takes the following form

$$N_1 \, d \ln a_1 + N_2 \, d \ln a_2 + \cdots + N_n \, d \ln a_n = 0 \qquad (4\text{-}11)$$

Ideal Solutions, Raoult's Law

If the activity a_1, relative to the pure component 1 as standard state, is equal to the mole fraction N_1 at all temperatures and pressures, the solution is said to be *ideal*. Inserted into Eq. (4-11) this gives, for a binary solution,

$$N_1 \, d \ln N_1 + N_2 \, d \ln a_2 = 0$$

Since $N_1 = 1 - N_2$, and $N_1 \, d \ln N_1 = dN_1 = -dN_2 = -N_2 \, d \ln N_2$, it follows that $a_2 = N_2$. Thus if a binary solution is ideal with respect to one component, it is ideal also with respect to the other.

From Eqs. (3-4) and (3-2) follows that the relative partial enthalpy and volume in an ideal solution are related to the activity by the following relations:

$$\Delta \bar{H} = \left(\frac{\partial (\Delta \bar{G}/T)}{\partial (1/T)} \right)_P = R \left(\frac{\partial \ln a}{\partial (1/T)} \right)_P$$

and

$$\Delta \bar{V} = \left(\frac{\partial \Delta \bar{G}}{\partial P} \right)_T = RT \left(\frac{\partial \ln a}{\partial P} \right)_T$$

Since the activity in an ideal solution is equal to the mole fraction at all temperatures and pressures, the partial derivatives above must be zero, which means that $\Delta \bar{H}$ and $\Delta \bar{V}$ are also zero. The relative partial entropy is then obtained from the relation

$$\Delta \bar{S} = \frac{\Delta \bar{H} - \Delta \bar{G}}{T} = -R \ln N \qquad (4\text{-}12)$$

Combined with Eq. (4-5) this gives

$$\Delta S_M = -R(N_1 \ln N_1 + N_2 \ln N_2) \tag{4-13}$$

From statistical thermodynamics, Sec. 4-6, it can be shown that this is the entropy of a completely random mixing of the different atoms.

Very few systems are known to show ideal behavior. Even a solution of metals as similar as silver and gold shows deviation from ideal behavior. Only if no experimental data exist can ideal behavior be assumed as a first approximation for solutions of chemically similar components or elements.

The relation $a = N$ is sometimes called *Raoult's law*. This name is used even in cases when the relation is obeyed only for a certain concentration range or at a certain temperature, and does not necessarily mean that the solution as a whole is ideal.

Nonideal Solutions

Most solutions deviate from ideal behavior. This deviation can be expressed by introducing the *activity coefficient*: $\gamma_1 = a_1/N_1$. If $\gamma_1 > 1$, the solution is said to show positive deviation from Raoult's law. If $\gamma_1 < 1$, the deviation is called negative. Usually the deviation has the same sign over the entire composition range. Some solutions are known, however, which show positive deviation in one composition range and negative in another. If the activity coefficient is inserted in the Gibbs–Duhem equation this becomes, for a binary solution,

$$N_1 \, d \ln \gamma_1 + N_2 \, d \ln \gamma_2 = 0 \tag{4-14}$$

This expression is particularly well suited for calculation of the activity of one component from the activity of the other.

On the basis of Eq. (3-4) it may be shown that the activity coefficient is a function of temperature

$$\left(\frac{\partial \ln \gamma}{\partial (1/T)} \right)_P = \Delta \bar{H}/R$$

For most solutions $\Delta \bar{H}$ and $\ln \gamma$ are either both positive or both negative. This means that with increasing temperature γ usually approaches unity, i.e., the solution approaches ideality.

The partial molar entropy $\Delta \bar{S}$ may either be larger or smaller than for an ideal solution. If ΔS is equal to the value for an ideal solution, i.e., equal to $-R \ln N$, the solution is said to be *regular*. It follows that in a regular solution $\Delta \bar{H} = RT \ln \gamma$. Very few solutions are known to exhibit regular behavior. On the other hand if the entropy of mixing is not known from experiments, regular behavior may be assumed as a first approximation for solutions of not too dissimilar components.

More fruitful than the regular solution concept is perhaps the empirical observation of a relation between the enthalpy and the entropy of mixing.

According to Kubaschewski et al.[1] who surveyed a large number of metallic solutions, a positive enthalpy of mixing is usually associated with an entropy larger than ideal, whereas a negative enthalpy is associated with an entropy less than ideal. According to this observation regular metallic solutions are almost as rare as ideal ones.

Excess functions Deviations from ideal behavior are sometimes expressed by the so-called *excess functions*. These give the difference between the various functions for the actual solution and the corresponding functions for an ideal solution. As $\Delta H = 0$ for an ideal solution this gives, for the relative integral excess functions,

$$\Delta S_M^{xs} = \Delta S_M + R(N_1 \ln N_1 + N_2 \ln N_2)$$

$$\Delta H_M^{xs} = \Delta H_M$$

$$\Delta G_M^{xs} = \Delta G_M - RT(N_1 \ln N_1 + N_2 \ln N_2)$$

and for the relative partial excess functions,

$$\Delta \bar{S}_1^{xs} = \Delta \bar{S}_1 + R \ln N_1$$

$$\Delta \bar{H}_1^{xs} = \Delta \bar{H}_1$$

$$\Delta \bar{G}_1^{xs} = \Delta \bar{G}_1 - RT \ln N_1 = RT \ln \gamma_1$$

All the interrelations that have been given for the total molar functions may be applied also for the excess functions. As the ideal contribution has been subtracted this allows a more accurate analysis of the deviations.

Integration of the Gibbs–Duhem Equation

Attempts to integrate Eq. (4-11) directly are hampered by the fact that when the concentration of, say, component 2 goes to zero, $\ln a_2$ goes to minus infinity. Instead, we base our integration of Eq. (4-14) which, for a binary system, may be rearranged and integrated

$$\int_1^2 d \ln \gamma_1 = - \int_1^2 \frac{N_2}{N_1} d \ln \gamma_2$$

The lower integration limit is here set by the pure component 1, i.e., for $N_2/N_1 = 0$ and $\ln \gamma_1 = 0$. The integration is carried out graphically as shown in

[1] O. Kubaschewski et al.: "Metallurgical Thermochemistry," Pergamon Press, London, 1979, p. 55.

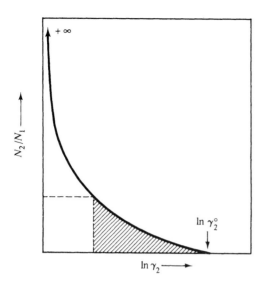

Figure 4-3 Graphical integration of the Gibbs–Duhem equation. The hatched area, which is negative, is equal to $-\ln \gamma_1$ for the chosen value of N_2/N_1.

Fig. 4-3, which applies to a case where $\ln \gamma_2 > 0$, i.e., component 2 shows positive deviation from ideality. As will be shown in Sec. 4-4, on decreasing value of N_2, that is, of the ratio N_2/N_1, $\ln \gamma_2$ approaches a finite limit, $\ln \gamma_2^\circ$. This makes it possible to measure the area under the curve, starting from $N_2/N_1 = 0$, and ending at any desired N_2/N_1 ratio. This area gives $-\ln \gamma_1$.

Notice that measuring toward the left, as in Fig. 4-3, the area under the curve has negative sign, corresponding to $\ln \gamma_1$ being positive, and component 1 showing positive deviation from ideality. Conversely, measuring toward the right, negative values for $\ln \gamma_1$ are obtained. In some rare systems $\ln \gamma_2$ may change from negative to positive. In such cases the net value for $\ln \gamma_1$ is the difference between the negative and positive areas under the curve.

Figure 4-3 shows that when N_1 goes to zero, the ratio N_2/N_1 goes to infinity, and the area under the curve can no longer be measured. This usually presents no problem. When $\ln \gamma_1$ has been evaluated up to a certain N_2/N_1 ratio, the remaining part may quite easily be obtained by a slight extrapolation.

Usually the Gibbs–Duhem equation is used to calculate the activity of one component from that of the other. In that case the known activity is converted to the corresponding activity coefficient. After integration the obtained activity coefficient is converted to the desired activity. A small uncertainty in the second activity coefficient at low concentration will, in that case, have little effect.

Some authors recommend an integration procedure based on the so-called α-function $= \ln \gamma/(1 - N)^2$. In the opinion of the present author this gives no advantage compared to the procedure outlined above.

Methods to integrate the Gibbs–Duhem equation for ternary and more complicated systems have been worked out, but these go beyond the scope of the present text.

4-4 DILUTE SOLUTIONS

If the activity coefficient γ_2 is plotted as a function of composition, it is found empirically that on increasing dilution it approaches a finite value which is called γ_2°. Thus at small concentrations the activity may approximately be expressed by the relation $a_2 = \gamma_2^\circ N_2$. This relation is known as *Henry's law*. It should be pointed out that Henry's law is only obeyed if the solute is present in the solution with the same molecular or atomic species as in the standard state. Thus Henry's law is not valid for components which undergo association or dissociation in the solution.

In the dilute range $d \ln \gamma_2$ has a finite value. If this is inserted into Eq. (4-14), it follows that when the ratio N_2/N_1 approaches zero, $d \ln \gamma_1$ approaches zero. Since $\gamma_1 = 1$ for $N_1 = 1$, the relation $da_1 = dN_1$ will be obeyed at decreasing concentration of component 2. Thus when Henry's law is valid for the solute, Raoult's law is valid for the solvent. It can furthermore be shown that Raoult's law for the solvent is obeyed more closely than Henry's law for the solute. Thus the curve for a_1 does not only approach the same slope as given by Raoult's law, but also the same curvature.

Even though Henry's and Raoult's laws are obeyed with increasing accuracy at increasing dilution, it is difficult to say at which composition the deviation becomes significant. For metallic solutions of not too dissimilar elements it may be assumed that Raoult's law is obeyed to within 10 percent for N_1 down to about 0.9, that is, that the activity at $N_1 = 0.9$ may be between 0.89 and 0.91. The deviation from Henry's law would, in that case, be within about 20 percent.

Change of Standard State

For many solutions we are mainly interested in the range of low concentration of one component. Thus in steel-making the concentration of most impurities rarely exceeds a mole fraction of 0.01, and the activity coefficients γ°, referred to the pure solute as standard state, may differ from unity by several orders of magnitude. It has been found convenient in such cases to introduce a new standard state which is more suitable for calculations of equilibria in the dilute range. The new standard state has the property that the activity, referred to this standard state, becomes equal to the concentration expressed in some convenient unit at infinite dilution. In steel-making the concentration unit is usually chosen to be one weight percent. Similarly in aqueous chemistry the concentration unit is usually a one molal solution, see Sec. 15-1.

For steel melts the activity relative to the new standard state may be expressed by the relation: $a_B' = f_B[\% B]$. Here f is a new activity coefficient (henrian coefficient) which approaches unity at infinite dilution, and which at moderate concentrations does not deviate much from unity. If the activity coefficient f had remained equal to unity up to a concentration of one weight percent, the new activity would be equal to unity at that concentration. This solution would therefore be the new standard state. If the activity coefficient f deviates from unity already below one weight percent, the standard state would be a hypothetical one percent solution with the same activity coefficient as the infinite dilution. This is illustrated in Fig. 4-4, which shows the activity of silicon in the Fe–Si system, and where the new standard state is indicated to the right.

If the partial Gibbs energy of B in the new standard state is denoted $G^\circ_{B(1\%)}$, and in the pure solute G°_B, the Gibbs energy change $\Delta G^\circ_B = G^\circ_{B(1\%)} - G^\circ_B$ may be expressed in terms of γ°_B by the following reasoning:

If the partial Gibbs energy at any concentration of B is $\bar{G}_B$, then

$$\bar{G}_B - G^\circ_B = RT \ln a_B = RT \ln \gamma_B N_B$$

and

$$\bar{G}_B - G^\circ_{B(1\%)} = RT \ln a'_B = RT \ln f_B [\% \; B]$$

This gives

$$\Delta G^\circ_B = RT \ln \frac{\gamma_B N_B}{f_B [\% \; B]}$$

This relation applies for any concentration and consequently also for infinite dilution. Here $\gamma_B = \gamma^\circ_B$, $f_B = 1$, and $N_B = [\% \; B] M_A / (100 \, M_B)$ where M_A and M_B are the molecular weights of A and B respectively. This gives

$$\Delta G^\circ_B = G^\circ_{B(1\%)} - G^\circ_B = RT \ln \frac{\gamma^\circ_B M_A}{100 M_B} \tag{4-15}$$

Hence if γ°_B is known, ΔG°_B may be calculated.

Similarly we may calculate the difference in enthalpy in the new and the old

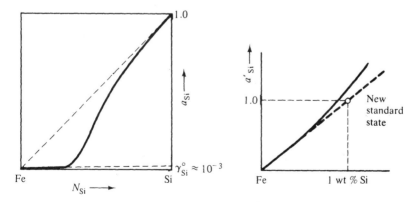

Figure 4-4 Activity of silicon in the iron–silicon system, (a_{Si}) relative to pure Si, and (a'_{Si}) relative to an idealized 1 weight percent solution.

standard states. If the enthalpy in the new standard state is denoted $H^\circ_{B\,(1\%)}$ and the old state H°_B we have, for any concentration

$$\bar{H}_B - H^\circ_B = R\,\frac{d\ln\gamma_B}{d(1/T)} \qquad \text{and} \qquad \bar{H}_B - H^\circ_{B\,(1\%)} = R\,\frac{d\ln f_B}{d(1/T)}$$

This gives

$$\Delta H^\circ_B = H^\circ_{B\,(1\%)} - H^\circ_B = R\left[\frac{d\ln\gamma_B}{d(1/T)} - \frac{d\ln f_B}{d(1/T)}\right]$$

At infinite dilution f is equal to unity at all temperatures, and consequently $d\ln f/d(1/T) = 0$. Hence

$$\Delta H^\circ_B = R\,\frac{d\ln\gamma^\circ_B}{d(1/T)} = \bar{H}^\circ_B - H^\circ_B$$

where $\bar{H}^\circ_B$ is the partial molar enthalpy of B at infinite dilution. The enthalpy of the new standard state is therefore equal to that of the infinitely diluted solution.

Example 4-1 The binary solution iron–nickel is assumed to be ideal, that is, $\gamma^\circ_{Ni} = 1$. The molecular weights of iron and nickel are 55.8 and 58.7 respectively. The change in standard Gibbs energy for the reaction $Ni(1) = \underline{Ni}\,(1\%)$ is then $\Delta G^\circ = RT\ln(0.558/58.7) = -38.7T$ J. Here $\underline{Ni}\,(1\%)$ denotes the new standard state.

Example 4-2 For the solution of silicon in liquid iron at $1600°C$, $\gamma^\circ = 10^{-3}$ and $\Delta H^\circ = -131\,400$ J[1]; this gives, for the reaction $Si(l) = \underline{Si}\,(1\%)$,

$$\Delta G^\circ_{1873} = RT\ln\frac{55.8\times10^{-3}}{28\times100} = -168\,600 \text{ J}$$

$$\Delta S^\circ_{1873} = \frac{\Delta H^\circ - \Delta G^\circ}{T} = \frac{-131\,400 + 168\,600}{1873} = 19.86 \text{ J/K}$$

and, since ΔH° and ΔS° are regarded as closely independent of temperature,

$$\Delta G^\circ = \Delta H^\circ - T\,\Delta S^\circ = -131\,400 - 19.86T \text{ J}$$

Gases in Metals

Gases like hydrogen, oxygen, and nitrogen are soluble in most liquid metals. In the solution the gases are present in the atomic state and are therefore not significantly different from other alloying elements. We may say that when a gas

---

[1] J. Chipman and R. Baschwitz: *Trans. AIME*, **227**: 473 (1963). E. T. Turkdogan, P. Grieveson, and J. F. Beisler: *Trans. AIME*, **227**: 1258 (1963). F. Woolley and J. F. Elliot: *Trans. AIME*, **239**: 1872 (1967).

is dissolved in a metal, it is no longer a gas. The solution of e.g., nitrogen in liquid iron, may be expressed by the equation $N_2(g) = 2\underline{N}$ (1%). In the diluted range, i.e., for $f_N = 1$, the equilibrium constant is then equal to

$$K = [\% \text{ N}]^2/p_{N_2} \quad \text{or} \quad [\% \text{ N}] = (Kp_{N_2})^{1/2} \tag{4-16}$$

This relation is known as *Sieverts' law* and it is illustrated in Fig. 4-5. At a nitrogen pressure of 1 atm liquid iron dissolves about 0.04 percent N, whereas at $\frac{1}{4}$ atm the nitrogen concentration is 0.02 percent.

Also for the solution of oxygen in liquid iron Sieverts' law is obeyed, but only up to an oxygen pressure of about 10^{-3} Pa. At this oxygen pressure a new liquid phase, liquid iron oxide, is formed. Similarly the solubility of nitrogen may be limited if the melt contains elements which form stable high-melting nitrides such as zirconium or titanium.

If the equilibrium constant for the solution of nitrogen in liquid metal is inserted into Eq. (3-4), the variation of the nitrogen solubility with temperature at constant nitrogen pressure is obtained

$$\frac{d \ln [\% \text{ N}]}{d(1/T)} = \frac{-\Delta H^\circ}{2R}$$

Here ΔH° is the enthalpy of solution of one mole of N_2 in the metal at infinite dilution. The enthalpy of solution depends on the chemical forces that act between the metal and the gas atoms. If these forces are relatively weak such as for the solution of hydrogen and nitrogen in liquid iron, the solution is endothermic, and the solubility increases with increasing temperature. If the forces are stronger such as for the solution of nitrogen in titanium or zirconium, the solution is exothermic, and the solubility decreases with increasing temperature.

At higher concentrations deviations from Sieverts' law may occur. Thus in the system Cr–N deviation is observed at about 1 percent N. At 5 percent N the

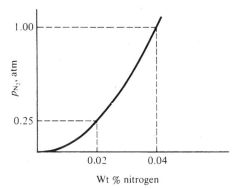

Figure 4-5 Pressure of N_2 in the iron–nitrogen system.

N_2 pressure is about ten times as high as given by Sieverts' law. Further discussion on diluted solutions and gases in liquid steel will be given in Sec. 13-2.

4-5 THERMODYNAMICS AND PHASE DIAGRAMS

Equilibrium between two or more phases requires that the partial Gibbs energy of each component is the same in all phases. For a binary system the composition of the coexisting phases may be conveniently derived from a plot of the relative molar integral Gibbs energy versus composition. This is shown in Fig. 4-6(a) for a binary system which forms a simple eutectic phase diagram. If the two pure liquids are chosen as standard states, the Gibbs energy curve for the liquid solution is illustrated by curve I. Relative to the same standard state, and at temperature T_1 intermediate between the melting points of the pure components and the eutectic temperature, the Gibbs energy of the two solid phases is illustrated by curves II and III. The compositions of the phases which are in equilibrium with each other may now be obtained by drawing common tangents to the Gibbs energy curves. At the points where these touch the curves, the partial Gibbs energy of each component is the same in the two phases. If similar curves are drawn for other temperatures, the phase diagram shown in Fig. 4-6(b) is obtained.

Similarly diagrams which show complete solid solubility as well as intermediate phases could be constructed if the Gibbs energy of the pertinent phases were known. Thus we may say as a general rule that if the Gibbs energy for all phases were known as function of composition and temperature, the phase diagram could be constructed.

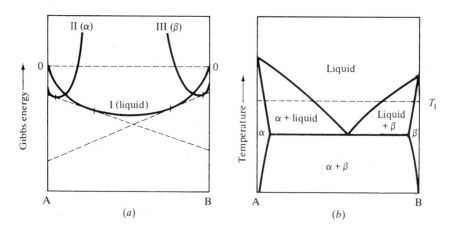

Figure 4-6 (a) Integral molar Gibbs energy of liquid and solid phases at temperature T_1, and (b) isobaric phase diagram for the A–B system.

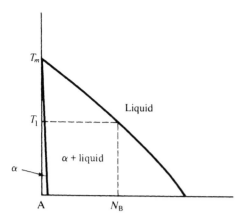

Figure 4-7 Freezing point depression for the addition of B to A.

We want next to discuss to what extent thermodynamic properties for the solution may be obtained from the phase diagram. At the temperature T_1 (Fig. 4-7) the practically pure solid phase A is in equilibrium with a melt with molar composition $N_A = 1 - N_B$. Under these conditions the partial Gibbs energy of A in the solution is equal to the partial Gibbs energy of A in the solid. In the absence of solid solubility the latter is equal to the Gibbs energy of the pure solid: $\bar{G}_A(l) = G_A^\circ(s)$. The partial Gibbs energy of A in the liquid is also related to the Gibbs energy of the pure supercooled liquid by the relation

$$\bar{G}_A(l) = G_A^\circ(l) + RT \ln a_A$$

Here $G_A^\circ(l)$ is the Gibbs energy of the pure supercooled liquid, and a_A is the activity referred to the pure supercooled liquid as standard state. Furthermore, by combination with Eq. (4-1) we get

$$RT \ln a_A = (\Delta H_m/T_m)(T - T_m) + \int\int_{T_m}^{T} \Delta C_P \, d \ln T \, dT \qquad (4\text{-}17)$$

If ΔC_P is independent of temperature this may be integrated

$$RT \ln a_A = [(\Delta H_m/T_m) - \Delta C_P](T - T_m) + \Delta C_P T \ln (T/T_m) \qquad (4\text{-}18)$$

Here ΔH_m is the enthalpy change at the melting point T_m, and ΔC_p is the change in heat capacity on melting. For temperatures not too far from the melting point the second term in Eq. (4-17) is usually small and may be disregarded.

The above relation may be further simplified for small concentrations of B in A. In that case $a_A \approx N_A$ and $\ln N_A \approx (N_A - 1) = -N_B$. Furthermore $T \approx T_m$. This gives the relation between the freezing point depression $T_m - T = \Delta T$ and the concentration of B.

$$\frac{\Delta T}{N_B} \approx \frac{RT^2}{\Delta H_m} \qquad (4\text{-}19)$$

This relation becomes exact when the concentration of B approaches zero. It should be emphasized, however, that Eq. (4-19) is only correct if Raoult's law is valid for the solvent. This again is dependent on whether the molar concentration of the solute is expressed in terms of the particles that are actually present in the melt. Conversely, if the molecular weight is unknown, this may be derived from the freezing point depression.

In the above derivations solid A was assumed to show no solid solubility. If this were not the case, the left-hand side of Eq. (4-17) would have the form $RT \ln (a_A/a_A^*)$, where a_A^* is the activity of A in the saturated solid solution relative to pure solid A. If the solid solubility is relatively small, Raoult's law may be assumed to be valid for the solid solution even though it is not valid for the liquid. It can also be shown that for solid solubility Eq. (4-19) becomes

$$\frac{\Delta T}{N_B - N_B^*} \approx \frac{RT^2}{\Delta H_m}$$

where N_B^* is the solubility of B in the solid phase.

The above derivation shows that knowledge of the phase diagram makes it possible to calculate the activity of A *along the liquids line*, i.e., at decreasing temperature. In order to obtain the activity curve at some constant temperature we need to know the temperature dependence of the activity. This cannot be derived from the phase diagram. In the absence of other information we could assume that the melt behaves as a regular solution. In that case $RT \ln \gamma_A = \Delta \bar{H}_A$. Since $\Delta \bar{H}_A$ is essentially independent of temperature, this expression may be used to calculate γ_A at any temperature. As was mentioned in Sec. 4-3, the assumption of regular solution behavior is not always justified, however.

We may conclude that whereas the phase diagram may be derived from a knowledge of the thermodynamics of the system, the opposite is not strictly the case. Nevertheless the phase diagram may give valuable indication regarding the activity relations. This is illustrated by the system iron–copper which is shown in Fig. 4-8. This system is characterized by a rather flat liquidus curve. Inserted into Eq. (4-15) this means that the system exhibits strong positive deviation from ideality and is actually at the point of separation into two liquid phases. Activity measurements have confirmed this behavior.

Intermediate Phases

If in a system A–B there is an intermediate compound A_xB_y this has a standard Gibbs energy of formation relative to the pure solids. Thus for the reaction $xA(s) + yB(s) = A_xB_y(s)$

$$\Delta G^\circ = -RT \ln \frac{a_{A_xB_y}}{(a_A)^x(a_B)^y} \tag{4-20}$$

Here the activities are referred to the pure solids and the stoichiometric solid compound as standard states. At the eutectic point between A and A_xB_y, and

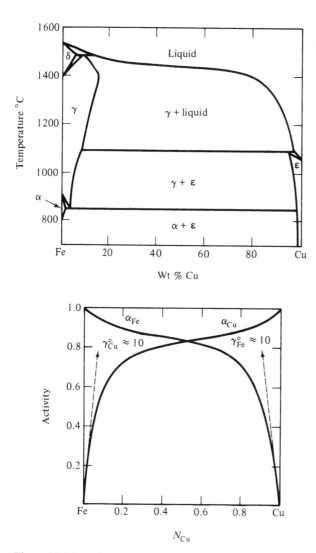

Figure 4-8 Phase diagram (above) and liquid activities (below) in the iron–copper system.

assuming no solid solubility, both a_A and $a_{A_xB_y}$ are unity. This gives the activity of B relative to solid B: $\ln a_B = \Delta G°/(yRT)$.

Similarly the activity of A may be calculated at the eutectic point between A_xB_y and B. Finally the activities may be converted to the pure liquids as standard state by means of Eq. (4-1). Usually we will find that if a system contains intermediate solid phases, the melt will exhibit negative deviation from ideality.

It is sometimes convenient to express the thermodynamics of solutions containing intermediate phases in terms of activities of these phases or compounds.

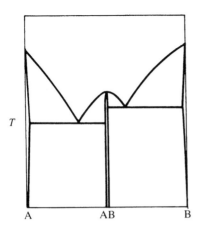 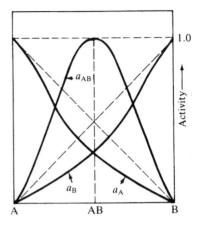

Figure 4-9 Phase diagram (left) and liquid activities (right) in a binary system with the intermediate compound AB.

Thus in the system shown in Fig. 4-9 the activity of A_xB_y is equal to unity at the stoichiometric composition. In the same way as shown for the activity of the pure components, the activity of the compound in the solution, relative to the pure supercooled compound, may be derived from the shape of the liquidus curve and the heat of fusion of the compound. One important difference should be pointed out. The activity of A_xB_y is related to the activity of A and B by Eq. (4-20). In the solution the activities of A and B are related by the Gibbs–Duhem equation, Eq. (4-11). In the *stoichiometric* melt $N_A = x/(x + y)$ and $N_B = y/(x + y)$. Hence

$$x \, d \ln a_A + y \, d \ln a_B = d \ln a_{A_xB_y} = 0$$

In other words: At the stoichiometric composition the activity of A_xB_y does not change by small additions of either A or B. The activity curve therefore has a horizontal tangent, and as a result also the liquidus curve has a horizontal tangent at the stoichiometric composition. It should be pointed out that this applies only for compounds which melt to form a mixture of A and B particles. If the compound is completely undissociated in the liquid state the binary system formed by the compound and any one of the two components may be treated as any other simple binary system.

Gibbs Energy of Unstoichiometric Compounds

Unstoichiometric compounds, such as wustite, $Fe_{1-x}O$, and pyrrhotite, $Fe_{1-x}S$, may in principle be treated like any other liquid or solid solution. Their total and partial relative molar Gibbs energies, as well as the activities of the various components, may be measured and plotted as function of composition. This is illustrated by Fig. 4-10, which shows the oxygen pressure as well as the iron activity, both on a logarithmic scale, for part of the iron–oxygen system at 1200°C. Of these, the oxygen pressure has been measured experimentally, and the iron activity has been calculated by means of the Gibbs–Duhem equation, Eq. (4-11).

It will be noticed that, in this system, the wustite range does not include the stoichiometric composition FeO. The question then arises: Which standard state should be used for wustite, what is the free energy of formation of that standard state, and how does the activity, referred to that standard state, vary with wustite composition? Since activities are to be used in stoichiometric equilibrium expressions, the most convenient *entity* for the standard state is stoichiometric FeO. This still leaves us with a choice of standard *state*. We could

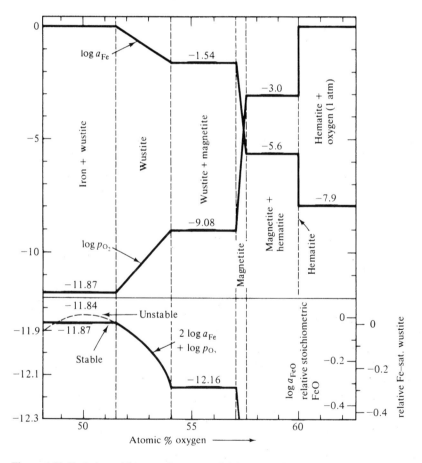

Figure 4-10 Variation of log a_{Fe}, log p_{O_2} and 2 log a_{Fe} + log p_{O_2} with composition in the iron–oxygen system at 1200°C. Right-hand scales give a_{FeO} relative to stoichiometric FeO as well as to iron-saturated wustite as standard state. (*Data from L. S. Darken and R. W. Gurry: J. Am. Chem. Soc., vol. 68, 1946, p. 798.*)

either choose FeO at the stoichiometric, but unstable composition, or we could choose FeO dissolved in iron-saturated wustite. This difference appears from the lower part of Fig. 4-10, which gives the expression $2 \log a_{Fe} + \log p_{O_2}$ as function of composition. This expression is related to the standard Gibbs energy of formation of 2 moles of FeO by the relation (from Eq. (3-6)):

$$\Delta G^\circ = RT(2 \ln a_{Fe} + \ln p_{O_2} - 2 \ln a_{FeO})$$

It is evident that the value of ΔG° depends on the choice of standard state for FeO, that is, on the choice of composition where the activity of FeO is taken to be unity. If we choose stoichiometric FeO, the curve for $2 \log a_{Fe} + \log p_{O_2}$ has to be extrapolated to the stoichiometric composition, remembering that its slope must be zero at that composition. This gives for the reaction

$$2Fe(s) + O_2 = 2FeO \text{ (stoichiometric)} \tag{1}$$

$$\Delta G_1^\circ = 19.144 \times 1473(-11.84) = -333.9 \text{ kJ}$$

Alternatively we could choose a_{FeO} equal to unity at iron saturation. This gives for the reaction

$$2Fe(s) + O_2 = 2FeO \text{ (in Fe-sat. wustite)} \tag{2}$$

$$\Delta G_2^\circ = 19.144 \times 1473(-11.87) = -334.7 \text{ kJ}$$

Since for the reaction $Fe(s) = Fe$ (in Fe-sat. wustite) the Gibbs energy change is zero, ΔG_2° also applies for the reaction

$$2(1-x)Fe(s) + O_2 = 2Fe_{1-x}O \text{ (Fe-sat. wustite)}$$

This is the equation most commonly used for the formation of "FeO", and is the one which is used for the standard free energy in App. C. Above the melting point of wustite the standard state is FeO dissolved in Fe-saturated wustite melt.

The activity of FeO, relative to the two chosen standard states, may now be read from the right-hand scale in Fig. 4-10. We notice that the two activities differ by a factor of about 1.04. For FeO dissolved in liquid slags the activity commonly refers to Fe-saturated wustite melt as standard state. This is, for example, the case for the activity curves shown in Figs. 11-6 and 11-7. Notice that in all cases the activity is that of the stoichiometric entity FeO, and not that of the unstoichiometric $Fe_{1-x}O$.

Sometimes we want to calculate the Gibbs energy of formation of higher oxides, for example, Fe_3O_4, from the elements. If the lower oxides had been stoichiometric, this could be done by simple addition of the Gibbs energy for each oxidation step. With the lower oxides being unstoichiometric this is no longer the case. But if the oxygen pressure and the metal activity as function of composition are known, the calculation is easily done, as for the reaction

$$\tfrac{3}{2}Fe(s) + O_2 = \tfrac{1}{2}Fe_3O_4(s) \tag{3}$$

$$\Delta G_3^\circ = RT(\tfrac{3}{2} \ln a_{Fe} + \ln p_{O_2} - \tfrac{1}{2} \ln a_{Fe_3O_4})$$

In the two-phase region wustite–magnetite the activity of Fe_3O_4 is unity, and the last term in the above expression is zero. This gives for the Gibbs energy of reaction (3) at 1200°C

$$\Delta G_3^\circ = 19.144 \times 1473(-2.31 - 9.08) = -321.2 \text{ kJ}$$

Referred to this expression, the activity of Fe_3O_4 may be calculated through the magnetite solid solution range. As this range is rather narrow, the activity at saturation with hematite will be only slightly less than unity.

4-6 ELEMENTS OF STATISTICAL THERMODYNAMICS[1]

The relation between the thermodynamic properties of melts and solutions and the corresponding structures may be discussed in terms of statistical thermodynamics. According to Boltzmann, the entropy of a system or a substance may be expressed by

$$S = k \ln W$$

Here k, the Boltzmann constant, is equal to R/N, that is, the gas constant divided by Avogadro's number. W is a measure of the *probability* of the system or substance in question, and is the number of different microscopic states which give rise to one and the same macroscopic state, i.e., it is the number of different arrangements which the atoms in the substance may have without changes in the macroscopic properties of the substance. It can be shown that the substance which can be realized by the largest number of microstates also is the most probable.

For crystalline materials, and with some modification also for liquids and gases, it may be convenient to divide the possible microstates in two groups:

1. Configurational states
2. Vibrational states

The entropy of the substance may then be divided corespondingly

$$S = S_{conf} + S_{vibr} = k \ln W_{conf} + k \ln W_{vibr}$$

and

$$W_{tot} = W_{conf} \cdot W_{vibr}$$

By W_{conf} we understand the number of possible ways the atoms or molecules may be distributed over the available lattice positions, whereas W_{vibr}

---

[1] For a more detailed treatment see: J. D. Fast: "Entropy," McGraw-Hill Book Co. Inc., New York, 1962. R. A. Swalin: "Thermodynamics of the Solid State," John Wiley and Sons, Inc., New York, 1962.

denotes the number of possible ways they may vibrate around these positions. We shall first consider configurational entropy of a pure crystalline substance.

One mole of the substance contains a total of $N = 6.02 \times 10^{23}$ atoms or molecules, and we would think that these could be distributed in a very large number of ways. This is not the case, however. Since for a pure substance all atoms or molecules are equal, they cannot be distinguished even on an atomic scale. For a pure substance with a perfect crystal structure the atoms or molecules may be arranged in only one way. This gives

$$S_{\text{conf}} = k \ln 1 = 0$$

For a pure substance the entropy is therefore only determined by the vibrational contribution

$$S = S_{\text{vibr}} = k \ln W_{\text{vibr}}$$

The same is the case for a compound with a fully ordered crystal structure.

The vibrational probability W_{vibr} may be studied further spectroscopically. We shall limit ourselves to stating that at the absolute zero the vibrational probability is equal to unity, i.e., $S_{\text{vibr}} = 0$ in accordance with the Third Law of Thermodynamics. At higher temperatures

$$S_{\text{vibr}} = \int_0^T C_P \, d \ln T$$

in accordance with classical thermodynamics.

We shall now discuss a solid solution or a compound with a disordered crystal structure. If the solution or compound consists of the atoms A and B, these are no longer indistinguishable on an atomic scale. If we have N (Avogadro's) number of lattice positions, and n_A atoms of A and n_B atoms of B, where $n_A + n_B = N$, these may be arranged in W_{conf} different ways which all give the same macroscopic properties. For a relatively ordered structure W_{conf} is relatively small, whereas for a completely disordered structure (statistical distribution) W_{conf} has its maximum value. It may be shown that in this case

$$W_{\text{conf}} = \frac{N!}{n_A! \, n_B!} = \frac{N!}{n_A!(N - n_A)!}$$

Since N is of the order of 10^{23}, it is clear that in this case W_{conf} is a very large number. For such large numbers we may use the *Sterling approximation*

$$\ln N! = N \ln N - N$$

This gives

$$S_{\text{conf}} = k \ln W_{\text{conf}} = Nk \left(\frac{n_A}{N} \ln \frac{N}{n_A} + \frac{n_B}{N} \ln \frac{N}{n_B} \right)$$

$$= -R(N_A \ln N_A + N_B \ln N_B) \tag{4-21}$$

Here N_A and N_B denote the atomic fractions of A and B. Since the configurational entropy of the pure components is zero, this means that the above expression is also equal to ΔS_{conf} of mixing. We see that this is equal to the value previously known from classical thermodynamics for an ideal solution, Eq. (4-13).

For solutions with a certain degree of order the configurational probability, and consequently the configurational entropy of mixing, is less than for the statistical distribution and becomes equal to zero for a completely ordered distribution of the atoms or for complete separation in two immiscible phases. In a later paragraph it will be shown how the entropy may be calculated as a function of the degree of order.

We know from actual measurements that for some mixtures the entropy of mixing may be considerably larger than for ideal mixing, whereas in other cases it may even be negative. Thus in the systems, Mg–Sb and Mg–Bi the integral entropies of mixing show a minimum at $N_{Mg} = 0.6$ for Mg–Sb with $\Delta S = -8$ J/K and at $N_{Mg} = 0.7$ for Mg–Bi with $\Delta S = -4$ J/K. The explanation for this may be found in the vibrational contribution to the entropy of mixing. The vibrational entropy increases with increasing amplitude of vibration and decreases with increasing binding forces between the atoms. Expressed in terms of classical thermodynamics a positive vibrational entropy of mixing means that the mixture has a larger heat capacity than the weighted mean of the heat capacities of the components. Conversely, a low vibrational entropy, as in the Mg–Sb and Mg–Bi alloys, means that the amplitude of vibration and the heat capacity are less than for the pure metals.

For melts and to an even larger extent for gases, it may be difficult to distinguish between configurational and vibrational entropy. Nevertheless, it may be useful, as a mental aid to consider how the atoms of the melt are distributed at any given moment over certain "lattice" positions and also how they vibrate around these positions. The same considerations which were valid for solid solutions would then apply for the liquid. For gases the vibrational contribution to the entropy of mixing is almost zero and the mixing is close to ideal.

Models for Metallic Solutions

Most metallic solutions deviate from ideal behavior, and the question arises how these deviations may be described in terms of statistical thermodynamics, and what is the relation between the thermodynamics and the structure of the melt. Through the years various models have been suggested in which the thermodynamics for metallic solutions is derived from some characteristic properties of the metals involved. In this text only the *regular solution model* and the *quasichemical model* will be presented.

Both models are based on the assumption that between neighboring atoms in a metal chemical forces exist, the strength or energy of which is dependent on the nature of the respective atoms. Thus in an alloy of components A and B there will be A–A, B–B, and A–B bonds with binding energies v_{A-A}, v_{B-B}, and v_{A-B} respectively. By convention we give these energies positive sign.

By the mixing of components A and B, the mixing enthalpy ΔH will be equal to the energy of the broken bonds minus the energy of the new bonds formed. As an idealized case we may consider a solid solution where the atoms of the two components have about the same size, and where the coordination number is constant throughout. If the coordination number is denoted Z (for close packed structures $Z = 12$), then one mole of metal or alloy contains $\frac{1}{2}ZN$ bonds. For the pure metal the total binding energy, which again corresponds to the enthalpy of vaporization, will be

$$\Delta H_{v(A)} = \tfrac{1}{2}ZNv_{A-A}$$

and

$$\Delta H_{v(B)} = \tfrac{1}{2}ZNv_{B-B}$$

The regular solution model It is here assumed that the atoms are statistically distributed. In that case every A or B atom will, on the average, be surrounded by A and B atoms in the same ratio as in the alloy as a whole. Of the total number of bonds a fraction N_A^2 will be A–A bonds, a fraction N_B^2 will be B–B bonds, and a fraction $2N_A N_B$ will be A–B bonds, where $N_A + N_B = 1$. The enthalpy of vaporization for the alloy is then

$$\Delta H_{v(A, B)} = \tfrac{1}{2}ZN[N_A^2 v_{A-A} + N_B^2 v_{B-B} + 2N_A N_B v_{A-B}]$$

For the formation of the alloy we therefore get

$$\Delta H = N_A \,\Delta H_{v(A)} + N_B \,\Delta H_{v(B)} - \Delta H_{v(A, B)}$$

which gives

$$\Delta H = \tfrac{1}{2}\, ZNN_A N_B(v_{A-A} + v_{B-B} - 2v_{A-B}) = \tfrac{1}{2}\, ZNN_A N_B w$$

Here $w = v_{A-A} + v_{B-B} - 2v_{A-B}$ is the energy required to break one A–A bond and one B–B bond and to form two A–B bonds. We see that if $w < 0$, then $\Delta H < 0$, and if $w > 0$, then $\Delta H > 0$, whereas if $w = 0$, the mixing takes place without any evolution of heat.

If we now want to develop an expression for the Gibbs energy of mixing, we need to know the entropy of mixing. With a statistical distribution we assume an ideal entropy. This gives

$$\Delta G = \tfrac{1}{2}ZNN_A N_B w + RT(N_A \ln N_A + N_B \ln N_B) \qquad (4\text{-}22)$$

It can easily be shown that, for the relative partial free energy,

$$\Delta \bar{G}_A = \tfrac{1}{2}ZNN_B^2 w + RT \ln N_A = RT(\ln \gamma_A + \ln N_A)$$

and hence

$$\ln \gamma_A = \frac{Zw}{2kT}\, N_B^2 = \frac{b}{RT}\, N_B^2 = \alpha N_B^2$$

where $b = \tfrac{1}{2}ZNw$.

The quasichemical model Further analysis of Eq. (4-22) shows that this can only be approximately correct. If $w < 0$ it is obvious that there will be a tendency to form as many A–B bonds as possible at the expense of A–A and B–B bonds. Conversely if $w > 0$ less A–B bonds will be formed than would correspond to statistical distribution. We will have a tendency towards "short-range ordering" and "clustering" respectively. This will affect the value for ΔH as well as for ΔS, the latter no longer being ideal. A revised expression was given by Guttman[1]

$$\Delta G = \tfrac{1}{2}ZN(N_A - x)w + \tfrac{1}{2}RTZ[x \ln x + 2(N_A - x) \ln (N_A - x)$$

$$+ (1 - 2N_A + x) \ln (1 - 2N_A + x)]$$

$$- RT(Z - 1)[N_A \ln N_A + (1 - N_A) \ln (1 - N_A)] \qquad (4\text{-}23)$$

Here the symbols are the same as above. In addition the symbol x is introduced, and is a measure of the *degree of order* and equal to the ratio between the number of A–A bonds and the total number of bonds. For a statistical distribution this ratio becomes equal to N_A^2, and the expression for ΔG reduces to the one given in Eq. (4-22). For complete immiscibility in two phases $x = N_A$ whereas for complete order $x \to 0$, corresponding to every A atom being surrounded by only B atoms and vice versa. Taking an alloy of a given composition, the atoms will find the distribution which gives the lowest Gibbs energy. The corresponding value for x is found by taking $(d\Delta G/dx) = 0$. This gives

$$\ln \frac{(N_A - x)^2}{x(1 - 2N_A + x)} = -w/kT$$

For $w < 0$, x will be somewhat smaller than N_A^2, corresponding to short range ordering, whereas for $w > 0$, x will be larger than N_A^2, corresponding to clustering. We furthermore see that the deviation from statistical distribution increases with decreasing temperature. Having calculated the equilibrium value of x, we may insert this into the expression for ΔG, which then is obtained as an explicit function of w only.

There is a question, however, how fruitful such calculations are. In the above discussion only the configurational entropy was considered, whereas we know that in many alloys the vibrational term may be significant. Furthermore, we have assumed the energy of the various interatomic bonds to have constant values independent of the nature of the neighboring bonds, i.e., independent of the composition of the alloy. This assumption is probably not correct. Finally we have assumed the atomic sizes and the coordination number to be constant. For solid solutions the coordination number is constant, but if the atoms have different sizes, additional strain energy terms are introduced. For liquid solutions of atoms of different sizes it is probable that the coordination number

[1] L. Guttman: *Trans. AIME*, **175**: 178 (1948).

changes with composition. Whereas the idealized expression for ΔG is symmetrical around the 50/50 composition, the additional factors mentioned above may cause the ΔG curve to be asymmetrical. We know from actual alloy systems that the ΔG curve often is asymmetrical.

Nevertheless, from a purely qualitative viewpoint we should expect a relation between the thermodynamics of the alloy system and the degree of short range order or clustering. For some solid systems such relations have been found experimentally by x-ray diffraction analysis.

The quasichemical theory gives no information about the nature of the interatomic binding forces, and it is unable to predict the strength of these forces or the properties of the mixture from the properties of the components. A satisfactory theory for metallic solutions would have to consider the electronic structure of the metals, their difference in electronegativity as well as their different atomic sizes. Even though some empirical correlations have been found between these properties and the thermodynamics of the mixture, a general theory for metallic solutions is still lacking, however.

4-7 FUSED SALTS AND SILICATES

Our knowledge of the structure of molten salts derives primarily from our knowledge of the solids. The melts are more disordered, however. In the same way as in the solid we expect every cation to be surrounded by only anions and every anion to be surrounded by only cations. We may say that the cations and the anions occupy two independent "lattices," mutually interwoven. Within each "lattice" the ions are free to move more or less independently of each other, as is demonstrated by the fact that the molten salts are good electrical conductors and that the current is carried by electrically charged ions.

Since the individual particles in a molten salt or salt mixture are electrically charged ions, their thermodynamic functions cannot be measured. We can only measure thermodynamic functions for electrically neutral groups, such as ionic pairs. These may be referred to the pure salts as standard states. Thus in a mixture of KCl and NaBr we cannot measure the activities of K^+ or Cl^-, but only of the neutral groups KCl and NaBr as well as of KBr and NaCl, the last two being referred to the corresponding pure salts as standard states.

We shall now see how thermodynamic functions for molten salts may be understood in terms of the structure of the melt and statistical thermodynamics. The integral as well as the partial Gibbs energy are composed of an enthalpy and an entropy term, and we shall first consider the entropy term. For a salt mixture we may consider all the cations distributed over the cation lattice, and all the anions over the anion lattice. The partial entropy in each lattice may be calculated from the Boltzmann relation $S = k \ln W$. If we assume the ions to be statistically distributed over their respective lattices, and if we only consider the configurational entropy, we get for the cation A^+ (from Eq. 4-21)

$$\bar{S}_{A^+} = -R \ln N_A$$

and for the anion X^-

$$\bar{S}_{X^-} = -R \ln N_X$$

Here N_A and N_X denote the fractions of A ions in the cation lattice and of X ions in the anion lattice, respectively. Since the configurational entropy in the pure salts is equal to zero, the above expressions also give the relative partial entropies. For an ion–pair AX the relative partial entropy then becomes

$$\Delta \bar{S}_{AX} = -R(\ln N_A + \ln N_X) \qquad (4\text{-}24)$$

The above derivation was first given by Temkin.[1]

The *enthalpy* of mixing is determined primarily by the electrostatic attraction between ions of opposite charge and repulsion between ions of similar charge. For ions of closely the same size the enthalpy change on mixing will be small, and may, as a limiting case, be taken as equal to zero. This gives for the relative partial Gibbs energy of a $1:1$ salt

$$\Delta \bar{G}_{AX} = RT \ln a_{AX} = RT(\ln N_A + \ln N_X)$$

which gives

$$a_{AX} = N_A N_X$$

If this expression is used for mixtures with either a common cation or a common anion, the ionic fraction of that ion is equal to unity, and the activity of the salt becomes equal to the ionic fraction of the noncommon ion, i.e., equal to the molar fraction of the salt in question. The same will be the case for mixtures like $NaCl$–$CaCl_2$ or $NaCl$–Na_2SO_4, which have one common ion.

For a mixture where both cations and anions are different, such as KCl–NaBr, we get

$$a_{KCl} = N_K N_{Cl} = (N_{KCl})^2$$

and

$$a_{NABr} = N_{Na} N_{Br} = (N_{NaBr})^2$$

Furthermore

$$a_{KBr} = N_K N_{Br} = N_{KCl} N_{NaBr}$$

$$a_{NaCl} = N_{Na} N_{Cl} = N_{NaBr} N_{KCl}$$

Thus for a 50/50 mixture of the two salts the activity of each of the four ionic pairs becomes equal to 0.25.

If the ions also have different charge, such as $NaCl$–$CaBr_2$, the Temkin theory gives the following activities:

$$a_{NaCl} = N_{Na} N_{Cl} = N^2_{NaCl}/(2 - N_{NaCl})$$

$$a_{CaBr_2} = N_{Ca}(N_{Br})^2 = 4N^3_{CaBr_2}/(1 + N_{CaBr_2})^2$$

[1] M. Temkin: *Acta Physicochimica URSS*, **20**: 411 (1945).

The above derivation has some interesting consequences with respect to freezing point depressions in salt mixtures. As was shown in Eqs. (4-17) and (4-18) the freezing point depression ΔT_m of a solvent A is essentially proportional to $-RT \ln a_A$. Thus for a binary Temkin mixture with a common ion the freezing point depression for AX is proportional to $-RT \ln x_{AX}$ as for a molecular mixture. For mixtures where all four ions are different it will be larger, however. Thus at high concentration of NaCl the freezing point depression in the system NaCl–KBr is twice, and in the system NaCl–CaBr$_2$ three times, that for the NaCl–KCl system.

The above expressions correspond to so-called "ideal mixing" of salts. In most cases we will have deviations from ideal behavior. The deviations may be either in the enthalpy or in the entropy term. Thus for the mixing of ions of somewhat different sizes the electrostatic binding energies do not cancel each other completely. Also, the interatomic bindings may not be purely electrostatic, but may be superimposed by covalent binding forces. These effects may cause the enthalpy of mixing to be different from zero.

These deviations may be described by an *activity coefficient*. Thus for an ion–pair AX

$$a_{AX} = N_A N_X \gamma_{AX}$$

For a mixture with a common ion, like AX–BX, and assuming ideal entropy of mixing γ_{AX} may be estimated from the differences between A–A, B–B, and A–B interaction energies and in the same way as shown for the regular solution model for metallic mixtures the relation $RT \ln \gamma_{AX} = bN_B^2$ may be derived. Here the interaction energy term is usually small as it is derived from interactions between cations that are next-nearest neighbors in the melt. Thus for the system NaCl–KCl: $b = -2$ kJ and for the system LiCl–KCl: $b = -13$ kJ at about 1000 K.[1]

For mixtures where all four ions are different, the interaction energy is larger, and may cause significant deviation from ideal Temkin behavior. Thus for a AX–BY mixture Flood and coworkers[2] have shown that the activity coefficient of, for example, AX in the mixture may approximately be given by

$$RT \ln \gamma_{AX} = N_B N_Y \, \Delta G_{ex}^{\circ} \qquad (4\text{-}25)$$

Here ΔG_{ex}° is the *standard* Gibbs energy change for the exchange reaction

$$AX(l) + BY(l) = AY(l) + BX(l)$$

[1] J. Lumsden: "Thermodynamics of Molten Salt Mixtures," Academic Press, London, 1966, p. 69.

[2] H. Flood, T. Förland, and K. Grjotheim: *Z. Anorg. Allg. Chem.* **276**: 289 (1954). See also H. Flood and K. Grjotheim: *J. Iron Steel Inst.* **171**: 64 (1952).

As an example, for the reaction $NaCl(l) + KF(l) = NaF(l) + KCl(l)$, $\Delta G°$ at 1000°C is about -21.7 kJ[1]. For a 50/50 mixture of NaCl and KF this gives $\gamma_{NaCl} = \gamma_{KF} = 0.6$, and $\gamma_{NaF} = \gamma_{KCl} = 1.67$. The corresponding activities will be 0.15 and 0.42 respectively.

The above equation applies for mixtures of ions of equal charge. If the ions are of different charge, for example, Na^+, Ca^{2+}, Cl^-, O^{2-}, the relation becomes

$$RT \ln \gamma_{CaO} = N'_{Na} N'_{Cl} \Delta G°_{ex}$$

where

$$N'_{Na} = \frac{n_{Na}}{n_{Na} + 2n_{Ca}} \quad \text{and} \quad N'_{Cl} = \frac{n_{Cl}}{n_{Cl} + 2n_O}$$

and similarly

$$N'_{Ca} = \frac{2n_{Ca}}{n_{Na} + 2n_{Ca}} \quad \text{and} \quad N'_O = \frac{2n_O}{n_{Cl} + 2n_O}$$

If the melt had contained a number of different cations and anions, then the activity coefficient for each ion pair $M_1 X_1$ would be

$$RT \ln \gamma_{M_1 X_1} = \sum_{i, j}^{i \neq 1, j \neq 1} N'_{M_i} N'_{X_j} \Delta G°_{M_i X_j} \tag{4-26}$$

Here $\Delta G°_{M_i X_j}$ is the change in Gibbs energy when one mole of $M_1 X_1$ is reacted reciprocally with $M_i X_j$, where $M_i X_j$ denotes all other ion pairs in the melts.

In general we see that if $\Delta G°_{ex}$ is negative, the activity coefficient of the ion pair $M_1 X_1$ will be less than one, i.e., the activity of the less stable pairs will be less than for the ideal ionic mixture.

The above derivations were based on the assumption of ideal entropy of mixing in the ionic sublattices. As in the quasichemical model for metallic solutions deviations from ideal entropy are to be expected, however, and will result in additional terms in the activity coefficient.

If one of the interatomic bonds is particularly strong, e.g., due to the presence of a considerable covalent contribution, a high degree of local order may occur. This is for example the case in the system $KCl-MgCl_2$. Here the Mg–Cl bonds are very strong, and each Mg cation is surrounded by four Cl anions with a tendency to form the complex anion $(MgCl_4)^{2-}$. In this case the melt may be regarded as consisting of $(MgCl_4)^{2-}$ anions and K^+ cations with additional Cl^- anions or Mg^{2+} cations depending on composition. This phenomenon is even more pronounced in sulfate melts where we have the sulfate ion $(SO_4)^{2-}$, whereas simple S^{6+} and O^{2-} ions hardly exist.

---

[1] J. Lumsden: loc. cit. p. 155.

Silicate Melts

An important example of complex ions is in the silicate melts, the structure of which may be understood from the structure of the solid silicates. In these melts each Si atom is surrounded by four O atoms forming a tetrahedral anion $(SiO_4)^{4-}$, the bonds between Si and O beeing essentially covalent. The $(SiO_4)^{4-}$ tetrahedra may again be bonded together by shared oxygen atoms to give double anions $(Si_2O_7)^{6-}$. These again may be linked into anions such as $(SiO_3)_n^{2n-}$, where n is some large number, corresponding to the anion being a long chain or a closed ring. These one-dimensional chains may again be linked together by further oxygen bridges to form two- and three-dimensional groups. For every oxygen bridge formed the negative charge per SiO_4 tetrahedron becomes less, and finally, for a complete three-dimensional network, this becomes neutral and has the composition $(SiO_2)_n$. Thus solid as well as liquid silica may be regarded as a gigantic three-dimensional molecule.

When basic oxides such as CaO are added to molten silica, this will have the effect of breaking the oxygen bridges, and at the same time free Ca^{2+} ions are formed

$$-Si-O-Si- + CaO = -Si-O^- \quad O^- -Si- + Ca^{2+}$$

By further addition of CaO more and more oxygen bridges are broken until finally the melt consists of single orthosilicate anions, $(SiO_4)^{4-}$, and Ca^{2+} cations. By further addition of CaO free oxygen ions, O^{2-}, may be formed.

Actually, in a melt of any composition we would expect there to be an equilibrium between the various types of anions as, for example,

$$O^{2-} + (Si_2O_7)^{6-} \rightleftarrows 2(SiO_4)^{4-}$$

and similarly for other groups. According to a theory by Masson[1] these equilibria, i.e., the *concentration* of the various anionic species, are strongly influenced by the nature of the cation. Thus for a lime silicate melt the above equilibrium is shifted strongly to the right, i.e., toward the formation of $(SiO_4)^{4-}$ ions, whereas for an iron silicate melt it is shifted to the left, to a mixture of oxygen and polysilicate anions. This is in agreement with calcium orthosilicate, Ca_2SiO_4, being much more stable than iron orthosilicate, Fe_2SiO_4.

If the chemical activity of calcium oxide is calculated from the Temkin theory, $a_{CaO} = N_{Ca} N_O$, then for a pure $CaO-SiO_2$ melt where Ca^{2+} is the only cation $N_{Ca} = 1$, whereas N_O and consequently a_{CaO} will decrease from unity for pure molten CaO to practically zero at the orthosilicate composition. Conversely the activity of SiO_2 is very small in the same range, but increases rapidly as the SiO_2 content increases beyond the Ca_2SiO_4 composition.

For an iron silicate melt, on the other hand, due to the predominance of highly polymerized anions N_O will be appreciable even in melts of high silica

[1] C. R. Masson: *Proc. R. Soc. London,* **A287**: 201 (1965).

content, and the activity of FeO is only slightly less than its mole fraction, through most of the system.

For melts in the system $CaO-Ca_2SiO_4-Fe_2SiO_4-FeO$ the activity of FeO and CaO may be calculated by means of Eq. (4-26). In accordance with the higher stability of Ca_2SiO_4 the activity of FeO becomes much higher than its mole fraction. This is in agreement with experimental measurements by Taylor and Chipman.[1] Further discussion on the thermodynamics of slag systems will be given in Chap 11.

PROBLEMS

4-1 For the binary liquid system zinc–tin the zinc vapor pressure at 700°C is measured as follows:

N_{Zn}	0.231	0.484	0.495	0.748	1.000
p_{Zn} (atm)	0.0246	0.0458	0.0470	0.0620	0.0788

Calculate the activity a_{Zn} and the activity coefficient γ_{Zn} for the above compositions, and plot these as functions of alloy composition. Estimate γ_{Zn}° for infinite dilution of zinc in tin.

4-2 For the binary liquid system silver–aluminum the following relative partial functions are listed for 900°C:

N_{Al}	0.1	0.3	0.5	0.7	0.9
$\Delta \bar{G}_{Al}$, J	−49 800	−22 100	−7570	−3180	−962
$\Delta \bar{H}_{Al}$, J	−28 900	−5060	+6190	+3220	+420

Calculate the relative partial entropy as well as the activity a_{Al} and the activity coefficient γ_{Al} at 900°C and 1100°C for the above compositions. Does this system obey Raoult's law?, Henry's law?, and, if so, in what range? ΔC_P on mixing is taken to be zero.

4-3 For the binary liquid system silver–lead, the following relative integral molar functions are listed for 727°C:

N_{Pb}	0.1	0.3	0.5	0.7	0.9
ΔH, J	1256	3245	61 230	3730	1885
ΔS, J/K	3.52	7.20	8.46	7.24	4.06

Calculate the relative integral molar Gibbs energy of mixing. By means of the tangent intercept method (Eq. (4-10)) calculate the relative partial Gibbs energies and the corresponding activities for each component.

[1] C. R. Taylor and J. Chipman: *Trans. AIME*, **154**: 228–246 (1943).

4-4 An *athermal* solution is defined as having zero enthalpy of mixing, whereas the entropy and hence the activity may deviate from ideal behavior. Derive an expression for the variation of the activity with temperature for constant composition for such a solution.

4-5 A given solution shows negative deviation from ideality. Also the change in volume on mixing is negative. How is the activity affected by an increase in pressure? The activity is referred to the pure component at the same temperature and pressure as standard state.

4-6 A *symmetrical* binary solution is defined as having the relative integral enthalpy of mixing obeying the relation $\Delta H = b N_A N_B$, where b is a constant independent of composition. Derive an expression for the relative partial enthalpy of each component in the solution.

4-7 In a dew point experiment a Cu–Zn alloy is placed in one end of an evacuated and closed silica tube, and is heated to 900°C. When the other end is cooled to 740°C zinc vapor starts to condense. Calculate the activity of zinc in the alloy, relative to pure liquid zinc, when the vapor pressure of pure liquid zinc is 75 mmHg at 740°C, and 750 mmHg at 900°C.

4-8 Figure 4-11 shows the austenite field of the stable iron–graphite phase diagram.

(a) Assuming Henry's law to be valid within the α and γ phases, estimate the chemical activity of carbon, relative to graphite at 800 and 1000°C as function of composition. (At low concentrations mole fraction may be taken proportional to weight percentage.)

(b) A steel with 0.5 percent C is to be bright annealed at 800°C in a CO–CO_2 atmosphere. Estimate the gas ratio $(p_{CO})^2/p_{CO_2}$ which would be in equilibrium with the steel, when for the reaction $C(\text{graph}) + CO_2 = 2CO$ the equilibrium constant at 800°C is 6.0. Estimate the gas composition if $p_{CO} + p_{CO_2} = 0.2$ atm.

(c) Calculate the Gibbs energy change for the reaction $C(\text{graph}) = \underline{C}(1\%)$ at 800 and 1000°C under the above assumption, and calculate the activity a'_C (relative to 1 percent solution) at graphite saturation.

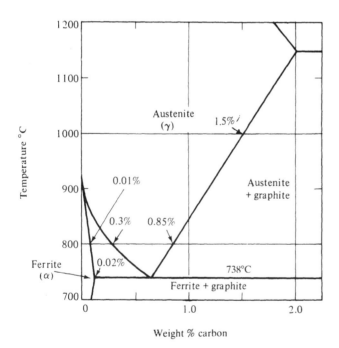

Figure 4-11 Stable iron–graphite phase diagram. Notice the expanded scale for the ferrite (α) field. (*After R. P. Smith: J. Am. Chem. Soc., vol. 68, 1946, p. 1163.*)

(d) Actually carbon in austenite shows positive deviation from Henry's law. In which direction will this affect the values calculated under point (c)?

(e) The addition of silicon is known to increase the activity coefficient of carbon dissolved in iron. How will the addition of silicon affect the solubility limit for graphite in austenite?

(f) For the reaction C(graph) = C(diam) $\Delta G^\circ_{1273} = +7.33$ kJ. Estimate the solubility of diamond in austenite at this temperature if graphite is not formed.

4-9 (a) By means of the Temkin model (Sec. 4-7) and assuming no enthalpy of mixing, calculate the activity of NaCl for 90 and 50 mole percent NaCl in the following binary molten salt mixtures: NaCl–KCl, NaCl–KBr, NaCl–CaBr$_2$. In all cases the number of cation and anion sites is taken equal to the number of cations and anions respectively.

(b) The melting point of NaCl is 801°C and the enthalpy of fusion is 28 000 J. Calculate the freezing point depression for NaCl with 10 mol percent of respectively KCl, KBr, and CaBr$_2$. There is no solid solubility, and the change in heat capacity on melting is disregarded.

4-10 For the system Ag–Si the partial pressure of Ag at 1500°C has been measured as follows (Robinson and Tarby, *Met. Trans.*, **2**: 1347 (1971)):

N_{Si}	p_{Ag}(mmHg)	N_{Si}	p_{Ag}(mmHg)
0.0000	6.68	0.530	3.23
0.0238	6.59	0.633	2.74
0.0546	6.23	0.736	2.29
0.111	5.58	0.843	1.68
0.292	4.43	0.928	0.755
0.358	3.94		

Calculate the activity of Ag at that temperature, relative to pure Ag, and by means of the Gibbs–Duhem equation calculate the activity of Si.

4-11 The liquidus temperature in the system Ag–Si has been measured as follows (J. P. Hager: *Trans. AIME*, **227**: 1000 (1963)):

wt % Si	T K	wt % Si	T K
0.0	1234	6.1	1318
1.0	1194	8.1	1386
2.1	1154	11.8	1480
3.0	1113 (min)	23.1	1562
3.5	1142	43.3	1608
4.0	1174	64.4	1641
4.8	1212	100.0	1686
5.9	1291		

By means of Eq. (4-18) calculate the activity of Ag and Si along the two branches of the liquidus curve. There is no solid solubility. Assuming a regular solution and $\Delta \bar{H}$ independent of temperature, calculate the corresponding activities at 1773 K, and by means of the Gibbs–Duhem equation calculate the activity of the other component. Compare the result with that of Prob. 4-10. For melting of the pure components the following molar values apply: Ag: $\Delta H_m = 11\,945$ J, $\Delta C_P = -0.8$ J/K; Si: $\Delta H_m = 50\,668$ J, $\Delta C_P = -1.7$ J/K.

BIBLIOGRAPHY

Hildebrand, J. H., and R. L. Scott: "The Solubility of Non-Electrolytes," Reinhold Publ. Co., New York, 1950. "Regular Solutions," Prentice-Hall, Inc., Englewood Cliffs, N.J., 1962.

Institution of Mining and Metallurgy: "The Physical Chemistry of Melts," London, 1953.

Kubaschewski, O., and J. A. Catterall: "Thermochemical Data of Alloys," Pergamon Press, London, 1956.

Lumsden, J.: "Thermodynamics of Alloys," The Institute of Metals, London, 1952.

Lumsden, J.: "Thermodynamics of Molten Salt Mixtures," Academic Press, London, 1966.

National Physical Laboratory, Symposium No. 9: "The Physical Chemistry of Metallic Solutions and Intermetallic Compounds," Her Majesty's Stationery Office, London, 1959.

Newton, R. C., A. Navrotsky, and B. J. Wood (eds.): "Thermodynamics of Minerals and Melts," Springer Verlag, New York, 1981.

Richardson, F. D.: "Physical Chemistry of Melts in Metallurgy," 2 vols., Academic Press, London and New York, 1974.

Sundheim, B.: "Fused Salts," McGraw-Hill Book Co. Inc., New York, 1964.

Turkdogan, E. T.: "Physical Chemistry of High Temperature Technology," Academic Press, New York, 1980.

Wagner, C.: "Thermodynamics of Alloys," Addison-Wesley Publ. Co. Inc., Reading, Mass., 1952.

CHAPTER
FIVE

REACTION KINETICS

From thermodynamics the final, equilibrium state for a reaction may be predicted, but thermodynamics gives no information about *the rate* at which this equilibrium is approached. At room temperature many reactions are very slow, as illustrated by the fact that most base metals may be used in air, even though they would be converted into oxides if chemical equilibrium were established. Thus kinetics is responsible for the fact that the products of metallurgy may be used at all. Kinetics also affects the processes of metal extraction. In most cases a fast reaction is preferred. Thus in open-hearth steel-making the time required to bring the refining process to completion is of great importance for the economy of the process. In other cases the outcome of a metallurgical process may depend on the rate being of moderate magnitude. A typical example is the iron blast furnace, where high efficiency is dependent on the coke being not too reactive.

In kinetics we distinguish between *homogeneous reactions* and *heterogeneous reactions*. The former term refers to reactions which take place within one single phase, e.g., between molecules in a gas or in a solution. A reaction is called *heterogeneous* if more than one phase is involved. Many reactions which apparently are homogeneous are actually heterogeneous inasmuch as they take place on the walls of the container or on the surfaces of a second phase. Heterogeneous catalytic reactions belong to this group. Almost all metallurgical processes are based on heterogeneous reactions, e.g., between gases and solids as in the case of roasting and reduction, or between two liquid phases as in slag–metal reactions. That homogeneous reactions are of little direct interest to metallurgists, is illustrated by the fact that a good metallurgical example is hard to find. Nevertheless, the treatment of homogeneous kinetics represents a model which is useful also for the discussion of heterogeneous reactions.

5-1 HOMOGENEOUS REACTIONS

In a phase of a given volume V there are n_A moles of A, n_B moles of B, and n_C moles of C. The molar concentrations of the three species are then $C_A = n_A/V$, $C_B = n_B/V$ and $C_C = n_C/V$. If a chemical reaction takes place among the three species, e.g., according to the equation

$$lA + mB = nC \tag{5-1}$$

the rate of the reaction may be expressed as

$$r_C = \frac{dn_C}{V\,dt} = \frac{dC_C}{dt} = -\frac{n}{l} \cdot \frac{dC_A}{dt} = -\frac{n}{m} \cdot \frac{dC_B}{dt} \tag{5-2}$$

where t denotes the time.

The rate may be regarded as the *flux* of the atoms in molecules A and B through the reaction per unit volume and time. It is found experimentally that the rate is a function of the concentrations of A, B, and C, and of the temperature. We may tentatively assume that the reaction is the result of a simultaneous collision of l molecules of A and m molecules of B to form n molecules of C, and also that n molecules of C may collide to form l and m molecules of respectively A and B (reverse reaction). In that case we would expect the rate to depend on the composition in the following way

$$r_C = \frac{dC_C}{dt} = k_1\,C_A^l\,C_B^m - k_2\,C_C^n \tag{5-3}$$

The constants k_1 and k_2 are called the *rate constants* in the forward and in the reverse direction, and are expected to be functions of the temperature. When stable equilibrium is established, the forward and the reverse rates are equal, and the net rate is zero. Hence *at equilibrium*

$$\frac{(C_C^*)^n}{(C_A^*)^l(C_B^*)^m} = \frac{k_1}{k_2} = K_C \tag{5-4}$$

Here C_A^*, etc., denote concentrations at equilibrium, and K_C is the equilibrium constant for the reaction expressed in terms of concentrations. As k_1 and k_2 are assumed independent of concentration it follows that $k_1 = k_2 K_C$ at any concentration.

The exponents l, m, and n are called the *order* of the reaction with respect to A, B, and C. The total order in the forward direction is given by $l + m$ and in the reverse direction by n. In some cases, e.g., for the reaction $H_2 + I_2 = 2HI$, experiments have shown that the reaction rate may actually be described by Eq. (5-3), where the order is equal to the *molecularity*. This indicates strongly that the reaction takes place by collision of single H_2 and I_2 molecules, and is reversed by collision of two HI molecules. In other cases, e.g., for the reaction $H_2 + Br_2 = 2HBr$, and particularly for reactions where a large number of molecules participate, the observed order of the reaction is not equal to the molecularity. This is not surprising if one realizes what a rare coincidence a simultaneous collision of a large number of molecules must be. The observed order is usually smaller than the molecularity, and it is doubtful whether total orders higher than three have ever been observed. In many cases the observed order is even fractional.

To account for this discrepancy it is believed that such reactions proceed over a number of *intermediate steps*. Thus for the reaction $H_2 + Br_2 = 2HBr$ the overall rate has been found to obey the relation

$$\frac{dC_{HBr}}{dt} = \frac{kC_{H_2}(C_{Br_2})^{1/2}}{m + C_{HBr}/C_{Br_2}}$$

where k and m are constants. As we see the order of reaction with respect to Br_2 is less than one, and the order with respect to HBr is a variable quantity.

This relation may be accounted for by the following assumed reaction mechanism:

$$(1) \qquad Br_2 = 2Br$$

$$(2) \qquad Br + H_2 = HBr + H$$

$$(3) \qquad H + Br_2 = HBr + Br$$

$$(4) \qquad H + HBr = H_2 + Br$$

Here reaction 4 is simply the reversion of reaction 2, and is assumed to have a rate constant of similar magnitude to those of reactions 2 and 3. On the other hand, the equilibrium constant for reaction 3 is assumed very large, making the reversion of that reaction insignificant.

The rate of formation of HBr would then be given by

$$\frac{dC_{HBr}}{dt} = k_2 \, C_{Br} \, C_{H_2} + k_3 \, C_H \, C_{Br_2} - k_4 \, C_H \, C_{HBr}$$

where k_2, k_3, and k_4 are the forward rate constants for reactions 2, 3, and 4.

Furthermore, since the concentration of monatomic H is very small,

$$\frac{dC_H}{dt} = k_2 \, C_{Br} \, C_{H_2} - k_3 \, C_H \, C_{Br_2} - k_4 \, C_H \, C_{HBr} = 0$$

Finally, it is assumed that the rate constant for reaction 1 is much higher than for the other reactions, causing the concentration of monatomic Br at any time to be in virtual equilibrium with Br_2

$$(C_{Br})^2 = (K_C)_1 \, C_{Br_2}$$

Here $(K_C)_1$ is the equilibrium constant for reaction 1. Eliminating C_H and C_{Br} from the above expressions, we get

$$\frac{dC_{HBr}}{dt} = \frac{[2k_2\,k_3(K_C)_1^{1/2}/k_4]C_{H_2}(C_{Br_2})^{1/2}}{k_3/k_4 + C_{HBr}/C_{Br_2}}$$

Putting $2k_2\,k_3(K_C)_1^{1/2}/k_4 = k$ and $k_3/k_4 = m$, the empirical equation is obtained.

In general we may look upon a reaction $A \to D$ where A represents all the reactants and D all the products, as proceeding over a number of steps connected in series, viz.,

$$A \underset{K_1}{\overset{k_1}{\longrightarrow}} B \underset{K_2}{\overset{k_2}{\longrightarrow}} C \underset{K_3}{\overset{k_3}{\longrightarrow}} D$$

Here B and C represent intermediate reaction products, k_1, k_2, and k_3 represent the forward reaction rates for the three steps, and K_1, K_2, and K_3 represent the corresponding equilibrium constants. The rate of each step will be given by its rate constant, its molecularity, and the concentration of reactants and products before and after the step. As material cannot accumulate in significant amounts between the intermediate steps the intermediate products B and C must be consumed as fast as they are formed. If the rate constant for one of the steps is significantly smaller than for the others, these other steps will each be, essentially, in chemical equilibrium, and the overall rate of reaction will be determined by the smaller rate constant and the equilibrium constants for the remaining steps together with the corresponding molecularities. We say that the step with the smallest rate constant is the *rate determining* one. We may also say that this step offers the greatest *resistance* to the reaction.

Occasionally there may be alternative reaction paths leading from the reactants to the products, viz.,

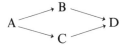

In that case the path which offers the *least* resistance will be the preferred one in the same way as in an electrical circuit the current will mainly go through that of several parallel conductors which offers the smallest resistance.

From experiments an empirical rate expression may be obtained. This may be compared with the rate expression which can be derived from a hypothetical sequence of intermediate reactions, i.e., an assumed *reaction mechanism*. In this way one may rule out those mechanisms which do not agree with the observed expression. On the other hand, since several different mechanisms may result in the same rate expression, it is not always possible from experiments to decide unambiguously which mechanism is actually operating.

Example 5-1 If the reaction rate is directly proportional to the concentration of one of the reacting species, we have a *first-order reaction*. Examples are radioactive decay, and the reaction of iodine with a large surplus of hydrogen under conditions where the reverse reaction is insignificant. In such cases we have: $-dC_A/dt = kC_A$, which integrated gives: $\ln C_A/C_A^\circ = -kt$. Here C_A° is the initial and C_A the instantaneous concentration of A. The variation of C_A with time for such a reaction is illustrated by curve (a) in Fig. 6-7.

The Effect of Temperature on Reaction Rates

The rate constants k_1, k_2, etc., increase with increasing temperature. The Swedish chemist Svante Arrhenius found that this increase could be described in the following way:

$$k = k_0\, e^{-E/RT} \tag{5-5}$$

This is an expression analogous to the van't Hoff equation for the temperature dependence of the equilibrium constant Eq. (3-8). The quantity E is called the *activation energy*, and is determined from the slope of a plot of log k vs. $1/T$. The first term k_0 is usually called the *frequency factor*.

A slightly different expression has been proposed by Eyring.[1] This will be illustrated for a simple reaction $A \rightarrow B$. According to Eyring's theory A* represents an *activated* compound formed intermediately. The activated compound is in chemical equilibrium with the reactant A. Hence $a_A^* = K^* a_A$, where a_A^* and a_A are the chemical activities of A* and A respectively, and may be regarded as proportional to their concentrations. K^* is the equilibrium constant for the reaction $A \rightarrow A^*$. The decomposition of A* into B takes place at a *universal* rate, independent of the nature of the reaction and only dependent on the concentration of A*

$$\frac{-dC_A^*}{dt} = \frac{RT}{Nh}\, C_A^*$$

Here h is the Planck's constant and N is Avogadro's number. This expression, therefore, gives the rate of the overall reaction of $A \rightarrow B$. Since $C_A^* = C_A \exp(-\Delta G^*/RT)$, where ΔG^* is the standard Gibbs energy for the reaction $A \rightarrow A^*$, the rate becomes equal to

$$\frac{-dC_A}{dt} = C_A \frac{RT}{Nh}\, e^{-\Delta G^*/RT}$$

or, for the rate constant,

$$k = \frac{RT}{Nh}\, e^{-\Delta G^*/RT} = \frac{RT}{Nh}\, e^{\Delta S^*/R} e^{-\Delta H^*/RT}$$

$$= k_0\, T e^{-\Delta H^*/RT} \tag{5-6}$$

---

[1] S. Glasstone, K. J. Laidler, and H. Eyring: "The Theory of Rate Processes," McGraw-Hill Book Co. Inc., New York, 1941, pp. 195–197.

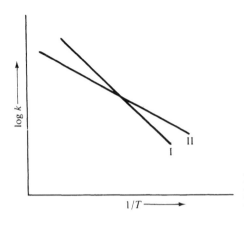

Figure 5-1 Variation of rate constant with temperature for two reactions I and II of different activation energy.

Here $\Delta G^* = \Delta H^* - T\Delta S^*$, and the expression $R/(Nh) \cdot \exp(\Delta S^*/R)$ is denoted k_0, and is regarded as independent of temperature. Also ΔH^* may be regarded as independent of temperature. As we see, the main difference between this expression and the Arrhenius expression is the addition of a linear temperature term. Unfortunately, most kinetic measurements are carried out over too small a temperature range to detect the difference between the Arrhenius and Eyring expressions, and for most purposes the former is adequate even though the latter may have a sounder theoretical basis.

The activation energy differs greatly for different types of reaction, but may be of the order of 50 to 500 kJ for chemical reactions. Also physical phenomena such as viscous flow and diffusion have temperature dependencies which may be expressed in terms of activation energies, usually up to about 100 kJ. It should be borne in mind, however, that strictly speaking the term "activation energy" should be used only for well-defined reaction steps. For more complex phenomena it will always remain an empirical entity with little physical reality.

If in a sequence of reactions in series different steps have different activation energies, an increase in temperature will make the step with the lowest activation energy rate controlling. Conversely for parallel reactions the step with the highest activation energy will dominate at high temperature, see Fig. 5-1.

5-2 DIFFUSION

If a concentration gradient is present within a single phase, diffusion will tend to smooth it out. Diffusion is a process of mass transfer on an atomic or molecular scale. In gases and liquids the individual molecules or atoms are always in motion. In solids they oscillate around their lattice positions. Once in a while thermal oscillation causes two atoms to change places, or one atom squeezes through the opening between two neighbors into an interstitial position or into a vacant lattice position. In this way a new vacant position is created, into which a new atom may squeeze. Also more complicated motions, such as the simultaneous movement of a number of atoms in a ringlike fashion, have been suggested.

How can the molecules or atoms "know" that there is a gradient to be levelled out? The answer is simply: They don't. Since there are more particles of a given kind in a region where the concentration is high than in a region of low concentration, more particles will, in their erratic movements, move from the region of high concentration than in the opposite direction. The result will be a net transfer from the high to the low concentration region. In spite of the apparently erratic nature of diffusion on an atomic scale, the result on a macroscopic scale may be expressed as a simple relation between the *concentration gradient* and the *flux* of a given particle A through a unit area perpendicular to the concentration gradient

$$J_A = \frac{dn_A}{dt} = -D_A \frac{dC_A}{ds} \tag{5-7}$$

At a given location s, see Fig. 5-2, the flux J_A is the product of the concentration gradient and a *diffusivity constant* D_A. This relation is called *Fick's first law*. The negative sign shows that the flux is positive in the direction of decreasing concentration. The flux J_A has the dimension of mole $\cdot$ time^{-1} length^{-2}, and C_A has the dimension of mole $\cdot$ length^{-3}. This makes the dimension of D equal to length$^2 \cdot$ time^{-1}, and it is usually expressed in square centimeters per second.

If a given concentration gradient is maintained by continuous supply of A to the high concentration area and removal from the low concentration area, the flux will remain constant with time. Such a stationary state is particularly well suited for experimental determination of D_A. Thus the diffusivity of hydrogen through palladium may be determined by diffusing hydrogen through a

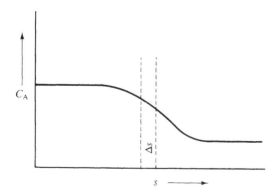

Figure 5-2 Variation of concentration of component A with distance during diffusion.

palladium foil of a given thickness and area, and with fixed hydrogen pressures on each side.

If the diffusing element is not supplied or removed, the concentration gradient will change with time as a result of the diffusion itself. For a volume element, Fig. 5-2, of unit cross section and of thickness Δs the entering flux will be $-(D_A \, dC_A/ds)_s$, and the effluent flux $-(D_A \, dC_A/ds)_{s+\Delta s}$. The difference between the fluxes is the increase in the amount of A within the volume element

$$\frac{dn_A}{dt} = \Delta s \frac{dC_A}{dt} = -\left(D_A \frac{dC_A}{ds}\right)_s + \left(D_A \frac{dC_A}{ds}\right)_{s+\Delta s}$$

When Δs goes to zero

$$\frac{dC_A}{dt} = \frac{d}{ds}\left(D_A \frac{dC_A}{ds}\right) \tag{5-8}$$

This is *Fick's second law*. In the special case where D_A is independent of s the law has the simpler form

$$\frac{dC_A}{dt} = D_A \frac{d^2 C_A}{ds^2} \tag{5-9}$$

Usually the diffusion coefficient is a function of composition. This is illustrated very clearly from an experiment carried out by Darken.[1] Figure 5-3

[1] L. S. Darken and R. W. Gurry: "Physical Chemistry of Metals," McGraw-Hill Book Co. Inc., New York, 1953, p. 462. See also: L. S. Darken, *Trans. AIME*, **180**: 430 (1949).

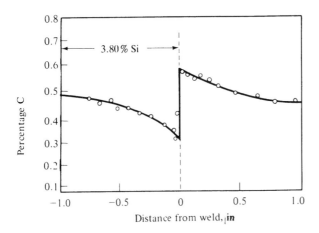

Figure 5-3 "Uphill" diffusion of carbon from Fe–Si–C alloy (3.8 percent Si) to Fe–C alloy. Treatment: 13 days at 1050°. (*From L. S. Darken: Trans. AIME, vol. 180, 1949, p. 430.*)

shows a diffusion couple which consists of two steel bars of approximately the same carbon content. The one bar is a plain carbon steel, whereas the other contains 3.8 percent silicon. Although originally there had been no carbon concentration gradient, carbon had during the experiment diffused from the silicon steel into the carbon steel. The diffusion had taken place against the carbon concentration gradient, which in itself was created by the diffusion (uphill diffusion). This phenomenon is attributed to the *chemical potential* of carbon in the silicon steel being higher than in the carbon steel, and shows that the driving force behind diffusion is the Gibbs energy which is seeking toward its minimum.

It has been suggested that the diffusion coefficients should be expressed in terms of chemical activity gradients rather than concentration gradients. Although this would more clearly emphasize the nature of the driving force in diffusion, it is found that even on this basis the diffusion coefficients are functions of composition. Obviously other factors, such as interatomic interactions, affect the diffusion rate. Since there is very little gained by converting concentrations into activities, and still ending up with a variable "constant," the original way of expressing diffusion coefficients in terms of concentrations is retained. This may have the result, however, that the diffusivity may become negative as in the junction between a silicon and a plain carbon steel.

If we know the variation of D_A with composition, or if D_A is constant, Fick's second law may in principle be integrated to show the composition at any time and at any location within the body under consideration. Since Fick's laws refer to unit cross sections, the integration depends on the relation between the cross section and the distance s, that is, of the geometry of the body. Because of its complexity a mathematical integration will not be given here. Instead concentration profiles as function of time for a few typical geometries are shown in Fig. 5-4(*a*) and (*b*). A more detailed treatment will be given in the discussion of nonsteady heat flow, which is a phenomenon quite analogous to diffusion (Sec. 6-5).

Diffusion in Gases

If two gases, A and B, diffuse into each other, the flux of A in one direction must be equal to the flux of B in the opposite direction. If this were not the case, a gas pressure gradient would build up, and this would cause the gas body to flow in such a way as to establish equal and opposite fluxes of A and B. Hence according to Fick's first law

$$J_A = -J_B = -D_{AB} \frac{dC_A}{ds} = D_{AB} \frac{dC_B}{ds} \tag{5-10}$$

Here D_{AB} denotes the diffusivity of A in B or vice versa. The concentrations are given in moles per unit volume and are related to the partial pressures by

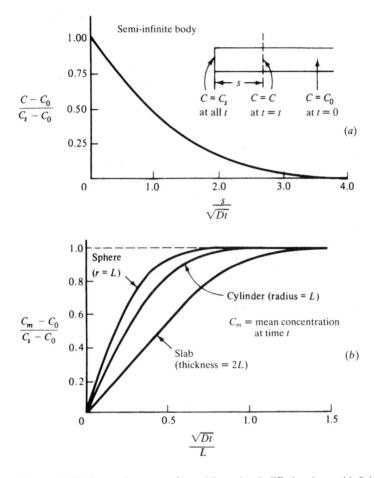

Figure 5-4 (a) Penetration curve for unidimensional diffusion in semi-infinite body. C_0 = initial uniform concentration, C_s = constant surface concentration, C = variable concentration, s = distance from surface, t = time. (b) Mean fractional concentration in slab, cylinder, and sphere of initial concentration C_0 and surface concentration C_s. (*From L. S. Darken and R. W. Gurry: "Physical Chemistry of Metals." Copyright 1953. Used with permission of McGraw-Hill Book Co. Inc., New York.*)

the expression $C_A = p_A/RT$ (perfect gas behavior is assumed). On the basis of the kinetic gas theory one may show that

$$D_{AB} = \frac{K_{AB}}{P_{tot}} T^{3/2} \tag{5-11}$$

Here K_{AB} is a constant which is independent of P and T, and $p_{tot} = p_A + p_B$.

If this expression is combined with Eq. (5-10), and the partial pressure inserted, we get

$$J_A = -J_B = \frac{-K_{AB}}{R p_{tot}} (T)^{1/2} \frac{dp_A}{ds} \qquad (5\text{-}12)$$

Since the composition of the gas is usually expressed in terms of partial pressures rather than in terms of concentration this expression is more useful than Eq. (5-10). Equation (5-12) shows that the rate of diffusion is proportional to the partial pressure gradient and to the square root of the absolute temperature. In this respect is differs from diffusion in solids and liquids, which increases exponentially with temperature in accordance with the Arrhenius equation. The square root relation gives a temperature dependence which, at say 1000 K, is equivalent to that of an activation energy of about 5 kJ/mol, i.e., considerably less than for solid and liquid diffusion.

The constant K_{AB} is a function of the properties of the gases. Empirically the following relation has been found:[1]

$$K_{AB} = \frac{0.0043}{(v_A^{1/3} + v_B^{1/3})^2} \left(\frac{1}{M_A} + \frac{1}{M_B} \right)^{1/2}$$

Here M_A and M_B are the molar weights of A and B respectively, and v_A and v_B are their molar volumes in condensed state. To give some examples, these volumes are: 14.3 cm^3 for hydrogen, 25.6 cm^3 for oxygen and 34.0 cm^3 for carbon dioxide, i.e., they do not change greatly when we go from lighter to heavier molecules. For organic molecules they may be larger, however. The numerical coefficient 0.0043 applies to K_{AB} having the dimension $cm^2 \cdot atm \cdot sec^{-1}(K)^{-3/2}$.

Diffusion plays an important role in heterogeneous reaction kinetics. If a chemical reaction takes place at the interface between two phases, the reactants have to be transferred to the interface, and the reaction products have to be removed. In the bulk of liquids and gases the mass transport is mainly by convection and turbulence. Near an interface, however, the flow is believed to be laminar in a thickness called the *boundary layer*. Through this boundary layer mass transfer takes place by diffusion. If the thickness of the boundary layer is denoted δ, the flux per unit area by diffusion will be

$$J_A = \frac{D_A}{\delta} (C_A - C_A^*)$$

For diffusion in gases this becomes $(p_A - p_A^*)D_{AB}/(\delta RT)$, where C_A and p_A denote the concentration and partial pressure on the interface and C_A^* and p_A^* denote the corresponding values outside the boundary layer.

---

[1] E. R. Gilliland: *Ind. Eng. Chem.*, **26**: 681–685 (1934). See also: L. Andrussow: *Z. Elektrochem.*, **54**: 566–571 (1950).

Usually we do not know δ, the thickness of the boundary layer. In analogy with the expressions used in heat transfer (Chap. 6) we may write

$$J_A = k_g(C_A - C_A^*) = k_g'(p_A - p_A^*) \qquad (5\text{-}13)$$

Here k_g and k_g' are the *mass transfer coefficients* in terms of concentration and partial pressure, respectively, and are functions of the physical properties of the gas or fluid and of the thickness of the boundary layer. The latter is again a function of the gas velocity parallel to the surface and the geometry and dimension of the surface. The relation between mass transfer coefficient and these parameters is analogous to the relation between the *heat* transfer coefficient and the same parameters. These relations will be discussed in Sec. 6-6. With the knowledge of these physical parameters the mass transfer coefficient and the rate of diffusion may be calculated.

5-3 HETEROGENEOUS REACTIONS

Different cases of heterogeneous reactions are shown in Fig. 5-5. Most heterogeneous reactions are composed of the following steps:

1. Supply of the reactant in the direction of fluid flow
2. Diffusion of reactant to the interface
3. Interface reaction (intrinsic rate)
   a. Adsorption of reactants
   b. Reaction proper
   c. Desorption of products
4. Diffusion of products from the interface
5. Removal of products in the direction of fluid flow

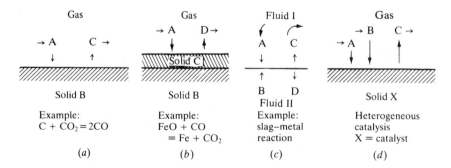

Figure 5-5 Types of heterogeneous reactions. (*a*) solid + gas → gas, (*b*) solid + gas → solid + gas, (*c*) diss. (phase I) + diss. (phase II) → diss. (phase I) + diss. (phase II). (*d*) gas + gas + catalyst → gas.

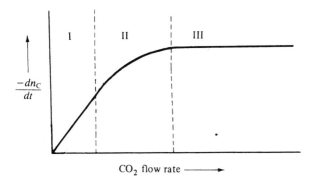

Figure 5-6 Rate of heterogeneous reaction as function of fluid flow rate (schematic).

These steps operate in series and may in principle be treated in the same way as shown for consecutive homogeneous reactions. The steps 1 and 5 are coupled, and may be treated as one step. The same applies to the steps 2 and 4. The reaction rate for each step may be expressed in terms of concentration or chemical activity before and after each step, and reaction constants for the forward and the reverse reaction. For the diffusion steps the reaction constant is replaced by the mass transfer coefficient. Since material cannot accumulate significantly in any of the intermediate stages, the flux must be the same through all steps. If the mechanism for each step is known we may work out a rate expression where the concentrations at the intermediate stages are eliminated, and which expresses the overall reaction rate in terms of the concentrations of the initial reactants and final products, and of the reaction constants and molecularities of the various steps, in the same way as was shown for homogeneous reactions.

It is obvious that a complete calculation of the reaction rate is difficult if all steps have rate constants of similar magnitude. Usually the rate constant for one of the above steps is significantly smaller than for the others, and this step is rate controlling, whereas the other steps are close to equilibrium. If we look upon a heterogeneous reaction between a solid and a gas, as for example the reaction $CO_2 + C = 2CO$, the controlling steps will depend on the gas flow rate as well as on the temperature. At very low flow rates, assuming a given temperature, the supply of CO_2 in the gas flow may be rate controlling. This means that diffusion to and from the carbon surface as well as the chemical reaction at the surface are sufficiently fast to bring the effluent gas in chemical equilibrium with the carbon. The reaction rate, i.e., the consumption of carbon per unit time, is, in that case, obviously proportional to the gas flow rate. This is illustrated by range I in Fig. 5-6.

At higher flow rates the diffusion to and from the carbon surface is no longer able to make the effluent gas equal to the gas on the carbon surface. Diffusion is now the rate controlling step, whereas chemical equilibrium is still established on the carbon surface. The diffusion rate, i.e., the mass transfer coefficient, increases with increasing gas flow rate corresponding to an increased rate of carbon consumption. This is illustrated by range II in Fig. 5-6.

If the gas flow is increased further, diffusion becomes so fast that the chemical reaction on the surface is no longer able to maintain chemical equilibrium. The surface reaction then becomes rate controlling. The surface reaction rate is independent of the gas flow as illustrated by range III in Fig. 5-6.

The rate constants for the different steps have different temperature coefficients. Thus the chemical reaction rate increases much faster with temperature than the diffusion to and from the surface. For a given gas flow rate, diffusion therefore becomes rate controlling at higher temperatures. Finally the supply of reactant may become rate controlling in the temperature range where the equilibrium constant is small.

The surface reaction is also called the *intrinsic* reaction. For reactions on solid surfaces this is believed to take place on *active points* on the surface. The intrinsic reaction rate will therefore increase with increasing concentration of these points. The reaction may be divided into several steps: 3*a*, adsorption on active points; 3*b*, reaction on the surface to form adsorbed products; and 3*c*, desorption of products. Detailed studies of intrinsic reaction rates have been made particularly for heterogeneous catalysis. For metallurgical processes the experimental material is very limited. A case of intrinsic rates will be shown in Example 5.2.

In order to study intrinsic reaction rates it is necessary to work at gas flow rates which make the gas composition on the surface equal to the composition in the main gas flow. For liquids the same result is obtained by stirring. The reaction rate is then measured as function of the significant parameters: gas composition, interface area, and temperature. Usually an empirical relation is obtained. This may be compared with the rate which is derived from assumed reaction mechanisms. As mentioned in connection with homogeneous kinetics, agreement between observed and calculated rate expressions does not prove that the assumed reaction mechanism is correct, however.

Similarly, for the range where the reaction is diffusion controlled this may be studied as function of composition, interface area and geometry, and fluid flow or stirring rate. In this way the mass transfer coefficient may be evaluated and, by means of dimensional analysis, may be estimated for other interface dimensions.

It should be borne in mind that even though the intrinsic reaction rate may be studied in the laboratory, this is no guarantee for this being the controlling step in an industrial process. But knowing the overall mechanism, it should be possible to affect the reaction velocity. Thus if the intrinsic rate is controlling, nothing is gained by increased gas flow rate or increased stirring of liquids. The rate may then be increased by increasing the surface area or one may increase the concentration of active points by addition of catalytic components to the surface. Also, the rate may be increased by increasing the temperature. The temperature will have little effect, however, if diffusion to and from the surface or the supply of reactants are rate controlling.

Example 5-2 For the reaction $CO_2 + C = 2CO$ the *intrinsic rate* per unit weight or surface area of carbon has been found experimentally to obey the relation[1]

$$\frac{-dn_C}{dt} = \frac{kp_{CO_2}}{lp_{CO_2} + mp_{CO} + 1}$$

Here k, l, and m are constants, independent of the partial pressures of CO_2 and CO. This rate equation may be accounted for by the following assumed *reaction mechanism*:

(1) $\quad CO_2(g) + C_f \xrightarrow{k_1} CO(g) + C_o$

(2) $\quad C_o + CO(g) \xrightarrow{k_2} CO_2(g) + C_f$

(3) $\qquad\qquad C_o \xrightarrow{k_3} CO(g) + nC_f$

Here C_f represents a free, active carbon site, and C_O a carbon site to which one oxygen is adsorbed. The number n is close to unity, and is, in the following, taken equal to unity. Whereas for reactions 1 and 2 the rate constants k_1 and k_2 are assumed to be of similar magnitude, the rate constant k_3 is much larger than its reverse, and the latter may be disregarded.

The rate of gasification of carbon is then given by $-dn_C/dt = k_3 C_{C_o}$, where C_{C_o} is the concentration of occupied sites on the carbon surface. Under steady state conditions the net rate of formation of occupied sites must be zero

$$\frac{dC_{C_o}}{dt} = k_1 p_{CO_2} C_{C_f} - k_2 p_{CO} C_{C_o} - k_3 C_{C_o} = 0$$

Expressing the total number of sites by $C_{C_t} = C_{C_o} + C_{C_f}$, and solving for C_{C_o}, we get:

$$C_{C_o} = \frac{k_1 C_{C_t} p_{CO_2}}{k_1 p_{CO_2} + k_2 p_{CO} + k_3}$$

This gives for the rate of gasification:

$$\frac{-dn_C}{dt} = \frac{k_3 k_1 C_{C_t} p_{CO_2}}{k_1 p_{CO_2} + k_2 p_{CO} + k_3}$$

If $k_1 C_{C_t}$ is denoted k, k_1/k_3 denoted l, and k_2/k_3 denoted m, the empirical rate equation is obtained.

---

[1] J. Gadsby et al.: *Proc. R. Soc. London*, **193**: 357–400 (1948). See also R. F. Strickland-Constable: *J. de Chim. Phys.*, **47**: 356–360 (1950), and S. Ergun: *U.S. Bur. of Mines Bull.*, **598** (1962).

Under most experimental conditions $k_1 p_{CO_2} > k_3$ and $k_2 p_{CO} > k_3$, i.e., reaction (3) is rate controlling. This gives the simplified equation:

$$\frac{-dn_C}{dt} = \frac{k_3 k_1 C_{C_t} p_{CO_2}}{k_1 p_{CO_2} + k_2 p_{CO}} = \frac{k_3 K_1 C_{C_t}}{K_1 + p_{CO}/p_{CO_2}}$$

where $K_1 = k_1/k_2$ is the equilibrium constant for reaction 1.

5-4 GAS–LIQUID AND LIQUID–LIQUID REACTIONS

Typical examples of gas–liquid reactions are absorption of a gas in, or desorption of a gas from, a molten metal, as illustrated by absorption or desorption of nitrogen in or from liquid steel, and the reaction between a gas and a liquid metal to give a gaseous product, as illustrated by the reaction between oxygen gas and carbon in liquid steel. In all cases the reaction will involve diffusion or mass transfer in the gaseous and in the liquid phase, as well as a chemical or interface reaction on the gas–liquid interface.

If the rate of gas flow or the equilibrium constant for the reaction is small, or if the mass transfer or reaction rates are high, the effluent gas flow may, in some cases, be in chemical equilibrium with the melt. In such cases the observed rate of reaction will be proportional to the gas flow rate and independent of the gas–liquid interface area, as already mentioned for gas–solid reactions (range I in Fig. 5-6). One example is oxygen blowing of liquid steel, where the effluent gas is virtually free of unreacted oxygen, corresponding to the rate of oxidation being proportional to the rate of oxygen flow. Another example is the reaction between sulfur in liquid iron and a flow of hydrogen according to the equation

$$\underline{S} + H_2 = H_2S$$

If the activity of sulfur in the melt is assumed to follow Henry's law, the equilibrium gas ratio $p_{H_2S}/p_{H_2} = K[\% \; S]$, where the equilibrium constant K is of the order of 10^{-3}. If the effluent gas is in equilibrium with the melt the rate of sulfur removal will be proportional to the rate of hydrogen flow and to the sulfur percentage in the melt. Thus for a constant hydrogen flow the rate of sulfur removal, expressed as $-d[\% \; S]/dt$ will be proportional to $[\% \; S]$ like for a first-order reaction. This example shows that if a first-order relation is observed experimentally, this does not necessarily mean that the reaction is kinetically controlled.

In other cases, like in the removal of nitrogen from liquid steel by argon purging or vacuum treatment, chemical equilibrium is not established, and the rate of reaction is determined by the kinetics of the reaction. In this case we may have the following reaction steps:

$$2\underline{N} \xrightarrow{\;k_1\;} 2N^* \xrightarrow{\;k_2\;} N_2^* \xrightarrow{\;k_3\;} N_2(g)$$

Here N^* and N_2^*, respectively, represent monatomic and diatomic nitrogen on the liquid–gas interface, whereas $\underline{N}$ and $N_2(g)$, respectively, represent the same species in the liquid and gas bulks. The mass transfer coefficients to and from the interface are given by k_1 and k_3 respectively, whereas k_2 is the interface reaction rate constant. For a sufficiently high argon flow rate or for vacuum treatment the partial pressure of $N_2(g)$ will be very small, and as the mass transfer coefficient k_3 is usually relatively large, the pressure or concentration of N_2^* will also be small. The rate controlling step may then either be mass transfer to the interface, given by k_1, or the interface reaction, given by k_2. In the first case the rate will be directly proportional to [% N], corresponding to a first-order reaction, whereas in the second case the rate will be proportional to $[\% \text{ N}]^2$, corresponding to a second-order reaction. In both cases the rate will be proportional to the liquid–gas interface area.

The question of which step is actually controlling in practice has been much debated, and rather conflicting experimental data exist. One reason for these discrepancies may be that the reaction is affected strongly by third elements, like oxygen or sulfur, which are concentrated on the interface. These elements seem to reduce drastically the interface reaction rate, making this the controlling step and the reaction a second-order one. At low concentration of these elements a transition toward mass transfer control and a first-order reaction seems to occur. For *absorption* of nitrogen *into* steel very much the same reasoning applies, with corresponding reaction rates and rate controlling steps.

Liquid–liquid reactions. These have been studied in particular at room temperature in connection with *solvent extraction*, which will be discussed in Sec. 15-5. In pyrometallurgy a typical example would be desulfurization of liquid steel by means of a basic slag, as given by the reaction:

$$\underline{S} + O^{2-}(\text{slag}) = \underline{O} + S^{2-}(\text{slag})$$

In such cases the reaction involves diffusion or mass transfer to and from the interface as well as the interface reaction. For slag–metal reactions the interface reaction appears to be fast, as it mainly involves an electron transfer, whereas diffusion to and from the interface appears to be the controlling steps. Accordingly the rate of reaction can be increased by vigorous stirring of the two liquids as well as by an increase in the interface area. A typical example is the Perrin process for refining of liquid steel (Sec. 13-5), where the steel is poured into a ladle which contains the premelted slag. Another example is the silicothermic reduction of a chromite slag (Sec. 14-2).

5-5 REACTIONS ON SPHERES AND BROKEN SOLIDS

So far we have expressed the reaction rate per unit surface area. This is a useful reference if one discusses reactions, e.g., between liquid metals and slags. In dealing with reactions between gases and lumps of solids or between droplets and the surrounding phase we are more interested in the rate per unit mass or per unit volume of the reacting substance. For a sphere the surface area is $A = \pi d^2$, the volume $V = (1/6)\pi d^3$, and the mass $m = (1/6) \times \pi d^3 \rho$, where d is the diameter and ρ the density of the sphere. The surface area per unit volume is then $A/V = 6/d$, and per unit mass $A/m = 6/(d\rho)$. In a bed of packed spheres the area per unit bed volume V' is $6(1 - \varepsilon)/d$ where ε is the void fraction. For close packing of spheres $\varepsilon = 0.26$, independent of the sphere diameter. If the rate per unit area is called $R_A = -(dn/dt)/A$, the rate per unit mass is then $R_m = 6R_A/(d\rho)$ and per unit bed volume $R_{V'} = 6R_A(1 - \varepsilon)/d$. For nonspherical particles the same expressions may be used if d is replaced by the mean diameter $d_m = 6m/(A\rho)$, where A/m is the surface area per unit mass. See Sec. 6-1.

For an evaluation of how the *rate* of a heterogeneous reaction is affected by the *degree* of reaction it is necessary to consider the geometry of the reacting interface. For the reaction between a lump of solid and a fluid phase as, for example, the burning of carbon in air or the reaction of a lump of ore with a gas, the reacting interface area will depend on whether the solid is dense or porous. Second, the rate of reaction will depend on whether the reaction product is dissolved in the fluid and is carried away with it, as for the combustion of carbon, or forms a solid layer as for the reduction of an oxide ore.

Dense reactants

For dense reactants the reaction interface is relatively sharp. As the reaction proceeds the interface will move from the external surface of the lump toward its center forming a sharp boundary between the reacted and the unreacted part. This gives rise to the so-called *shrinking core* or *topochemical model*. The rate of reaction will now depend on whether reaction on the interface or diffusion of the fluid phase through the layer of reaction products is rate controlling. We shall first consider the case where the interface reaction is the controlling one. This would be the case if the reaction product is either very porous or if it is soluble in the fluid. For a spherical lump the rate of reaction may be expressed in terms of the original volume V_0 in the following way:

$$\frac{-dn}{dt} = \pi d^2 R_A = \frac{6d^2 R_A V_0}{d_0^3}$$

Here R_A denotes the rate per unit area, and is a function of the composition of the fluid. The original and actual diameters are denoted d_0 and d respectively. If d^3/d_0^3 is denoted $(1 - X)$ where X is the fraction of solids which has reacted, then $n = V_0 \rho(1 - X)/M$, and

$$\frac{dX}{dt} = \frac{-dn}{dt} \cdot \frac{M}{V_0 \rho} = \frac{6R_A M}{d_0 \rho}(1 - X)^{2/3} \qquad (5\text{-}14)$$

Here M is the molecular weight of the reacting solid, and ρ its density. We see that the fraction which reacts per unit time is inversely proportional to d_0, the original lump diameter. Equation (5-14) may be integrated to give the time required to attain a certain value of X

$$t_X = \frac{d_0\,\rho}{2R_A M}\,[1 - (1 - X)^{1/3}] \tag{5-15}$$

The time required for complete reactions, i.e., for $X = 1$, is then

$$t_{\text{tot}} = \frac{d_0\,\rho}{2R_A M}$$

The equations developed above are also valid for a growing lump, for example in the precipitation of a solid or evolution of a gas bubble. In this case the sign of R_A is reversed, and the resulting values for X will be negative.

Solid reaction products By the oxidation of metals or metal sulfides in air, or by the reduction of metal oxides with reducing gases, solid reaction products are obtained. These products form a layer between the initial solid and the reacting gas. In order for the reaction to proceed, material has to be transported through this layer by diffusion. The diffusion may either be solid diffusion in the crystal lattice or gaseous diffusion through pores in the reaction product. The latter is usually the case when the volume of the reaction product is less than that of the initial solid reactant. This gives a porous layer, as for example in the reduction of metal oxides or in the oxidation of alkali metals. For the oxidation of most other metals the oxide occupies a larger volume than the initial metal. This gives a dense layer, and diffusion takes place through the crystal lattice. Thus for the oxidation of iron or copper in air oxygen has either to diffuse inward or the metal outward. Experiments have shown that in most cases the metal atoms diffuse, in the form of metal ions and electrons, outward through vacant metal positions in the oxide lattice. In some cases the oxidation of a metal may result in the formation of internal cavities.

In any case the diffusion acts in series with the reaction proper, and the net reaction is the result of a combination of reaction and diffusion rates and of the thickness of the produced layer. Within a certain range of temperature or product thickness, however, one of the two steps is usually significantly slower than the other, and is therefore controlling. For a porous reaction product of limited thickness and below a certain temperature diffusion is the faster, and chemical reaction at the reacting interface is controlling. In this case the diffusion step may be ignored, and the reaction rate will be given by Eqs. (5-14) and (5-15). This is the case, for example, for the oxidation of alkali metals in air, and is often the case for reduction of metal oxides with reducing gases.

In other cases the diffusion step is controlling, and the conditions at the reacting interface correspond to chemical equilibrium. The rate per unit area is then proportional to the concentration gradient and the diffusivity of the diffusing substance in the product layer. If the diffusing substance is denoted B, the flux, which is equal to the reaction rate, is given by Fick's first law

$$J_B = \frac{dn_B}{dt} = D_B \frac{dC_B}{ds}\,A$$

Here s denotes distance within the reaction product layer, and A the cross-sectional area, which may be a function of s. If A is independent of s, which is the case for a plane surface, the flux becomes proportional to $\Delta C_B/L$, where ΔC_B is the difference between the concentrations of B at the external surface and at the reacting interface, and L is the thickness of the product layer. The thickness of the product layer is again related to the amount of B which has reacted by the relation: $n_B = LA\rho_B/M_B$, where ρ_B is the density of B in the solid reaction product, and M_B its molecular weight. If B is not present in the reaction product itself, M_B/ρ_B is the volume of the reaction product formed by the action of one mole of B. This gives

$$\frac{dL}{dt} = \frac{D_B \Delta C_B M_B}{\rho_B L} \tag{5-16}$$

This expression may be integrated, and since $L = 0$ at $t = 0$:

$$t = \frac{L^2 \rho_B}{2D_B \Delta C_B M_B} = k \frac{L^2}{\Delta C_B} \tag{5-17}$$

This relation is known as the *parabolic law*, and shows that for constant values of D_B, ΔC_B, and ρ_B the thickness of the reaction product is proportional to $t^{1/2}$ for a plane surface.

For reactions with lumps of solid substances the area is no longer independent of s. For a sphere of diameter d_0 the relation between A, s, and L is shown on Fig. 5-7. In this case the cross-sectional area is $A = \pi(d_0 - 2s)^2$. At any time the flux is then equal to

$$\frac{dn_B}{dt} = \frac{D_B \Delta C_B}{\int_0^L \dfrac{ds}{A}} = \frac{\pi D_B \Delta C_B}{\int_0^L \dfrac{ds}{(d_0 - 2s)^2}} = \frac{2\pi D_B \Delta C_B}{1/d - 1/d_0}$$

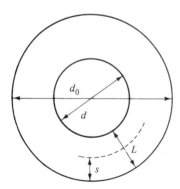

Figure 5-7 Topochemical reaction of a sphere, $d_0 = $ initial diameter, $d = $ diameter of unreacted part. $L = $ thickness of reacted layer, $s = $ distance within layer.

Here d is the diameter of the sphere formed by the reacting interface. The volume of the reacted layer is again related to the amount of B by the relation $n_B = \pi \rho_B (d_0^3 - d^3)/(6M_B)$. If we define $X = (d_0^3 - d^3)/d_0^3$, that is as the ratio between the reacted volume and the original volume, we get

$$\frac{dX}{dt} = \frac{12 \cdot D_B \, \Delta C_B \, M_B}{\rho_B \, d_0^2 [(1 - X)^{-1/3} - 1]} \tag{5-18}$$

This expression shows that, for a given value of X, the fraction of product formed per unit time is inversely proportional to d_0^2. This expression may be integrated from $t = 0$, $X = 0$ to give the time needed for a given degree of reaction

$$t_X = \frac{d_0^2 \, \rho_B}{4 D_B \, \Delta C_B \, M_B} \left[\frac{1 - (1 - X)^{2/3}}{2} - \frac{X}{3} \right] \tag{5-19}$$

Finally, the time for complete reaction, i.e., for $X = 1$, becomes equal to

$$t_{tot} = \frac{d_0^2 \, \rho_B}{24 D_B \, \Delta C_B \, M_B}$$

The two last expressions show that for a given degree of reaction the time is proportional to d_0^2. Thus if this relation is found experimentally, it indicates that the reaction is controlled by diffusion rather than by interface reaction.

Equations (5-18) and (5-19) have been developed under the assumption that d_0 is constant, i.e., that there is no macroscopic swelling or shrinking during the reaction. This assumption is closely satisfied for reduction of metal oxides or for roasting of metal sulfides. For the oxidation of metals in air the volume usually increases. This requires the introduction of a correction term, which is usually small and does not alter greatly the relations given above.

If the reaction is controlled by diffusion of fluid through a porous product layer the concentration difference ΔC_B is equal to the difference between the concentration of B in the external fluid and the equilibrium concentration at the interface. The pore diffusivity D_B is usually different from the diffusivity in open space, however. First of all the pores are not usually radial, and second some diffusion may take place on the pore surfaces. In general the pore diffusivity may not easily be predicted but has to be calculated from observed reaction rates.

For diffusion through dense solid phases the concentration gradient of the diffusing species is not necessarily equal to the concentration gradient for the corresponding gaseous diffusion. This will be illustrated by an example. For the oxidation of copper Wagner[1] has shown that this is controlled by the diffusion of copper ions and electrons over vacant copper positions in the crystal lattice

---

[1] See, e.g., K. Hauffe: article in "Progress in Metal Physics," vol. 4; pp. 71–104. Pergamon Press, London, 1953. K. Hauffe: "Oxidation von Metallen und Metallegierungen," Springer Verlag, Berlin, 1956.

of $Cu_{2-x}O$, and from the metal to the oxide surface. This may again be regarded as diffusion of copper ion vacancies and "electron holes" in the opposite direction. By electron holes we understand divalent copper ions that have one electron less than the normal monovalent ions. The relation between the concentrations of vacant copper positions, electrons, and electron holes is given by the equilibrium:

$$4Cu^+ + 4e^- + O_2(g) = 2Cu_2O + 4\square + 4\oplus$$

Here $\square$ denotes copper ion vacancies, and $\oplus$ electron holes. Since the crystal lattice is mainly one of Cu^+, e^-, and O^{2-}, the concentrations of Cu^+ and e^- are large and do not change much by small variations in stoichiometry. Also the activity of Cu_2O is practically constant and unity. The ideal law of mass action then gives:

$$C_\square^4 \cdot C_\oplus^4 = K p_{O_2}$$

According to the stoichiometry of the reaction $C_\square = C_\oplus$ which gives $C_\square = C_\oplus = (K p_{O_2})^{1/8}$. In order to obtain electroneutrality, the diffusion rates for the ion vacancies and electron holes must be equal. Thus if the rate of diffusion is given by the concentration gradient for any one of them, it becomes proportional to $(p_{O_2})^{1/8} - (p_{O_2}^*)^{1/8}$, where $p_{O_2}^*$ is the oxygen pressure at the metal–oxide interface. The latter is usually small compared to the outer oxygen pressure p_{O_2}, and the diffusion rate, which is equal to the oxidation rate, becomes closely proportional to $(p_{O_2})^{1/8}$. This relation has been essentially confirmed by the experiments by Wagner and Grünewald,[1] who found the oxidation rate of copper proportional to the seventh root of the oxygen pressure. This example shows that the rate of diffusion through solid phases cannot be derived directly from the pressure gradient in the gas without knowledge of the exact diffusion mechanism.

The above treatment applies to conditions where Cu_2O is the only oxide phase. This is the case at high temperature or at low oxygen pressure. At lower temperatures or higher oxygen pressure Cu_2O becomes covered with a layer of a second oxide, CuO. In order for the oxidation to proceed, material has to be transported through both the Cu_2O and the CuO lattices. The diffusivity in the CuO lattice is very much less than in the Cu_2O lattice. For a given flux the thickness of the CuO layer is therefore much less than that of the Cu_2O layer, and, as long as metallic copper is present, the oxidation is mainly one where Cu_2O is formed. Since the oxygen pressure at the interface Cu_2O–CuO corresponds to equilibrium between the two oxides, the oxygen gradient over the Cu_2O layer is constant and independent of the outer oxygen pressure. As a consequence the oxidation rate is also independent of the oxygen pressure. When all the copper has been converted to Cu_2O, the rate of further oxidation

[1] C. Wagner and K. Grünewald: *Z. phys. Chem.*, **B40**: 455–475 (1938).

to CuO will be low, but will now once more be a function of the oxygen pressure.

A somewhat analogous case occurs in the reduction of hematite with hydrogen. In this case successive layers of magnetite, wustite, and iron are formed. The reduction of hematite and magnetite to wustite takes place by solid diffusion, mainly by iron ions and electrons inward through metal vacancies in the wustite and magnetite lattices. The reduction of wustite to iron takes place by diffusion of gases through pores in the iron layer and chemical reaction at the wustite–iron interface. Thus all the oxygen in the hematite is removed from this last interface, and the rate of reduction is wholly determined by the reduction of wustite. Only in the case of an atmosphere which is unable to reduce wustite will the partial reduction of hematite to magnetite or wustite be rate controlling.

This discussion has so far been restricted to cases where the interface reaction is fast, and diffusion through the reaction product is controlling. If the rates of the two steps are of the same order of magnitude, the net rate will be a combination of the two. In most cases the two reaction rates will be of similar magnitude only within a limited temperature or dimensional range. Since diffusion, solid or gaseous, usually shows a much smaller temperature dependence than the interface reaction, the latter will be controlling below a certain temperature. Also since the two reaction rates, expressed as dX/dt, show different dependencies on X and d_0, each will dominate for certain values of X and d_0. Thus for small values of X, that is, for initial reaction, the interface reaction will be controlling. The same is the case for d_0 below a certain value. For the reduction of dense hematite and magnetite pellets 1 cm diameter with hydrogen, McKewan[1] has found that the interface reaction is controlling. For pellets of about 3 cm diameter, on the other hand, experiments by Bogdandy and Janke[2] indicate the diffusion through pores in the reduced layer to be controlling. For hematite cylinders 1 × 1 cm Warner[3] found a mixed reaction control, partly by diffusion and partly by interface reaction. In any case the value of the critical diameter is likely to be a function of the properties of the oxide.

Reaction on Porous Substances

So far it has been assumed that the reacting solid is dense, and with a defined surface area. Most heterogeneous reactions of metallurgical importance take place with porous solids, however. Typical examples are gaseous reduction of oxide ores and the combustion of coal and coke. Even the reaction between apparently dense graphite and carbon dioxide is not confined to the external graphite surface, but also takes place on the "internal surface," i.e., the walls of

[1] W. M. McKewan: *Trans. AIME*, **212**: 791–793 (1958); **221**: 140 (1961).
[2] L. v. Bogdandy and W. Janke: *Z. Elektrochem.*, **61**: 1146–1153 (1957).
[3] N. A. Warner: *Trans. AIME*, **230**: 163–176 (1964).

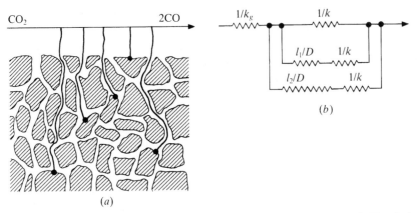

Figure 5-8 (a) Reaction of CO_2 on external and internal surfaces of graphite. (b) Electrical analogy.

the very fine network of pores that penetrate the solid. We shall first consider the case where no solid product is formed as illustrated by reactions like

$$C(s) + CO_2 = 2CO$$

$$ZnO(s) + CO = Zn(g) + CO_2$$

In such cases the reaction path may be illustrated by Fig. 5-8(a) and schematically by the electrical analogy Fig. 5-8(b). Here $1/k$ represents the chemical reaction resistance per unit area, on the external as well as on the internal surface, whereas l_1/D, l_2/D, etc.—where l_1, l_2, etc., are distances to successively deeper unit areas on the internal surfaces, and D is the pore diffusivity—represent the diffusion resistance. In addition there is the boundary layer resistance $1/k_g$ which, for the time being, will be disregarded. As seen from the electrical analogy model, reaction on the external surface and diffusion to and reaction on the internal surfaces represent parallel reaction paths. If diffusion and chemical reaction both are assumed to be of first order with respect to the reacting gas (Ohm's law relation) the combined effective reaction rate constant, or "conductivity," per external unit area becomes:[1]

$$k_{eff} = k + \frac{1}{l_1/D + 1/k} + \frac{1}{l_2/D + 1/k} + \cdots$$

It is evident that by increasing the value of l, the contribution from reaction on the internal surface decreases to an extent which depends on the relative magnitude of the pore diffusivity D and the rate constant k. These again will be functions of the porosity of the solid and of the temperature.

We remember that the activation energy for chemical reaction is significantly larger than for gas diffusion, which causes the rate constant to increase

---

[1] A mathematically more stringent treatment is given by H. Y. Sohn in H. Y. Sohn and M. E. Wadsworth: "Rate Processes in Extractive Metallurgy," Plenum Press, New York, 1979, pp. 24–30.

more strongly with increasing temperature than the diffusivity. At high temperature, therefore, l/D will dominate in the denominator, and the contribution to the effective reaction constant will decrease rapidly with increasing distance l, that is, the depth of the reaction zone will decrease. Conversely, at low temperature the depth of the reaction zone will increase, and in some cases the reaction may take place through the entire cross section of the solid body. This is illustrated by Fig. 5-9, which shows the reaction rate between carbon dioxide and graphite cylinders as function of temperature, as well as the variation in the partial pressure of CO_2 with distance from the external surface.

At low temperature, zone I, the chemical reaction rate is sufficiently low to make the gas composition in the pores equal to that of the external atmosphere. The reaction takes place throughout the entire volume of the graphite cylinder, and the activation energy is equal to that for the intrinsic chemical reaction. Experimental studies on the intrinsic reaction, as was discussed in Example 5-2, should, therefore, be done within temperature zone I.

At higher temperatures, zone II, the CO_2 pressure decreases with increasing distance from the external surface, and will at some depth reach equilibrium with graphite, i.e., be close to zero. As this depth decreases with increasing

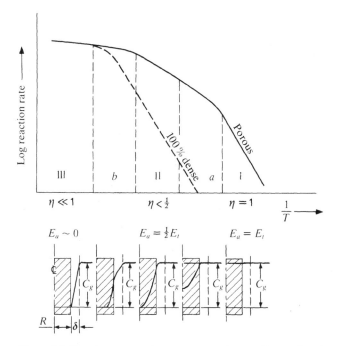

Figure 5-9 Reaction rate and gas concentration profiles as function of temperature for reaction between porous graphite cylinders and CO_2. (*After P. L. Walker et al.: Gas reaction on carbon in "Advances in Catalysis," vol. 11, 1958 p. 165, Academic Press, New York.*) E_t = true, E_a = apparent activation energy, $\eta = E_a/E_t$. Dashed line gives expected rate for 100 percent dense graphite.

temperature the effective reaction rate will increase more slowly than in zone I, and it can be shown[1] that the apparent activation energy in zone II is about half of that for the intrinsic reaction. Also, the *order* of the reaction with respect to CO_2 will be the average of the order for the intrinsic reaction and for diffusion.

At very high temperature, zone III, the chemical reaction will be sufficiently fast to establish chemical equilibrium at the external surface, and there will be no reaction inside the body. On the other hand, mass transfer to the surface, i.e., boundary layer diffusion, may now be rate controlling, corresponding to a very small apparent activation energy. For intermediate temperature ranges between the zones I, II, and III intermediate situations with mixed control exist.

For given temperatures within zones II and III the rate per unit weight of solid, as expressed by dX/dt, will be proportional to the external surface area of the solid reactant. Thus in these zones the dependency on initial lump size and on the degree of reaction will be the same as for dense solids, and Eqs. (5-14) and (5-15) may be applied. In zone II the apparent chemical rate constant will be higher than for dense solids, however, as the reaction takes place in a zone of some thickness. In zone III the rate constant is to be replaced by the mass transfer coefficient k_g as given in Sec. 5-2. In zone I the reaction rate per unit weight of solid will be independent of lump size as the reaction takes place throughout the entire volume of the lump (volume control). It will depend, however, on the porosity and chemical reactivity of the solid.

If the graphite had been 100 percent dense the chemical reaction would be confined to the external surface, and we would expect the rate to follow the dashed line in Fig. 5-9. At sufficiently high temperature the reaction will be controlled by boundary layer diffusion as for porous graphite.

Solid reaction products For reduction of, for example, iron ores with reducing gases similar relations as for the graphite reaction will apply. For relatively large lumps or pellets of relatively dense ores the reduction may be confined to a relatively narrow zone, which moves in a topochemical fashion from the external surface to the center of the lump. In this case the rate may, in the same way as for dense oxides, be controlled either by the interface reaction or by pore diffusion through the product layer, and either Eqs. (5-14) and (5-15) or Eqs. (5-18) and (5-19) may be applied. It should be emphasized that in the first case the interface reaction takes place in a reaction zone of some width and will also include diffusion within this zone. The apparent rate constant will, therefore, be larger than for a dense oxide.

For more porous lumps or pellets below a certain size the reaction may, in the same way as for graphite, take place throughout the entire volume of the lump, and the reaction rate per unit weight or volume will be independent of the lump size. It will still depend on the nature of the ore, its porosity, and

[1] H. Y. Sohn: loc. cit.

reactivity, however. The ore usually consists of relatively dense subgrains separated by the pore network, and the reaction probably occurs topochemically in each subgrain. As the subgrains as well as the pores may vary greatly in size within each lump of ore it will be difficult to formulate an expression which relates reduction rate dX/dt to the degree of reduction X. This relation will have to be determined experimentally for each type of ore.

In summary the variation of the reduction rate, dX/dt, with the initial pellet diameter, d_0, for a constant value of X may be illustrated by Fig. 5-10, which describes dense as well as porous oxide ores. For dense ores below a certain pellet diameter the rate will be inversely proportional to the initial diameter, as expressed by Eq. (5-14), corresponding to interface reaction. At larger diameters pore diffusion through the product layer will be controlling, and the rate will be inversely proportional to d_0^2 in accordance with Eq. (5-18).

For porous ores the same dependencies on d_0 apply above certain pellet sizes, but the rates will be higher. For pellets below a certain critical size the rate will be independent of d_0, as the reaction takes place through the entire cross section of the pellet. As seen, the curves for dense and porous ores will intersect at some small d_0 value, which corresponds to the size of the subgrains in the porous pellet.

For iron-making in the blast furnace, which will be further discussed in Chap. 9, one is interested in a high reduction rate. On the other hand the ore particles should not be too small as this will impair the flow of gases. The optimum pellet size would, therefore, be close to that for transition between interface and volume control, which usually corresponds to 10 to 15 mm diameter.

The transitions between pore diffusion, interface, and volume control are functions of temperature as well as the degree of reduction. As pore diffusion has a lower activation energy than chemical reaction, pore diffusion control becomes increasingly dominating at high temperature, whereas volume control

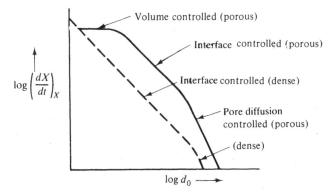

Figure 5-10 Fractional reaction rate dX/dt for given value of X as function of lump diameter d_0.

dominates at low temperature. Also, interface and volume control may dominate at low degrees of reaction (low X values), and pore diffusion may dominate at high degrees of reaction for a given pellet diameter.

5-6 NUCLEATION AND BUBBLE FORMATION

Some heterogeneous reactions are characterized by the formation of a new phase. Typical examples are crystallization, condensation, and evolution of gases like carbon monoxide in the refining of molten steel. In order for the new phase to grow the interfacial energy between the two phases must be overcome. This may in some cases affect the rate of the reaction. We will first look at the thermodynamics of nucleation.

If, in a homogeneous phase, a second phase is to be formed, the Gibbs energy difference between the new and the old phase is composed of two terms: (1) the difference in Gibbs energy between the two phases in macroscopic quantities (usually called the volume term), and (2) the difference in Gibbs energy caused by interfacial energy (called the surface term). If expressed for one single *spherical* particle or one bubble[1]

$$\Delta G = \tfrac{4}{3}\pi r^3 \, \Delta G_{vol} + 4\pi r^2 \sigma$$

Here ΔG_{vol} denotes the change in Gibbs energy associated with the formation of one unit volume of the second phase in macroscopic quantities, and σ denotes the interfacial energy per unit area. The second term is always positive and increases with r^2. The first term is negative if ΔG_{vol} is negative, and is proportional to r^3. As shown in Fig. 5-11 the net change in Gibbs energy always increases for small values of r. At a certain radius, called the critical radius r^*, the Gibbs energy has its maximum and will then decrease. The critical radius is obtained by differentiation

$$\frac{d\Delta G}{dr} = 4\pi r^2 \, \Delta G_{vol} + 8\pi r\sigma = 0$$

which gives

$$r^* = -\frac{2\sigma}{\Delta G_{vol}} \quad \text{and} \quad \Delta G^* = \frac{16\pi\sigma^3}{3\Delta G_{vol}^2}$$

where ΔG^* is the Gibbs energy for the formation of one critical sphere. Particles of less than the critical radius are unstable and will dissolve, whereas particles of critical and supercritical size may grow, and will do this under a decrease in Gibbs energy.[2]

---

[1] According to J. H. Hollomon: article in "Thermodynamics in Physical Metallurgy," American Society for Metals, Cleveland, Ohio, 1950, pp. 161–177.

[2] A somewhat different, and theoretically perhaps more correct, derivation of the thermodynamics of nucleation was given by J. P. Hirth et al. (*Met. Trans.*, **1**: 939 (1970)), without this affecting significantly the conclusions given above.

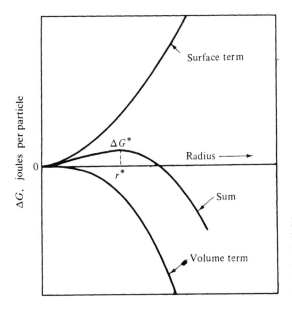

Figure 5-11 Gibbs energy of formation of a spherical particle from a matrix phase. (*From J. H. Holloman: Heterogeneous Nucleation in "Thermodynamics in Physical Metallurgy," Am. Soc. for Metals, Cleveland, 1950.*)

The question then arises: How is it possible for a new phase to be formed at all? There are two possible mechanisms. The first is *homogeneous nucleation*. In any system there will always be local fluctuations in atomic arrangements. Once in a while these fluctuations give rise to an atomic arrangement equal to the new phase and of critical size or more. The frequency of such spontaneous nuclei increases rapidly with increasing supersaturation of the system with respect to the new phase. Thus Turnbull and Fisher[1] give the following expression for the number of critical nuclei per unit volume and time:

$$I = (nkT/h) \exp \left[-(\Delta G^* + \Delta G_A)/kT \right]$$

Here n is the number of atoms or molecules per unit volume of the new phase, and ΔG_A is the *Gibbs energy of activation* for the transfer of one atom or molecule across the interface. The frequency of nucleation increases very rapidly with decreasing value for ΔG_{vol}, that is, by increasing supersaturation. Eventually a significant number of stable nuclei are formed within the size of the system. Thus for the freezing of water in a one liter volume, the number of stable nuclei becomes large at about $-39°C$. Homogeneous nucleation of gas bubbles in a liquid may also be enhanced by vigorous stirring of the liquid. In this case vortices are formed with large negative pressures that favor the gas formation.

[1] D. Turnbull and J. C. Fisher: *J. Chem. Phys.*, **17**: 71–73 (1949).

More important than homogeneous nucleation is *heterogeneous nucleation,* particularly under conditions of engineering interest. Heterogeneous nucleation is caused by a third phase, which may be the container or may be a solid impurity. If the interfacial energy between the third phase and the product is less than between the reactant and the product phase, nucleation of the product will be enhanced. Thus for the precipitation of a solid from a fluid phase precipitation may be initiated by the addition either of seeds of the product or of another solid which has similar interatomic dimensions as the precipitate. In the latter case the interfacial energy between the two solids is a minimum. This principle is used in artificial rainmaking where silver iodide crystals may be used to precipitate snow from a supersaturated atmosphere.

For the evolution of a gas from a liquid, as for example in the evolution of carbon monoxide from liquid steel, nucleation frequently takes place on rough surfaces. Rough surfaces always contain cavities where nucleation of the gas is favorable. Once a bubble is formed inside a cavity, it will grow and break loose, but will leave a small bubble behind, which therefore acts as a continuous nucleus.

To what extent nucleation is rate controlling in metallurgical reactions has been a subject of discussion. There is no doubt that lack of nucleation may slow down a reaction appreciably. Thus in open-hearth steel-making the boil sometimes stops, even though final equilibrium is not reached. But once nucleation is started the rate of carbon removal seems not to be affected by the nucleation rate but rather by the rate at which carbon and oxygen diffuse to the rising bubbles and make these grow.

Specific nucleation phenomena may alter the expected mechanism of reaction. Thus if silver sulfide is reduced by hydrogen according to the reaction $Ag_2S + H_2 = 2Ag + H_2S$ one would expect the metal to form a layer on the sulfide surface. Instead, silver is formed at a few local spots where it grows into long threads. Apparently the reduction itself takes place at the entire surface, but since silver diffuses easily through the sulfide, it diffuses to a few local spots where nucleation is easy. Also for the reduction of wustite to iron a similar phenomenon has been observed.

5-7 METASTABLE PRODUCTS AND PARTIAL EQUILIBRIUM

Depending on the rate of reaction we may get three types of conditions.

1. The reaction is completed in a time which is short compared to the duration of the experiment or process. In this case virtual equilibrium is established and the composition of the product may be calculated from thermodynamics.
2. For intermediate reaction rates, incomplete reactions are obtained. If the reaction mechanism is simple, the product will be a mixture of the reactants and the equilibrium product. If the mechanism involves simultaneous or consecutive steps, various intermediate products may be obtained. In some chemical industries these may represent the wanted product.

3. If the reaction towards equilibrium is very slow, *metastable* products may be obtained. A typical example is the nitriding of iron with ammonia at red heat according to the reaction $4Fe + NH_3 = Fe_4N + \frac{3}{2}H_2$. The phase Fe_4N is under these conditions unstable, and has a decomposition pressure, $Fe_4N = 4Fe + \frac{1}{2}N_2$, of about one thousand atmospheres. The decomposition is slow, however, and Fe_4N is a product with considerable lifetime.

The fact that certain reaction steps may reach equilibrium whereas others are very slow, is an example of *partial equilibrium,* a phenomenon which has been discussed particularly by Darken.[1] By partial equilibria in general we understand situations where equilibrium is established with respect to certain variables but not to others. Thus a steel container may admit thermal equilibrium between the inside and the surroundings, whereas chemical equilibrium and pressure equilibrium is not established. Another example was illustrated by the diffusion of carbon from silicon to plain carbon steel, Fig. 5-3. If enough time were allowed, equilibrium between the plain carbon steel and the silicon steel would be established with respect to carbon. This would result in two different carbon concentrations. The two steels would not, however, be in equilibrium with respect to silicon. If such equilibrium were established after a sufficiently long time, the carbon would diffuse back and establish equal concentration in the two bars.

In the past the attention of the extractive metallurgist has been focused mainly on the total equilibrium of a reaction. The concept of partial equilibrium may in the future prove to be very useful in the development of processes which would have been impossible if total equilibrium were established. An example of such a possibility will be discussed in connection with desulfurization of liquid steel (Sec. 13-5).

5-8 REACTION RATES AND HEAT SUPPLY

All our discussions so far have been referred to rates of reaction at a given temperature, but we have not been concerned about how this temperature is obtained. As was shown in Chap. 2, the temperature is the result of heat supplied and removed. In the case of reactions with significant heat effects the reaction itself will affect the temperature. This was discussed in Sec. 2-4 for exothermic reactions where the actual reaction temperature is determined by the enthalpy balance and heat losses.

Endothermic reactions are dependent on the necessary heat being supplied from the surroundings. Typical examples are the boiling of water in a pan, the calcination of limestone in a kiln, or the reduction of zinc oxide with carbon in a retort. In these cases the rate is determined by the rate of supply of the

[1] L. S. Darken, see J. F. Elliot: "The Physical Chemistry of Steelmaking," M.I.T. Technology Press, 1958, pp. 101, 233.

necessary heat. If heat is not supplied at the rate needed, the reactants will cool until a balance is established between the rate of reaction and the heat supplied. This effect is shown clearly in the lime boil in the open-hearth steel-making process. Here limestone is covered with hot metal, the latter having a temperature of about 1300°C. At this temperature limestone has a decomposition pressure of about one thousand atmospheres, and we should expect a vigorous explosion. This does not occur, however. The decomposition is strongly endothermic, and since burned lime is a poor heat conductor, the heat leaks only slowly into the limestone which never gets much above 900°C, at which temperature the decomposition pressure is one atmosphere.

Heat transfer from the surroundings into a reacting body follows very much the same laws as mass transfer. The heat transfer controlling step may be either heat transfer by convection and radiation to the surface of the body or heat conduction through the body. Thus for the calcination of limestone it seems that heat conduction through the layer of burned lime is the rate controlling step, whereas chemical reaction on the reacting interface as well as diffusion or flow of carbon dioxide into the atmosphere are fast. Heat transfer by conduction follows equations analogous to those which apply for diffusion, and the movement of the reacting interface will be described by equations analogous to Eqs. (5-18) and (5-19). Thus the time needed for complete calcination of a lump of limestone will be proportional to the square of the lump diameter if other parameters are constant. Further discussion on heat transfer will be given in Chap. 6.

PROBLEMS

5-1 The rate of the reaction $H_2 + I_2 = 2HI$ is given by the expression

$$\frac{dC_{HI}}{dt} = k_1 C_{H_2} C_{I_2} - k_2(C_{HI})^2$$

where at 400 K $k_1 = 6.3 \times 10^{-8}$ and $k_2 = 8.0 \times 10^{-11}$ cm^3/(mol·s).

(a) Calculate the apparent equilibrium constant for this reaction and compare it with the one obtained from $\Delta G°$ (App. C).

(b) Express in moles per cubic centimeter the concentrations in an initial mixture of equal parts of $H_2(g)$ and $I_2(g)$ at 400 K and 1 atm pressure, and calculate the corresponding equilibrium concentrations.

(c) Calculate the rate of reaction for the above initial mixture, as well as for a mixture where half of the H_2 and I_2 have already reacted.

(d) Make a semiquantitative plot of how you expect the concentrations of all three gases to vary with time. (For each question you utilize the results obtained for the preceding questions.)

(e) Make a similar plot for the case that the initial mixture had contained $\frac{2}{3}H_2$ and $\frac{1}{3}I_2$.

5-2 For the reaction $H_2 + I_2 = 2HI$ the Arrhenius activation energy in the forward direction is 163 kJ/mol. Calculate the initial rate of reaction at 300 and 500 K for an initial mixture of equal parts of H_2 and I_2 at 1 atm pressure, using the rate constant for 400 K as given in Prob. 5-1.

5-3 Hydrogen dissolves in palladium by the reaction $H_2(g) = 2H(diss)$. At 300°C and 1 atm H_2 pressure the solubility is 164 cm^3 (STP) per 100 g Pd.

(a) Applying Sieverts' law calculate the solubility at the same temperature for $p_{H_2} = 0.1$ atm.

(b) A membrane of Pd of 300°C separates two large hydrogen volumes, one with $p_{H_2} = 1$ atm, the other with $p_{H_2} = 0.1$ atm. Calculate the diffusion flux in cubic centimeters H_2 (STP) per square centimeter and hour, when the foil thickness is 0.1 mm, the density of Pd is 11.9 g/cm³, and the diffusivity of hydrogen in Pd is 3.8×10^{-5} cm²/s. (*Hint*: In order to get the right dimension the solubility of hydrogen should be converted into cubic centimeters H_2 (STP) per cubic centimeter Pd.)

5-4 The solubility of CO_2 in rubber at 20°C and $p_{CO_2} = 1$ atm is reported to be 0.9 cm³ CO_2/cm³ rubber, and its diffusivity to be 1.1×10^{-6} cm²/s. We want to keep CO_2 in a rubber balloon of diameter 40 cm and wall thickness 0.03 cm. The CO_2 pressure inside the balloon is taken equal to 1 atm, and outside equal to zero. Calculate the amount of CO_2 which diffuses out during 1 h, and express this as fraction of the initial amount.

5-5 The end of a semi-infinite rod of pure iron is brought into equilibrium with graphite at 1000°C, corresponding to a surface concentration of 1.5 wt % carbon. The diffusivity of carbon in iron at 1000°C is taken equal to 3×10^{-7} cm²/s.

(a) Using the curve in Fig. 5-4(a), calculate the carbon concentration 1 mm from the end after 1 h and after 100 h.

(b) Repeat the calculation for the case that the iron initially had contained 0.5 wt % carbon.

5-6 The surface of a cylindrical rod of pure iron with 10 mm diameter is brought into equilibrium with graphite at 1000°C. The surface concentration and diffusivity of carbon are as given in Prob. 5-5.

(a) Using the curves given in Fig. 5-4(b), calculate the mean carbon content of the rod after 1 h and after 100 h.

(b) Repeat the calculation for the case that the rod diameter had been 2 mm.

5-7 A bed of powdered wustite is reduced in a flow of hydrogen according to the reaction FeO + H_2 = Fe + H_2O. The gas flow is sufficiently slow to make the reaction go to virtual equilibrium.

(a) Calculate the rate of reduction, expressed in moles Fe per minute as function of the gas flow (mol H_2/min), and equilibrium constant K. Insert for 800°C: K = 0.5.

(b) Do you expect this rate to be dependent on the lump size or of the amount of wustite in the bed?

(c) Make a plot of how you expect the degree of reduction to vary throughout the bed in the direction of gas flow.

5-8 A flow of CO_2 reacts with a bed of graphite at p = 1 atm. The flow is sufficiently slow to make the reaction CO_2 + C = 2CO go to virtual equilibrium. Derive an expression for the reaction rate, expressed in moles of carbon consumed per minute as function of the equilibrium constant, K, and the gas velocity, expressed in moles of CO_2 introduced per minute, and insert for 900°C: K = 30.

5-9 At high gas flow rates the reaction between dense graphite and CO_2 is controlled by the surface reaction. For the reaction with pure CO_2 at 1 atm and 900°C the rate of reaction is found: $dn_{CO}/dt = 0.5 \times 10^{-3}$ mol/(min·cm²).

(a) Calculate the rate of reaction per gram of graphite if this is present as spheres of diameter 1 cm and the density of graphite is 2.25.

(b) Calculate the rate of reaction per cubic meter of a closed packed bed of the above spheres.

(c) Utilizing Eq. (5-15) calculate the time needed for half the amount of graphite to be consumed.

(d) At 900°C and 1 atm the equilibrium gas composition is 97 percent CO and 3 percent CO_2. Show qualitatively how you expect the rate of reaction to vary with gas composition between zero and 100 percent CO.

5-10 A spherical, sintered pellet of dense NiO is reduced with hydrogen according to the reaction NiO + H_2 = Ni + H_2O. For a pellet diameter of 0.5 cm and high gas velocity, the rate controlling reaction occurs at the interface between reduced and unreduced oxide, and the rate is found equal to 2×10^{-3} mol/(cm²·min) at 600°C and 1 atm H_2.

(a) By means of Eq. (5-15) calculate the time needed for 50 percent and 100 percent reduction when the density of the NiO pellet is 7.4 g/cm³.

(*b*) Calculate the corresponding times if the diameter had been 1.0 cm.

(*c*) For 3 cm diameter the time for 75 percent reduction is found equal to 60 min, and the reaction is assumed to be controlled by pore diffusion. Calculate the pore diffusion coefficient in square centimeters per second when the hydrogen concentration at the interface is taken equal to zero.

5-11 A shaft furnace is fed with NiO pellets at the top and pure hydrogen at the bottom. The molar ratio of H_2 to NiO is 1.5. The furnace temperature is constant, 600°C, corresponding to the equilibrium constant for the reduction being very large. Estimate the change in composition of the pellets and the gas along the height of the shaft. The feeding rate is sufficiently slow to make the reduction of NiO in the shaft complete.

BIBLIOGRAPHY

Bogdandy, L. von, and H.-J. Engell: "The Reduction of Iron Ores," Springer Verlag, Berlin, 1971.

Benson, S. W.: "The Foundation of Chemical Kinetics," McGraw-Hill Book Co. Inc., New York, 1960.

Hauffe, K.: "Reaktionen in und an festen Stoffen," Springer Verlag, Berlin, 1955.

Hinshelwood, C. N.: "The Kinetics of Chemical Change," Clarendon Press, Oxford 1940, Oxford University Press, 1949.

Hougen, O. A., and K. M. Watson: "Chemical Process Principles," Vol. III, John Wiley and Sons, Inc., New York, 1947.

Laidler, K. J.: "Chemical Kinetics," 2d ed., McGraw-Hill Book Co. Inc., New York, 1965.

Smith, J. M.: "Chemical Engineering Kinetics," McGraw-Hill Book Co. Inc., New York, 1956.

Sohn, H. Y. and M. E. Wadsworth: "Rate Processes in Extractive Metallurgy," Plenum Press, New York, 1979.

Turkdogan, E. T.: "Physical Chemistry of High Temperature Technology," Academic Press, New York, 1980.

SIX

REACTOR DESIGN

Metallurgical processes are carried out in reactors of various kinds. Into the reactors are fed raw materials such as ore, coke, oil, and air, as well as mechanical and electrical energy. Out of the reactors come products, which may be molten metal and slag, or solid materials such as a roasted or reduced product, and gases of various kinds. In hydrometallurgical and electrometallurgical processes the input may be crude or roasted ores or impure metals and aqueous solutions, and the output may be aqueous or organic solutions, solid precipitates, or electrodeposited metals. For the calculation of the size and capacity of the various reactors, from which capital costs of the process may be derived, it is necessary to know not only the chemical equilibria and the rate of the chemical reactions, but also how the rate is affected by the design of the reactor. It is furthermore necessary to know the mechanical aspects of the transport of material through the reactors as well as the transport and transfer of heat between the various reactants and products including heat losses to the surroundings.

This whole field of fluid dynamics and heat and mass transfer, from which reactors may be designed and dimensioned, is an important part of every type of process engineering, and it is treated extensively in many textbooks of mechanical, chemical, and metallurgical engineering, some of which are listed in the bibliography.

This chapter intends primarily to give a qualitative understanding of the factors which are of importance for the design and dimensioning of metallurgical reactors with emphasis on their application to pyrometallurgical processes.

For actual design purpose and a more quantitative treatment the reader is referred to specialized literature. Although it is not strictly necessary, some general background in fluid dynamics, heat and mass transfer, as well as dimensional analysis, is desirable.

6-1 METALLURGICAL REACTORS

A metallurgical reactor is defined as a piece of equipment designed to carry out a certain type of reaction, a certain unit process. As almost all metallurgical processes are heterogeneous the main purpose of the reactor is to bring the reacting phases together under conditions where the chemical reactions are favorable, to supply the necessary energy, and finally to allow a separation of the product phases.

Even though an abundance of reactors exist most of these may be classified in certain types which have essential features in common. A reactor may be either *continuous* or *discontinuous*, i.e., either fed continuously with raw materials and have the products continuously withdrawn, or charged with raw materials which are processed for a certain time whereupon the products are withdrawn (batch process). Some reactors may consist of several smaller units which are connected either in series or in parallel, and with the whole set of units operating together to carry out the desired process. Typical examples are the use of flotation cells in ore-dressing and of mixer-settlers in hydrometallurgy.

The most important pyrometallurgical reactor types are shown schematically in Fig. 6-1(*a–h*), and will be described briefly.

Fixed-bed reactors Figure 6-1(*a*) illustrates a fixed-bed reactor. Here granular material is placed in the reactor and a fluid, which may be a gas or a liquid, is passed through the bed. A pyrometallurgical example is grate sintering, which will be discussed in greater detail in Sec. 8-4. A hydrometallurgical example is so-called vat, or percolation, leaching where a liquid solvent is passed through a bed of granular ore, and which will be discussed in Sec. 15-4.

In grate sintering the grate may be made part of a moving belt to give the continuous Dwight–Lloyd sintering machine. Such a traveling grate reactor, although continuous, has most physical features in common with the fixed-bed reactor.

Shaft furnaces The shaft furnace shown in Fig. 6-1(*b*) is one of the most common metallurgical reactors. It is used in the burning of limestone and firing of iron ore pellets as well as in the production and smelting of iron and nonferrous metals. In the iron blast furnace as well as in shaft furnaces for nonferrous metals the products are molten metal and slag, whereas in lime-burning and pellet-firing solid products are discharged from the reactor.

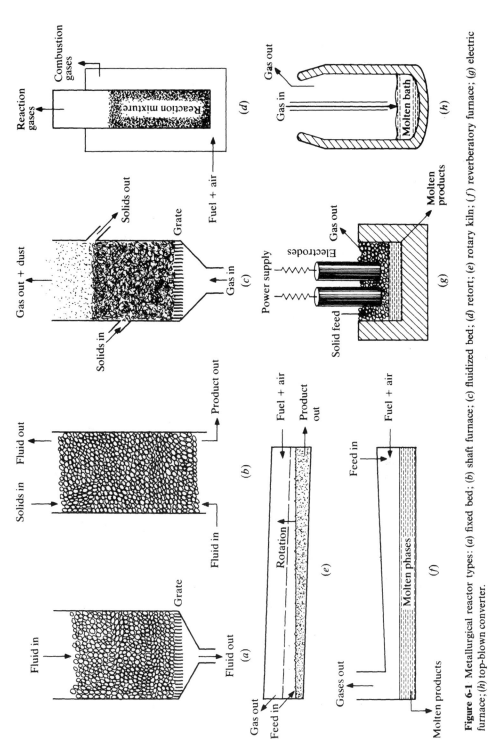

Figure 6-1 Metallurgical reactor types: (a) fixed bed; (b) shaft furnace; (c) fluidized bed; (d) retort; (e) rotary kiln; (f) reverberatory furnace; (g) electric furnace; (h) top-blown converter.

137

In most shaft furnaces the solid burden moves counter-current to the gases, which may be hot combustion gases or chemical reactants. In this way an efficient heat and mass transfer is achieved, and shaft furnaces usually have a high production capacity. In a few cases con-current reactors may be considered, or the gas may be made to pass perpendicular to the burden flow.

In addition to their use in pyrometallurgical processes the principle of the shaft furnace is used also in continuous solvent extraction columns as well as in distillation columns.

Fluidized-bed reactors These are illustrated in Fig. 6-1(c) and are used in the roasting of sulfide concentrates and in the burning of limestone, and have also been tried for the reduction of iron oxides. In a fluidized bed the solid particles are carried by the upward flow of gas, in such a way that the bed of solids behaves like a liquid (see Sec. 6-4). Most fluidized-bed reactors are continuous with continuous introduction of solid feed and withdrawal of the product either as an overflow or by separation of dust from the effluent gas flow.

Fluidized-bed reactors have their counterpart in hydrometallurgical equipment for agitation leaching of fine-grained concentrates, which operate either mechanically or by pneumatic agitation (Pachuca tanks), and very much the same theoretical treatment may be applied.

Retorts By retorts, as shown in Fig. 6-1(d), we understand reactors where the solid charge is placed in a more or less gas-tight chamber, and where the necessary heat is introduced by conduction through the chamber walls and into the charge. Retorts are used in cases where contact with hot oxidizing combustion gases would be harmful for the charge. Typical examples are retorts for the coking of coal, and retorts for production of zinc and magnesium by, respectively, carbothermic and silicothermic reduction of the oxide.

As will be shown in Sec. 6-5 the time needed for the heat to reach the center of the retort increases with the square of the distance to the center. This puts a limit to the size of the individual retort, and retort processes are usually carried out in furnaces which contain a large number of retorts. Retorting is therefore a relatively expensive process.

The term "retort" is sometimes used for other types of closed reactors, where combustion or reacting gases are made to flow through the inside charge. Such reactors should more correctly be classified as fixed-bed reactors.

Rotary kilns These are illustrated in Fig. 6-1(c) and are used for the burning of limestone, and the production of cement clinker, as well as for reduction or treatment of iron and nonferrous ores. The kiln has the shape of a slightly slanting cylinder which rotates slowly around its axis. The feed is introduced in the one end, and the product is discharged in the other. Heat is supplied by fuel and air introduced through a burner in the discharge end, and the combustion gases are withdrawn from the charge end of the kiln. Heat is transferred from the hot gases to the charge partly by direct radiation from the flame, partly by reflected radiation from the kiln lining, and partly by radiation and convection to the lining and further by the rotating action of the kiln into the charge.

Compared to shaft and fluidized-bed reactors the rotating kiln has the advantage that it can handle material with variable particle size, and even partly molten material. However, heat and mass transfer between the furnace gases and the charge is poorer and, due to its large surface area, heat losses to the surroundings are usually large.

Reverberatory furnaces These, which are illustrated in Fig. 6-1(f), are used for the smelting of iron–copper matte as well as for open-hearth steel-making. The furnace has the shape of a refractory lined chamber with a relatively shallow bath of molten metal or matte and slag. Heat is supplied from fuel which is burned above the bath and, like in the rotary kiln, it is transferred to the molten bath partly by direct radiation from the flame and partly by reflected radiation from the furnace roof (reverberate = to reflect).

In some matte smelting processes the sulfide concentrate is used as fuel, and is burned with preheated or oxygen-enriched air (flash smelting). In these cases the furnace is often given a shape which facilitates easy uptake of the partly oxidized concentrate into the slag and matte melts, see Sec. 12-3.

Electric furnaces These, Fig. 6-1(g), are used for matte smelting and the smelting of iron and ferroalloys, as well as for steel-making. For these purposes electric energy is introduced through carbon electrodes, which either dip into the slag that acts as the main resistor, or terminate above the bath causing electric arcs to play between the electrode and the bath. An intermediate case is the submerged arc furnace, which is used for iron-making and ferroalloy production, and where heat is generated partly by arcs and partly by resistance in the solid and semimolten charge (Sec. 9-4).

Electric resistance and arc furnaces have the advantage, compared with reverberatory furnaces, that heat is generated *inside* the charge or slightly above it and with good heat transfer. Furthermore there are no combustion gases, and the effluent furnace gases are limited to what is produced by the process, with correspondingly smaller heat losses.

Pneumatic reactors These are reactors where a gas is brought to react with a molten charge. In this group we find the important steel-making converters, which may be either top-blown, as in Fig. 6-1(h), or bottom- or side-blown, where air or oxygen is introduced through tuyeres directly into the bath. The gas may also be introduced through lances which dip into the bath.

Pneumatic processes may either be purely oxidizing, like in steel-making converters, or they may be blown with reducing gases, whereby reduction and deoxidation of the bath is obtained. Also slag-forming constituents, like lime, may be blown into the bath.

During oxidizing processes, like in steel-making, the reaction with the bath is usually sufficiently exothermic to supply the necessary heat. For reducing processes the necessary heat may be obtained by introducing some oxygen together with the reducing gas, or a mixture of coal powder and oxygen-enriched air may be used. One example is slag-fuming, Sec. 10-7.

A metallurgical plant may also contain other furnaces and equipment such as holding furnaces, tunnel kilns, heat exchangers, etc. These cannot be classified as reactors, however, as they are not used to carry out a chemical process.

A special type of reactor is the *electrolytic cell* type, which will be discussed in Chap. 16.

6-2 RETENTION TIMES IN REACTORS

In order to calculate the degree of reaction obtained in a given reactor it is, in addition to kinetic data, necessary to know how long the material has stayed within the reactor, the so-called *retention time*. For batch processes, such as a fixed-bed or a steel converter, this is an easy task, as all material has stayed within the reactor for the time of the process. For continuous reactors, like a shaft furnace or a fluidized-bed furnace, it becomes more difficult. We can in such cases distinguish between two extreme situations, that of *perfect displacement* and that of *perfect mixing*.

Perfect displacement, also called *plug flow*, is approximately realized in a shaft furnace. In the ideal shaft furnace all the material moves parallel and with the same velocity, and the retention time is the same for each piece of material. It is easily seen that if the feeding rate is m kg/h and the mass inside the reactor is M kg, then the retention time for the material in the reactor is M/m h. In practice the displacement in a shaft furnace is not perfect. The material along the walls will, due to friction, tend to descend more slowly than in the middle. In that case the retention time, as calculated above, represents an average value t_m, around which the actual retention times are distributed. Even larger deviations from perfect displacement occur in the rotary kiln, where the larger lump sizes move through the kiln faster than the fine material, as well as in flow systems, where some *back-mixing* occurs.

The extreme case of *perfect mixing* is illustrated by fluidized-bed reactors (Fig. 6-1(c)) and by agitation leaching tanks. As soon as the raw ore enters the fluidized bed it is mixed completely with the bed content. At the same time the bed overflows continuously, and the composition of the overflow is the same as that of the bed. Also in this case we may define the mean retention time as the ratio between bed mass and feeding rate. Since the mass of the ore may change during the reaction, the bed mass as well as the overflow should be given on the same basis, e.g., in terms of the corresponding raw ore masses.

We shall now calculate the actual retention time for each ore particle relative to the mean retention time. We assume that at a given time $t = 0$ a small quantity, n_0, of trace ore is introduced, and we will follow this material. After a certain time t, the quantity remaining in the reactor is n and its concentration is n/M. The rate at which it overflows is $-dn/dt = (n/M)m$, where m, the total rate of overflow, is expressed in terms of and is equal to the feeding rate. If this expression is integrated we get

$$-\int \frac{dn}{n} = \frac{m}{M} \int dt = \frac{1}{t_m} \int dt$$

Inserting the limits $n = n_0$ at $t = 0$, and $n = n$ at $t = t$ we get

$$\frac{n}{n_0} = e^{-t/t_m} = e^{-\theta} = r \qquad (6\text{-}1)$$

In this expression r gives the fraction of the original material which remains in the reactor at the time t and thus has a retention time equal to or larger than t, and t/t_m is the *relative retention time*, which will be denoted θ.

Furthermore, we see that the fraction of material which has a retention time between t and $t + dt$ will be

$$-dr = \frac{1}{t_m} e^{-\theta} \, dt = E \, dt \qquad (6\text{-}2)$$

where E is called the *age distribution function*, which for perfect mixing is equal to $e^{-\theta}/t_m$. It follows from Eq. (6-2) that $-dr/dt$ will have its largest value, $1/t_m$, for $t = 0$, and will decrease with increasing time. This means that more material has a retention time within the first time interval than within any subsequent interval of the same length. We also see that about 10 percent of the material has a retention time less than $t_m/10$, and about 10 percent has a retention time more than $3.2t_m$. A small amount of material will have a retention time which approaches infinity.

For other types of reactors we may still define an age distribution function $E = -dr/dt$ where $E \, dt$ is the fraction of material in the reactor with retention time between t and $t + dt$. Some typical age distribution curves are shown in Fig. 6-2. Here curve (a) represents the case of perfect mixing, curve (c) an almost perfect displacement, whereas curve (b) represents an intermediate case such as illustrated by a rotary kiln.

As all particles in a reactor must have some retention time, it is evident that

$$\int_0^\infty E \, dt = 1 \qquad (6\text{-}3)$$

that is, the area under the curves in Fig. 6-2 will be unity in all cases. For the case of an ideal plug flow, therefore, E would be infinite at $t = t_m$ and zero at any other t values. In practice there will be a spread as shown by curve (c).

Occasionally the material flow may not utilize the entire reactor volume. Thus in a shaft furnace some of the material may stick to the wall to form accretions and have an infinite retention time. Similarly, if in a fluidized-bed reactor the feeding and the overflow are placed too close together the material

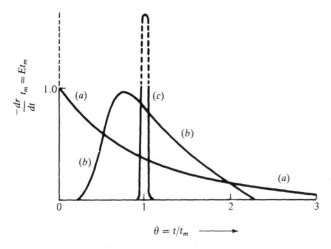

Figure 6-2 Age distribution function for various reactors, (a) fluidized bed, (b) rotary kiln, (c) shaft furnace.

may not mix completely with the entire reactor volume before it is discharged. In such cases the area under the observable part of the curves in Fig. 6-2 will be less than unity, corresponding to the mean retention time for the observed material being less than that calculated from the feeding rate and the total mass inside the reactor. We may say that the reactor has some dead volume or that it is more or less *short-circuited*.

Even if the age distribution function is known for the material as such this may not be sufficient to describe the reactor or to calculate the degree of reaction obtained in the product. For granulated material of nonuniform particle size the different size fractions may have different retention times. Thus in a fluidized bed reactor the retention time will be shorter for the fine-grained than for the coarse material. In the rotary kiln the opposite will be the case. In order to calculate the degree of reaction obtained in a given reactor, or the size of the reactor for a desired degree of reaction, it may be necessary to know the age distribution function as well as the chemical reaction rate for the different fractions in the feed. This will be discussed in Sec. 6-6, but first some words about the nature of the feed.

6-3 GRANULATED MATERIAL

The ideal feed for a reaction between a fluid and a solid would be spherical particles of uniform size. A bed of such particles would offer the least resistance to fluid flow. Furthermore, all particles would react at the same rate and they would have the same age distribution function in a given reactor. Unfortunately, metallurgical feed comes usually as broken solids or granulated material of nonuniform size and shape. In order to describe such material we need to know their average *size* and *shape* and their *size distribution.*

The average size and shape may be described by two quantities, the *particle diameter* and the *mean diameter.* The particle diameter d_p is equal to the diameter of the *sphere* which has the same *volume*, V, as the particle. This gives

$$d_p = (6V/\pi)^{1/3} \qquad (6\text{-}4)$$

The mean diameter d_m is a quantity defined by the equation

$$d_m = 6V/A \qquad (6\text{-}5)$$

where A is the surface area of the particle. We see that for spherical particles both expressions will be equal to the sphere diameter. The ratio between the two expressions is called the *shape factor*

$$\phi_p = d_m/d_p = \pi^{1/3}(6V)^{2/3}/A \qquad (6\text{-}6)$$

The shape factor corresponds to the ratio between the surface area of a sphere with the same volume as the particle, and the actual surface area of the particle, and it is unity for spherical and less than unity for nonspherical particles. For most crushed ores it will have a value around 0.75, but may be smaller for particles of odd shape, such as needles.

As in a given bed of broken solids of different particle sizes the total surface area of all particles is proportional to their total volume, it follows that the mean d_m for all particles in the bed may be calculated from the total particle volume and total surface area. If the shape factor is known the *mean* particle diameter $\bar{d}_p$ may be derived.

In many cases it will be necessary to know the *size distribution* of the feed. This may be determined by *screening*, Sec. 7-2, and the fractions of the feed with particle size between screen sizes are recorded. Alternatively the screen analysis could be given on a *cumulative* basis by giving the percentage of material which either passes, or is stopped by, each screen size.

The void fraction, ε, in a bed of broken solids expresses the fraction of the bed which is not occupied by solids, and may be calculated from

$$1 - \varepsilon = \frac{\text{volume of solids}}{\text{volume of bed}} = \frac{(m/\rho_s)}{\text{volume of bed}}$$

where m is the mass and ρ_s is the density of the solid.

The void fraction in a bed of solids will depend on the particle shape and the size distribution. For close packing of spheres of uniform size $\varepsilon = 0.26$, independent of the sphere diameter, whereas for loosely packed beds of most broken solids ε is between 0.4 and 0.6, and may be as high as 0.7 for very small or crooked particles. Smaller void fractions may be obtained by mixing two fractions of greatly different diameter. In that case the small particles will fill the voids between the larger particles. The effect of different size and mixing ratios

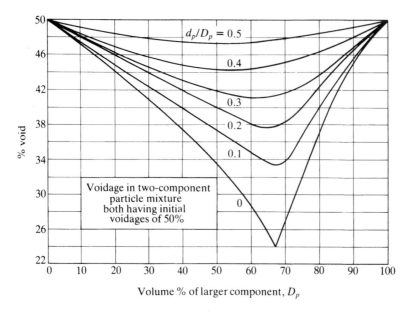

Figure 6-3 Effect of size and mixing ratios on voidage in two-component particle mixture. (*From "Fluidization and Fluid-Particle Systems" by F. A. Zenz and D. F. Othmer. Copyright © 1960 by Van Nostrand Reinhold Company. Reprinted by permission of the publisher.*)

on the void fraction of a bed of broken solids is shown in Fig. 6-3.[1] For most metallurgical reactors based on fixed or moving beds a large void fraction is desired. This is obtained by using material of uniform particle size. If the available material has a wide particle distribution it may be advantageous first to separate it into different fractions by screening, and then to charge each size fraction as separate layers in the bed.

For the production of ceramics or refractory furnace linings from granulated material the void fraction should be as small as possible. This is obtained by mixing two or even three size fractions of optimum sizes as shown in Fig. 6-3. By combination of accurate sizing, mixing and stamping void fractions of less than 0.1 may be obtained.

6-4 FLUID DYNAMICS IN REACTORS

Fixed-, moving-, and fluidized-bed reactors are based on the flow of the fluid, which may be a gas or a liquid, through the bed of granulated material. Of particular interest is the pressure drop in the fluid, which is of importance for

---

[1] C. C. Furnas, quoted by F. A. Zenz and D. F. Othmer, "Fluidization and Fluid-particle Systems," Reinhold Publ. Co., New York, 1960, p. 112.

the energy needed for the flow, for the distribution of the fluid over the cross section of the bed, and for the behavior of the solid particles under the influence of the flow of fluid.

For small particle sizes, low fluid velocity and high viscosity of the fluid, the flow will be *laminar* or *viscous*, whereas for larger particle sizes, high fluid velocity and less viscous fluids, the flow will be *turbulent*. In actual cases there may be a combination of both types. For metallurgical applications the equation given by Ergun[1] has been found particularly useful

$$\frac{\Delta P}{L} = 150 \frac{\mu v (1 - \varepsilon)^2}{d_m^2 \varepsilon^3} + 1.75 \frac{\rho v^2 (1 - \varepsilon)}{d_m \varepsilon^3} \tag{6-7}$$

Here $\Delta P/L$ is the pressure drop per unit height of bed, μ and ρ are the viscosity and density of the fluid, and v is the *superficial velocity* of the fluid, i.e., velocity referred to the empty bed. As everywhere else it is important to use consistent units. To get the pressure drop in newtons per square meter, the diameter d_m and the bed height L must be given in meters, the velocity v in meters per second, the density ρ in kilograms per cubic meter, and the viscosity μ in newton·seconds per square meter. Notice that $1 \text{ N} \cdot \text{s/m}^2 = 1 \text{ kg/(s} \cdot \text{m)} = 10$ poise.

The first term in Eq. (6-7) refers to the viscous, the second term to the turbulent, pressure drop and we see that both increase strongly with decreasing void fraction ε.

For a bed of constant void fraction Eq. (6-7) may be simplified

$$\frac{\Delta P \rho}{L} = a \frac{\mu G}{d_m^2} + b \frac{G^2}{d_m} \tag{6-8}$$

Here $G = v\rho$ is the *mass velocity* of the fluid, i.e., the mass of fluid which passes through a unit cross section per unit time. We see that for constant void fraction the second term in Eq. (6-8) becomes small compared to the first for small values of G and d_m, corresponding to the flow being essentially laminar. Conversely for large values of G and d_m and small values of μ the second term will dominate, corresponding to turbulent flow.

Furthermore, we see that, if for a given mass velocity the density ρ is increased, the pressure drop decreases. Conversely, for a given pressure drop the mass velocity may be increased by increasing the density. This relation may be utilized in the iron blast furnace where, by increasing the total pressure, the mass velocity, i.e., the driving rate, may be increased without increase in the pressure drop and, as will be shown in Eqs. (6-11) and (6-12), without increase in the dust losses from the furnace.

We see furthermore that, for a given pressure drop and gas density, the mass velocity in the laminar region is proportional to d_m^2 whereas in the turbulent region it is proportional to $d_m^{1/2}$. This has the implication that if the burden

[1] S. Ergun: *Chem. Eng. Prog.*, **48**: 89 (1952); see also ibid., p. 227.

in a shaft is unevenly distributed, the mass velocity will be higher in those channels where the particle size is larger, resulting in an uneven gas distribution. The effect will, however, be much more pronounced in the laminar than in the turbulent region.

As shown below, the Ergun equation may be used to calculate the necessary fan power as well as the mean diameter and void fraction of a bed of solids.

Example 6-1 Through a bed of solids, 100 cm high, air of 0°C and of 1 atm mean pressure is passed. The following pressure drops were observed:

$$v_1 = 1 \text{ m/s} \qquad \Delta P_1 = 160 \text{ mmH}_2\text{O} = 1570 \text{ N/m}^2$$

$$v_2 = 2 \text{ m/s} \qquad \Delta P_2 = 630 \text{ mmH}_2\text{O} = 6180 \text{ N/m}^2$$

The viscosity of air is $1.7 \times 10^{-5} \text{ N} \cdot \text{s/m}^2$ and its mean density is 1.29 kg/m^3.

(a) Calculate the necessary theoretical fan power in watts per square meter in the two cases.

(b) Calculate the mean diameter d_m and the void fraction ε in the bed.

SOLUTION (a) Remembering that $1 \text{ W} = 1 \text{ N} \cdot \text{m/s}$, the fan power $\Delta P \cdot v$ is, (for case 1) 1570 W/m^2 and (for case 2) $6180 \times 2 = 12\,360$ W/m^2.

(b) Equation 6-7 may be written in the form

$$\frac{\Delta P}{Lv} = a\mu + bG$$

where $G_1 = 1.29 \text{ kg/(m}^2\text{s)}$ and $G_2 = 1.29 \times 2 = 2.58 \text{ kg/(m}^2\text{s)}$. Inserting the viscosity and the two sets of measurements we get

$$1570 = 1.7 \times 10^{-5}a + 1.29b$$

$$3090 = 1.7 \times 10^{-5}a + 2.58b$$

These equations are solved with respect to a and b to give

$$a = \frac{150(1 - \varepsilon)^2}{d_m^2 \varepsilon^3} = 2.94 \times 10^6 \text{ m}^{-2}$$

$$b = \frac{1.75(1 - \varepsilon)}{d_m \varepsilon^3} = 1178 \text{ m}^{-1}$$

These equations are solved with respect to d_m and ε to give

$$d_m = 0.022 \text{ m} = \underline{2.2 \text{ cm}} \text{ and } \varepsilon = \underline{0.35}.$$

Fluidization and Pneumatic Transport

By further increase in the fluid velocity the bed becomes *fluidized*, i.e., each particle will "float" in the fluid and the mixture of particles and fluid will behave like a liquid. The pressure drop per unit height, $\Delta P/L$, at fluidization is equal to the weight of the bed per unit volume, i.e., equal to $(\rho_s - \rho)(1 - \varepsilon)g$, where ρ_s is the density of the solid and g is the acceleration of gravity. Within the *laminar* flow region this gives

$$v_{mf} = \frac{(\rho_s - \rho)g\varepsilon_{mf}^3 d_m^2}{150\mu(1 - \varepsilon_{mf})} \tag{6-9}$$

Here v_{mf} is the fluid velocity based on an empty bed, and ε_{mf} is the void fraction at minimum fluidization. Similarly we get for the *turbulent* region

$$v_{mf} = \left[\frac{(\rho_s - \rho)g\varepsilon_{mf}^3 d_m}{1.75\rho}\right]^{1/2} \tag{6-10}$$

We see that for constant values of ρ_s, ρ, μ, and ε_{mf} the gas velocity is proportional to d_m^2 within the laminar, and proportional to $d_m^{1/2}$ within the turbulent region. Most fluidization processes occur within the laminar region; for large particle sizes turbulent conditions may be noticeable. The void fraction ε_{mf} for most types of beds is in the range 0.5 to 0.6, but may be as high as 0.7 for particles of odd shape.

As the fluid velocity is increased beyond v_{mf} Eqs. (6-9) and (6-10) still apply, but the bed will expand in such a way that the pressure drop remains equal to the weight of the bed. The *maximum fluidization* is reached when the void fraction has increased to practically unity, i.e., when the particles are far apart. The gas velocity is now equal to the *terminal settling velocity* of the particle in a stagnant fluid. This is, for the laminar region, given by *Stokes' law*

$$v_t = \frac{(\rho_s - \rho)gd_m^2}{18\mu} \tag{6-11}$$

For the turbulent region *Newton's* or *Rittinger's law* is approximately obeyed

$$v_t \simeq \left[\frac{3(\rho_s - \rho)gd_m}{\rho}\right]^{1/2} \tag{6-12}$$

Once more we see the proportionality to d_m^2 and $d_m^{1/2}$ respectively. This relation is shown in Fig. 6-4 which applies for given values of g, ρ_s, ρ, μ and ε_{mf}. The transition from a fixed to a fluidized bed is given by the curve marked v_{mf}, and the transition from fluidized bed to *pneumatic transport* by the curve marked v_t.

Figure 6-5 shows the pressure drop across the bed as well as the void fraction of the bed as function of fluid velocity. Within the fixed-bed region the void fraction remains essentially constant, but the pressure drop increases with increasing fluid velocity. In the fluidized-bed region the pressure drop is essentially constant, but the void fraction increases from ε_{mf} at minimum fluidization to unity in the pneumatic transport range. The small hump in the pressure drop which is often observed at initial fluidization is caused by the breaking up of interlocking grains of irregular shape. In the pneumatic transport region the pressure drop falls off as material is blown out of the bed.

Granular material usually consists of particles of different size. As long as the size differences are not too large the mean diameter of all particles may be

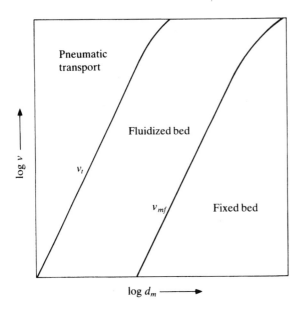

Figure 6-4 Minimum, v_{mf}, and maximum, v_t, fluid velocity for fluidization as function of mean diameter of particles for given values of g, ρ_s, ρ, μ, and ε_{mf}.

used to calculate the fluid velocity for minimum fluidization. On the other hand, if there are large differences in particle size we may have a situation where the smaller particles are fluidized and the larger are not. Terminal fluidization and pneumatic transport, however, will occur at different fluid velocities for different particle sizes. Thus in order to calculate the fluid velocity at which a given particle size fraction is blown out of the bed, the d_m value for that fraction must be used.

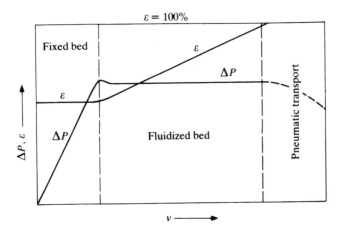

Figure 6-5 Pressure drop, ΔP, and voidage, ε, in a bed of solid particles as function of fluid velocity.

The bubble model At fluid velocities between v_{mf} and v_t the bed may behave in one of two different ways:

1. *Homogeneous or particular fluidization.* Here the particles are distributed rather evenly in the bed. Homogeneous fluidization is obtained mostly for fluids of high density, such as liquids or gases at high pressure. This type is of minor interest to the metallurgist.
2. *Heterogeneous or aggregative fluidization.* This type is usually obtained by fluidization with gases of normal density. One distinguishes between (*a*) *bubble formation,* (*b*) *channeling,* and (*c*) *slugging.* Of these bubble formation is the normal phenomenon and will be discussed below. Channeling may occur if the gas is introduced through orifices that are far apart, whereas slugging may occur if the size of the bubbles becomes equal to the cross section of the bed.

According to the *bubble model* the fluidized bed will, at fluid velocities above approximately $2v_{mf}$, separate into two "phases," one *emulsion phase,* which has a density and void fraction close to that of minimum fluidization, and a *bubble phase,* which is practically free of solid particles. The relative amount of the bubble phase increases with increasing gas velocity, i.e., an increasing part of the gas passes the bed through the bubbles. As a result, the average gas velocity in the bed may be increased beyond the terminal velocity which applies for the solid particles, without excessive dust losses. Whereas the gas velocity in the emulsion phase in average corresponds to v_{mf}, the rising velocity of the bubbles is much higher and, as a result, there is a continuous exchange of gas between the bubble phase and the emulsion phase. This exchange represents an important vehicle in the transfer of heat and mass between the gas and the solid particles in the bed, which will be discussed further in the following sections.

6-5 HEAT TRANSFER IN REACTORS

Heat transfer is a result of the second law of thermodynamics which states that heat will flow from high to low temperature until equal temperatures are obtained. The *rate* of heat transfer is determined partly by the temperature *difference* between the source and the sink, and partly by the *resistance* to heat flow. The latter may again be dependent on the mechanism of heat transfer. The main heat transfer mechanisms are: (1) *conduction,* (2) *convection,* and (3) *radiation.* In a given case all three mechanisms may operate, and may either be working in *series,* as illustrated by convection from a fluid to a solid surface followed by conduction into the solid, or in *parallel,* as illustrated by simultaneous convection and radiation to or from a solid surface.

In the present discussion on metallurgical reactors heat transfer is of importance partly in the calculation of heat losses through reactor walls to the surrounding atmosphere, and partly in the transfer of heat from, for example, hot combustion gases to a reaction mixture. In the present text the main emphasis will be put on the problems connected with heat transfer to a reacting mixture.

Heat Conduction

For the flow of heat through a body, the *Fourier equation* applies

$$\dot{q} = -kA\left(\frac{dT}{ds}\right) \tag{6-13}$$

Here $\dot{q}$ is the quantity or *flux* of heat per unit time which flows through an area A, when the temperature gradient is dT/ds. The coefficient k is called the *thermal conductivity* of the material. The minus sign shows that heat flows toward the lower temperature. It follows from Eq. (6-13) that thermal conductivity has the dimension joules per meter · second · kelvin which is the same as watts per meter · kelvin. The thermal conductivity is a property of the material. Examples: silver, 427; copper, 400; aluminum, 236; steel, about 50; stainless steel, about 15; brick, 0.5 to 1.0; sand, about 0.6; asbestos and other insulating materials, about 0.1; stagnant air, 0.025; all in watts per meter · kelvin at room temperature. The thermal conductivity is usually a function of temperature, and for porous materials it increases with increasing temperature, whereas it decreases with increasing temperature for dense materials and metals.

By integration of Eq. (6-13) we obtain

$$\dot{q} \int_{s_1}^{s_2} \frac{ds}{A} = -\int_{T_1}^{T_2} k \, dT$$

If A and k are independent of s and T respectively, the following simple relation is obtained:

$$\dot{q} = -kA \frac{\Delta T}{\Delta s} \tag{6-14}$$

If k is a function of temperature, as for example, $k = a + bT$, then the mean value, $k_{mean} = a + b(T_1 + T_2)/2$ may be used. In addition, for different geometric shapes A may be a function of s. Thus for radial heat flow in a cylindrical body $A = 2\pi rL$, where r is the distance from the cylinder axis and L is the length of the cylinder. This gives

$$\int_{s_1}^{s_2} \frac{ds}{A} = \int_{r_1}^{r_2} \frac{dr}{2\pi rL} = \frac{\ln (r_2/r_1)}{2\pi L} = \frac{2.3 \log (r_2/r_1)}{2\pi L}$$

If k is independent of T, or by applying a mean value for k, the heat flow becomes

$$\dot{q} = -k \frac{2\pi L \Delta T}{2.3 \log (r_2/r_1)}$$

or, as $2\pi L = (A_2 - A_1)/\Delta r$ and $r_2/r_1 = A_2/A_1$, where A_1 and A_2 are the areas at r_1 and r_2 respectively

$$\dot{q} = -k\,\frac{(A_2 - A_1)}{2.3\,\log\,(A_2/A_1)}\,\frac{\Delta T}{\Delta r}$$

By comparison with Eq. (6-14) it is seen that the second term on the right-hand side represents the mean area for the heat flow. This is called the *logarithmic mean area*. For other geometric shapes other mean areas apply. Thus for heat flow between concentric spherical surfaces $A_{mean} = \sqrt{A_2 A_1}$, that is, the *geometric mean area*. For more complicated shapes the mean area may have to be determined experimentally.

Heat Convection

By convection we understand transfer of heat between a solid surface and a fluid, which may be a gas or a liquid, or between two fluid phases. In these cases the interface between the two phases gives a certain resistance to heat transfer. The heat flux $\dot{q}$ will be a function of the interface area A, the temperature difference ΔT, and the *heat transfer coefficient by convection* h_c, viz.,

$$\dot{q} = h_c\,(\Delta T)A \qquad (6\text{-}15)$$

It follows that the heat transfer coefficient has the dimension watts per square meter · kelvin. We distinguish between *free convection* where the motion of the fluid near the interface is caused by temperature differences in the fluid, and *forced convection* where the fluid moves relative to the interface as a result of an external force.

Free convection This has been studied experimentally for a number of geometric cases, temperature differences, and properties of the fluid. In the present context it is of particular interest for calculation of heat losses from, e.g., the outer shell of a metallurgical reactor to the surrounding atmosphere. For such cases, i.e., for convection between plane or cylindrical surface and air or similar diatomic gases at moderate temperatures, 0 to 200°C, the following expressions may be used:[1]

Vertical planes and cylinders

$$h_c = 1.4(\Delta T/d)^{0.25}\ \text{W}/(\text{m}^2 \cdot \text{K}) \qquad (6\text{-}16)$$

Horizontal cylinders

$$h_c = 1.3(\Delta T/d)^{0.25}\ \text{W}/(\text{m}^2 \cdot \text{K}) \qquad (6\text{-}17)$$

---

[1] Converted to SI units from expressions given by: W. H. McAdams: "Heat Transmission," 3d ed., McGraw-Hill Book Co. Inc., New York, 1954, pp. 173–180. See also 2d ed., p. 240.

In these expressions the characteristic dimension d is either the vertical dimension of the plane or the diameter of the cylinder in meters. These expressions are particularly useful for values of $d \leq 0.3$ m. For values of $d > 1$ m the dimension is of minor importance, and the following simplified expressions may be used:

Horizontal surfaces facing up

$$h_c = 2.4(\Delta T)^{0.25} \text{ W}/(\text{m}^2 \cdot \text{K}) \qquad (6\text{-}18)$$

Vertical surfaces

$$h_c = 1.7(\Delta T)^{0.25} \text{ W}/(\text{m}^2 \cdot \text{K}) \qquad (6\text{-}19)$$

Horizontal surfaces facing down

$$h_c = 1.2(\Delta T)^{0.25} \text{ W}/(\text{m}^2 \cdot \text{K}) \qquad (6\text{-}20)$$

These expressions apply for heat transfer *from* the surface. For heat transfer *to* the surface the expressions for surfaces facing up and down are exchanged. For surface temperatures above 200°C the numerical coefficients in the above expressions have somewhat smaller values. For polyatomic gases such as CO_2 and SO_2 somewhat larger values apply.

The interesting thing about the above expressions is that the heat transfer coefficient is proportional to $(\Delta T)^{0.25}$, that is, the heat flow is proportional to $(\Delta T)^{1.25}$ This is in contrast to conduction where the heat flow is directly proportional to the temperature difference.

Example 6-2 For a horizontal cylindrical pipe of 1 m diameter in air and for $\Delta T = 100$°C we calculate $h_c = 4.1$ W/(m$^2 \cdot$ K). For a heating wire with 1 mm diameter, we calculate for the same ΔT: $h_c = 23$ W/(m$^2 \cdot$ K). For a large vertical surface as, for example, a furnace shell, and for $\Delta T = 200$°C, we calculate $h_c = 6.5$ W/(m$^2 \cdot$ K). In most cases of natural convection in gases h_c is between 1 and 20 W/(m$^2 \cdot$ K).

Forced convection As already mentioned forced convection is of greater importance than free convection in metallurgical reactions. The forced convection heat transfer coefficient is a function of the size and geometric shape of the interface, the temperature and physical properties of the fluid, and the velocity of the fluid relative to the interface. The heat transfer coefficient will also depend on whether the fluid flow is *laminar* or *turbulent*. For most cases of metallurgical interest the flow velocity and surface dimensions are sufficiently large to make the flow turbulent. In other cases transition to laminar flow occurs. For *turbulent* flow through *pipes*, i.e., for Re > 2100 the following relation between Nusselt's, Prandtl's, and Reynold's numbers has been found:

$$\text{Nu} = 0.023\text{Re}^{0.8}\text{Pr}^{0.33}$$

In this equation $Nu = h_c d/k$, $Re = Gd/\mu$ and $Pr = c_p \mu/k$ where k is the heat conductivity of the fluid and d the diameter of the pipe. For *air and similar diatomic gases* $Pr \simeq 0.70$ rather independent of temperature. Inserting this and solving for h_c we get:

$$h_c = 0.020kd^{-0.2}(G/\mu)^{0.8} \qquad (6\text{-}21)$$

We see that the heat transfer coefficient increases with increasing heat conductivity and mass velocity of the gas and decreases with increasing pipe diameter and gas viscosity. Both the heat conductivity and the viscosity of gases increase slightly with increasing temperature, and values which apply for the average solid–gas temperature should be applied.

Example 6-3 For air at atmospheric pressure and an average temperature of 500°C the heat conductivity is 0.054 W/(m · K) and the viscosity $\mu = 3.6 \times 10^{-5}$ kg/(m · s). 0.01 kg of air per second is blown through a circular pipe with 0.1 m diameter.

(a) Calculate the mass velocity, and calculate the heat transfer coefficient per square meter of pipe as well as per meter of pipe length.

(b) Do the same calculation if the same amount of air per unit time is blown through a pipe of 0.01 m diameter.

SOLUTION (a) The pipe cross section is $\frac{1}{4}\pi d^2 = 0.00785$ m^2, which makes the mass velocity $G = 0.01/0.00785 = 1.274$ kg/(m^2 · s). This gives Reynold's number $Re = 1.274 \times 0.1/(3.6 \times 10^{-5}) = 3538$, which means that Eq. (6-21) may be used. This gives for the heat transfer coefficient

$$h_c = 0.020 \times 0.054 \times 0.1^{-0.2}[1.274/(3.6 \times 10^{-5})]^{0.8} = 6.2 \text{ W/(m}^2 \cdot \text{K)}$$

Expressed per meter of pipe length the heat transfer coefficient is $6.2 \times \pi \times 0.1 = 1.95$ W/(m · K).

(b) If the same amount of air is blown through a pipe of 0.01 m diameter the mass velocity increases by a factor of 100 and the heat transfer coefficient by a factor of $100^{0.8}/0.1^{0.2} = 63$, giving $h_c = 390$ W/(m^2 · K) and per meter of pipe length 12.25 W/(m.K). We see that the combined effect of increased gas velocity and decreased pipe diameter more than compensates for the decrease in surface area, and that the most favorable heat transfer is obtained by forcing the gas at high velocity through narrow openings.

For heat convection between fluids and a single solid particle or a bed of broken solids the following relations have been given:[1]

$$Nu = 2 + 0.6Pr^{1/3}Re^{1/2} \qquad \text{(single particles)} \qquad (6\text{-}22)$$

$$Nu = 2 + 1.8Pr^{1/3}Re^{1/2} \qquad \text{(fixed bed)} \qquad (6\text{-}23)$$

---

[1] D. Kunii and O. Levenspiel: "Fluidization Engineering," John Wiley and Sons Inc., New York, 1969, pp. 210–211.

Here $Nu = h_c d_p/k$, $Pr = c_p \mu/k$ and $Re = Gd_p/\mu$ where the mass velocity G as before refers to the *empty bed*, i.e., with the particles removed. In these equations the first term, 2, gives Nusselt's number for stagnant fluid, and becomes insignificant at higher fluid velocities. Disregarding the first term and inserting the value $Pr \simeq 0.70$ for air and similar diatomic gases we solve for h_c

$$h_c = 0.532k\left(\frac{G}{d_p \mu}\right)^{1/2} \qquad \text{(single particles)} \qquad (6\text{-}24)$$

$$h_c = 1.60k\left(\frac{G}{d_p \mu}\right)^{1/2} \qquad \text{(fixed bed)} \qquad (6\text{-}25)$$

In these expressions h_c refers to unit area of the particle surface. For beds of broken solids it is sometimes more convenient to refer the heat transfer coefficient to unit volume or unit mass of solids. For this purpose h_c is multiplied by the surface area per unit volume or unit mass. These were given in Sec. 5-5 as $6(1 - \varepsilon)/d_m$ and $6/(d_m \rho_s)$ respectively, where ε is the void fraction and ρ_s the density of the particle. Thus the heat transfer coefficient per unit bed volume or bed mass becomes inversely proportional to $d_m^{1.5}$ provided other factors are constant.

Experimental studies on heat transfer coefficients have shown that Eqs. (6-22) to (6-25) are closely obeyed in gases for particle diameters larger than about 5 mm. These equations may, therefore, be used for, for example, iron-ore pellets in a blast furnace. For smaller diameters, however, the observed Nusselt's number, and consequently the heat transfer coefficient, is significantly *smaller* than expected. This is the case for fixed beds of fine particles as well as for fluidized beds, where the Nusselt number may be as low as one-thousandth of the limiting value of 2 given by Eqs. (6-22) and (6-23). This is shown in Fig. 6-6, which is composed from data given by Kunii and Levenspiel.

For fixed beds this deviation is believed to be caused by channeling, i.e., that the small particles form aggregations separated by channels through which the main flow goes. Similarly, for fluidized beds the deviation is believed caused by the formation of bubbles in the bed (Sec. 6-4), and the heat transfer is restricted by the exchange of gas between the bubble phase and the emulsion phase. Nevertheless, in spite of the Nusselt number being significantly smaller than expected, it still results in a fairly large heat transfer coefficient, because of the small particle diameter, and in general one may conclude that a high degree of thermal equilibrium exists between the gas flow and the solid particles during fluidization.

For heat transfer between the gas and the wall of a fluidized bed reactor the heat transfer coefficient is very high. Values between 100 and 500 W/(m² · K) have been reported. This is significantly *more* than for transfer between a solid surface and a gas flow of the same velocity, but without fluidized particles. The large transfer coefficient is believed to be caused by the particles destroying the stagnant gas film on the solid surface.

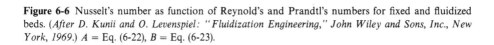

Figure 6-6 Nusselt's number as function of Reynold's and Prandtl's numbers for fixed and fluidized beds. (*After D. Kunii and O. Levenspiel: "Fluidization Engineering," John Wiley and Sons, Inc., New York, 1969.*) A = Eq. (6-22), B = Eq. (6-23).

Heat Radiation

For radiation from a blackbody surface into open space the *Stefan–Boltzmann law* applies

$$\dot{q} = KT^4 A \tag{6-26}$$

where A is the area of the surface in square meters and T is the temperature in kelvin. The value of KT^4 depends on the units of measure used. Thus in the SI system

$$\dot{q} = 5.67(T/100)^4 A \quad \text{W}$$

For nonblackbodies the above expression is to be multiplied by ε, the *emissivity* of the surface.

For radiation from one surface to another of lower temperature the heat flow will depend on the two temperatures, the relative size of the surfaces, their relative orientation, and the presence of intermediate reflecting surfaces, as well as their emissivities ε_1 and ε_2 and absorptivities α_1 and α_2. In most cases the emissivity and the absorptivity of a surface are the same. Thus the heat flow between two surfaces may be expressed by

$$\dot{q} = 5.67F\left[\left(\frac{T_1}{100}\right)^4 - \left(\frac{T_2}{100}\right)^4\right]A_1 \qquad \text{W} \qquad (6\text{-}27)$$

Here T_1 and T_2 are the temperatures in kelvin of the source and the sink, respectively, and A_1 is the area of the source in square meters. The coefficient F is a dimensionless *efficiency factor*, which is determined by the relative size, the orientation, and the emissivities of the two surfaces, as well as the effect of intermediate reflecting surfaces.

For evaluation of the efficiency factor F for various cases the reader is referred to textbooks listed in the bibliography. Here only a few simple cases will be given:

1. If the area A_1 is convex and small compared to A_2, the area of the sink, the efficiency factor, becomes equal to ε_1, the emissivity of the source. This would be the case for heat radiation from a relatively small object, e.g., a laboratory furnace or the outer shell of a metallurgical reactor into a large room of uniform temperature.
2. If the area A_1 is large compared to A_2, the efficiency factor becomes equal to $\varepsilon_2 A_2/A_1$ and the heat flow $\dot{q} = 5.67\varepsilon_2[(T_1/100)^4 - (T_2/100)^4]A_2$ W. This would be the case for the heating of a relatively small object, like a lump of ore, inside a furnace with uniform wall temperature.

Heat radiation from solid surfaces is of importance for calculation of heat losses from a furnace shell into the surroundings, where it works in parallel to heat convection, and for radiation from the heated brickwork and roof of a reverberatory furnace or rotary kiln to the molten or solid charge. In the latter cases direct radiation from the flames and hot combustion gases is usually of greater importance.

Luminous flames radiate heat mainly because of their content of unburned carbon particles. Such flames are obtained by the combustion of fuel oil or powdered coal. A large luminous flame can approximately be regarded as a blackbody. For small flames and for more complete combustion the emissivity can hardly be expressed generally, but may, with some experience, be estimated.

The heat radiation from *nonluminous* flames is of a different nature. Such flames contain no solid particles but molecules such as CO_2, H_2O, SO_2, in addition to N_2, H_2, and O_2. The three last gases have symmetrical molecules and do not radiate heat in the ordinary temperature range. The first three gases have asymmetrical molecules which result in a dipole moment and give rise to a radiation spectrum with significant emission bands in the infrared.

Since the emission bands occur at fixed wavelengths, the emissivity ε_g varies with the temperature of the gas. It is also a function of the partial pressure of the polar component and of the thickness of the gas layer. Furthermore, the *absorptivity* of the gas α_g is not equal to its emissivity, but is a function of the incident radiation wavelength, i.e., of the temperature of the surrounding surface. In order to calculate radiant heat exchange between polar gases and the surrounding surfaces it is necessary to know the emissivity and absorptivity of the gas as function of the various parameters: temperature, partial pressure, and gas thickness, as well as the emissivity of the surfaces. In most textbooks on heat transfer such data may be found, together with the necessary equations for their application.

It follows from Eq. (6-27) that heat radiation increases strongly with increasing temperature of the source. At around room temperature it is usually smaller than heat transfer by free convection, but may in some cases be significant. Similarly, for heat transfer between small gas thicknesses and solid surfaces, such as between a gas flow and a bed of broken solids, heat transfer by radiation usually is less important than by forced convection.

Total transfer coefficients As was shown in Eq. (6-15) heat transfer by convection may be given by a heat transfer coefficient multiplied by the temperature difference between the source and the sink. In comparison Eq. (6-27) shows that heat transfer by radiation is proportional to the difference between the fourth power of the two absolute temperatures. However, by rearrangement of Eq. (6-27) radiation transfer may also be expressed in terms of a heat transfer coefficient

$$\dot{q} = KF(T_1 + T_2)[(T_1)^2 + (T_2)^2](T_1 - T_2)A_1$$

Here the radiation transfer coefficient is

$$h_r = KF(T_1 + T_2)[(T_1)^2 + (T_2)^2]$$

For small temperature differences T_1 and T_2 may be approximated by T_{mean} which gives $h_r = 4KF(T_{\text{mean}})^3$ or, expressed in the SI system, $h_r = 0.227F(T_{\text{mean}}/100)^3$ W/(m$^2 \cdot$ K).

Heat transfer between a hot surface and its surroundings usually consists of both convection and radiation which work in parallel giving the *total coefficient*: $h = h_c + h_r$. The inverse of this quantity is called the *total transfer resistance*.

Unsteady Heat Flow

The preceding treatment of heat transfer mechanisms and heat transfer coefficients apply regardless of whether the temperatures of the source and the sink remain constant with time or are time functions. In the first case the flow is *steady* as illustrated by heat losses from a furnace shell to the surrounding atmosphere or by heat transfer between the gas and the solid burden at a *constant level* in a shaft furnace that works under stationary conditions. In other cases the heat flow is *unsteady*. This applies, for example, for the heating of the reaction mixture in a retort process or the cooling of a steel ingot in air. Unsteady heat flow also applies for the burden in a shaft furnace as it *descends* in counter-current with hot combustion gases.

In the heating of a solid body heat flows from the source, which may be the surrounding hot gas flow, and into the body of the solid. The resistance to this flow is composed of the resistance to heat transfer by convection and radiation *to* the outer surface of the body, and resistance to heat flow *into* the body. These two resistances are connected in series, and for a rigorous treatment both resistances should be considered. In many cases, however, one of these is significantly larger than the other which, therefore, may be disregarded. Thus in the heating or cooling of a steel ingot the heat conductivity is large, and the main resistance to heat flow is given by the heat transfer coefficient between the ingot and the surroundings. The same is the case for the heating or cooling of broken solids of ore or coke below a certain particle size, say 1 to 2 cm. If, in such a case, the heat flow from or to the surroundings are denoted $\dot{q} = dq/dt$, this causes an increase or decrease, as the case may be, in the temperature of the body by dT, and we have

$$\frac{dq}{dt} = hA(T_2 - T) = V\rho c_p \frac{dT}{dt}$$

Here A is the surface area, V the volume, ρ the density, and c_p the specific heat of the body. T_2 is the constant temperature of the surroundings, and T is the variable temperature of the body. This gives

$$\frac{dT}{dt} = \frac{hA}{V\rho c_p}(T_2 - T)$$

If the temperature of the body at $t = 0$ is T_0 and if h is assumed constant, integration gives

$$\ln\left[\frac{T - T_2}{T_0 - T_2}\right] = \frac{-hAt}{V\rho c_p} \tag{6-28}$$

This is called *Newton's law*, and is analogous to the change in composition during a first-order chemical reaction. It is equally valid for heating and for cooling of bodies provided the heat conductivity is large compared to the surface transfer coefficient. The ratio $(T - T_2)/(T_0 - T_2)$ is called *the relative temperature* of the body with respect to its initial and final temperature. It is unity for $t = 0$ and becomes zero when the body has reached the temperature of the surroundings. It follows from the equation that this requires infinite time.

For less conducting bodies such as refractory bricks or metallurgical ores it is necessary to take into consideration the heat conductivity. The extreme case is obtained when the surface resistance is very small compared to the conduction resistance as, e.g., in the heating or cooling of refractory bricks or the heating of a reaction mixture of low heat conductivity in a retort. In these cases the temperature follows equations analogous to *Fick's second law* for diffusion (Eqs. (5-8) and (5-9)), and may in the general case be given by

$$\left(\frac{dT}{dt}\right)_s = \frac{k}{c_p \rho}\left[\frac{d^2T}{ds^2} + \frac{d\ln A}{ds}\frac{dT}{ds}\right] \tag{6-29}$$

The expression $k/(c_p\rho)$ is called the *thermal diffusivity* of the body and is analogous to the diffusion coefficient D; s is the distance from the surface, and A the cross-section area at that distance and perpendicular to the heat flow. If A is independent of s, the second term inside the brackets becomes zero.

In the same way as shown for unsteady diffusion the above equation may be integrated for various geometric shapes, and the results are identical to those given in Fig. 5-4(*a–b*). Instead of the relative concentration the *relative temperature* $(T - T_0)/(T_s - T_0)$ or $(T_m - T_0)/(T_s - T_0)$ is plotted as ordinate, and instead of the diffusivity D, the thermal diffusivity is inserted in the abscissa. We see that for a given abscissa value and a given diffusivity the time is proportional to the square of the linear dimension. This means that for a given heating or cooling of a given material a doubling of the linear dimension causes an increase in the necessary time by a factor of four, etc. This again means that in the heating or cooling of solids of low heat conductivity as, for example, in the coking of coal or for carbothermic reduction of zinc oxide in a retort (Secs. 8-1 and 10-5), relatively small dimensions are necessary in order to complete the process within a reasonable time. This is in contrast to the heating or cooling of solids with high conductivity, where the time is proportional to V/A, that is, to the linear dimension, Eq. (6-28).

6-6 CHEMICAL REACTIONS IN FLOW SYSTEMS

The degree of reaction obtained in a given reactor will depend on the rate equation for the chemical reaction as well as on the retention time and age distribution function for the material in the reactor. For simplicity we shall first consider a *first-order reaction*. As shown in Sec. 5-1 a reaction $A + B \rightarrow C$ may be considered first-order with respect to A if there is a large surplus of reactant B and if the reverse reaction is insignificant. The change in concentration of reactant A is then given by

$$\frac{-dC_A}{dt} = kC_A$$

where k is the rate constant.

For a *batch process* this expression may be integrated to give the concentration of unreacted A as function of time

$$C_A = C_A^\circ e^{-kt} \tag{6-30}$$

Here C_A° represents the initial concentration of A in the feed.

The same equation will be valid also for a *perfect plug-flow reactor* where the reactant A has a unique retention time, provided the concentration of reactant B remains constant within the reactor.

It follows from Eq. (6-30) that the concentration of unreacted A will decrease with time, and that an infinite time is needed for a complete reaction. This is illustrated by Fig. 6-7 where curve (*a*) gives the *relative concentration* C_A/C_A° as function of retention time for a batch or a perfect plug-flow reactor.

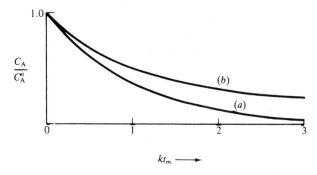

Figure 6-7 Degree of reaction as function of retention time for a first-order reaction, (a) batch process, (b) perfect mixing process.

We shall now consider the same first-order reaction carried out in a reactor with perfect mixing. The flow into the reactor is m kg/h with concentration C_A° of unreacted material. Referred to the feed the effluent flow and the reactor volume have an unreacted concentration C_A. We then have the following mass balance per unit time

$$C_A^\circ \cdot m = k C_A M + C_A \cdot m$$

This gives the relative concentration of unreacted material in the overflow

$$C_A/C_A^\circ = \frac{m}{kM + m} = \frac{1}{kt_m + 1} \tag{6-31}$$

In Fig. 6-7 curve (b) shows the value of C_A/C_A° as function of the product kt_m. Comparing the curves (a) and (b) for a given value of k, the concentration of A is seen to decrease faster with time in a batch process than in a perfect mixing process. For example, for 25 percent unreacted the mean retention time in the latter process must be about twice that of the batch process.

Most metallurgical reactions do not follow the first-order rate equation, however. Furthermore the concentration of both reactants A and B may change during the reaction. In the following we shall still consider the concentration of B to be constant but will consider the effect of various rate equations and age distribution functions on the degree of reaction.

Gas Solid Reactions

As discussed in Chap. 5 heterogeneous reactions between a fluid and a solid may involve diffusion through a stagnant fluid layer followed by diffusion through a product layer and chemical reaction on a reactant–product interface. The first of these steps is commonly called *mass transfer* between the fluid and the solid. It shows great similarity to heat transfer by convection, as discussed in Sec. 6-5. For mass transfer between a fluid and a single particle or a bed of broken solids, equations analogous to Eqs. (6-22) and (6-23) have been given

$$Sh = 2 + 0.6Sc^{1/3}Re^{1/2} \qquad \text{(single particles)} \qquad (6\text{-}32)$$

$$Sh = 2 + 1.8Sc^{1/3}Re^{1/2} \qquad \text{(fixed bed)} \qquad (6\text{-}33)$$

Here Sh (Sherwood's number) $= k_g d_p/D$ and Sc (Schmidt's number) $= \mu/(\rho D)$, where k_g is the *mass transfer coefficient* and D the *diffusivity* of the transferred species in the fluid. Strictly speaking this form of Sherwood's number applies only for equimolecular counterdiffusion, as for the reaction FeO + CO = Fe + CO_2. For unidirectional diffusion in the presence of an inert gas the expression for Sherwood's number should be multiplied by a factor y, which is the logarithmic mean of the contents of inert gas in the fluid flow and on the particle surface.

In the same way as for heat transfer, experiments have shown that Eqs. (6-32) and (6-33) are closely obeyed for particles with diameters larger than about 5 mm, whereas for smaller diameters significantly smaller values for Sherwood's number are found, corresponding to lower mass transfer coefficients. In the same way as shown for heat transfer this is believed caused by channeling for fixed beds, and bubble formation for fluidized beds, and Fig. 6-6 may be applied directly also to mass transfer by replacing Nusselt's with Sherwood's number and Prandtl's with Schmidt's.

Mass transfer between the fluid flow and the external surface of solid particles may be of importance for reactions where solid reaction products are not formed, such as volatilization reactions and for the gasification of solid carbon with CO_2. For most other metallurgical reactions, such as the roasting of sulfide concentrates or reduction of metal oxides, mass transfer to the external surface is normally not the rate-limiting step. For such reactions either diffusion through pores in the solid product layer or reaction at a solid reactant–product interface is usually rate-limiting. As discussed in Sec. 5-5 such reactions are often *topochemical*, i.e., they follow the *shrinking-core* model. Depending on whether interface reaction or pore diffusion is rate controlling we get different expressions for the reaction rate. If the fraction of reacted component A at time t is denoted X_A, and the time for complete reaction is denoted t_{tot}, Eqs. (5-15) and (5-19) may be written in the following forms:

1. *Interface control*

$$t/t_{tot} = 1 - (1 - X_A)^{1/3} \qquad \text{i.e.} \qquad (1 - X_A) = (1 - t/t_{tot})^3 \qquad (6\text{-}34)$$

Here $t_{tot} \propto d_0/R_A$ where d_0 is the initial diameter of the particle and R_A the rate of reaction per unit interface area.

2. *Pore diffusion control*

$$t/t_{tot} = 1 - 3(1 - X_A)^{2/3} + 2(1 - X_A) \qquad (6\text{-}35)$$

Here $t_{tot} \propto d_0^2/D$ where D is the diffusion coefficient in the product layer.

These expressions apply for each individual particle. If all particles were of the same size and all had the same retention time in the reactor, as in a perfect plug-flow reactor, the degree of reaction at any time t could be calculated if the time for complete reaction, t_{tot}, were known. In practice particles will usually be of different sizes, corresponding to different values for t_{tot}, and the age distribution function will differ from that for perfect plug flow. If the mean fraction of unreacted A in the combined product is denoted $(1 - \bar{X}_A)$ we have

$$(1 - \bar{X}_A) = \sum_{\substack{\text{all} \\ \text{particles}}} \begin{bmatrix} \text{unreacted fraction in par-} \\ \text{ticles with retention time} \\ \text{between } t \text{ and } t + dt \end{bmatrix} \begin{bmatrix} \text{fraction of product} \\ \text{with retention time} \\ \text{between } t \text{ and } t + dt \end{bmatrix}$$

or

$$(1 - \bar{X}_A) = \int_0^{t_{tot}} (1 - X_A)E \, dt \tag{6-36}$$

where E is the *age distribution function* (Sec. 6-1).

For a *perfect mixing reactor* $E = e^{-\theta}/t_m$ where t_m is the mean retention time, and $\theta = t/t_m$ is the relative retention time. If, furthermore, we assume all particles to be of the same size, corresponding to a unique value for t_{tot}, we have, for *interface control*,

$$(1 - \bar{X}_A) = \int_0^{t_{tot}} (1 - t/t_{tot})^3 (e^{-t/t_m}/t_m) \, dt$$

This expression may be integrated and given in expanded form[1]

$$(1 - \bar{X}_A) = \frac{1}{4}\left(\frac{t_{tot}}{t_m}\right) - \frac{1}{20}\left(\frac{t_{tot}}{t_m}\right)^2 + \frac{1}{120}\left(\frac{t_{tot}}{t_m}\right)^3 \cdots$$

For $t_{tot}/t_m < 0.2$ all terms except the first may be ignored, which gives

$$(1 - \bar{X}_A) \simeq \frac{1}{4}\frac{t_{tot}}{t_m} \propto \frac{d_0}{R_A t_m}$$

We see that in order to have, for example, 5 percent unreacted A in the product, the mean retention time t_m has to be $5t_{tot}$. Similarly for 1 percent unreacted $t_m = 25t_{tot}$, and so on.

If *pore diffusion* had been rate-controlling the fraction of unreacted A in the product is obtained by combination of Eqs. (6-35) and (6-36) to give in expanded form[2]

$$(1 - \bar{X}_A) = \frac{1}{5}\left(\frac{t_{tot}}{t_m}\right) - \frac{19}{420}\left(\frac{t_{tot}}{t_m}\right)^2 + \frac{41}{4620}\left(\frac{t_{tot}}{t_m}\right)^3 \cdots$$

---

[1] D. Kunii and O. Levenspiel: "Fluidization Engineering," John Wiley and Sons Inc., New York, 1969, p. 491.
[2] D. Kunii and O. Levenspiel: loc. cit.

Again for $t_{tot}/t_m < 0.2$ this may be simplified

$$(1 - \bar{X}_A) \simeq \frac{1}{5}\frac{t_{tot}}{t_m} \propto \frac{d_0^2}{Dt_m}$$

Thus for $t_m = 5t_{tot}$ we get $(1 - \bar{X}_A) = 0.04$, that is, 4 percent unreacted in the product. Similarly for 1 percent unreacted we need $t_m = 20t_{tot}$.

We see that for both versions of the shrinking-core model applied to a perfect mixing reactor, the mean retention time needed in order to get an almost complete reaction will have to be many times that needed to react one single particle to completion. The reason for this is obvious: In a perfect mixing reactor a fair fraction of the material has a retention time which is much shorter than the mean, and will be only partially reacted, whereas part of the material has retention times which approach infinity. This last part will be completely reacted, and will stay in the reactor for no useful purpose. Thus for fluidized-bed roasting of sulfide concentrates (Sec. 8-3) mean retention times of several hours may be needed even though some particle may have reacted completely after only a few minutes.

For gas reduction of porous oxide ores below a certain particle size other equations may apply. As mentioned in Sec. 5-4 such reactions do not follow the shrinking-core model, but take place throughout the entire volume of the particle. As a first approximation we may assume the reaction to be first order with respect to the oxide, and Eqs. (6-30) and (6-31) may be used, although it is unlikely that infinite time is needed for complete reaction.

It was assumed in the preceding treatment that all particles were of the same size, i.e., that their t_{tot} was the same, and also that all particles had the same age distribution function with the same mean retention time. In practice this will not necessarily be the case, and for a rigorous treatment it would be necessary to determine the t_{tot} value as well as the age distribution function for all different particle sizes in the material, and to insert these in Eq. (6-36) to get the overall fraction of unreacted material in the product.

Without going into detail it should only be mentioned that for fluidized-bed roasting the coarse particles will have longer mean retention time than the fine ones. On the other hand, they will also need a longer time for complete reaction. In total the two effects will compensate each other to give a reasonably thorough reaction of the effluent product. In the rotary kiln furnace, on the other hand, the coarser fractions, which react more slowly, will have a shorter mean retention time than the fine ones. This will tend to give a product which is rather unevenly reacted.

So far we have assumed the concentration of the other reactant, B, to be constant, and also that we were sufficiently far from chemical equilibrium to make the reverse reaction rate insignificant. This would approximately be the case for fluidized-bed roasting of sulfide concentrates, where there is considerable back-mix in the gas phase and where the composition of the gas is far removed from chemical equilibrium with the solids. In other cases, such as in the iron blast furnace, the composition of the gas changes as it ascends in *counter-current* with the burden, and chemical equilibrium with the solids is closely approached at some level in the shaft. In such cases the above treatment needs further adjustment. The counter-current principle is used also in other reactors and for other types of processes, and a generalized treatment will, therefore, be given in the following section.

6-7 COUNTER-CURRENT REACTORS

The *counter-current principle* is used in a number of chemical and metallurgical processes. Typical examples are solvent extraction (where an aqueous and an organic liquid move counter-current), continuous distillation (where a liquid and a gas move counter-current), and oxide ore reduction in a shaft furnace. The counter-current principle is used also for *heat exchange* between hot and cold fluid flows. In the following a generalized treatment will be given with application to different types of processes and, as before, we shall start with the simpler and proceed to the more complicated.

Heat Exchangers

Important examples of heat exchange are shown in Figs 6-8(*a–c*). Figure 6-8(*a*) illustrates a simple *heat recuperator*. In this the two flows are separated by a wall of temperature-resistant material, e.g., steel. Figure 6-8(*b*) illustrates the

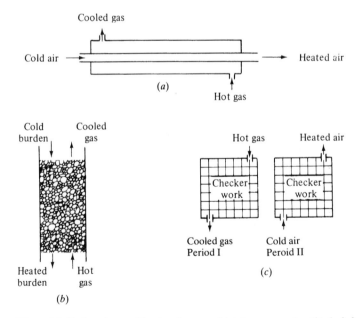

Figure 6-8 Various types of heat exchangers, (*a*) tube recuperator, (*b*) shaft furnace, (*c*) regenerator.

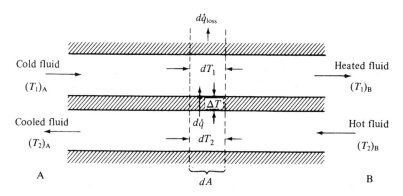

Figure 6-9 Heat flow and temperature in tube recuperator.

transfer of heat from a hot gas flow to a solid burden, e.g., in the iron blast furnace. Figure 6-8(c) illustrates a *heat regenerator*. In this the heat is first transferred from the hot gas to a checkerwork of refractory bricks. Subsequently these bricks are used to heat the cold gas flow. Regenerators are used if the temperatures involved are too high for common recuperator materials. Typical examples are the blast preheater for the iron blast furnace, which is heated with hot combustion gases, and the regenerator for the open-hearth steel-making furnace, where the hot gases from the furnace are used to preheat the combustion air.

The theoretical treatment of the three types of heat exchangers follows very much the same scheme, and will be illustrated for a simple tube recuperator shown in Fig. 6-9.

The cold flow is designated by the symbol 1 and the hot flow by symbol 2. The cold end of the recuperator is designated by A and the hot end by B. The masses flowing per unit time are designated by $\dot{w}$ and their specific heat by c. Thus the heat capacity per unit time is $\dot{w}c$.

Within the infinitesimal element with a wall area dA, the temperature difference $T_2 - T_1 = \Delta T$ and the heat transferred is $d\dot{q}$ per unit time. This gives rise to a temperature increase in the cold flow by dT_1 and a decrease in the hot flow by dT_2. We consider first the *heat balance*

$$\dot{w}_1 c_1 \, dT_1 + d\dot{q}_{loss} = \dot{w}_2 c_2 \, dT_2$$

If we disregard heat losses, the heat flow per unit time becomes

$$d\dot{q} = \dot{w}_1 c_1 \, dT_1 = \dot{w}_2 c_2 \, dT_2$$

This gives

$$dT_1 = \frac{d\dot{q}}{\dot{w}_1 c_1} \quad \text{and} \quad dT_2 = \frac{d\dot{q}}{\dot{w}_2 c_2}$$

Since $dT_2 - dT_1 = d\Delta T$ we get

$$d\dot{q} = \frac{d\Delta T}{1/(\dot{w}_2 c_2) - 1/(\dot{w}_1 c_1)} \tag{6-37}$$

Integrated from A to B we get, for the total heat transferred per unit time,

$$\dot{q} = \frac{\Delta T_B - \Delta T_A}{1/(\dot{w}_2 c_2) - 1/(\dot{w}_1 c_1)} \tag{6-38}$$

We see that if $\dot{w}_2 c_2 > \dot{w}_1 c_1$ then $\Delta T_B < \Delta T_A$ and vice versa, and that if $\dot{w}_2 c_2 = \dot{w}_1 c_1$, then $\Delta T_B = \Delta T_A$ and $dT_1 = dT_2$.

The value of ΔT is determined by the transferred heat $d\dot{q}$ and the *overall heat transfer coefficient* U, viz.,

$$d\dot{q} = U\Delta T\, dA$$

Here U is given by

$$U = \frac{1}{1/h_1 + s/k + 1/h_2}$$

where s is the thickness and k the heat conductivity of the separating wall, and h_1 and h_2 are the heat transfer coefficients on either side. Inserting this into Eq. (6-37) we get

$$\frac{d\Delta T}{\Delta T} = U\left(\frac{1}{\dot{w}_2 c_2} - \frac{1}{\dot{w}_1 c_1}\right) dA$$

Assuming U constant, this expression may be integrated over the entire recuperator

$$\ln \frac{\Delta T_B}{\Delta T_A} = U\left(\frac{1}{\dot{w}_2 c_2} - \frac{1}{\dot{w}_1 c_1}\right) A_{A-B}$$

where A_{A-B} is the total recuperator area. Finally, this expression may be combined with Eq. (6-38) to give the total heat transferred per unit time

$$\dot{q} = U \frac{\Delta T_A - \Delta T_B}{\ln(\Delta T_A/\Delta T_B)} A_{A-B} \tag{6-39}$$

Thus the transferred heat is the product of recuperator area, the logarithmic mean temperature difference, and the heat transfer coefficient. If $\Delta T_A = \Delta T_B = \Delta T$ the expression reduces to $\dot{q} = U\Delta T A_{A-B}$.

The *practical efficiency* of a heat exchanger is defined as the ratio between the transferred heat and the heat content of the incoming hot flow above $(T_1)_A$, and is therefore

$$E_p = \frac{\dot{w}_1 c_1 [(T_1)_B - (T_1)_A]}{\dot{w}_2 c_2 [(T_2)_B - (T_1)_A]}$$

If $\dot{w}_2 c_2 > \dot{w}_1 c_1$ this expression has its maximum value for $\Delta T_B = 0$. This value is called the *theoretical efficiency limit* of the heat exchanger and is equal to

$$E_t = \frac{\dot{w}_1 c_1}{\dot{w}_2 c_2}$$

We see that the theoretical efficiency limit reaches its maximum value of unity if the two flows have the same heat capacity per unit time. If $\dot{w}_2 c_2 < \dot{w}_1 c_1$ the theoretical efficiency limit will also be unity, corresponding to $(T_2)_A = (T_1)_A$, but the cold flow can not be heated to the same temperature as the incoming hot flow.

The *relative efficiency* is equal to the ratio between practical and theoretical efficiency, viz.,

$$E_r = \frac{(T_1)_B - (T_1)_A}{(T_2)_B - (T_1)_A} \quad \text{or} \quad \frac{(T_2)_B - (T_2)_A}{(T_2)_B - (T_1)_A}$$

Some examples of theoretical and practical efficiency are shown in Fig. 6-10(a–d).

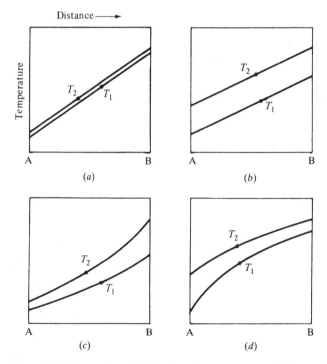

Figure 6-10 Temperature distribution in heat exchangers, (a) $\dot{w}_1 c_1 = \dot{w}_2 c_2$, $E_r \simeq 1$, (b) $\dot{w}_1 c_1 = \dot{w}_2 c_2$, $E_r < 1$, (c) $\dot{w}_1 c_1 > \dot{w}_2 c_2$, $E_r < 1$, (d) $\dot{w}_1 c_1 < \dot{w}_2 c_2$, $E_r < 1$.

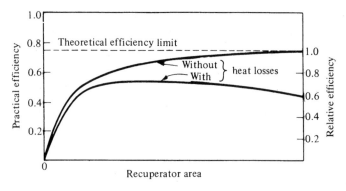

Figure 6-11 Practical and relative efficiency as function of recuperator area, with and without heat losses considered.

The above expressions may further be combined with Eq. (6-39) to give the recuperator area as a function of relative and practical efficiencies, viz.,

$$A_{A-B} = \frac{\dot{w}_2 c_2 E_t \ln\left[(1 - E_p)/(1 - E_r)\right]}{U(1 - E_t)} \qquad (6\text{-}40)$$

For $E_t = 1$, that is, $E_p = E_r$, the expression reduces to

$$A_{A-B} = \frac{\dot{w}_2 c_2 E_r}{U(1 - E_r)}$$

These relations are illustrated in Fig. 6-11 for a theoretical efficiency limit, E_t, of 0.75, and for the case of no heat loss as well as with heat losses considered. In the latter case the heat loss is assumed proportional to A_{A-B}. We see that without heat losses the relative efficiency approaches 100 percent asymptotically with increasing recuperator area. With heat losses the relative efficiency goes through a maximum. In practice, when also capital expenses have to be considered, the optimum recuperator area is below this maximum.

Other heat exchangers The transfer of heat from a gas to a solid burden in a shaft furnace is analogous to heat exchange in a recuperator, and the same equations may be applied. The recuperator area is here taken equal to the surface area of the burden. As shown in Sec. 5-5 this is related to the volume of the shaft by the relation: $A = 6V(1 - \varepsilon)/d_m$, where ε is the void fraction and d_m the mean lump diameter of the solid burden. If the surface resistance is large compared to the conduction resistance of the burden, the overall transfer coefficient, U, may be taken equal to the surface transfer coefficient, h. If the conduction resistance is significant, the overall transfer coefficient may approximately be taken equal to

$$U = \frac{1}{1/h + d_m/(6k)}$$

where d_m is the mean diameter and k the thermal conductivity of the solid burden. This equation implies that the heat *on the average* is conducted a distance of $d_m/6$, or that the entire heat capacity of the solid is regarded as concentrated in a shell with a diameter two-thirds of that of the solid lump. We see from the above relation that on decreasing lump size the conduction resistance of the solid plays a decreasing role.

In *regenerators* heat is transferred to refractory bricks which again give off heat to the cold flow. Also in this case we may assume the heat capacity of the bricks concentrated in a shell with a "diameter" two-thirds of that of the brick. For the total transfer of heat from hot to cold gas we then get

$$U = \frac{1}{1/h_1 + d/(3k) + 1/h_2}$$

where h_1 and h_2 are the transfer coefficients for the two flows. For a refractory brick we may take $d = 0.06$ m and $k = 1$ W/(m·K). Assuming further that the transfer coefficient to the brick is 60 and from the brick 25 W/(m²·K) we get

$$U = \frac{1}{0.017 + 0.02 + 0.04} = 13 \text{ W/(m}^2 \cdot \text{K)}$$

Of the total transfer resistance that of conduction amounted to 26 percent. In other cases between 15 and 30 percent have been reported.

The calculated value for U may be inserted into Eq. (6-39) for calculation of the transferred heat $\dot{q}$. Taken over the entire recuperator period of heating *and* cooling the average amount of heat transferred will be $\dot{q}/2$ W.

In calculating the heat transfer in a regenerator the heat capacity of the brickwork does not enter the picture, provided a steady state condition has been established. During the heating up of a cold regenerator the efficiency will be less, and heat will be stored in the brickwork. This heat is again given off when the regenerator is finally shut down.

Also for con-current and perpendicular heat exchangers similar equations may be derived. These types of heat exchangers are less efficient, however, and are of less interest to the extractive metallurgist.

Solvent Extraction

Some metal ions are soluble in aqueous as well as in organic liquids (Sec. 15-6). In *solvent extraction* the metal is transferred from the aqueous to the organic phase; in *stripping* the transfer is from the organic to the aqueous phase. The *equilibrium distribution* between the two phases is determined by the chemical composition of the two phases, and the temperature. If the molar concentration of the metal ion in the aqueous phase is denoted x and in the corresponding organic phase y, the equilibrium distribution for a given temperature may be given by curve I in Fig. 6-12. As we see this curve does not need to be linear.

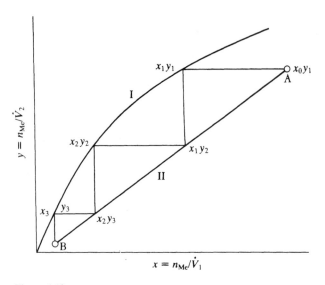

Figure 6-12 Equilibrium line, I, and operating line, II, for counter-current solvent extraction. Stepwise curve refers to mixer-settler, Fig. 6-13.

We shall now consider a reaction where the two phases flow counter-current to each other and will assume that there is no mixing within each phase in the direction of the flow (no back-mix), whereas each flow mixes completely perpendicular to the flow. Such a reactor is approximately realized by an extraction column where the aqueous phase sinks, and the lighter organic phase rises as droplets and overflows at the top. If the vertical distance in the column is denoted z we have the following *mass balance* per unit time:

$$\dot{V}_1 \, dx/dz = \dot{V}_2 \, dy/dz = dn_{Me}/dz$$

Here $\dot{V}_1$ and $\dot{V}_2$ are the volume flow rates of the aqueous and the organic phases respectively, dx and dy are the changes in concentration in the flows, and dn_{Me} is the amount of metal transferred across the liquid–liquid interface. If integrated over the entire length of the extraction column we get, for the total amount transferred per unit time,

$$n_{Me} = \dot{V}_1(x_A - x_B) = \dot{V}_2(y_A - y_B)$$

where A and B, respectively, represent the upper and lower ends of the column. Thus we have

$$\frac{x_A - x_B}{y_A - y_B} = \frac{\dot{V}_2}{\dot{V}_1}$$

Therefore, if corresponding values of x and y are known at one level in the column, together with the ratio of the volume flows, corresponding values for all other levels can be calculated. This relation is shown in Fig. 6-12 by curve II, which is called the *operating line*. The assumption was here made that the transferred amount of metal ion is sufficiently small to make each volume flow constant. If this were not the case, the operating line would have been slightly curved. If the two flows had moved *con-current* the operating line would have had a negative slope.

In Fig. 6-12 the operating line lies below the equilibrium line, corresponding to the metal ion being transferred from the aqueous to the organic phase. For *stripping* the operating line would be above the equilibrium line. Although the operating line describes the conditions along the height of the column, the concentrations x and y do not vary proportionally to the height, but are functions of the transfer rate.

The rate of transfer is given by the difference between the equilibrium line and the operation line and by the *mass transfer coefficient* at the interface. The latter may again be composed of diffusion transfer to and from the interface and chemical reaction at the interface. Without going into detail it should only be pointed out that the concentration in the two flows will change with column height in a way analogous to what was shown in Fig. 6-10 for the temperature change in a recuperator, and that the most efficient extraction is obtained for a system where $\dot{V}_1 x \simeq \dot{V}_2 y$.

Solvent extraction in a vertical column is not very common for large liquid flows. More common is a number of extraction steps carried out in apparatus called *mixer-settlers*. In these the two flows are first mixed thoroughly, whereupon the two phases are allowed to separate, and each phase is passed into the next step in counter-current. This is illustrated in Fig. 6-13 which shows three mixer-settlers arranged in counter-current. For the first extraction step, 1, the concentration of the entering aqueous phase and the effluent organic phase are given by point A on the operating line in Fig. 6-12 with the coordinates $x_0 y_1$. We shall now assume that equilibrium is established between the two phases in each step, corresponding to the composition of the *two effluent* flows lying on the equilibrium line. The concentration of the metal ion in the organic phase from extraction step 1 is, as already mentioned, given by the ordinate y_1. The corresponding *equilibrium* concentration in the effluent aqueous phase is given by the abscissa x_1 on the equilibrium line. This solution passes on to extraction step 2. The concentration in the organic phase coming from step 2 must, according to the mass balance, be given by the ordinate y_2 on the operating line.

Figure 6-13 Three mixer-settlers in counter-current (schematic).

The corresponding concentration in the aqueous phase from step 2 is given by the abscissa x_2. In the same way the concentration in the organic phase from step 3 is given by the ordinate y_3 and in the aqueous phase by x_3, and so on until the desired concentration x_n in the effluent aqueous phase at point B is obtained.

We see that for the case given in Fig. 6-12 a total of three extraction steps were needed. We see also that if a 100 percent extraction of the metal ion were desired, point B on the operating line would be at the origin, and an infinite number of extraction steps would have been necessary.

In practice complete equilibrium is not always reached in a given mixer-settler. From practical experiments one may, in such cases, construct an *effective equilibrium line* which describes the actual distribution ratios obtained, and the same stepwise procedure may be carried out to calculate the number of necessary steps.

Oxide Reduction in a Shaft Furnace

For simplicity we shall consider a metal which forms only one oxide, MeO, and with insignificant oxygen solubility in the metal phase. We shall consider the temperature constant, corresponding to a constant equilibrium constant for the reduction with a gas, which we shall take to be hydrogen. Pure metal oxide, e.g., as pellets, is introduced at the top of the shaft, and pure hydrogen in excess of what is needed to reduce all the oxide is introduced in the bottom. The equilibrium and operating lines for the process are shown in Fig. 6-14(a). In this figure $n_O/n_{H_2} = N_{H_2O}$ is chosen for the abscissa, and $n_O/n_{Me} = N_{MeO}$ is chosen for the ordinate, where N_{H_2O} and N_{MeO} give the mole fractions of H_2O in the gas and MeO in the solids, respectively. Furthermore, N_{H_2O} for equilibrium will be denoted a, and in the effluent top gas b. The slope of the operating line $n_{H_2}/n_{Me} = 1/b$ gives the number of moles of H_2 needed per mole of MeO.

We shall furthermore assume that the *rate of reduction* is proportional to N_{MeO} (i.e., first-order with respect to MeO), and proportional to $a - bN_{MeO}$ (i.e., to the distance from chemical equilibrium). This gives

$$-\frac{dN_{MeO}}{dt} = k(aN_{MeO} - bN_{MeO}^2)$$

where k is a rate constant. This rate expression is given in arbitrary units in Fig. 6-14(a), and we see that it has a maximum at some intermediate N_{MeO} value, and becomes zero for $N_{MeO} = 0$. The expression may be integrated and by inserting the limits $N_{MeO} = 1$ for $t = 0$ the time needed for a given degree of reduction is obtained

$$t = -\frac{k}{a} \ln \frac{(a-b)N_{MeO}}{a - bN_{MeO}}$$

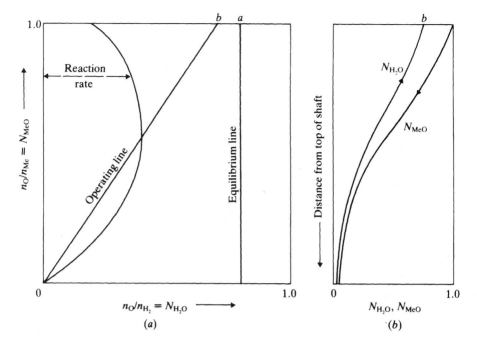

Figure 6-14 Counter-current reduction in shaft furnace, (a) equilibrium and operating lines and reaction rate, (b) variation of ore and gas composition with shaft height.

For an evenly descending charge in a shaft furnace the time is proportional to the distance from the top of the shaft, and in Fig. 6-14(b) N_{MeO} as well as N_{H_2O} are shown as functions of shaft height, the latter being given in arbitrary units. The actual size of the shaft will, in addition to the value for a and b, depend on the rate constant k and the feeding rate.

As we see from Fig. 6-14(b) N_{MeO} decreases with increasing distance from the top but, as a first-order reaction with respect to N_{MeO} was assumed, it will never be zero. As discussed in Sec. 6-6 oxide reduction does not necessarily follow a first-order rate equation, however. Furthermore, the temperature in a shaft furnace is usually not constant, but is a result of heat exchange between the gases and the solids, as well as of heats of reaction. A rigorous calculation of the degree of reaction and temperature as function of shaft height for given compositions and temperatures of the entering solids and gases is, therefore, a tedious task. Very often the necessary rate equations are not available, but they will be needed for future development and design of reduction processes.

The counter-current flow of solids and gases in the *iron blast furnace* will be further discussed in Chap. 9. The counter-current principle is used also in *continuous distillation*, such as the refluxing of zinc and purification of volatile metal halides. This will be discussed in Chap. 10.

PROBLEMS

6-1 A fluidized bed for roasting pyrite (FeS_2) has a volume corresponding to 5 tonnes of Fe. The feeding rate corresponds to 1 tonne of Fe per hour.

(a) Calculate the mean retention time, t_m.

(b) At time $t = 0$, five kilograms of Cu (as $CuFeS_2$) is introduced. Calculate the Cu/Fe ratio in the overflow immediately after this addition and after 10 h.

6-2 Air of 200°C and 1 atm (101.3 kPa) mean pressure is blown through a bed of spherical iron ore pellets of 1 cm diameter, the void fraction in the bed being $\varepsilon = 0.35$. The mass velocity of the air, based on an empty bed, is $G = 10^3$ kg/(m² · h), and the mean density and viscosity of the air are $\rho = 0.742$ kg/m³ and $\mu = 2.1 \times 10^{-5}$ kg/m · s) respectively.

(a) By means of the Ergun equation calculate the laminar, turbulent, and total pressure drop per meter of bed height, as well as the necessary fan power (in watts) per cubic meter of bed.

(b) Repeat the calculation for $P_{mean} = 2$ atm, assuming the same viscosity, and not including the fan power needed to compress the air from 1 to 2 atm.

6-3 A flat furnace lining consists of three courses of brick with the following thermal conductivities [all in W/(m · K)]:

Temperature °C	100	400	800	1200
1 High-alumina brick	2.9	2.3	2.0	1.7
2 Fireclay brick	1.0	1.2	1.4	1.6
3 Insulating brick	0.12	0.13	0.20	0.24

The thickness of each course is 25 cm. The inner temperature of the high-alumina brick is 1500°C, and the outer temperature of the insulating brick is 100°C. Calculate the heat flow in watts per square meter through the lining, and calculate the temperature at the interfaces between the courses. If necessary, use approximations and extrapolations of the data.

6-4 A flat furnace lining consists of one course of insulating brick, 25 cm thick, and with thermal conductivity as in Prob. 6-3, and a steel shell 0.5 cm thick and with $k = 60$ W/(m · K). The inner temperature is 1000°C, while the shell faces the surroundings which are at 20°C. The shell is vertical, more than 1 m high, and has an emissivity of 0.9. The area of the shell is small compared to that of the surroundings. Calculate the heat flow through the lining, as well as the part of this flow which is transferred by natural convection and by radiation respectively. Calculate the temperature drop across the steel shell.

Hint: Make a graph of $\dot{q}_{cond}$, $\dot{q}_{conv}$, and $\dot{q}_{rad}$ as functions of assumed outer temperatures of the steel shell.

6-5 A semi-infinite cylindrical laboratory tube furnace has a central heating tube, 5 cm diameter, with a temperature of 1000°C. The furnace shell is 20 cm diameter and at a temperature of 50°C. The annular space is filled with insulation with $k_m = 0.12$ W/(m · K).

(a) Calculate the heat flow through the lining in watts per meter of furnace length.

(b) In the above case the furnace shell was rusty sheet iron. What would happen to the heat flow and the shell temperature if it is replaced by bright aluminum?

6-6 A vertical steel pipe, 10 cm outer diameter, has a temperature of 120°C, whereas the surroundings are at 20°C. The emissivity of the surface is 0.5 and the area of the surface is small compared to that of the surroundings. Calculate the convective and radiant heat flow from the pipe, in watts per square meter as well as in watts per meter of pipe length.

6-7 Air at $P = 1$ atm and $t = 200$°C passes through a pipe of 10 cm internal diameter, with a linear velocity of 10 m/s.

(a) Taking its molecular weight equal to 28.8, calculate the density and mass velocity of the air.

(b) Applying Eq. (6-21) calculate the convective heat transfer coefficient if for air $k = 0.039$ W/(m · K) and $\mu = 2.6 \times 10^{-5}$ kg/(m · s), and calculate the amount of transferred heat in watts per meter of pipe length if the temperature of the pipe is 150°C.

6-8 A recuperator tube is made of stainless steel, 10 mm thick and with $k = 15$ W/(m · K). The heat transfer coefficient on the hot side is 60, on the cold side 25 W/(m² · K). Calculate the overall heat transfer coefficient, U, and calculate the heat flow per square meter if the temperature of the hot flow is 500, of the cold flow 300°C.

6-9 A semi-infinite rod of iron, initially at 20°C, is dipped at one end in a lead bath, which is kept at 500°C. Applying Fig. 5-4(a), calculate the temperature of the iron rod 1 m from the lead bath after one and after ten hours, when the heat conductivity for iron is 60 W/(m · K), the density 7800 kg/m³, and the specific heat 0.46 kJ/(kg · K). The heat transfer coefficient between the lead bath and the iron is very large, and the heat losses from the iron to the surroundings are very small.

6-10 Iron ore pellets (25°C) are heated in a shaft furnace (5 m high) in counter-current with hot combustion gases (1300°C). The pellet diameter is 15 mm and the feeding rate is 1000 kg/h. The heat capacity of the pellets is 0.90 kJ/(kg · K). The combustion gases are fed at a rate of 1500 kg/h and their heat capacity is 100 kJ/(kg · K).

(a) Show *qualitatively* in a sketch how you expect the temperature of the pellets and of the gas to vary along the height of the shaft.

(b) The feeding rate of the pellets is increased to 2000 kg/h. Discuss how this affects the temperature of the top gas, of the heated pellets, and the pressure drop in the shaft.

(c) The feeding rate of the pellets is 1000 kg/h and of the gas 750 kg/h. Discuss how this affects the temperature of the top gas, the heated pellets, and the pressure drop, all relative to case (a).

(d) The feeding rate of the pellets and gas are as for case (a), but the pellet diameter is 30 mm. Do the same discussion as for case (b) and (c).

(e) The conditions are as under case (a) but the shaft is 10 m high. Do the same discussion as above. In all cases heat losses are disregarded.

BIBLIOGRAPHY

Bird, R. B., W. E. Stewart, and E. N. Lightfoot: "Transport Phenomena," John Wiley and Sons Inc., New York, 1960.

Kitaev, B. I., Yu. G. Yaroshenko, and V. D. Suchkov: "Heat Exchange in Shaft Furnaces," Pergamon Press, London, 1967.

Kreith, F., and W. Z. Black: "Basic Heat Transfer," Harper and Row, Publishers, New York, 1980.

Kunii, D., and O. Levenspiel: "Fluidization Engineering," John Wiley and Sons Inc., New York, 1969.

Levenspiel, O.: "Chemical Reaction Engineering," 2d ed., John Wiley and Sons Inc., New York, 1972.

Szekely, J., and N. J. Themelis: "Rate Phenomena in Process Metallurgy," John Wiley and Sons Inc., New York, 1971.

Zenz, F. A., and D. F. Othmer: "Fluidization and Fluid-particle Systems," Reinhold Publ. Co., New York, 1960.

SEVEN

PHASE SEPARATION

Extractive metallurgy would be impossible without phase separation. This is illustrated by the following example. The reduction of iron oxide with carbon may in its simplest form be written: $Fe_2O_3 + 3C = 2Fe + 3CO$. If the carbon monoxide produced had dissolved completely in the iron, the composition of the product would have been identical to that of the reactants. Only because carbon monoxide is a gas, and because this gas can be separated physically from the iron, is it possible to produce iron by carbothermic reduction of the oxide.

Physical separation of different phases occurs in practically every unit process employed in the winning of the metals from their ores. Starting with the crude ore the first processes often involve ore-dressing. The crude ore contains a number of different solid phases, i.e., minerals. Some of these contain valuable elements, others are worthless rock, called gangue. In ore-dressing the crude ore is crushed and ground to the point where each mineral grain becomes essentially free. By means of various physical methods the mineral grains are separated in various fractions, some of which contain the valuable elements, whereas others are discarded as worthless. The principles of ore-dressing have found application also at later stages in metal production. One well-known example is the separation of the sulfides of copper and nickel by flotation of finely ground Cu–Ni matte (Sec. 12-4), another the recovery of copper from slag (Sec. 12-3).

Some of the principles employed in ore-dressing also find application in other phase-separating processes. Thus the principles of settling and filtration are utilized in hydrometallurgical processes. Settling is also an important step in the separation of liquid metals from slags. Finally, settling, filtration, and related processes are used in the removal of dust from metallurgical gases.

The present chapter will mainly be devoted to separation of solid phases from each other, which is the main object of ore-dressing, but will also cover processes for separating solids from fluids. The discussion will necessarily be brief, and the reader is referred to the bibliography for a more thorough treatment, particularly of the equipment used. Finally, methods for the removal of dust, etc., from gases will be discussed.

7-1 CRUSHING AND GRINDING

The ore from the mine may have lump sizes between 0.1 and 1 m, whereas the individual minerals may have grain sizes below 0.1 mm. The first part in any ore-dressing process will therefore involve the crushing and grinding, commonly called *comminution*, of the ore to a point where each mineral grain is free or practically free.

Crushing and grinding are usually carried out in a sequence of operations by which the lump size is reduced step by step. The first or primary crushing, down to about 100 mm, may be made either in *jaw crushers*, *gyratory crushers*, or *impact crushers*. For descriptions of these crushers the reader is referred to the bibliography.

Secondary and tertiary crushing from 100 mm to less than 10 mm lump size may take place in *cone crushers*, *roll crushers*, or *hammer mills*, which are all described in standard textbooks. In some cases the crushing in rolls and hammer mills is carried on to product sizes below 1 mm.

The final reduction in size usually takes place in *rod mills* or *ball mills* where the finely crushed ore is ground between rods or balls in a revolving drum. The latter may give a product which is less than 0.1 mm in diameter. These grinding operations may be done on dry material, but are usually carried out wet, i.e., the ore is mixed with enough water to give a thick pulp which is passed on to further ore-dressing processes. Extremely fine grinding to micron sizes is carried out by mutual attrition of the grains in a high-speed air or steam jet in machines called *jet mills* or *micronizers*.

A recent development in the comminution field is the *autogeneous mill* where coarse ore is reduced to fine product in one step by abrasion in a revolving mill drum. It is used extensively for iron ore grinding and is expanding into other fields.

Energy Requirements in Crushing

Crushing and grinding require large amounts of mechanical energy; the power cost may represent up to half the total processing costs, the remainder being for labor, supplies, and maintenance. Typical energy requirements for crushing and grinding are:

Coarse crushing	0.2–0.5 kWh/tonne ore
Fine crushing	0.5–2 kWh/tonne ore
Coarse grinding	1–10 kWh/tonne ore
Fine grinding	2–25 kWh/tonne ore
Micronizing	> 100 kWh/tonne ore

Assuming, for example, that a copper ore with 1 percent Cu needs a total of 20 kWh/tonne for crushing and grinding, this corresponds to 2000 kWh per tonne of copper, or about the same amount of energy as is used in all subsequent smelting and refining operations. With 0.5 % Cu in the ore, which is about the lower limit of what is mined economically today, the energy per tonne of copper for crushing and grinding would be twice as large. It is clear, therefore, that considerable attention has been devoted to the energy requirements and the theory of comminution.

From a thermodynamical viewpoint the main difference between a material before and after a crushing process is the increased surface area. Rittinger postulated in 1860 that the crushing work should be proportional to the surface area increase or equal to $\sigma \Delta A \propto \sigma(1/d_2 - 1/d_1)$, where σ is the specific surface energy, ΔA is the increase in surface area, and d_1 and d_2 are characteristic dimensions of the material before and after crushing.

Unfortunately our knowledge both of the surface area and of the surface energy of solids is rather limited. The surface area per unit mass of perfect spheres or cubes is $6/(d\rho)$, where d is the sphere diameter or the cube edge, and ρ is the density of the material. For actual ores the particles are neither spherical nor cubic, however, and the faces may be rough and even concave. It is usual to estimate the surface area of a crushed particle as 1.75 times the area of a cube with the same cross section. This does not include possible cracks inside the particle.

The surface energy of solids is not easily measured, and sometimes conflicting values have been reported. In general it seems that it is of the order of 0.1 to 1 J/m^2 for oxides and between 1 and 2 J/m^2 for metals. The theoretical energy required to grind a hard material like quartz to a fineness corresponding to a surface area increase of 10^5 m^2/m^3 should then be $\sigma \Delta A \simeq 10^5$ $J/m^3 \simeq 10^{-2}$ kWh/tonne. In actual grinding the energy consumed is in the order to 1 to 10 kWh/tonne, or between 100 and 1000 times as large. The reason for such a low efficiency is not clear. It has been demonstrated that parallel to the creation of new surface there are also cracks and defects inside the particles. These defects can only account for a minor part of the energy losses, however. Calorimetric investigations have shown that the major part of the expended energy is converted into sensible heat, probably created by the release of elastic stress which must precede the mechanical comminution. Whether this is a basic necessity, or whether it can be avoided by improved crushing technique, is not clear. The fact that widely different types of crushing equipment use roughly the same amount of energy for the same crushing operation may indicate that we are confronted with a fundamental principle which limits the "theoretical efficiency" of the process to some low value.

Even though there are order of magnitude differences between the actual grinding energy E and that which can be calculated from the increase in surface energy, the *Rittinger relation*, $E = C_R(1/d_2 - 1/d_1)$, seems to be obeyed, in particular for small particle sizes. For larger lumps, i.e., for crushing and coarse grinding, practical data agree more with the *Bond relation*, $E = C_B(d_2^{-1/2} - d_1^{-1/2})$. The size range where transition from the Rittinger to the Bond relation occurs depends on the nature of the ore. For comminution from large lumps to fine particles, i.e., for $d_1 \gg d_2$, the second term in the Rittinger or Bond relations becomes insignificant and can be omitted. Furthermore, the ground product will not all be of the same size, but may be described by a quantity k, the *size modulus*, which is a measure of the average particle size of the product.

For the total energy of such comminution the following relation has been suggested:[1] $E = Ak^{-m}$, where m is a characteristic exponent slightly less than unity. Thus for fine grinding of quartz, Brown et al.[2] found $m = 0.96$. For other materials values between 0.86 and 1.0 have been reported. The last value agrees essentially with the Rittinger relation. This is not surprising when one realizes that most of the energy is consumed in the fine grinding range.

The coefficient A is a measure of the resistance to grinding, that is, $1/A$ is the *grindability*. For grinding to extremely small particle sizes the energy consumption is found to exceed the value given by the above relation.

Closed and Open Circuits

The product from the crushing or grinding processes varies considerably in size. In order for the largest lump or particle to be below a certain size, there will be large amounts which are much smaller. In order to utilize efficiently the capacity of the equipment and avoid unnecessary overgrinding, some screening or classifying device may be operated together with the crusher or grinder. These devices separate out a fine fraction which goes on to the next step, whereas the coarser fraction either is returned to the previous step or is subjected to special treatment. Thus, after the primary crusher the product may be screened on, say, a 10 mm screen. The fine material goes directly to the ball mill, whereas the coarser material is subjected to secondary crushing. The product from the secondary crusher is returned to the same screen. In this way the crushing is used only to get the material through the screen without excessive overcrushing. Similarly the product from the ball-mill may be subjected to *classification*, a process which will be discussed in Sec. 7-3; the finer fraction passes on to further treatment, whereas the coarser fraction is returned to the ball mill. In both cases we say that the secondary crusher or the ball mill works in *closed circuit* with the screen or with the classifier. Conversely, a crusher or mill which works without such return is said to work in *open circuit*.

It may be that, after a certain amount of intermediate crushing or grinding, some of the ore or waste minerals have been essentially liberated. The product may then be subjected to a *primary concentration* process where these grains are separated out, whereas the remaining material is subjected to further grinding. In this way no energy is wasted to grind those grains which are already free.

Even after grinding to the point where each particle is smaller than the smallest mineral grain, complete liberation is not always obtained. Since the

[1] R. Schuhmann, Jr.: *Trans. AIME*, **217**: 22 (1960); *see also* G. E. Agar and J. H. Brown: *Can. Min. Metall. Bull.*, **57**: 147 (1964).

[2] J. H. Brown, S. R. Mitchell, and M. Weissmann: *Trans. AIME*, **217**: 203 (1960).

fractures do not necessarily occur along the mineral grain boundaries, some grains may contain fragments of several minerals. As a result, ore-dressing will never be perfect, there will always be a certain loss of valuable minerals in the discarded gangue, and a certain amount of gangue minerals in the ore concentrate. Often the process is arranged so that the nonliberated grains are collected in a *middling product* which is reground and reseparated.

The minerals in an ore often show large differences in grinding resistance and the softer minerals can to a large degree accumulate in the fine part of the ground product. This feature is often utilized for removal of constitutents, such as clay, sand, or slag.

7-2 SCREENING

Screening is mechanical separation of particles according to size. It is used industrially mainly in conjunction with ore crushing and grinding equipment, as described in Sec. 7-1. It is used also in cases where product specifications include rigid size limits as, for example, in the production and marketing of powders.

For the larger lump sizes coarse *grizzle bars* or *punched plate* screen may be used. For finer material, screens are usually made of woven metal wire. Various types of equipment are used. Plane screens with shaking or vibrating motion and revolving trommel screens are the most usual types. Also, screening may be either wet or dry.

It is easily seen that the capacity of screening decreases rapidly with decreasing screen opening. Take for example two screens where the linear dimensions of the openings, and therefore the linear dimensions of the passing particles, differ by a factor of d_1/d_2. The number of openings per unit area of screen will then differ by a factor of $(d_2/d_1)^2$, and the volume of each particle that just passes through the screen by a factor of $(d_1/d_2)^3$. If per unit time the same *number of particles* passes through the coarse and the fine screen openings, it is seen that the volume (and weight) of screened particles per unit screen area and time will differ by a factor of d_1/d_2. For this reason screening on an industrial scale is limited to relatively large particle sizes, down to 0.1 mm, occasionally less. For finer particle sizes industrial separation is made mainly by *classification*. In the size region 0.1 to 1 mm both methods are in use.

Screening is used also as a laboratory measuring technique to determine the relative amounts of various particle sizes in a given sample. For this purpose standardized screening scales have been developed. Most common is the American *Tyler screen scale*, where the number of meshes per linear inch (abbreviation: mesh), and the linear dimensions of the opening of successive standard screens differ by a constant factor of $\sqrt{2} = 1.41$, as shown in Table 7-1. For smaller particle sizes the dimensions are usually given in micrometers ($1 \ \mu m = 10^{-3}$ mm). Thus, 200 mesh corresponds to 74 μm, etc. Since screen

Table 7-1 Standard Tyler screen scale

Mesh number	Mesh opening		Wire diameter	
	in	mm	in	mm
—	1.050	26.67	0.148	0.375
—	0.742	18.85	0.135	0.343
—	0.525	13.33	0.105	0.266
—	0.371	9.423	0.092	0.233
3	0.263	6.680	0.070	0.178
4	0.185	4.699	0.065	0.165
6	0.131	3.327	0.036	0.091
8	0.093	2.362	0.032	0.081
10	0.065	1.651	0.035	0.089
14	0.046	1.168	0.025	0.063
20	0.0328	0.833	0.0172	0.0437
28	0.0232	0.589	0.0125	0.0317
35	0.0164	0.417	0.0122	0.310
48	0.0116	0.295	0.0092	0.0233
65	0.0082	0.208	0.0072	0.0183
100	0.0058	0.147	0.0042	0.0107
150	0.0041	0.104	0.0026	0.0066
200	0.0029	0.074	0.0021	0.0053
270	0.0021	0.052	0.0016	0.0041
400	0.0015	0.037	0.0010	0.0025

Note: The ASTM and the BS screen scales also use $\sqrt{2}$ as screen ratio with the 74 μm (200 mesh) screen as base, but some of the screen numbers are different. In France and Germany a screen scale with base 1 mm and screen factor $10^{1/10} = 1.26$ is standardized. The compromise ISO scale uses base 1 mm and factor $10^{3/20} = 1.41$.

standards differ somewhat from country to country, it is recommended for all sizes to give the linear dimension of the opening, and not the mesh number.

Although most laboratory screening usually stops at 52 μm, screens with openings down to 10 μm exist, but they are fragile and have low capacity. For laboratory sizing in the -74 μm range sedimentation and air-classification are often used.

The result of a screening analysis is given as the fraction of the sample which passes through one screen, but which is stopped by the subsequent one. Thus we can say that a certain percentage is $+26.67$ mm, i.e., coarser than the coarsest screen, another percentage is $-20 + 28$ mesh, i.e., it has dimensions between 0.833 and 0.589 mm, and finally that a certain percentage is -200 mesh, i.e., finer than 0.074 mm.

Screening can never be exact. This follows from the fact that the particles usually have odd shapes and their chances of passing through the screen depend on how they hit. Also these chances depend on the screening time. Therefore, standardized procedures are developed in order to give comparable results.

The result of the screening analysis is often given on a *cumulative* basis, i.e., for each screen the total percentage of material which either passes or is stopped by that screen is given. The result of the screen analysis is frequently plotted as a curve of cumulative percentage passing against the logarithm of the particle dimension. The standard screens will then appear with even intervals along the x axis. From such a curve the experienced observer can easily see the relative amounts of the different particle sizes in the material.

7-3 CLASSIFICATION

By classification we understand separation of particles according to their settling velocity in water, in air, or in other fluids. We remember from Sec. 6-4 that the terminal settling velocity, v, of a particle in the laminar range as derived from Stokes' law will be

$$v = \frac{d_m^2(\rho_s - \rho)g}{18\mu} \qquad (6\text{-}11)$$

In the turbulent region from Newton's or Rittinger's law

$$v \simeq \left[\frac{3d_m(\rho_s - \rho)g}{\rho}\right]^{1/2} \qquad (6\text{-}12)$$

Here ρ_s and d_m denote the density and mean diameter of the particle, and ρ and μ are the density and viscosity of the fluid. For settling of rock minerals in water the transition from laminar to turbulent settling occurs around 1 mm particle size, i.e., most ball-mill products will settle in the laminar range. The above relations show that the settling velocity depends not only on the diameter but also on the density of the particle as well as on the density and viscosity of the fluid.

These relations apply for *free-settling* conditions where the particles are sufficiently far apart not to affect each other, and the density of the fluid is that of the pure fluid, e.g., water. Assuming that the material contains two minerals of density ρ_1 and ρ_2 respectively, then there will be two diameters d_1 and d_2 for which the settling velocity is the same. Inserted into Stokes' law this gives, for the laminar range,

$$\frac{d_1}{d_2} = \left[\frac{\rho_2 - \rho}{\rho_1 - \rho}\right]^{1/2} \qquad (7\text{-}1)$$

This is called the *free-settling ratio* for the two minerals under laminar conditions. Similarly we get the free-settling ratio under turbulent conditions

$$\frac{d_1}{d_2} = \frac{\rho_2 - \rho}{\rho_1 - \rho} \qquad (7\text{-}2)$$

Assuming that the two minerals were quartz with $\rho_1 = 2.65$ and galena with

$\rho_2 = 7.5$ we get in the laminar region $d_1/d_2 = 1.985$ and in the turbulent region $d_1/d_2 = 3.94$.

In *hindered settling* the pulp is so thick that the particles interfere with each other. The relations between grain sizes, velocity, density, and viscosity are very complex in this case. When we only consider the behavior of the coarser particles in the pulp a practical approximation is to use the density of the pulp rather than that of the fluid in the velocity expressions. Thus, if we have quartz and galena in a pulp of density $\rho' = 1.5$, we get in the laminar range

$$\frac{d_1}{d_2} = \left[\frac{7.5 - 1.5}{2.65 - 1.5}\right]^{1/2} = 2.28$$

and in the turbulent range

$$\frac{d_1}{d_2} = \frac{7.5 - 1.5}{2.65 - 1.5} = 5.22$$

We see that these *hindered settling ratios* are larger than the corresponding free-settling ratios. The purpose of classification is primarily to separate the particles according to size and not according to density. We see that this requires settling ratios close to unity, i.e., it is more closely obtained in the laminar than in the turbulent range, and more closely in free settling than in hindered settling.

Industrial classification may be carried out in different types of classifiers. Basically they all operate according to the principle that the particles are suspended in water which has a movement relative to the particles. Particles below a certain size and density are carried away with the water flow, whereas the coarser and heavier particles will settle.

If a classifier works in closed circuit with a ball mill the settled part is continuously returned to the mill for further grinding, whereas the overflow passes for further treatment. It follows from the above discussion that after a steady state is established in a closed circuit operation, the heavy minerals in the overflow will have a smaller particle size than the lighter ones, but the overall composition of the solids in the overflow is still the same as in the crude ore.

Classification may also be carried out in air. This requires dry material, and is used, for example, for cement. Also it can be seen from Eqs. (6-11) and (6-12) that the settling velocities increase by increased gravitational force. This may be simulated by the use of centrifuges as will be discussed later in connection with cyclones.

7-4 CONCENTRATION

Concentration has as its purpose to separate the ground ore into different products, some of which may be high-grade concentrates which pass on to metallurgical treatment, some may be worthless tailing which goes to the dump, and some be middlings which may need further grinding and repeated concentration treatment. The most important concentration processes are:

1. Sorting
2. Gravity separation
3. Flotation
4. Magnetic and electrostatic separation

Sorting

Sorting is the oldest ore-dressing method. Some sorting was always done at the mine, where barren rock and submarginal ore were discarded. More important was *manual sorting* that was done on the ore after primary or secondary crushing. The crushed ore was conveyed on a moving belt, and submarginal lumps were picked out by hand and discarded. With increasing labor cost and decreased cost of subsequent ore-dressing operations manual sorting has become almost obsolete. In recent years, however, *mechanized sorting* has become an interesting separation method. As in manual sorting the material is spread out on a moving belt. As the belt moves along it is scanned by some detecting device, which may be a photocell or a sensor for magnetic, electric, or even radioactive properties. The sensor works together with a computer which records the position on the belt of those lumps which either meet or fail to meet the desired property. A signal is passed on to a device placed at the end of the belt, and the desired or undesired lumps are removed mechanically. Mechanized sorting is usually applied on lump sizes above 10 mm, but may occasionally be used on finer material. One example is in canneries for peas and beans, where the discolored ones automatically are removed from the flow.

Gravity Separation

Gravity separation separates the minerals according to their different densities, and is the second oldest concentration technique. It is used particularly for the concentration of very heavy or very light minerals within a wide range of grain sizes. A typical example is the panning of river sand for gold, which was used by the early prospectors. Both the efficiency and capacity falls off rapidly when the grain size decreases below 0.1 mm, however, owing to the large increase in drag effects in relation to inertia.

Gravity separation is often carried out in a thick pulp, thus making use of the hindered settling ratio, which is most sensitive to differences in particle density. If, in the ground ore, the heavy minerals also are the largest particles, a fairly good separation can be obtained directly by classification. This is illustrated by certain iron ores where the siliceous or clayey impurities are segregated into the finer sizes of the crushed ore, and may be washed away.

Another extreme case of gravity separation is *heavy media* separation where the density of the pulp is intermediate between that of the valuable minerals and the gangue. In that case the light minerals will float on top and the heavy minerals sink to the bottom of the pulp, independent of particle size. Most pulps will not satisfy this condition, however, but such a pulp may be obtained by the addition of a finely ground heavy component. This gives the so-called *sink-and-float* method. Ground ferrosilicon or galena may be used as heavy component, the particle size of which should be significantly smaller than that of the ore. After the valuable minerals and the gangue have been separated, the ferrosilicon or galena is recovered for reuse by magnetic separation or flotation respectively. Another example is in upgrading of coal. Here a slurry of fine-grained magnetite is used; the light coal floats on top and the heavier gangue sinks.

It should be mentioned that in the laboratory minerals may be separated according to their different densities by use of *heavy organic liquids*, like tetra-brom-ethane (sp. gr. 2.97). This procedure is usually too expensive for industrial ore-dressing, however.

Other methods of gravity separation utilize a combination of gravitational, inertial, frictional, and viscous effects. Commonly used separators are sluices, spirals, washing tables, and jigs, and between them they can handle a wide range of grain size. A description of these machines is outside the scope of this text and the reader is referred to standard ore-dressing textbooks. For dry gravimetric separations so-called air jigs and tables have been developed. They are generally less efficient than their wet counterparts.

Typical applications for gravimetric separators are concentrating coal from shale and separating heavy oxides like hematite, ilmenite, rutile, cassiterite, etc., from light rock minerals. Before the flotation process was developed gravimetric separation was also extensively used for recovering metal sulfides. Jigging and heavy-media separation are often used as a preconcentration stage for rather coarse ore to reduce the amount of material going on to the more expensive grinding and flotation processes.

Flotation

Separation by flotation is based on the ability or lack of ability of different surfaces to be wetted by water. The wettability of a surface is determined by the relative values of the interfacial energies: solid–water, solid–air, and water–air. If the first of these is large compared to the other two, water will not cling to the solid, and the solid is said to be *hydrophobic* (fear of water). Conversely if the solid–air interface energy is the larger, the surface is said to be *hydrophilic* (love for water) and is easily wetted. In flotation processes air is blown through the aqueous pulp of ground ore. Those minerals which are hydrophobic will cling to the air bubbles and rise with them, whereas the hydrophilic minerals will sink. In order to carry out flotation successfully various additions have to be made to the pulp, viz.:

1 Frothers In pure water the air bubbles will burst when they reach the surface, and any particles adhering to them will sink back. By the addition of certain surface-active organic compounds, commonly called frothers, a stable froth is formed, which may be skimmed off together with the floated particles. Pine oil (a turpentine fraction) and higher alcohols are typical flotation frothers.

2 Collectors Most ore minerals are hydrophilic; possible exceptions are graphite, sulfur, and molybdenite. By the addition of so-called collectors which are adsorbed to the mineral surface they may be made hydrophobic. Collectors are organic compounds, and different types of collectors are active for different types of minerals, e.g., xanthates for metal sulfides, fatty acids for metal oxides, hydroxides and carbonates, amines for silicates, etc.

3 Activators These are compounds which make the mineral surface more responsive to the collectors. Thus clean zinc sulfide (sphalerite) floats rather poorly with xanthate collectors. By the addition of small amounts of copper sulfate to the pulp, however, it is easily floated. It is believed that copper ions in the solution react with the zinc sulfide to give a mono-molecular coating of copper sulfide. The latter is readily made floatable by addition of xanthates.

4 Depressors These act in the opposite manner to the activators and counteract the collectors. Typical examples are chromate, cyanide, and hydrosulfide ions, which depress various metal sulfides.

5 Conditioners These are chemicals added to adjust the pH of the solution. The commonest one is lime which is used to make the solution slightly alkaline.

The amounts of additives are usually small. Thus for sulfide flotation the necessary amount of xanthate may be around 100 g/tonne of ore, increasing with decreasing particle size. Flotation found originally its greatest application in the concentration of sulfide minerals, but with recent improvements it can be used for practically any minerals, oxides as well as silicates and halides, sulfur, coal, etc. Thus it has found important application in the removal of silicates from iron ores, and is definitely the most versatile of all existing separating methods. It is applicable to the finest particle sizes which are not easily concentrated by gravity methods. Furthermore, by the use of combinations of different additions the flotation process can be made highly selective. Thus a complex sulfide ore may be floated either to give a bulk concentrate, which is a mixture of the various sulfide minerals, or it can be treated to give different flotation concentrates each containing mainly one mineral, as galena (PbS), chalcopyrite ($CuFeS_2$), sphalerite (ZnS), and pyrite (FeS_2). This is called *selective flotation*.

A flow sheet for selective flotation of a Cu–Pb–Zn–pyrite ore is shown in Fig. 7-1. Lime is added to the feed in the ball mill to raise the pH to 11.5,

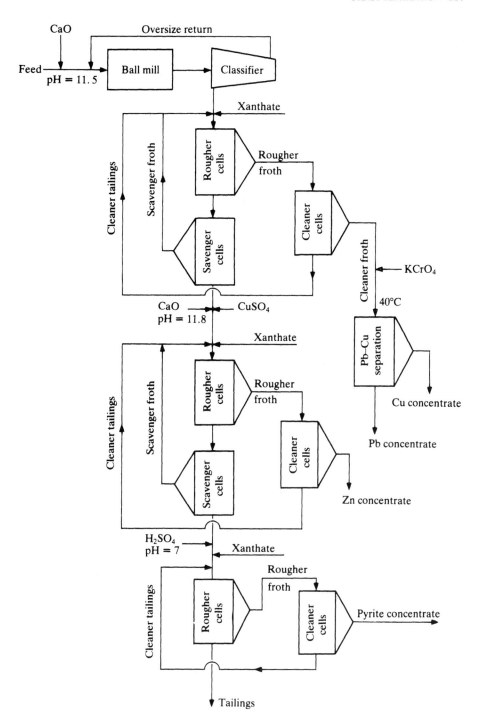

Figure 7-1 Flow sheet for selective flotation of a complex sulfide ore.

which is necessary to selectively float the copper and lead sulfides with xanthate. The flotation is carried out in a system of three banks of flotation cells. The concentrate (froth) from the *rougher* cells still contains a fair amount of unwanted minerals, which are separated out in the *cleaner* cells. On the other hand, the discharge (tails) from the rougher cells contains some of the desired copper and lead sulfides, which are recovered in the *scavenger* cells. The tails from the cleaner cells and the froth from the scavenger cells are *middlings* and are returned to the rougher cells, if necessary after further grinding. This represents an application of the *counter-current* principle discussed in Chap. 6.

The copper–lead concentrate from the cleaner cell is passed on to the *lead–copper separator* where the lead minerals are depressed by addition of $KCrO_4$ at about 40°C, whereas the copper minerals are floated off.

Copper sulfate is added to the tails from the scavenger cells, and the pH is raised further to 11.8, which is necessary to prevent pyrite from being floated together with the zinc sulfide. The zinc sulfide is now taken out in a second system of rougher, cleaner, and scavenger cells, and the tails are passed on for pyrite recovery. Here sulfuric acid is added to bring the pH down to about 7, which is necessary to float the pyrite and separate it from the gangue minerals.

Magnetic Separation

This method is based on the difference in magnetic properties of the minerals. A few minerals, such as metallic iron, magnetite, and some forms of pyrrhotite, are *ferromagnetic* and are attracted strongly by a magnet. A number of other minerals are weakly attracted, and are called *paramagnetic*. In this group we find ilmenite, hematite and other iron minerals, garnet, etc. Finally some minerals such as quartz and calcite are practically nonmagnetic or may even be *diamagnetic*, i.e., they are repelled by a magnet.

Magnetic separation has found its greatest application for the ferromagnetic minerals, in particular for magnetite. It is used dry mainly for coarse material from 150 mm down to 6 mm, whereas separation of finer materials, all down to micron size, is usually carried out wet, i.e., from an aqueous pulp. Low-intensity permanent magnets may be used. Magnetic concentration of magnetite is one of the cheapest ore-dressing methods, and it has made possible the economic exploitation of magnetite ores with less than 30 percent Fe, from which concentrates with up to 70 percent Fe may be produced. Low-intensity magnetic separation cannot be used on hematite ores, however, and during the years considerable effort has been made to convert such ores into magnetite by *magnetizing reduction* (see Sec. 9-2). Unfortunately magnetizing reduction is rather expensive, which makes the method less attractive.

For hematite ores *high-intensity* magnetic separation appears to be a more promising method. Here high-intensity electromagnets are used, which attract paramagnetic minerals. Another example of high-intensity separation is the removal of small amounts of iron silicates from quartz and feldspar for the porcelain industry. The separation was originally made on dry material, but wet separators are now common. The method may be used for particle sizes down to 10 μm. Because of heat produced in the electromagnets the energy consumption is relatively large.

Electrostatic Separation

Electrostatic separation is based on differences in electrical conductivity of the minerals. The method has to be applied to a dry feed. This is fed, in a thin layer, on to the top of a slowly revolving metal roll. The roll may either be highly electrically charged, or it may be grounded, and a highly charged electrode is placed on the outside of the feeding flow. In both cases an electrical charge, which is opposite to that of the roll, will be induced in the particle, and the particles will adhere to the roll. Conducting particles lose this charge rapidly and acquire the charge of the roll, whereupon they are repelled and collected in a suitable container. Nonconducting particles will retain their charge and will be carried around with the roll until they are removed by a brush and fall into a second container.

Electrostatic separation requires a dry feed within a narrow size range, say 1 to 0.1 mm. Also since the layer of particles on the roll must be quite thin, its capacity is small, and the method seems to have limited applicability. One common use is for concentrating rutile and ilmenite from beach sands.

7-5 SEPARATION OF SOLIDS FROM LIQUIDS

After flotation and other wet ore-dressing processes the concentrate may occur as a pulp of high water content. Also in many hydrometallurgical processes (Chap. 15) the valuable element may be obtained as a precipitate from an aqueous solution. In order to obtain the valuable element in a concentrated form, it is necessary to remove the water. This may be done by *thickening* and *filtration*. These processes are also used to prepare pure solutions, e.g., for electrolysis as well for the purification of water and other liquids in general.

The principle of *thickening* may be studied on an aqueous suspension of solids in a measuring cylinder, see Fig. 7-2. After some time we observe four different layers, listed from above: A, practically clear water, B, a diluted suspension, C, a more concentrated suspension, and D, an almost solid cake. We distinguish between the following thickening mechanisms:

1. Free settling in zone B
2. Hindered settling in zone C
3. Exudation of water in zone D

The first two mechanisms have been discussed in connection with classification. The only difference between classification and thickening is that in the latter process all the solids are allowed to settle. As the particle size decreases

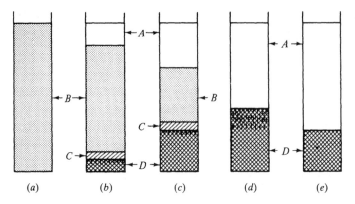

Figure 7-2 Batch sedimentation. (*From W. L. McCabe and J. C. Smith: "Unit Operations of Chemical Engineering." Copyright 1956. Used with permission of McGraw-Hill Book Co., New York.*) The sequence from (*a*) to (*e*) represents increasing sedimentation time.

the settling velocity decreases and would approach zero for the smallest particles. For such particles the velocity may be increased by *flocculation*, i.e., the individual particles are made to adhere to each other to form flocs of larger diameter. Flocculation is promoted by the addition of chemicals such as lime, alum, starch, and certain soluble polymers, whereas other chemicals such as alkali silicates, sulfides, and tannin are known to counteract flocculation.

In the hindered settling region the flocs settle to form a loose bed. As the height and weight of this bed increases, water, which had been present within the flocs, is excluded and oozes upward through the bed. After all the solids have settled by free and hindered settling further compression of the bed takes place by water exudation, and the volume and water content of the bed decreases.

Industrial thickening is usually carried out on a continuous scale. Most common in the metallurgical industry is the *Dorr thickener*. This is a wide, shallow cylindrical tank, equipped with a central shaft and rakes. The feed suspension, to which flocculants may be added, is introduced around the central shaft, whereas the clean overflow is taken out along the periphery. As the liquid moves slowly across the tank, the solid particles settle to the bottom. The rakes serve a double purpose: The slow action of the rakes promotes the exudation of water from the settled pulp, and the rakes move the settled pulp toward the center of the tank where it is withdrawn as a thickened underflow. Dorr thickeners may be made to work in series, and with over- and underflow in countercurrent.

Filtration may be applied either to the diluted suspension or to a thickened pulp. Cloths of various kinds are used as filter media, and suction or pressure is usually applied to force the liquid through the filter. Filtering differs from screening in one important respect. In screening particles smaller than the screen opening will go through. In filtration the particles form a bridge across the pores in the filter cloth. In that way a filter cake is built up with pore diameter much smaller than that of the filter cloth. Thus the filter is able to withhold particles which are much smaller than the pores in the filter cloth.

The capacity of a filter is determined by the rate at which the liquid can flow through the filter and the filter cake. The flow through the filter cake, being mainly laminar, may approximately be given by

$$J = k \cdot \Delta P \cdot A/(L \cdot \mu) \tag{7-3}$$

Here ΔP is the pressure difference across the filter, A the area and L the thickness of the filter cake, and μ is the viscosity of the liquid. The *permeability*, k, of the filter cake is determined mainly by the particle size and the void fraction in the cake. Of these the void fraction, and consequently the permeability, again decreases with increased pressure and increased filter-cake thickness. The permeability, and its dependence on pressure and cake thickness, may be measured on a small scale in the laboratory, and the result applied to calculate the capacity of an industrial filter.

Industrial filters usually work continuously. One example is the *rotary-drum filter*, shown in Fig. 7-3. This has the form of a cylindrical drum, the cylindrical surface being the filter cloth. The interior of the drum is divided radially in several sector chambers. These chambers are connected through a valve system to a suction and air pressure system. The axis of the drum is horizontal and the drum is partly immersed into the pulp. Suction is applied to those chambers which correspond to the immersed part, making a filter cake deposit on the filter cloth. As the drum rotates, new surface is continuously immersed in the

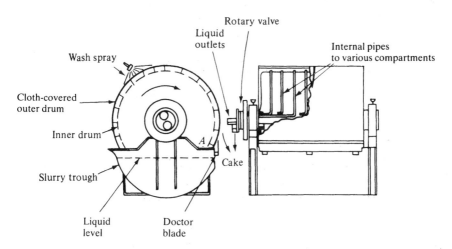

Figure 7-3 Continuous vacuum rotary-drum filter. (*From W. L. McCabe and J. C. Smith; "Unit Operations of Chemical Engineering." Copyright 1956. Used with permission of McGraw-Hill Book Co., New York.*)

pulp and the corresponding chamber is connected to the suction system. As the surface emerges from the pulp it is first washed with clean water and is then sucked relatively dry. The corresponding chamber is then shifted over to a slight overpressure, and before the filter drum has completed one rotation, the washed filter cake is released from the cloth and falls into a bin.

Other filters operate more or less according to the same principle. If the pulp is difficult to filter or if its volume is large, an overpressure of several atmospheres may be applied to the filter. Such *filter presses* have a high capacity, and they give a low water content in the filter cake. With recent developments they can be made almost continuous, and are gaining increasing importance.

After filtration the cake still contains considerable amounts of water, usually in the range 10 to 40 percent by volume (for filter presses about half that much). In some cases this wet product may be charged directly to metallurgical smelting furnaces. In other cases the product may have to be dried by evaporation of the water, as will be discussed in Sec. 8-2.

7-6 GAS CLEANING

The purpose of gas cleaning in the metallurgical industry is twofold: (1) to prevent air pollution, and (2) to recover valuable elements. The impurities in the gas may be solids, liquids, or gases. Gaseous impurities, for example SO_2, are taken out by absorption in suitable solvents, as will be discussed briefly in Sec. 8-3. Solid and liquid impurities are usually divided into *dust*, *smoke* (fume), and *fog*. By dust we understand solid particles, e.g., from a blast furnace burden, which is carried along with the gas. Smoke is solid particles and fog is liquid droplets formed by chemical reactions or by cooling of the gas. Typical examples of smoke are ZnO, formed by oxidation of zinc vapor, Fe_2O_3, formed by oxidation of iron vapor in the oxygen blown converter, and SiO_2, formed in the ferrosilicon process. Fog may be droplets of sulfuric acid formed by the cooling of roast gases. In Fig. 7-4 the size ranges for various particles in industrial gases, as well as the ranges covered by various gas cleaning devices, are shown. We see that the particles cover a large size spectrum, and it is easily understood that the difficulties of gas cleaning increase with decreasing particle size.

Gas cleaning equipment may be divided into the following groups:

1. Settling chambers
2. Cyclones
3. Filters
4. Scrubbers
5. Electrostatic precipitators (electrofilters)

Settling is analogous to the settling of solids in liquids, and follows Stokes' law. Since the settling velocity is proportional to the particle diameter it is easily seen that settling can be applied only on relatively coarse particles, larger than

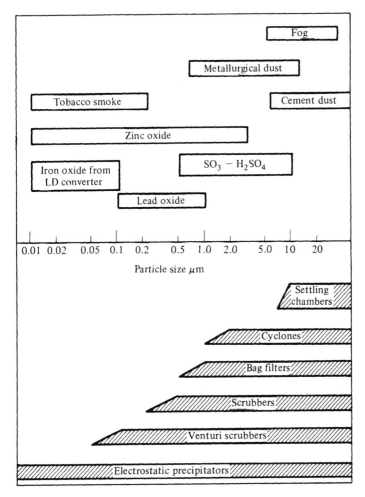

Figure 7-4 Size ranges for various types of dust, smoke, and fog (above), as well as working ranges for various types of gas cleaning equipment (below).

10 μm. Settling chambers may have the shape of a simple rectangular box (Fig. 7-5(a)), or they are constructed in such a way that the direction of the gas flow is altered (Fig. 7-5(b)). In the latter case the settling is promoted by the *inertia* of the particles.

Cyclones (Fig. 7-6) are also based on Stokes' law, but with the modification that, instead of the gravitational acceleration, acceleration due to centrifugal forces applies. This is equal to $v^2/r = \omega^2 r$, where v is the linear tangential velocity and ω the angular velocity (rad/s) of the particle, and r its radius of rotation, the latter being a function of the radius of the cyclone. In a cyclone

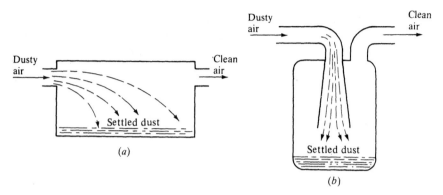

Figure 7-5 Two types of settling chambers.

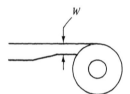

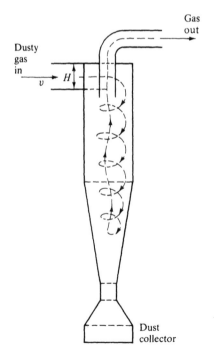

Figure 7-6 Cyclone.

the centrifugal acceleration may be many hundred times that of gravity, with a corresponding increase in settling velocity and efficiency. For a given gas flow the linear velocity increases with decreasing radius of rotation, thus making the acceleration inversely proportional to the radius raised to some power between 2 and 2.4. Thus a small cyclone is more efficient than a large one. On the other hand the pressure drop over the cyclone, and consequently the necessary fan power, increases with decreasing cyclone radius. For efficient cleaning of large gas quantities it has been found better to use a large number of small cyclones in parallel (multiclones) than one large cyclone.

The separation of dust in a cyclone is never perfect. In order to describe the efficiency of a cyclone, one uses the *critical particle diameter* d_c, corresponding to the particle size of which 50 percent is retained by the cyclone. This is again related to the dimensions of the cyclone and of the gas flow rate. For a typical cyclone, like the one shown in Fig. 7-6, the following empirical expression has been developed:[1]

$$d_c = \left[\frac{9\mu W}{10\pi v(\rho_s - \rho)} \right]^{1/2} \qquad (7\text{-}4)$$

Here μ is the viscosity and v the entering velocity of the gas, W is the width of the gas inlet, and should be about one-fifth of the cyclone diameter, ρ_s and ρ are the densities of the solid and the gas respectively.

When the critical particle diameter has been established the degree of retention for other particle sizes is more or less given. This is shown in Fig. 7-7,

[1] C. E. Lapple in J. H. Perry (ed.): "Chemical Engineers' Handbook," 4th ed., McGraw-Hill Book Co. Inc., New York, 1963, pp. 20–71.

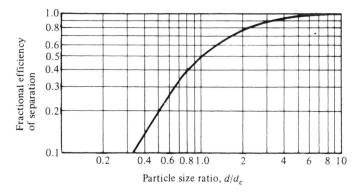

Figure 7-7 Efficiency vs. particle size for cyclones. (*From W. L. McCabe and J. C. Smith: "Unit Operations for Chemical Engineering." Copyright 1956. Used with permission of McGraw-Hill Book Co., New York.*)

which gives the fractional retention (efficiency) as function of the ratio between actual and critical particle diameter. We see that between the particles that are almost completely retained and those which are barely caught there is a size ratio of about 100.

Cyclones have the advantage of being robust, without moving parts, and they may be used for gases at high temperatures. On the other hand the power requirements are relatively large, and they are limited to particles larger than 1 μm.

Closely related to cyclones are *mechanical centrifuges*, where the gas is given a circular velocity by means of, for example, a rotating disc or cone. As the energy needed for the circular velocity is introduced mechanically instead of by the gas flow the pressure drop in the gas across the mechanical centrifuge is much less than across the cyclone, which may be an advantage in some situations.

Filters are based on the same principle as already mentioned for liquids. Most common is the bag filter. It consists of a large number of cloth bags or stockings, suspended vertically, and with the openings facing downward. The gas enters the bags through the openings and passes out through the fabric. The dust forms a filter cake inside the bags, and this makes it possible to catch particles which are much smaller than the openings in the fabric. At regular intervals the bags are shaken, the filter cake loosens and falls down through the bag openings and may be removed. Bag filters may be used for particle sizes down to 0.1 μm, and are particularly used to recover ZnO, $PbSO_4$, and SiO_2 smoke from metallurgical gases. The pressure drop and consequently the power requirements are small. Their main disadvantage is that they are limited to low temperatures. Cotton bags may be used up to 80°C and wool bags up to 95°C, whereas some modern synthetic fibers may be used up to 200°C. For higher temperatures bags of Teflon, glass fiber, or other inorganic material may be used, but they are expensive and easily embrittled.

Scrubbers are based on the principle that the gas flow is sprayed with a shower of water droplets. If the dust particles are wetted by the water drops, they will be brought down with them, and may be drained away. The process is somewhat analogous to flotation, but with the difference that the dust is brought down with the water droplets instead of up with the air bubbles. Ordinary, so-called mechanical scrubbers, remove dust down to about 0.2 μm. For finer particle sizes brownian movement, which is a result of collision with gas molecules, makes contact between the dust particles and the water droplets difficult. In order for the dust particles to hit the water droplets it is necessary that they are given a high velocity relative to the droplets. This may be achieved in the *Venturi scrubber*. Here the velocity of the gas flow is accelerated in a venturi throat, and a spray of water is introduced perpendicular to the gas flow. Venturi scrubbers may remove dust down to 0.05 μm, or about 500 Å diameter.

Electrostatic precipitators, also called *Cottrell filters*, are used to remove even smaller dust particles. Here the gas flow passes through a strong electric field, about 50 000 V dc. In this field the particles become negatively charged and are attracted to the positive electrode. Here they give off their charge and flocculate to coarser particles which settle in the bottom of the filter. In order for the electrofilter to work, it is necessary for the particles to conduct electricity sufficiently to become discharged. This is the case for many ore particles, which therefore may be removed from a dry gas. For nonconducting particles it is necessary to add enough water to the gas to make them conducting (wet Cottrell).

It should finally be pointed out that the efficiency of all gas-cleaning equipment decreases with decreasing particle size. Thus the size limits indicated in Fig. 7-4 represent the point where the equipment starts functioning, but with a low efficiency. Complete removal of the dust will be achieved only for particle sizes about 10 times the minimum ones.

7-7 OTHER PHASE SEPARATIONS

Of importance in extractive metallurgy is the separation of liquid slag from liquid metal, mattes, and speisses. The settling of heavier droplets in molten slags usually follows Stokes' law, and the settling velocity is enhanced by low slag viscosity and large difference in density between the phases. Particularly for the settling of mattes (Chap. 12), where the density is only slightly higher than that of the slag, the settling is slow. As a result much of the copper lost in matte smelting slags is in the form of matte droplets. By lowering the viscosity of the slag, or by increasing the time of separation, for example, by the use of a fore-hearth, the copper losses in the slag can be significantly reduced. Also in the iron blast furnace slag small amounts of iron droplets are found, although without affecting the economy of the process.

As for gases, settling in liquids may be enhanced by centrifugal forces. One well-known example is the milk separator, where fat droplets (cream) are separated from the aqueous (skimmed milk) phase. Centrifugal separation may be used in hydrometallurgical processes at room temperature (Chap. 15), and has been used for laboratory experiments on separation of molten metals and slags at high temperatures. It could possibly also be adapted to industrial smelting processes.

Of particular interest to steel-making is the rise and separation of oxide inclusions, formed during the deoxidation process (Sec. 13-6). These inclusions are often very small, around 10 μm or less. In a quiescent steel bath their upward velocity will be very low, making the separation difficult. Most steel baths show some turbulence, however, and the rate of separation of the inclusions is larger than what can be calculated from Stokes' law. Even higher rates are obtained by vigorous stirring of the bath. According to present theory the inclusions become trapped when they hit the slag layer or the refractory lining of the steel ladle. Transport of the inclusions to these interfaces occurs through a stagnant boundary layer, and the thickness of this layer is decreased when the bath is stirred. Thus separation of inclusions from molten steel resembles mass transfer by boundary layer diffusion.

In some cases inclusions in metallic melts may be removed by gas purging. Apart from the fact that this also causes stirring of the melt some particles are believed to adhere to the gas bubbles and to be carried to the surface. This will be an application of the principle of *flotation*. As in flotation a necessary prerequisite is that the interface energy between the particle and the gas is less than between the particle and the melt. This may possibly be obtained by the use of suitable gas mixtures.

PROBLEMS

7-1 A certain mill uses 2 kWh/1000 kg to grind a certain rock from about 10 mm diameter to an average particle size of 1 mm. (*a*) Using the Rittinger relationship, calculate the amount of energy required to grind the rock further to 0.05 mm average particle sizes. (*b*) The surface energy of the rock is 1 J/m^2. Calculate the "efficiency" of the grinding process if creation of new surface is regarded as the only useful work. The density of the rock is 2.5 g/cm^3.

7-2 A ground ore shows the following screen analysis: $-3 + 6$ mech: 18 percent, $-6 + 10$ mesh: 50 percent, $-10 + 20$ mesh: 25 percent, -20 mesh: 7 percent. The ore is to be screened at a rate of 100 tonnes/h and the screens have a capacity of 40 tonnes/(h $\cdot$ m^2 $\cdot$ mm opening). Calculate the necessary area of each of the above four screens (Tyler screen scale is assumed.)

7-3 In a screening test on a 1000 g sample the following fractions were obtained: $-14 + 20$ mesh, 58 g; $-20 + 28$ mesh, 72 g; $-28 + 35$ mesh, 90 g; $-35 + 48$ mesh, 120 g; $-48 + 65$ mesh, 150 g; $-65 + 100$ mesh, 160 g; $-100 + 150$ mesh, 140 g; $-150 + 200$ mesh, 100 g; -200 mesh, 110 g. Make a cumulative plot of the results against the logarithm of the screen openings in micrometers. (Tyler scale screens are assumed.)

7-4 Sphalerite has a density of 4.0 and galena a density of 7.5 g/cm^3. (*a*) Calculate the free-settling ratio both in the Stokes' and in the Newton's law ranges. (*b*) Calculate the corresponding hindered settling ratios if the pulp density is 1.5 g/cm^3.

7-5 Ground ferrosilicon with a density of 6.5 g/cm^3 is mixed with water to make a pulp of density 2.7 g/cm^3. What would be the relative amounts of ferrosilicon and water in the pulp?

7-6 A ferric hydroxide slurry was filtered under a constant pressure difference of 3 atm and the following rates of filtrate recovery were recorded:

Time, min	Total recovery, liters
0.5	0.50
2.0	0.96
4.0	1.35
9.0	2.02

(*a*) Calculate the average permeability, in suitable units, for each of the three time intervals.
(*b*) Estimate the time needed to filter a total of 10 liters without stopping.
(*c*) Estimate the time to filter 10 liters if the pressure difference had been 1 atm.

7-7 A settling chamber for dust-laden air (Fig. 7-5a) consists of a rectangular box 1 m high, 2 m wide, and 20 m long. In order to prevent the air stream from lifting the particles from the floor the average linear velocity is set at 3 m/s. The temperature of the air is 100°C and $P = 1$ atm. (a) Calculate the volume flow of the air in cubic meters per second when the air enters and leaves at opposite short ends of the chamber. (b) The air carries dust of various particle sizes. Assuming that Stokes' law can be applied, calculate the minimum particle size that will settle (pass from ceiling to floor) when the density of the particles is 2 g/cm^3, which is large compared to the density of the air, and the viscosity of air at 100°C is 1.75×10^{-5} kg/(m·s). (c) How long would the chamber have to be for particles 20 μm diameter to settle?

7-8 We want to design a cyclone to handle a gas flow of 0.1 m^3/s and to give a critical particle diameter of 10 μm. With reference to Fig. 7-6 the H/W ratio of the inlet is 2. (a) By means of Eq. (7-4) calculate the inlet width W as well as the entering gas velocity v. The density of the particles is 3 g/cm^3 which is large compared to that of the gas. The viscosity of the gas is 2×10^{-5} kg/(m·s). (b) The screen analysis of the entering dust is: 74 percent > 5 μm; 64 percent > 10 μm; and 43 percent > 20 μm. By means of Fig. 7-7 calculate the screen analysis of the retained dust as well as the total percentage of dust retained.

BIBLIOGRAPHY

Brown, J. H.: "Unit Operations in Mineral Engineering" International Academic Services, Ltd., Kingston, Ontario, 1979.
Gaudin, A. M.: "Principles of Mineral Dressing," McGraw-Hill Book Co. Inc., New York, 1939.
Lapple, C. E.: "Fluid and Particle Mechanics," University of Delaware, Newark, Delaware, 1956.
Liddell, D. M. (editor): "Handbook of Nonferrous Metallurgy," vol. I, McGraw-Hill Book Co. Inc., New York, 1945.
Michell, S. J.: "Fluid and Particle Mechanics," Pergamon Press, London, 1970.
Orr, Clyde, Jr.: "Particulate Technology," The Macmillan Company, New York, 1966.
Poole, J. B., and D. Doyle: "Solid-Liquid Separation," Her Majesty's Stationery Office, London, 1966.
Strauss, W.: "Industrial Gas Cleaning," Pergamon Press, London, 1966.
Taggart, A. F.: "Elements of Ore Dressing," John Wiley and Sons Inc., New York, 1951.

CHAPTER
EIGHT

FUEL AND ORE PREPARATION

In this chapter we shall discuss a number of processes which apply to the preparation of raw materials for metallurgical use. This includes the coking of coal, gasification of coke, coal, and oil, calcination, roasting of sulfide ores, and agglomeration and sintering of oxide and sulfide ores, as well as the treatment of roast gases.

8-1 COAL, COKE, OIL, AND GAS

Coal, coke, oil, and gas represent important raw materials for metallurgical processes. They are used as fuels and also as reducing agents. Chemically coal, mineral oil, and natural gas are made up of carbon and hydrogen with minor amounts of other elements: nitrogen, sulfur, and oxygen. The atomic ratio n_H/n_C in coal may be about one or less, in mineral oils between two and three. The highest n_H/n_C ratio, four, is found in methane which is the major constituent in natural gas.

Coals, and to a lesser extent oil, contain varying amounts of ash, consisting of metal oxides and silicates. When the coal or the oil is burned the ash remains behind unreacted. Ashes are usually regarded as a burden on the fuel and lower its economic value. Most coals and oils contain sulfur, which may occur partly as inorganic sulfides, e.g., pyrite, and partly chemically bonded to the hydrocarbons. When the coal is burned, sulfur enters the combustion gases as SO_2. When coal or coke is used as a reducing agent sulfur may contaminate the metal. It is therefore regarded as harmful.

For a discussion of the reactions which occur among the various constituents in fuels and hydrocarbons we shall look at their Gibbs energy of formation, referred to one mole of carbon as standard states. This is shown in Fig. 8-1. The standard state chosen for carbon is graphite, which is the stable modification at atmospheric pressure, and its Gibbs energy is represented by the horizontal zero line.

We see that the only hydrocarbons which are stable at room temperature are methane, CH_4, and ethane, C_2H_6. All other hydrocarbons including coal, oil, and most organic compounds and all living organisms including ourselves are thermodynamically unstable at room temperature, and have a tendency to decompose into simpler molecules. Only the slowness of the reactions, and in our own case the continuous supply of fresh organic food, prevents this decomposition.

At temperatures above 500°C even methane becomes unstable and decomposes into graphite and hydrogen. If the original material contains oxygen this will react with carbon and hydrogen to form mixtures of CO_2 and H_2O, and at higher temperatures CO. Nitrogen is present in some coals and organic compounds. On heating it will partly be expelled as ammonia, but since ammonia

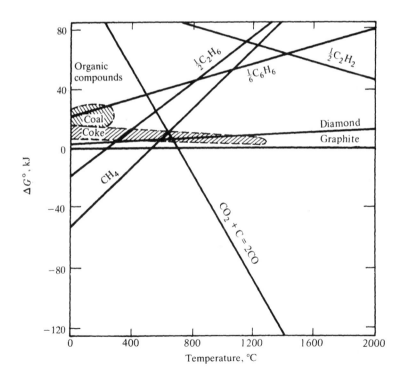

Figure 8-1 Standard Gibbs energy of formation, per mole of carbon, of various carbon compounds relative to graphite.

in itself is unstable the final product will be nitrogen gas. Sulfur in coal will, on heating in the absence of oxygen, be converted into hydrogen sulfide or it may remain as metal sulfides.

The composition of coal, coke, and oil may be expressed by their *proximate* and their *ultimate* analysis. The proximate analysis gives the percentage of "moisture," "volatile matter," "fixed carbon," and "ash." Each of these is determined by standardized procedures, and different values would be obtained if different procedures were used. The ultimate analysis gives the percentage of the individual elements, carbon, hydrogen, nitrogen, sulfur, and oxygen, as well as the ash-content, all expressed on moisture-free basis. The percentage of moisture is given separately, since this may vary with the storage conditions.

The *heat of combustion* of the fuel may be estimated from the ultimate analysis by the *Dulong formula*

$$NCP = 338C + 1423(H - O/8) + 92S - 24.4(9H + M)$$

Here *NCP* means the *net calorific power* (Sec. 2-4) in kilojoules per kilogram. The symbols C, H, O, S, and M represent the weight percentage of the four elements and moisture in the fuel. If the *gross calorific power* is required the last term in the Dulong formula is omitted.

The Dulong formula implies that oxygen in the fuel is present as H_2O, and that all other elements have the same heat of combustion as in the pure state, i.e., that the heat of formation of the fuel from the elements is negligible. The heat of combustion of carbon is taken equal to 406 kJ/mol = 33 830 kJ/kg, which is the value that applies for amorphous carbon, and which is 12.5 kJ/mol more than for crystalline graphite. As will be shown in the following section the assumption of a negligible heat of formation of the fuel from the elements is only approximately correct, however.

Coking

Coking is a process where coal is heated in the absence of air in such a way that the volatile constituents are expelled. We distinguish between *low-temperature* and *high-temperature* coking. Low-temperature coking is carried out at temperatures up to about 500°C. We get a low-temperature coke, "char," which still contains considerable amounts of hydrogen. The gas evolved is rich in hydrocarbons and tar, and contains relatively little free hydrogen. Low-temperature coke has found some application as smokeless household fuel, and as a reducing agent in ferrosilicon production, but on the whole it is of limited metallurgical interest.

High-temperature coking is carried out at around 1000°C. The gas is rich in hydrogen and methane, whereas higher hydrocarbons and tars to a great extent have been decomposed (cracked) under the influence of the high temperature. Furthermore the gas contains some CO, and minor amounts of CO_2, H_2O, C_2H_6, H_2S, and NH_3. The resulting coke is low in volatile compounds, about

1 to 2 percent. For metallurgical use high-temperature coke is used almost exclusively.

The coking process is usually carried out in retorts by the so-called "by-products method." The retorts are narrow, vertical chambers which are heated from the outside. The width of the chamber is about 0.5 m and differs little between the different coking plants. The other dimensions may vary considerably from one plant to another. A height of about 5 m and a length of about 10 m are not unusual. The reason for the narrow width is that heat has to be supplied from the outside by conduction through the chamber walls and through the coking charge. The time needed for the heat to penetrate the charge is, as discussed in Sec. 6-5, proportional to the square of the chamber width. The chambers are therefore made as narrow as is compatible with a satisfactory mechanical operation.

A *coke oven* consists of a battery of a large number of retorts separated by combustion chambers. The chamber walls are made of highly refractory bricks, usually silica. The heat may be supplied by the burning of some of the coke oven gas, but other fuels may also be used. After leaving the coke oven the hot combustion gases are passed through regenerative chambers for recovery of their heat content and preheating of the combustion air. In this way a good heat economy is obtained.

Not all coals are suitable for coking. Even though the volatile constituents may be expelled, the product may be weak and fall into powder. In order to obtain a mechanically strong coke it is necessary to use so-called "coking coals." On heating, these first go through a partly fluid state. On further heating the gases are expelled and the mushy fluid solidifies into solid coke. Even in this case the coke may be of varying mechanical strength, depending on the composition of the coal. By mixing different coal qualities various cokes may be obtained. For the iron blast furnace the coke has to be mechanically strong, for other metallurgical processes a somewhat weaker coke may be used.

The ashes in the coal are retained exclusively in the coke. Of other constituents the sulfur content is of particular interest. Of the total sulfur content in the coal about one-third is expelled in the gas and two-thirds is retained in the coke. Since the weight of the coke is about two-thirds of the weight of the coal this means that the sulfur percentage in the coke is about the same as in the coal. For iron reduction the sulfur content should be as low as possible, preferably less than 1 percent.

During isothermal coking the fuel passes from a state of higher to a state of lower enthalpy, i.e., the coking process as such is exothermic. This is confirmed by accurate calorimetric measurements which show that the combined calorific power of all coking products is about 200 kJ/kg *less* than that of the original coal. The actual heat used for coking, about 2000 kJ/kg, is needed to cover heat losses and the physical heat of the coking products. Even after coking the carbon has not reached its stable graphite form. This is illustrated by Fig. 8-1 where the Gibbs energy of coke is indicated as being 10 to 15 kJ/mol more positive, and the heat of combustion correspondingly larger, than for graphite.

Gasification of Coke and Coal

Whereas during the coking process, gas is produced only to the extent that the coal contains volatile compounds, it is also possible to convert the entire carbon content of the coal into gaseous products. This may be done by partial combustion with air, oxygen, steam, or even with carbon dioxide. Most common is gasification in the *gas producer* (Fig. 8-2(a)). This is a low shaft furnace which may be charged with either coke or, most usually, coal. Through a grate in the bottom of the furnace air on air + steam, in certain cases only steam, is introduced. The ashes are taken out through a waterlock or they may melt and be tapped, as in a blast furnace. Between carbon and oxygen the following reactions occur

$$C + O_2 = CO_2 \qquad \Delta G_1^\circ = -394\,100 - 0.8T \text{ J} \qquad (1)$$

$$2C + O_2 = 2CO \qquad \Delta G_2^\circ = -223\,400 - 175.3T \text{ J} \qquad (2)$$

The so-called "Boudouard reaction" is obtained by subtraction

$$CO_2 + C = 2CO \qquad \Delta G_3^\circ = +170\,700 - 174.5T \text{ J} \qquad (3)$$

The above ΔG° values refer to graphite as standard state. By the use of coke the first term, which represents the ΔH° of the reaction, becomes about 12 kJ more negative per mole of carbon.

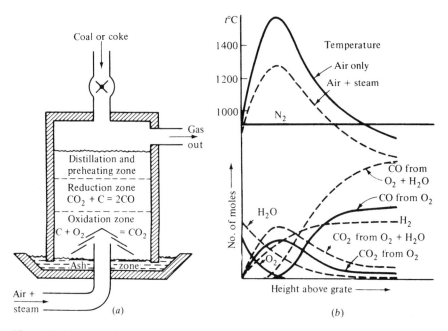

Figure 8-2 (a) Gas producer (schematic), (b) temperature and gas composition as functions of height above grate for gasification of coke with air or with air + steam.

We see from the above equations that reaction (1) which is the most exothermic, will dominate at low temperatures and high O_2 concentrations, whereas reactions (2) and (3) will take over at higher temperatures and by O_2 deficiency. It has, furthermore, been found experimentally that under standardized conditions the chemical rate is about 100 times higher for reaction (1) than for reaction (3).

In a gas producer there will be a high concentration of oxygen immediately above the grate, and mainly CO_2 will be formed. Higher up the oxygen will be used up, and CO_2 will react according to reaction (3). The extent of this reaction depends on the temperature and the time of contact. Thus the composition of the gas will vary with the distance from the grate as shown in Fig. 8-2(*b*).

Reaction (1) is strongly exothermic. If the gas producer is blown with air, very high temperatures, in excess of 1500°C, are obtained. Reaction (3) is strongly endothermic and the temperature in the gas drops to about 1100°C. By further cooling of the gas in counter-current with the cold charge the temperature drops to about 800°C before the gas leaves the producer. If the producer is charged with coal instead of coke this upper part will act as a coke-oven and the volatiles in the coal will mix with the furnace gas.

The gas obtained by blowing with air has a high nitrogen content and consequently a low calorific value. Furthermore, the high maximum temperature is harmful to the grate. In order to obtain a richer gas and to lower the temperature above the grate, steam is usually added to the blast. Steam reacts above the grate according to the reaction

$$H_2O(g) + C = H_2 + CO \qquad \Delta G_4^\circ = +134\,700 - 142.5T \text{ J} \qquad (4)$$

In addition some CO_2 is formed. We see that reaction (4) is strongly endothermic, and lowers the maximum temperature to about 1300°C. Since steam is free of nitrogen, the gas obtained will be richer in combustibles and have a higher calorific value than without steam. Also the temperature of the top gas will be lower. If we disregard heat losses it is clear that the calorific value of the coal must reappear as physical and chemical enthalpy in the producer gas. By steam addition some of the physical enthalpy in the gas is converted into useful chemical enthalpy.

An even richer gas would be obtained by further increase of the steam addition. One would then get to the point where the necessary temperature could no longer be maintained. The gas producer can then be operated alternatingly: First it is blown with air until the maximum temperature is reached. The gas obtained during this period is of poor quality but may be used in the steam boiler. The gas producer is then switched to steam and is blown until the temperature has dropped to about 1000°C. The gas is now practically pure $CO + H_2$. This gas, *water gas*, has a high calorific value. In a subsequent catalytic step it may be converted into pure hydrogen which again may be used, e.g., for the production of ammonia or as a reducing agent.

Another way to obtain a better heat economy is to blow the gas producer with steam and preheated air, or with steam and oxygen. In the latter case the gas obtained will be free of nitrogen. A third possibility would be to supply the necessary heat electrically, e.g., by making the coke bed the resistor in an electric circuit. In this way the producer may be blown continuously with steam or even with carbon dioxide. Due to the relatively high cost of electrical energy these possibilities have so far been used only under very special circumstances.

Partial Combustion of Oil and Natural Gas

Oil and natural gas find increasing application in the chemical and metallurgical industries, as raw materials for the production of hydrogen and ammonia, as reducing agents in iron- and steel-making, and as protective atmospheres in the heat treatment of steel and other metals.

Since hydrocarbons on heating will decompose and deposit soot they first have to be converted into simple gases, CO and H_2. This is achieved by partial combustion with air or oxygen. Reaction with steam or carbon dioxide is also used. For the partial combustion of methane with oxygen or air we have

$$2CH_4 + O_2 = 2CO + 4H_2 \qquad \Delta H_{298} = -71 \text{ kJ}$$

This reaction is only slightly exothermic. In order to obtain a temperature above 1000°C, which is necessary to obtain a good conversion and prevent sooting, the air must be preheated, or pure oxygen may be used. For the gasification of oil the danger of sooting is even larger and special burners have been developed for this purpose. Reforming of natural gas or oil by means of steam would be of the type

$$CH_4 + H_2O = CO + 3H_2 \qquad \Delta H_{298} = +206.3 \text{ kJ}$$

This reaction is endothermic, and extra heat has to be supplied, e.g., by electricity or by regenerative heating. The same is the case for reforming with carbon dioxide. The reforming is usually carried out at around 1000°C. In order to obtain a fast reaction and to avoid sooting, a catalyst, usually nickel sponge, is used. See, for example, the Midrex process (Sec. 9-6).

If natural gas is reacted with oxygen at temperatures around 2000°C acetylene may be formed

$$2CH_4 + \tfrac{3}{2}O_2 = C_2H_2 + 3H_2O$$

In order to obtain the high temperature, additional gas has to be burned to CO and H_2O, or electric energy must be supplied. As is seen from Fig. 8-1 acetylene is the most stable hydrocarbon at high temperature but becomes increasingly unstable at lower temperatures. Therefore, in order to prevent it from decomposing the gas, after reaction, has to be shock-cooled to room temperature. The fact that acetylene is unstable at room temperature (its enthalpy of formation from the elements is about $+220$ kJ/mol) results in a correspondingly large heat of combustion. An oxyacetylene torch, therefore, gives a higher flame temperature than can be obtained with other fuels.

Some other High-Temperature Reactions

When carbon monoxide, hydrogen, or hydrocarbons are burned with oxygen, as in a welding torch, the following equilibria are established at high temperature:

$$2CO + O_2 = 2CO_2 \qquad \Delta G° = -564\,800 + 173.6T \text{ J}$$

$$2H_2 + O_2 = 2H_2O \qquad \Delta G° = -492\,900 + 109.6T \text{ J}$$

At temperatures below 2000°C, that is, in combustion with air, these reactions are completely shifted to the right. At higher temperatures, for example, by combustion with pure oxygen, a certain equilibrium is established, the equilibrium constant being unity at about 3000°C for the first and at about 4000°C for the second reaction. This again means that at these temperatures CO_2 and H_2O will be partly dissociated even in the presence of oxygen, the dissociation being most pronounced for CO_2. These dissociations, which are endothermic, put a practical limit to the maximum temperatures which can be obtained by the burning of hydrocarbons with pure oxygen, this limit being about 3000°C, the highest temperature for acetylene.

Low-Temperature Reactions

The most important low-temperature reaction is the conversion of water gas into hydrogen, also called the *shift reaction*. This is based on the equilibrium

$$H_2O + CO = H_2 + CO_2 \qquad \Delta G° = -36\,000 + 32.01T \text{ J}$$

The equilibrium constant for this reaction is unity at about 900°C, and the reaction shifts to the right with decreasing temperature. At 300°C the equilibrium constant is about 40, but at this temperature the reaction rate is very low. In order to obtain the best possible conversion a temperature of about 450°C as well as a catalyst, usually iron oxide, and excess of steam are used. After cooling of the gas mixture CO_2 is removed, for example, by absorption in water under pressure or by compression and condensation to give liquid CO_2. The remaining small amounts of CO may be absorbed in an ammoniacal cuprous solution.

Today hydrogen is used mainly in the production of ammonia for the fertilizer industry, and to a smaller extent as reducing agent to make iron powder. In the future it may be used to run fuel cells for the production of electric power. In this connection it is of interest that modern gasification and conversion of hydrocarbons into hydrogen can be carried out with a high efficiency, the calorific value of the hydrogen produced being about 80 percent of that of the original fuel.

At temperatures below 400°C and in the presence of special catalysts a number of reactions may be carried out between carbon monoxide, hydrogen, and other reactants to give various hydrocarbons (synthetic gasoline) and even organic compounds (methanol). These processes represent a large and important chapter in petrochemical technology, but are of minor interest to the metallurgist.

8-2 DRYING AND CALCINATION

Drying refers to the removal of water from other substances such as ore or coke by evaporation. The necessary requirement for evaporation is that the vapor pressure of the water is higher than the partial pressure of water in the surrounding atmosphere. In order to obtain a sufficiently high rate of evaporation, however, it is preferred to have the vapor pressure higher than the total atmospheric pressure. Drying may, therefore, be accomplished either at atmospheric pressure by heating the substance above the normal boiling point of water, or under reduced pressure where the atmospheric pressure is brought below the vapor pressure of the water at the temperature in question. The latter technique is sometimes used for organic substances which would decompose above 100°C, but it is of minor interest to the metallurgist.

Evaporation of water is an endothermic process; for the reaction $H_2O(l) = H_2O(g)$, $\Delta H_{298} = 43.9$ kJ. This means that in addition to the heat needed to bring the substance to the drying temperature, the heat of evaporation must be supplied at that temperature.

Drying is usually accomplished by passing hot combustion gases through or above the substance. In most integrated metallurgical plants there are available hot gases of a few hundred degrees Celsius which have no other use. If not, extra fuel has to be burned, but this does not need to be of the highest quality. Drying may be carried out in a number of different types of furnaces, e.g., in a rotary kiln or in a fixed or fluidized bed. Since the temperature is relatively low there are usually no serious material problems.

By contrast with drying, *calcination* involves the removal of water, carbon dioxide, or other gases, which are chemically bound as, for example, hydrates or

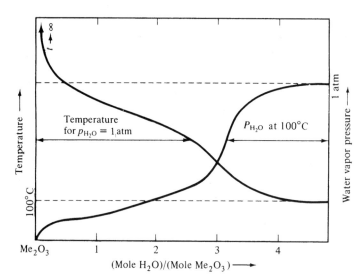

Figure 8-3 Water vapor pressure at constant temperature (right) and temperature for constant water vapor pressure (left) for the system Me_2O_3–H_2O (schematic).

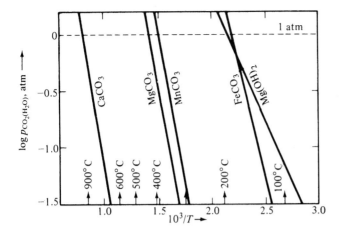

Figure 8-4 Decomposition pressure (logarithmic) of various carbonates and hydrates as function of inverse temperature.

carbonates. It is worth noticing that there is no sharp distinction between free and chemically bound water. As the free water is removed from a substance we get into the range of adsorbed water, then to relatively loosely bound and finally to more strongly bound water. This is for example the case for clays and also for hydroxides of iron and aluminum. For such substances, the vapor pressure of water decreases continuously from that of free water for a moist material to zero for the completely dehydrated oxide, with a corresponding increase in the temperature necessary for dehydration. This behavior is illustrated in Fig. 8-3. Thus, in order to remove the last traces of water from Al_2O_3 temperatures in excess of 1000°C are required. Also, the enthalpy of evaporation increases with decreasing water content.

In Fig. 8-4 is shown the decomposition pressure of some well-defined carbonates and hydrates as a function of temperature. We see that the temperature necessary for the decomposition pressure to reach 1 atm varies considerably from one substance to the other. For $FeCO_3$ and $Mg(OH)_2$ temperatures below 200°C are sufficient; for $MgCO_3$ about 400°C, and for $CaCO_3$ almost 900°C are needed. Even higher temperatures are required to decompose $BaCO_3$ and Na_2CO_3.

Calcination is more endothermic than drying. Thus for the reaction $CaCO_3 = CaO + CO_2$, $\Delta H_{298} = 177.8$ kJ, and the heat must be supplied at a relatively high temperature.

The *rate* of calcination seems to be governed primarily by the supply of the necessary heat of decomposition. This follows among other things from work by Wuhrer[1] and Satterfield and Feakes.[2] According to these authors the *chemical*

---

[1] J. Wuhrer: *Chemie-Ing.-Techn.*, **30**: 19–30 (1958).
[2] C. N. Satterfield and F. Feakes: *Am. Inst. Chem. Eng. J.*, **5**: 115–122 (1959). See also: A. W. D. Hills: *Chem. Eng. Sci.*, **23**: 297–320 (1968).

rate of decomposition of limestone at 900°C is relatively high, and the rate of calcination is governed by the supply of the necessary heat by *conduction* through the layer of already burned lime. The rate of decomposition is in that case governed by the same equations which apply for diffusion controlled reactions (Eqs. 5-18, 5-19), i.e., it is inversely proportional to the square of the diameter, and the time needed to complete the calcination is proportional to the square of the diameter. In order to obtain a reasonable calcination time for lump limestone it is necessary to operate with a relatively large temperature excess in the heating gas. This again means that the outer layer of the burned lime will be unnecessarily overheated, which again may result in loss of chemical reactivity for subsequent applications.

For industrial application calcination may be carried out in a number of different furnaces. For the burning of coarse limestone the *shaft furnace* is most useful. For material of mixed particle size or for lumps which disintegrate during the process, a *rotary kiln* may be used. Finally for material of uniform, small particle size the calcination may be done in a *fluidized bed*. Depending on what is available, the fuel may be gas, oil, or coke. In order to illustrate the distribution of heat during the process, the calcination of limestone with coke in a shaft furnace will be treated below.

Heat Balance for Shaft Furnace Calcination

The shaft furnace is illustrated schematically in Fig. 8-5(*a*). The furnace is charged with a mixture of limestone and coke. Air is introduced at the bottom of the furnace, where also the burned lime is withdrawn. The furnace may be regarded as divided into three zones:

1. *The preheating zone.* In this zone the solid charge is preheated to 800°C in counter-current with the hot furnace gases, but there is no calcination taking place, and for practical purposes, not yet any combustion of the coke.
2. *The reaction zone.* In this zone burning of the coke and decomposition of the limestone take place, and the burned lime is overheated to an estimated average temperature of 1000°C. Furthermore we assume that the gases which leave the reaction zone have a temperature of 900°C, that is, 100°C higher than the entering solid material.
3. *The cooling zone.* In this zone the burned lime is cooled in counter-current with cold air. We assume that the burned lime is cooled to 100°C.

In order to ensure complete combustion of the coke we assume that 25 percent air excess is needed. Also, for simplicity, we assume the limestone to be 100 percent $CaCO_3$, and the coke to be 100 percent carbon. Any deviation from these assumptions can easily be corrected for. Also, for simplicity, we disregard heat losses through the furnace walls.

The question as to how much carbon (coke) is required for the calcination of 1 kg of limestone may then be handled in the following way:

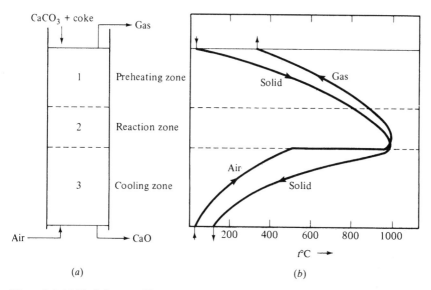

Figure 8-5 (*a*) Shaft furnace, (*b*) temperature of gas and solids during calcination of limestone in a coke-fired shaft furnace.

(1) The heat requirements for the calcination of 1 mol of $CaCO_3$ in the reaction zone, and (2) the available heat from the combustion of 1 mol of carbon are calculated. From these values the number of moles or weight of carbon per mole or unit weight of $CaCO_3$ is obtained. The calculations follow essentially the procedure outlined in Sec. 2-5 and thermochemical data from Apps. A and B are used. See Tables 8-1 and 8-2.

It will be noticed that in Table 8-2 the combustion air is assumed preheated to 500°C. This value is derived from the subsequent heat balance for the cooling zone, and actually has to be obtained by approximation. From the above calculations follows that $183/328 = 0.56$ mol of carbon is required per mole of $CaCO_3$. We may now work out the total heat balance for the process separately for the three zones, and with 1 kg = 10 mol of $CaCO_3$ as basis. See Table 8-3.

Table 8-1 Heat requirement per mole of CaCO₃

Reaction: $CaCO_3$ (800°C) = CaO (1000°C) + CO_2 (900°C)

Input		kJ	Output		kJ
$CaCO_3$	H_{298}^{1073}	87.0	CaO	H_{298}^{1273}	50.2
Heat deficiency		183.3	CO_2	H_{298}^{1173}	42.3
			$CaCO_3 = CaO + CO_2$	ΔH_{298}	177.8
Sum		270.3	Sum		270.3

Table 8-2 Heat available per mole of carbon

Reaction: C (800°C) + (1.25O_2 + 4.7N_2) (500°C) = (CO_2 + 0.25O_2 + 4.7N_2) (900°C)

Input		kJ	Output		kJ
C	H_{298}^{1073}	13.4	CO_2		42.3
1.25O_2	H_{298}^{773}	18.4	0.25O_2	H_{298}^{1173}	7.1
4.7N_2		64.9	4.7N_2		125.5
C(coke) + O_2 = CO_2	$-\Delta H_{298}$	405.8	Available heat		327.6
Sum		502.5	Sum		502.5

Comments on heat balance The theoretical requirement of 5.6 mol corresponds to 0.067 kg of carbon per kilogram of $CaCO_3$. In practice a total amount of about 0.1 kg is needed in order also to cover heat losses through the furnace wall. The heat balance for the preheating zone shows a surplus of 461 kJ, which

Table 8-3 Heat balance for calcination shaft

1. Preheating zone

Input		kJ	Output		kJ
10$CaCO_3$	25°C	0	15.6CO_2		
5.6C	25°C	0	1.4O_2	340°C	461
15.6CO_2		661	26.3N_2		
1.4O_2	H_{298}^{1173}	42	10$CaCO_3$	H_{298}^{1073}	870
26.3N_2		703	5.6C		75
Sum		1406	Sum		1406

2. Reaction zone

Input		kJ	Output		kJ
10$CaCO_3$	H_{298}^{1073}	870	10CaO	H_{298}^{1273}	502
5.6C		75	15.6CO_2		661
7O_2	H_{298}^{773}	105	1.4O_2	H_{298}^{1173}	42
26.3N_2		364	26.3N_2		703
5.6(C(coke) + O_2			10($CaCO_3$		
= CO_2)	$-\Delta H_{298}$	2272	= CaO + CO_2)	ΔH_{298}	1778
Sum		3686	Sum		3686

3. Cooling zone

Input		kJ	Output		kJ
10CaO	H_{298}^{1273}	502	10CaO	H_{298}^{373}	33
7O_2	25°C	0	7O_2	H_{298}^{773}	105
26.3N_2			26.3N_2		364
Sum		502	Sum		502

gives a top gas temperature of about 340°C. The heat balance for the cooling zone shows that, for a product temperature of 100°C, the air will be preheated to about 500°C, the value which was used in the preliminary calculations. If the calculation for the cooling zone had not balanced, the preliminary calculation would have had to be adjusted.

We can now draw a temperature profile for the furnace, as shown in Fig. 8-5(b). We see that in the preheating zone the gas temperature is always higher than that of the solids, in the cooling zone the opposite is the case. In the preheating zone the temperature difference increases toward the top of the furnace in accordance with the gas flow having the higher heat capacity. Similarly, in the cooling zone the burned lime of 1000°C is able to preheat the air only to about 500°C. A better heat exchange and consequently a better heat economy would be possible with smaller gas volume, e.g., with stoichiometric air or oxygen-enriched air. The heat balance tells us nothing about the height or capacity of the furnace, i.e., of the retention time of the burden. These are determined by the overall heat transfer coefficients between the gas and the solids, as discussed in Sec. 6-5. An increased retention time, as would be possible with a higher shaft or a decreased feeding rate, would result in a smaller difference between gas and solid temperatures, and a smaller *theoretical* fuel requirement. On the other hand the heat losses through the furnace walls and the capital expenses would increase.

Other Calcination Furnaces

Other calcination furnaces operate on much the same principles as the coke-fired shaft furnace. In the *oil-fired shaft furnace* oil is burned in combustion chambers adjacent to the shaft, and the hot combustion gases are passed into the shaft. In the *rotary kiln* oil, gas, or pulverized coal is burned with air, and the hot combustion gases are passed through the kiln in counter-current with the solid charge. In the *fluidized-bed furnace* heavy oil is injected into the bed, where the combustion and the calcination take place simultaneously. In order to utilize the heat content in the hot gases these are passed through a number of fluidized beds thus preheating the cold limestone. Similarly, the hot burned lime is passed through one or more beds for preheating of the combustion air.[1]

An interesting type of furnace is the old-fashioned *ring kiln*, which may also be used for the firing of bricks. The kiln consists of a number of stationary chambers, Fig. 8-6, arranged in a ring around a common chimney. Fuel and combustion air may be introduced to any of these chambers, and the flue gases may be withdrawn from any of them. The operation follows a cycle: At any time one chamber (1) is being fed with raw limestone. The subsequent chambers (2–6) which contain unburned limestone, are preheated by hot combustion gases

---

[1] F. S. White and E. L. Kinsalla: *Mining Eng.*, **4**: 903–905 (1952).

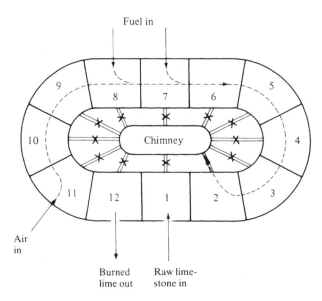

Figure 8-6 Ring kiln. Dashed line = gas flow. Flues marked × are closed.

before these go into the chimney. Then follow a few chambers (7–8) where fuel is introduced and where the calcination takes place. Finally there follows a number of chambers (9–11) where combustion air is being preheated and the burned lime is being cooled. In the very last chamber (12), which is adjacent to the first, cooled lime is removed from the kiln. After the first chamber has been filled and the last emptied, the inlets for fuel and air and the outlet for flue gas are shifted clockwise one step, the filled chamber is being preheated, the emptied one is once more filled, and a new chamber is being opened and emptied. Thus, even though the solid charge always remains in a stationary position, the operation of the kiln is analogous to the shaft furnace where the gases move counter-current to the charge.

In addition to thermal decomposition of hydrates and carbonates calcination may serve other purposes. One of these is to obtain a certain grain growth to make the oxide chemically resistant. Thus magnesium oxide and dolomite, which are to be used for refractories, are calcined at high temperature in order to make them more resistant against atmospheric moisture and carbon dioxide. Another application, particularly of the rotary kiln, is in the production of *portland cement*. Here a mixture of limestone and some clay-material is heated to about 1350°C whereby a slight fusion occurs, and calcium silicates and aluminosilicates are formed. The product, *cement clinker*, is subsequently ground to make cement.

A final application of calcination kilns is in the firing of iron-ore pellets which will be discussed in Sec. 8-4.

8-3 ROASTING OF SULFIDES

Roasting is the oxidation of metal sulfides to give metal oxides and sulfur dioxide. Typical examples are

$$2ZnS + 3O_2 = 2ZnO + 2SO_2$$

$$2FeS_2 + 5.5O_2 = Fe_2O_3 + 4SO_2$$

In addition other reactions may take place: formation of SO_3 and metal sulfates and formation of complex oxides such as $ZnFe_2O_4$.

Typical ores which are roasted are the sulfides of copper, zinc, and lead. In these cases the main purpose is to convert the metals, partly or completely, into oxides for subsequent treatment. The sulfur dioxide is then a by-product. When pyrite is roasted the gas is the main product, and iron oxide (pyrite cinder) is a by-product.

Roasting is usually carried out below the melting points of the sulfides and oxides involved, usually below 900 to 1000°C. On the other hand, in order for the reactions to occur with sufficient velocity the temperature has to be above 500 to 600°C. Thus the temperature range of interest is between 500 and 1000°C.

Thermodynamics of Roasting

The necessary conditions for the formation of the various roasting products may be illustrated by the equilibrium relations in a system which contains metal, sulfur, and oxygen. We have here three components, and according to the phase rule, we can have a maximum of five phases, i.e., four condensed phases and one gas phase. If the temperature is fixed, we may have a maximum of three condensed phases and a gas phase. The gas phase contains normally SO_2 and O_2; but also SO_3 and even S_2 may be present. Among these gaseous compounds the following equilibria exist:

$$S_2 + 2O_2 = 2SO_2 \tag{1}$$

$$2SO_2 + O_2 = 2SO_3 \tag{2}$$

For a given temperature the composition of the gas mixture is defined by the partial pressure of any two of the gaseous components. Also, for a fixed gas composition the composition of the condensed phases is fixed. Thus the phase relations in the ternary system at *constant temperature* may be described in a two-dimensional diagram where the two coordinates are the partial pressures of two of the gaseous components. Such a diagram is shown in Fig. 8-7 for an arbitrary metal Me, and as coordinates the partial pressures of SO_2 and O_2 are chosen.

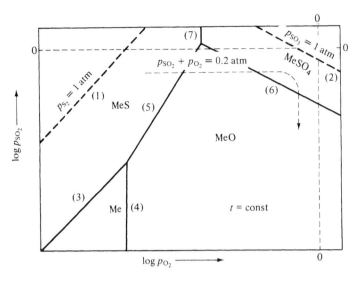

Figure 8-7 Equilibria and predominance areas at constant temperature (Kellogg diagram) for the system Me–S–O (schematic).

The lines that describe equilibrium between any two condensed phases are given by equations such as

$$Me + SO_2 = MeS + O_2 \tag{3}$$

$$2Me + O_2 = 2MeO \tag{4}$$

$$2MeS + 3O_2 = 2MeO + 2SO_2 \tag{5}$$

$$2MeO + 2SO_2 + O_2 = 2MeSO_4 \tag{6}$$

$$MeS + 2O_2 = MeSO_4 \tag{7}$$

If the metal had formed several sulfides and oxides, additional equations would have to be considered for the formation of MeS_2, Me_2O_3, and $Me_2(SO_4)_3$, etc. Also basic sulfates such as $MeO \cdot MeSO_4$ may exist.

For the above reactions and for all condensed phases in their standard states the equilibria are given by the expressions

$$\log p_{O_2} - \log p_{SO_2} = \log K_3$$

$$\log p_{O_2} = -\log K_4$$

$$2 \log p_{SO_2} - 3 \log p_{O_2} = \log K_5$$

$$2 \log p_{SO_2} + \log p_{O_2} = -\log K_6$$

$$2 \log p_{O_2} = -\log K_7$$

We notice that for a given stoichiometry of reaction the form of the equilibrium expression is the same for all metals, i.e., the slope of the corresponding curves in Fig. 8-7 is the same. Only the value for the equilibrium constants K_3, K_4, etc., may differ from one metal to the other. This means that the position of the equilibrium lines may change, and consequently the size and the position of the areas between the lines (compare Fig. 3-3).

These areas are called the *predominance area* for that particular phase. We notice that as long as only one condensed phase exists the partial pressure of SO_2 and O_2 may be changed independently of each other, i.e., the system at constant temperature has two degrees of freedom. Along the lines for equilibrium between two condensed phases the system has one degree of freedom. Finally when three condensed phases are present the system is nonvariant at constant temperature. In Fig. 8-7 are also given lines to describe reactions (1) and (2), that is, for the formation of S_2 and SO_3. These are given by the expressions

$$2 \log p_{SO_2} - 2 \log p_{O_2} = \log K_1 + \log p_{S_2}$$

$$2 \log p_{SO_2} + \log p_{O_2} = -\log K_2 + 2 \log p_{SO_3}$$

It follows that for fixed values of K_1 and K_2, the relation between $\log p_{SO_2}$ and $\log p_{O_2}$ depends also on the partial pressures of S_2 or SO_3. In Fig. 8-7 the lines are drawn for p_{S_2} and p_{SO_3} being one atmosphere. For other pressures the lines are to be shifted up and down in accordance with the above expressions. We notice that the partial pressure of S_2 becomes large when the partial pressure of O_2 is small and that of SO_2 is large. The partial pressure of SO_3 becomes large for large values of SO_2 and O_2. Diagrams of the kind shown in Fig. 8-7 are often called Kellogg diagrams.[1]

When roasting is carried out in air the sum of the partial pressure of SO_2 and O_2 is about 0.2 atm. This means that the conditions during roasting will be described by the dotted line in Fig. 8-7. First the sulfide is converted into the oxide by reaction (5). Then the oxide may be converted into sulfate, which by prolonged heating in air at constant temperature again may decompose to give the oxide.

Since the predominance areas for the different metals have different locations, the reactions for a mixed sulfide ore will not occur simultaneously for the different metals, and some reactions may not occur at all. Thus for a mixed Fe–Cu sulfide (see Fig. 8-8), iron sulfide will first oxidize to form Fe_3O_4 (assuming $p_{SO_2} + p_{O_2} = 0.2$ atm). Copper will then still be present as Cu_2S. Further oxidation converts Fe_3O_4 into Fe_2O_3 and Cu_2S into Cu_2O and further into CuO. For temperatures below about 750°C and for about equal amounts of SO_2 and O_2, CuO may be converted into $CuO \cdot CuSO_4$ and $CuSO_4$, whereas Fe_2O_3 will not form sulfate under these conditions.

The effect of temperature on the roasting equilibria may be given in a two-dimensional diagram for constant values of p_{SO_2}. In Fig. 8-9 this is done in a plot which describes $\log p_{O_2}$ against the inverse absolute temperature, for

[1] H. H. Kellogg and S. K. Basu: *Trans. AIME*, **218**: 70–81 (1960).

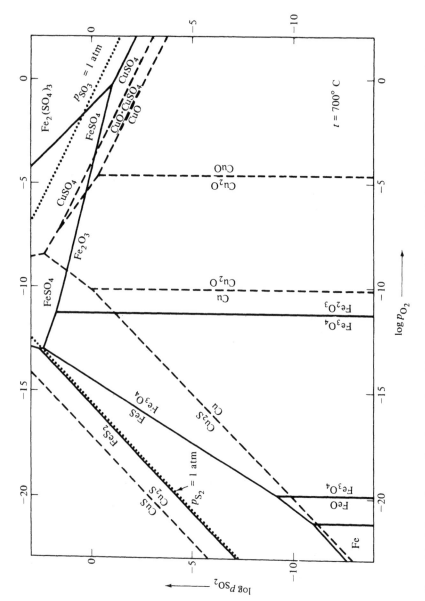

Figure 8-8 Kellogg diagram for the roasting of iron and copper sulfides at 700°C. Solid lines = Fe compounds, dashed lines = Cu compounds, dotted lines = gases. Ternary compounds are disregarded.

218

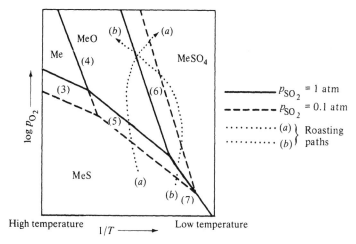

Figure 8-9 Equilibria and predominance areas at constant SO_2 pressure as function of temperature for the system Me–S–O. The lines (a)–(a) and (b)–(b) indicate two different roasting paths (schematic).

$p_{SO_2} = 1$ atm and 0.1 atm. We see that, with the exception of reactions (4) and (7), $\log p_{O_2}$ is also a function of $\log p_{SO_2}$, and shifts up or down with changing values for the SO_2 pressure.

The slopes of the curves are, in this case, given by the enthalpy of the reactions, all expressed in terms of one mole of oxygen, see Eq. (3-8). Even though the enthalpy of reactions (3) to (7) may vary considerably from one metal to another, their relative values are usually about the same, and the diagrams obtained will usually be of the type given in Fig. 8-9.

We see from Fig. 8-9 that, whereas MeS and MeO may coexist in an intermediate temperature range, they will react to give Me and SO_2 at a sufficiently high temperature. The temperature for this reaction differs considerably from one metal to another. For metals such as copper, lead, and the noble metals the reaction occurs below 1000°C, that is, it may occur under roasting conditions. For the less noble metals such as nickel, iron, and zinc the reactions require much higher temperatures and are, for all practical purposes, impossible.

The oxidation of a metal sulfide to the corresponding metal is called a *roast-reaction*, as it was believed earlier that some sulfide first had to be converted into oxide, which afterwards reacted with the remaining sulfide to give the metal. The principle of the roast-reaction is used in matte smelting and converting of copper sulfide ores, and to a minor extent in lead smelting, both of which will be discussed in Chap. 12. In comparison, a scheme where the sulfide is first roasted to oxide, which subsequently is reduced with, for example, carbon, is called *roast-reduction*.

We see furthermore that metal sulfate is formed at low temperatures and high partial pressures of SO_2 and O_2. Also the temperature for sulfate formation differs considerably from one metal to another. The highest sulfating temperature, corresponding to high temperature for sulfate decomposition, is found for lead and zinc. Lower temperatures are found for copper and nickel, whereas iron oxide is converted into sulfate only at temperatures below 600°C. Direct conversion from sulfide to sulfate requires even lower temperatures.

In Fig. 8-9 two possible paths for the roasting process are indicated. Usually the path is as given by the dotted line (a). The roasting mainly occurs from sulfide to oxide with an increase in temperature. Toward the end of the process the temperature drops and some sulfate may be formed. In other cases the process may be according to path (b). First the sulfate is formed at low temperature. By means of additional heat the temperature is raised until the sulfate is decomposed.

So far we have considered the reactions for each metal sulfide independent of other metals. In the roasting of complex sulfide ores additional reactions may occur. First of all the different sulfides may form solid solutions and even complex sulfides. Second the oxides produced may react among themselves to make complex oxides. This is illustrated by Figs. 8-10(a) and (b) which show the phase equilibria in the system Zn–Fe–S–O for a constant SO_2 pressure of one atmosphere.[1] Figure 8-10(a) shows the oxygen potential, log p_{O_2}, for the various equilibria at 891°C as function of the molar fraction of iron, $N_{Fe} = n_{Fe}/(n_{Fe} + n_{Zn})$, in the solids. We see that at low oxygen potential ZnS is able to take up to 35 mole percent of FeS in solid solution forming the sphalerite, (Zn,Fe)S, phase. At higher iron contents the pyrrhotite phase, "FeS", is also found. On oxidation of the sulfides a spinel phase $(Zn,Fe)Fe_2O_4$, is formed. Starting with a two-phase mixture (Zn,Fe)S + "FeS" the first oxidation will take place at log $p_{O_2} \simeq -11.7$ and the oxidation product will be almost pure Fe_3O_4. As iron sulfide is oxidized the remaining sphalerite will be richer in zinc, and spinels of higher zinc content will be formed until at log $p_{O_2} \simeq -10.5$ almost pure ZnS is in equilibrium with almost pure $ZnFe_2O_4$. At this oxygen potential the remaining sphalerite will oxidize to a mixture of ZnO and $ZnFe_2O_4$. At oxygen potentials above log $p_{O_2} \simeq -7$, Fe_3O_4 in the spinel phase will oxidize to Fe_2O_3, and at log $p_{O_2} \simeq -2.1$, ZnO will react to give the basic sulfate $ZnO \cdot 2ZnSO_4$. Finally at log $p_{O_2} \simeq -1.6$, $ZnFe_2O_4$ will be converted to basic sulfate, which again reacts to give the normal sulfate, $ZnSO_4$, at log $p_{O_2} \simeq -0.5$, all values for 891°C.

The effect of temperature on all three-phase equilibria in the Zn–Fe–S–O system is shown in Fig. 8-10(b), and we see once more that the sulfate phases are favored by low temperature and high oxygen potential and that $ZnSO_4$ may coexist with Fe_2O_3 within a wide temperature range. At even lower temperature ferric sulfate, $Fe_2(SO_4)_3$, will be formed.

Figures 8-10(a) and (b) refer to 1 atm of SO_2. At lower SO_2 pressures, such as encountered in industrial roasting processes the oxygen potentials for the sulfide–oxide equilibria will shift to lower values, and for the oxide–sulfate

T. Rosenqvist: *Met. Trans.*, **9B**: 337 (1978); see also: J. M. Skeaff and A. W. Espelund: *Can. Metall. Q.*, **12**: 445 (1973); A. W. Espelund and H. Jynge: *Scand. J. Metall.* **6**: 256 (1977).

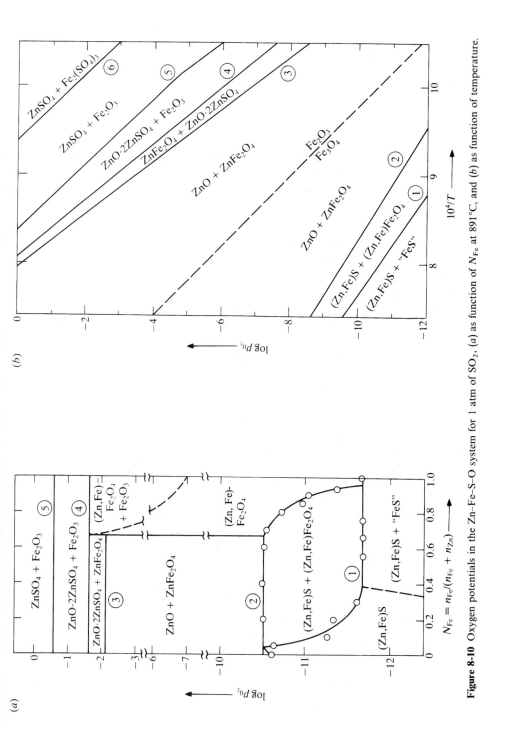

Figure 8-10 Oxygen potentials in the Zn–Fe–S–O system for 1 atm of SO_2, (a) as function of N_{Fe} at 891°C, and (b) as function of temperature.

221

equilibria to higher values in accordance with the stochiometry of the corresponding reactions. Similar but even more complicated phase relations apply for the copper–iron–sulfur–oxygen system.[1]

Kinetics of Roasting

Roasting is a process where a solid reacts with a gas to form another solid and a gas. Thus the process is analogous to the gaseous reduction of oxides described in Sec. 5-5. In the roasting of FeS, concentric layers of Fe_3O_4 and Fe_2O_3 are formed. In the roasting of Cu_2S there will be concentric layers of Cu_2O and CuO and even $CuSO_4$. Thus in a partly roasted ore we may have unreacted sulfide as well as sulfate present.

The oxide layer is relatively dense and the rate of roasting is usually controlled by diffusion of oxygen in, and sulfur dioxide out through the oxide layer. This means that roasting will follow Eqs. (5-18) and (5-19), and a rather long time is needed to oxidize the last traces of sulfides. Sulfate formation is even slower than the roasting as such. This follows from the fact that the sulfate occupies a larger volume than does the oxide; the sulfate layer will be quite dense, and diffusion of the gases through this layer correspondingly slow.

The formation of *ferrites* depends also on the physical nature of the original ore. If the ore consists of separate phases of FeS and ZnS these will oxidize to form separate phases of Fe_2O_3 and ZnO, and ferrite formation will take some time. If, on the other hand, the original ore were a solid solution of FeS in ZnS the formation of ferrites will be greatly enhanced. Furthermore, it is found by experience that the formation of zinc ferrite is less pronounced if the roasting is carried out very rapidly and at a high temperature, whereas slow roasting at lower temperature increases the amount of zinc ferrite in the product. With reference to Fig. 8-10(a) a possible explanation for this phenomenon would be that the first oxidation product, which is almost pure Fe_3O_4, is oxidized further to Fe_2O_3 before it has a chance to form the zinc ferrite.

A kinetic phenomenon of rather special character is the so-called "kernel roasting," (German: Kern Röstung).[2] If coarse Fe–Cu sulfide is roasted slowly and at low temperature we will find that the copper diffuses toward the center where it combines with the remaining sulfur, and iron diffuses toward the surface and is oxidized. In extreme cases we get a core which is almost pure Cu_2S, whereas the outer oxide layer is practically free of copper. This phenomenon may easily be understood from the thermodynamics of roasting. Since iron oxides are more stable than copper oxides, i.e., for the reaction Cu-oxide + FeS = Fe-oxide + Cu_2S the change in Gibbs energy is negative, the Gibbs energy for a partly roasted ore will be the lowest if copper is present as sulfide and iron as oxide. Since the rate of diffusion of the cations (copper and iron) is

[1] T. Rosenqvist and A. Hofseth: *Scand. J. Metall.*, **9**: 129 (1980).
[2] See e.g.: T. A. Henderson: *Trans. Inst. Min. Metall.*, **67**: 437–462; 497–520 (1957–58); **68**: 193–200 (1958–59).

higher than that of the anions (sulfur and oxygen), copper diffuses to the center which is rich in sulfur. Thus kernel roasting is a special case of "uphill diffusion" as described in Sec. 5-2. It has been found that kernel roasting is greatly affected by the nature of the original ore, and it may be accelerated by the addition of small amounts of alkali salts. These probably increase the rate of diffusion of the metal ions by forming eutectic melts along the grain boundaries.

Technology of Roasting

Roasting may be carried out in a number of different furnaces. In the old days it was done in stationary heaps, later in hand-rabbled furnaces. The mechanically rabbled *multiple-hearth furnace* was for a long time dominant for the roasting of most sulfide ores. Later flash-roasting was developed. For descriptions of these furnace types reference is made to standard textbooks in extractive metallurgy. In recent years *fluidized-bed roasting* has become the dominant technique (see Fig. 8-11).

Roasting is a strongly exothermic process:

$$FeS_2 + 2.75O_2 = 0.5Fe_2O_3 + 2SO_2 \qquad \Delta H_{298} = -833 \text{ kJ}$$

$$ZnS + 1.5O_2 = ZnO + SO_2 \qquad \Delta H_{298} = -442 \text{ kJ}$$

$$Cu_2S + 1.5O_2 = Cu_2O + SO_2 \qquad \Delta H_{298} = -384 \text{ kJ}$$

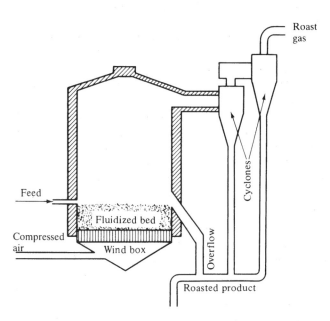

Figure 8-11 Fluidized-bed reactor for roasting of sulfide concentrates.

We see that the heats of reaction are of the same order of magnitude as for combustion of carbon. If the heats of reaction are combined with the heat capacity of the reaction products the roasting of, say ZnS with air, produces an adiabatic temperature of about 1700°C. This is above the melting point of most metal sulfides and oxides, and if roasting were carried out without heat losses it would give a more or less molten product. Usually the heat losses are sufficiently large to keep the temperature down. The heat losses, and consequently the available heat, depend on the roasting rate, and if the rate is very low a relatively large part of the heat is lost to the surroundings. The rate of roasting depends on the type of the equipment used, but also on how completely the roasting is carried out. Thus a copper concentrate may be roasted autogeneously in a multiple hearth furnace provided the sulfur is not completely eliminated. If, on the other hand, a so-called "dead roast" is wanted, additional heat may have to be supplied.

In modern flash-roasting and fluidized-bed roasting the roasting velocity is very high, and heat losses correspondingly small. In these cases cooling may be necessary in order to prevent the charge from sintering.

The heat balance for roasting will be illustrated by the roasting of pyrite with 50 percent air excess in a fluidized bed. See Table 8-4.

We see that the process has a heat surplus of about 214 kJ, which is more than the normal heat loss, and in order to keep the temperature down, additional cooling is necessary. In some types of furnaces this is done by feeding water directly into the fluidized bed. In others water-cooled tubes are placed in the reactor wall. We see further that the roast gas has a heat content of about 554 kJ, which is more than half the total heat of roasting. In most modern plants this is used for the production of steam, and more than 1 tonne of steam may be produced per tonne of pyrite.

The capacity of a fluidized-bed roaster is determined partly by the chemical roasting rate, partly by the dynamics of the fluidized bed. For coarse-grained ore, say 1 to 6 mm, the roaster may be operated at a high air rate without extensive dust losses. In this case there is always enough air available, and the necessary retention time for the ore is determined by the chemical rate. (See

Table 8-4 Heat balance for pyrite roasting

Reaction: $(FeS_2 + 4.13O_2 + 15.5N_2)(25°C) = (0.5Fe_2O_3 + 2SO_2 + 1.38O_2 + 15.5N_2)(900°C)$

Input		kJ	Output		kJ
FeS$_2$	25°C	0	0.5Fe$_2$O$_3$		65
4.13O$_2$ + 15.5N$_2$	25°C	0	2SO$_2$	H_{298}^{1173}	86
Fe + 0.75O$_2$			15.5N$_2$		429
= 0.5Fe$_2$O$_3$	$-\Delta H_{298}$	411	1.38O$_2$		39
2S(rh) + 2O$_2$			FeS$_2$ = Fe + 2S(rh)	ΔH_{298}	172
= 2SO$_2$	$-\Delta H_{298}$	594	Heat surplus		214
Sum		1005	Sum		1005

Sec. 6-6 and Fig. 6-7). For fine-grained material the chemical rate is higher and the capacity is limited by the rate at which air can be supplied. In such cases the roaster is usually operated with large dust losses, and in some processes the entire production is delivered by the cyclone.

In addition to "dry roasting," roasting may be carried out also at temperatures where a sintered or even molten product is obtained. Roasting combined with sintering will be discussed in Sec. 8-4, and roasting combined with smelting will be discussed in Sec. 12-3.

Chloridizing Roast

The purpose of what is normally called chloridizing roast is to bring the nonferrous metals in, for example, pyrite cinder into a water-soluble form. Pyrite cinder which contains various amounts of Cu, Zn, S, etc., is mixed with about 10 percent NaCl and is roasted with air at 500 to 600°C. The process is often described by the reaction

$$MeS + 2NaCl + 2O_2 = Na_2SO_4 + MeCl_2$$

In addition, various amounts of SO_2, HCl, and Cl_2 are formed. Most likely a low-melting mixture of sulfates and chlorides is formed, penetrating the pores of the ferric oxide. In such an ionic salt mixture we cannot say that one metal is present as a chloride, another as a sulfate. Thus the process may equally well be called a sulfating roast where sodium chloride is present as a catalyst. The roasted product is subsequently leached in acid to recover the nonferrous metals.

Another type of chloridizing roast is used to convert the nonferrous metals into volatile chlorides. Such a process must be carried out at a relatively high temperature, and it may be carried out either by means of calcium chloride or elemental chlorine. In the first case the reactions are of the type

$$MeO + CaCl_2 = MeCl_2(g) + CaO$$
$$MeS + CaCl_2 + 1.5O_2 = MeCl_2(g) + CaO + SO_2$$

In the second case the reactions are

$$MeO + Cl_2 = MeCl_2(g) + \tfrac{1}{2}O_2$$
$$MeS + Cl_2 + O_2 = MeCl_2(g) + SO_2$$

The first reactions require a temperature of about 1250°C. For the second reactions 900 to 1000°C is sufficient. Iron oxide is not easily chloridized, the reaction of the type

$$3MeO + 2FeCl_3 = 3MeCl_2 + Fe_2O_3$$

being strongly shifted to the right.

The Calcium–Sulfur–Oxygen System

This system has received attention for the various reasons:

1. Desulfurization of metals and metal sulfides
2. Fixation of sulfur from combustion and roast gases
3. Utilization of gypsum and anhydrite

As seen from Fig. 8-12 the stable phases that can be formed are CaO, CaS, and $CaSO_4$. For unit activity of CaO and 1 atm of SO_2 the oxygen potentials for the equilibrium CaO–CaS and CaO–$CaSO_4$ are single functions of temperature.

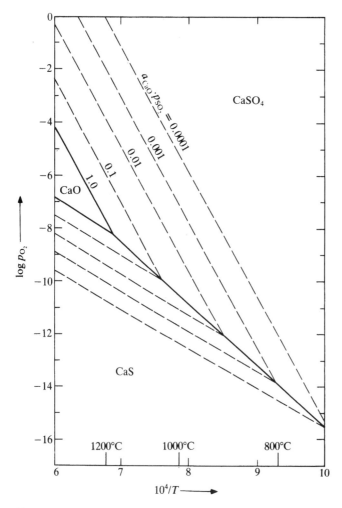

Figure 8-12 Oxygen potential in the Ca–S–O system.

For SO_2 pressures of less than 1 atm or for CaO activities less than unity the oxygen potential for the CaO–CaS equilibrium shifts to lower, and for the CaO–CaSO$_4$ equilibrium to higher values. This is shown by the dashed lines which give the oxygen potential for various values of the product $a_{CaO} \cdot p_{SO_2}$. In contrast, the line which gives the equilibrium between CaS and CaSO$_4$ is independent of this product.

It follows that CaO can react with sulfur in two different ways: under reducing conditions it can react to form sulfide; under oxidizing conditions it can form sulfate. The formation of sulfide is the basis for the removal of sulfur from metals and metal sulfides. Thus lime or a lime-rich slag is used to remove sulfur from liquid iron and steel as well as from solid sponge iron (Chaps. 9, 11, and 13). Lime has also been proposed as a means to produce metals directly from their sulfide ores by reduction with hydrogen

$$MeS + CaO + H_2 = Me + CaS + H_2O$$

Whereas reduction of most metal sulfides with only hydrogen is virtually impossible due to unfavorable equilibrium constants the presence of lime, which ties up the sulfur, helps to shift the reaction to the right. Instead of hydrogen other reducing agents such as CO or even coal can be used. Calcium oxide may also be used to remove impurity sulfur directly from hot gaseous or solid fuels.

Whereas formation of CaS is most favorable at high temperatures CaSO$_4$ is preferably formed at lower temperatures. The lower temperature limit is set by the kinetics of the reaction. As already mentioned sulfation is a slow process, and temperatures above 800°C are needed in order to get an acceptable rate. Even then the efficiency of the lime rarely exceeds 25 percent. The reaction to form CaSO$_4$ has been particularly studied as a way to remove SO_2 from hot combustion gases, e.g., in steam boilers. It has also been suggested as a way to remove SO_2 from metallurgical roast gases. In most metallurgical plants the roast gases are utilized to produce sulfuric acid. However, in localities where there is no need for sulfuric acid, and where emission of the roast gases to the atmosphere is prohibited, absorption in lime is a possibility. This absorption may be done in a separate reactor, but could also be done directly in the roaster. Thus it has been proposed to roast chalcopyrite with lime to obtain the reaction[1]

$$CuFeS_2 + CaO + \tfrac{17}{4}O_2 = CuSO_4 + \tfrac{1}{2}Fe_2O_3 + CaSO_4$$

The roasted product could be leached in acid to extract its copper values.

Whereas in the fixation of sulfur from hot combustion and roast gases SO_2 is tied down as sulfate which is deposited, the opposite reaction is also possible. Thus in the 1950s when there was a shortage of sulfur on the world market some SO_2 for the production of sulfuric acid was made from naturally occurring gypsum or anhydrite. As is seen from Fig. 8-12 decomposition of

[1] R. W. Bartlett and H. H. Haung: *J. Met.*, December 1977, p. 28.

$CaSO_4$ to give SO_2 requires high temperatures and mildly reducing conditions. It is also favored by a low activity of CaO in the product. In the processes which were in use, some siliceous or clayey material was added to the sulfate, and the mixture was fired in a rotary kiln. In this way portland cement was obtained as an additional product.

The decomposition of $CaSO_4$ is a strongly endothermic reaction. As it also requires high temperatures and slightly reducing conditions the fuel consumption will necessarily be large. Under present market conditions where fuel is expensive, and sulfur from natural deposits or as by-products from the metal and petroleum industry is abundant, the use of $CaSO_4$ as raw material for sulfur is uneconomical. However, when today's deposits of cheap sulfur are depleted it may once more be attractive to attack the large deposits of gypsum and anhydrite that exist in nature.

Other Treatments of Roast Gases

Gases with various concentrations of SO_2 are produced in a number of processes for sulfide ores, such as roasting, sintering, smelting, and converting. If the SO_2 concentration is more than 5 percent the gases may be used directly for the production of sulfuric acid. For lower concentrations and in localities where the gases cannot be emitted to the atmosphere the sulfur dioxide may be removed in different ways. One method is reaction with limestone and water in an absorption tower, whereby $CaSO_4$ is formed and is washed away. Another possibility is absorption connected with upgrading. In that case an aqueous solution with a high buffering capacity in the pH range 3 to 6 is used, for example dimethylaniline or sodium citrate. Such a solution has the ability to absorb SO_2 at lower temperature and to give it off at higher temperature. By letting the solution circulate between a low-temperature absorption tower and a high-temperature desorption tower a high concentration of SO_2 gas can be obtained. Alternatively the desorption can be made by use of reduced pressure. The gas thus obtained can either be used for production of sulfuric acid or compressed to form liquid sulfur dioxide for other uses.

8-4 AGGLOMERATION

By agglomeration we understand processes where fine-grained materials, e.g., ore concentrates, are converted into coarser lumps. Agglomeration is used particularly if the ore is to be smelted in a shaft furnace where fine-grained material would plug up the gas passage. Agglomeration may be of the following types:

1. Briquetting
2. Nodulizing
3. Grate sintering
4. Pelletizing

Briquetting and Nodulizing

Briquetting is carried out in presses usually with the addition of a binder. The binder may be inorganic: lime, cement, clay, or metal-salts; or it may be organic: oil, tar, or pitch. Also sulfite liquor from wood pulp factories may be used. When lime or cement is used as a binder the briquettes may be hardened at room temperature, in the case of lime by treatment with carbon dioxide. When clay is used as a binder the briquettes are hardened by firing, e.g., in a tunnel kiln. Briquettes may also be made without binder and should then be hardened by firing. Organic binders are used for the production of zinc oxide–coke briquettes for the New Jersey vertical zinc retort process (Sec. 10-5). In this case the hardening takes place by heating in the absence of air whereby the binder is converted into coke.

Due to the heavy wear on the briquetting press, briquetting is a relatively expensive process, and with the exception of the above-mentioned zinc process it has not found wide application in extractive metallurgy.

Nodulizing is analogous to the production of cement clinker described in Sec. 8-2. The process is used, for example, to convert precipitated zinc hydroxide or zinc oxide fume into lumpy oxide, but on the whole it has limited application.

Grate Sintering

This is the process which probably has the largest industrial application, and many million tonnes of iron ore concentrate are annually converted into a spongy coke-like sinter by this method. The principle of the process is shown in Fig. 8-13.

The grate may either be stationary (grate sintering) or it may be part of a moving belt (Dwight–Lloyd sintering). The principle in both cases is the same.

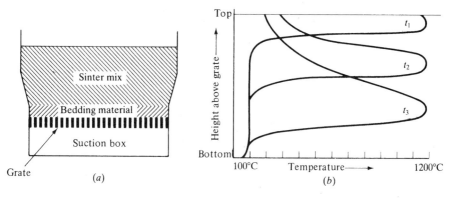

Figure 8-13 (*a*) Sintering grate (schematic). (*b*) Temperature distribution in grate sintering as function of time and height above grate, t = time.

In the sintering of a magnetite concentrate, this is mixed with about 5 percent coke powder and 5 to 10 percent water. The mixture is placed on the grate to a height of about 30 cm. In order to protect the grate material, usually cast iron, against the high temperature, a layer of previously sintered material is placed in the bottom of the bed. A suction of about 0.1 atm (1 m water column) is applied, and the sinter mix is ignited at the top by means of a burner. Due to the suction, the combustion zone moves downward through the bed, and when it reaches the bottom the sintering is finished. This is illustrated in Fig. 8-13(*b*) which shows the temperature distribution through the bed at three different times. Notice the similarity to Fig. 8-5(*b*).

The conditions in the bed may be described as follows: In the lower part of the bed the mix is still moist, and hot gases are coming from above. Farther up the bed all the water has evaporated, and the temperature increases to the ignition temperature of the coke powder. The coke burns rapidly, and the temperature increases to between 1200 and 1300°C, which is the maximum temperature of the sintering zone. At this point a certain reduction of Fe_3O_4 to FeO may take place near the coke particles. Above the combustion zone the temperature is still high, and the atmosphere is essentially pure air. Here FeO and Fe_3O_4 are oxidized to Fe_2O_3 under evolution of additional heat. Finally in the upper part of the bed the hot sinter is cooled by the air which conversely is being preheated. As the combustion-sintering zone moves down through the bed all the material in the bed goes through the above sequence. We may say that, relative to the sintering zone, the solid charge and the air move in counter-current to each other.

The agglomeration of the ore particles takes place in the zone of highest temperature. It is caused partly by crystal growth and recrystallization, and partly by the formation of small amounts of molten phase. If the ore contains silica, low-melting iron silicate is formed. This gives a very strong sinter, but the glazy iron silicate which penetrates the sinter makes it less reactive during the subsequent reduction in the blast furnace. Usually a balance will have to be found between the strength of the sinter and its reducibility. Sometimes limestone or other basic material is added to the sinter mix. Instead of iron silicates we get lime silicates which have a higher melting point. Sometimes sufficient limestone is added to make the sinter "self-fluxing," i.e., no further addition of lime is needed in the blast furnace. Self-fluxing sinter usually has a high reducibility, and the formation of lime ferrites helps to make it mechanically strong.

The capacity of the sintering grate is determined primarily by the rate at which air may be sucked through the bed. As an approximate rule we may say that the sintering zone reaches the bottom of the bed when the amount of air sucked through has reached the same heat capacity as the solid charge. The reason for this is as follows: As was shown in Sec. 6-7 the best heat exchange is obtained if two fluids in counter-current have the same heat capacity per unit time. In a sintering grate the heat capacity of the charge per unit time is given by the velocity of the sintering zone, whereas the heat capacity of the air per unit time is given by the suction rate. If, for some reason, the sintering zone had moved more slowly than what corresponded to equal heat capacity per unit time, the temperature difference between the hot gas and the cold sinter mix would increase. This would cause the sinter mix to heat up more rapidly until balance was again established between the heat capacity of the air and of the charge.

The rate of suction is determined by two factors: (1) the applied suction, and (2) the gas permeability of the charge. The suction may be increased to a maximum of about 0.2 atm, but with a rapid increase in the necessary fan power. At larger suction the pressure on the sinter bed becomes so large that the bed collapses. The permeability may be increased by making the sinter mix as porous as possible, Sec. 6-4, and this is the method most commonly used. The porosity depends mainly on the grain size of the ore concentrate, and sintering is particularly well suited for rather coarse-grained materials, say +100 mesh. Concentrates finer than 200 mesh may have to be subjected to a pretreatment, *micropelletizing*, before they can be sintered satisfactorily on the grate.

The water addition also affects the permeability. At moderate water contents the grains will stick to each other and form bridges, giving a high porosity. At higher water contents the bridges will collapse, and the porosity will again decrease. The highest permeability is obtained usually, for about 10 percent water in the mix. The water also retards the heating of the charge. This makes the sintering zone narrower, and with a correspondingly higher temperature. Thus the water content may actually be instrumental in reducing the fuel requirements.

The quality of the coke powder is also important. If the coke is too coarse or unreactive the combustion is retarded and the desired maximum temperature is not reached. If, on the other hand, the coke is too reactive the combustion zone may run away from the sintering zone, again resulting in a poorly sintered product.

Grate sintering is a very efficient process. Due to the counter-current principle the air and the solid charge are both preheated before they reach the combustion-sintering zone. This results in a low fuel consumption, about 5 percent for magnetite. Here the oxidation of magnetite to hematite gives additional heat. The sintering of hematite concentrates requires more fuel and, in general, hematite is more difficult to sinter.

Sintering of Ores with Sulfur or Arsenic

Sulfur may be present in ores as sulfides but also as sulfates. Thus some iron oxide ores contain small amounts of pyrite. The pyrite oxidizes rapidly and completely during the sintering process, and the evolved heat helps to reduce the fuel requirements. The author has made experiments where an iron ore was sintered with the addition of 9 percent pyrite, that is, 5 percent sulfur and without any coke. The product was of satisfactory quality and its sulfur content was about 0.1 percent.

Sintering is also applied to nonferrous sulfide ores, particularly zinc and lead sulfides. In that case the sintering becomes a roasting process as well. If pure sulfide concentrates were sintered, the temperature would be too high, and complete melting would occur. The ore may then either be preroasted by traditional methods to about 5 to 10 percent sulfur, or it may be mixed with large amounts of return sinter. The latter method is much used in the sintering of zinc sulfide for the carbothermic zinc process. In this case the raw sulfide ore is mixed with about five times as much return sinter. After sintering, the product is screened, and $\frac{5}{6}$ of the product is returned, whereas $\frac{1}{6}$ is passed on to the reduction plant.

In the sintering of lead sulfide ores for subsequent smelting in the blast furnace the highest possible elimination of sulfur is attempted. For this purpose a high temperature is needed, and in addition to lead oxide some metallic lead will be formed. Since, however, PbS, Pb, and PbO have low melting points, special precautions must be made to prevent the bed from collapsing or liquid lead from running into the grate. In addition to return sinter some limestone is added to the sinter mix. Calcination of limestone around 900°C helps to absorb the excess heat of the process and the burned lime forms a refractory skeleton which carries the partly molten phases. Furthermore, lead sintering is often made by *updraft*: After a thin initial layer of sulfide has been ignited by suction the remaining sinter mix is added, and air is blown from below through the bed. Thus the sintering zone moves upward instead of downward, and the updraft prevents the liquid phases from running into the grate.

In the sintering of sulfide ores one attempts to get a gas with sufficiently high concentration of SO_2 for subsequent production of sulfuric acid. If the sintering is done on a moving grate (Dwight–Lloyd machine), this is usually constructed to allow several passes of the gas through the bed. The gas from the last part of the bed, which is low in SO_2, is returned to the first part in order to become further enriched before it is passed on to the sulfuric acid plant.

Some ores contain sulfates, for example, $BaSO_4$ in some iron oxide ores. This may decompose at high temperature according to the reaction

$$BaSO_4 = BaO + SO_2 + \tfrac{1}{2}O_2$$

As already mentioned for the decomposition of gypsum the reaction is promoted by slightly reducing conditions which exist around the coke particles, and also by the addition of acid oxides, such as silica, which forms stable barium silicates. Since the reaction is endothermic, the temperature should be as high as possible.

Analogous to the decomposition of sulfates is the removal of arsenic. Arsenic is present in some ores as a stable arsenate, for example, $Ca_3As_2O_8$. This may be reduced to form volatile As_2O_3

$$Ca_3As_2O_8 + 2CO = 3CaO + As_2O_3(g) + 2CO_2$$

Also in this case the reaction is promoted by high temperature and the presence of acid oxides. Thus for both the decomposition of sulfates and arsenates the requirements are: (1) high temperature, (2) slightly reducing conditions, (3) addition of acid oxides. In practice the two first requirements are met by increased coke addition.

Pelletizing

This method is particularly well suited for very fine grained iron ore concentrates which are not easily sintered on the grate. The process consists of two steps: (1) rolling of pellets at room temperature, and (2) firing of pellets at 1200 to 1300°C. The rolling is carried out with the addition of water and some binder, which may be clay mineral (bentonite), lime, salts, or organic material. The rolling is based on the principle that water in the pores between the ore particles exhibits a capillary force which binds the particles together. This force is proportional to the total surface area of the particles, and thus increases with decreasing particle size. The rolling is carried out either in a drum or on a rotating disc, and resembles the rolling of snowballs on a hillside. The pellets are rolled, usually, to a diameter between 10 and 20 mm. In some cases micropellets of about 3 mm diameter are also made.

The raw (green) pellets have sufficient strength to be handled and moved. During the firing process water evaporates, and the pellets would collapse if it were not for the binders. These give the pellets sufficient strength to withstand the heating from room temperature to between 1200 and 1300°C, which is necessary for the pellets to harden by sintering.

The firing of the pellets is carried out either on a sintering grate or in a shaft furnace. Firing on a Dwight–Lloyd machine is most common, particularly for large tonnages. The fuel may be coal dust, oil, or gas. When coal dust is used it is usually rolled as a separate layer on the outside of the pellets. The firing in that case is very similar to direct sintering with coke powder on the grate; only due to the size of the pellets, the heat requires a longer time to penetrate the pellets and the sintering zone is not as sharp as for sintering of concentrates.

When the firing is by oil or gas these are burned in a combustion chamber above the grate, and the hot combustion gases are sucked through the bed of pellets. In order to utilize the heat content in the gas, this is usually made to pass several times through the bed. A layer of previously burned pellets is usually placed in the bottom of the bed to absorb the heat content of the gases before they leave the grate, and protect the grate from overheating. A bed of pellets has a much higher permeability than a bed of concentrate. Therefore a larger gas flow and a higher bed are possible, resulting in a larger sintering capacity.

The firing of pellets in a shaft furnace is very similar to the calcination of limestone described in Sec. 8-2. In the lower part of the shaft the combustion air is preheated in counter-current with hot pellets, and in the upper part the green pellets are preheated in counter-current with the hot combustion gases. Contrary to calcination shown in Fig. 8-5(b), the heat capacity of the gas flow is less than that of the solid flow. In the upper, preheating zone, therefore, the combustion gases cool rapidly, whereas the cooling of the fired pellets is less effective. Owing to good heat exchange good heat economy is obtained, and fuel values down to about 10 to 15 kg of oil per metric ton (tonne) of pellets have been reported. This applies to magnetite ores. For hematite ores the fuel consumption is about twice as large.

PROBLEMS

8-1 A coal has the following ultimate analysis (on moisture-free basis): $C = 79.86$ percent, $H = 5.02$ percent, $S = 1.18$ percent, $N = 1.86$ percent, $O = 4.27$ percent, ash $= 7.81$ percent, moisture $= 3$ percent.

(a) By means of the *Dulong formula* calculate its net and gross calorific power in kilojoules per kilogram (moisture-free basis).

(b) Calculate the adiabatic flame temperature if the coal ($25°C$) is burned with 125 percent of the stoichiometric amounts of air ($25°C$).

8-2 Pure amorphous carbon is blown with air ($25°C$) in a gas producer to give a gas of only CO and N_2. Calculate the volume of air and producer gas (in normal cubic meters) per kilogram of carbon, and calculate the calorific power of the gas. Compare the calorific power of the gas with that of the carbon, and explain the difference.

8-3 In a laboratory gas producer charcoal, assumed to be pure amorphous carbon ($25°C$), is reacted with a mixture of CO_2 and O_2 ($25°C$) to make pure CO. The produced CO is assumed to leave the reaction zone at $1000°C$ and to preheat the charcoal in counter-current to $800°C$. Heat losses through the walls are compensated by electric heating, and may be ignored. Calculate the ratio of O_2 to CO_2 which is required to make the process autogeneous, and calculate the temperature of the effluent CO.

8-4 Lump limestone is calcined in an atmosphere of $1000°C$ and with $p_{CO_2} = 0.5$ atm. The heat and mass transfer resistances at the lump surface are assumed to be insignificant. After some time a layer of burned lime is formed.

(a) Assuming that the rate constants for chemical reaction at the carbonate–oxide interface and for gas diffusion through the oxide layer are large compared to the heat conductivity through the oxide (all expressed on a common basis), make a plot of how you expect the CO_2 pressure and temperature to vary through the lump.

(b) Make a similar plot assuming the rate constants for chemical reaction and diffusion each to be of similar magnitude to the heat conductivity. The *equilibrium* decomposition pressure of limestone is 0.5 atm at $860°C$ and 1.0 atm at $900°C$. Be careful in giving the temperature and pressure scales on your plot.

8-5 By means of data given in App. C calculate the temperatures at which the decomposition pressure $(p_{SO_3} + p_{SO_2} + p_{O_2})$ is one atmosphere for the following sulfates: $Fe_2(SO_4)_3$, $CuSO_4$, $ZnSO_4$, $CoSO_4$. Remember that the atomic ratio n_O/n_S in the gas phase must be three.

8-6 Calculate the lowest temperature at which Cu_2S may be roasted *with air* to give metallic copper and SO_2.

8-7 Calcium oxide may be used to remove SO_2 from hot flue gases. Calculate the equilibrium pressures of SO_3 and SO_2 in the presence of CaO and $CaSO_4$ at $800°C$ if $p_{O_2} = 0.01$ atm.

8-8 ZnS is roasted with 50 percent air excess beyond what is needed for the reaction: $ZnS + 1.5O_2 = ZnO + SO_2$.

(a) Calculate the roast gas composition in volume percent.

(b) Make a heat balance for the process (basis 1 kg ZnS) if the reactants are introduced at 25°C, and the products are withdrawn at 900°C.

(c) The roasting is carried out in a cylindrical fluidized-bed furnace with internal dimensions 2 m diameter, and 5 m high. The cylindrical surface and the ends have firebrick linings, 0.25 m thick, with mean heat conductivity $k = 1.5$ W/(m·K). The inner temperature of the lining is 900°C, the outer 100°C. The feeding rate of ZnS is 5000 kg/h. Calculate the surplus heat available for steam production in the furnace in kilojoules per hour.

8-9 FeS_2 is dead roasted with air in a fluidized-bed reactor. The mean diameter of the bed particles is 1 mm and their density is 3.35 g/cm^3.

(a) By means of information given in Sec. 6-4, calculate the minimum and maximum linear and mass velocities of the fluidizing gas, when its density and viscosity at roasting temperature are 0.34 kg/m^3 and 4×10^{-5} kg/(m·s) respectively. The void fraction at initial fluidization is taken equal to 0.5.

(b) In practice air is introduced with a mass velocity, referred to the bed area, of 1 kg/(m^2·s). Calculate the feeding rate of FeS_2 in kilograms per square meter·hour when 100 percent air excess is used.

(c) For the conditions given under (b) a mean retention time of 5 h is required to ensure a dead roast. Calculate the height of the fluidized bed when the void fraction is 0.6.

(d) For the conditions given under (b)–(c) calculate the pressure drop in the bed in pascals as well as millimeters of mercury. ($\rho_{Hg} = 13.55$ g/cm^3).

(e) Discuss the effect on gas velocity, feeding rate, retention time, and bed height if the mean particle size is reduced to 0.1 mm, keeping other parameters constant.

8-10 Magnetite concentrate is sintered on a fixed-bed sintering grate (Sec. 8-4) whereby it is oxidized to hematite.

(a) Calculate the theoretical amount of air, in normal cubic meters per kilogram sinter, required for the sintering zone to move through the bed.

(b) In practice 25 percent air surplus is used. The pressure drop in the bed is 0.1 atm ($= 10.13$ kPa). Calculate the theoretical fan energy in watthours per kilogram, if the fan temperature is 60°C, and the fan operates against atmospheric pressure. Only friction energy in the bed is considered.

8-11 Magnetite pellets with 10 percent moisture are to be fired in an oil-fired shaft furnace, whereby Fe_3O_4 is oxidized to Fe_2O_3. The fired pellets and the combustion gases are assumed to leave the firing zone at 1200°C. The green pellets (25°C) are dried and preheated in counter-current with the combustion gases, cooling these to 100°C, and the combustion air is assumed preheated to 1000°C in counter-current with the hot pellets before it is reacted with the oil. The oil contains 85 percent C and 15 percent H, and the air excess for the combustion of oil *and* magnetite is 50 percent.

(a) Using the procedure given in Sec. 8-2 calculate the oil consumption per metric ton of produced pellets, as well as the temperatures of the preheated magnetite and of the cooled hematite pellets. Heat losses are disregarded. The net calorific power of the oil is 42 000 kJ/kg.

(b) If heat losses are considered, in which direction will this affect the oil consumption and the temperatures of the preheated magnetite and of the cooled hematite pellets?

BIBLIOGRAPHY

Ball, D. F., J. Dartnell, J. Davison, R. Grieve, and R. Wild: "Agglomeration of Iron Ores," Heinemann Educational Books Ltd., London, 1973.
Griswold, J.: "Fuels, Combustion and Furnaces," McGraw-Hill Book Co. Inc., New York, 1946.
Gumz, W.: "Gas Producers and Blast Furnaces," John Wiley and Sons Inc., New York, 1950.
Knepper, W. A. (Editor): "Agglomeration," Interscience Publishers, New York, 1962.
Meyer, K.: "Pelletizing of Iron Ores," Springer Verlag, Berlin, 1980.
Trinks, W.: "Industrial Furnaces," vol II, 3d ed., John Wiley and Sons Inc., New York, 1955.

NINE

REDUCTION OF METAL OXIDES

A large number of metals are produced from oxide raw materials. This is for example the case for iron, manganese, chromium, and tin, which are produced almost exclusively from oxide ores. In other cases, as for example lead and zinc, the sulfide ores are first roasted to give oxides which subsequently are reduced to metal.

Only the oxides of the more noble metals may be converted to the metal by simple thermal decomposition. This is the case for silver oxide which, at temperatures above 200°C, decomposes according to the reaction $2Ag_2O = 4Ag + O_2$. Similarly, PtO decomposes above about 500°C and PdO above about 900°C.

All other oxides are decomposed by means of a reducing agent. This may be carbon, carbon monoxide, hydrogen, or in special cases some other metal which has a higher affinity for oxygen. Carbon, carbon monoxide, and hydrogen are the reducing agents which are of the greatest industrial and economic importance, and may be produced from raw materials such as coal, oil, or natural gas.

9-1 THERMODYNAMICS OF OXIDE REDUCTION

Like every metallurgical reaction the reduction with carbon and carbon monoxide is governed by the prevailing chemical equilibria and by reaction kinetics. From thermodynamic data the standard Gibbs energy change and the equilibrium constant may be calculated for the various reactions of the type MeO + CO = Me + CO_2. The corresponding gas ratio p_{CO_2}/p_{CO} is shown in Fig. 9-1 for a number of metals of industrial importance. As long as the metal and the oxide coexist as condensed phases, i.e., with their activity equal to or close to unity, this gas ratio has a definite value and is a function only of temperature. It is seen that the ratio varies between about 10^5 for the reduction of Cu_2O to Cu and Fe_2O_3 to Fe_3O_4, to values of the order of 10^{-5} or less for the reduction of MnO and SiO_2 to the corresponding metals. Even lower gas ratios would be found for the reduction of Al_2O_3 and MgO.

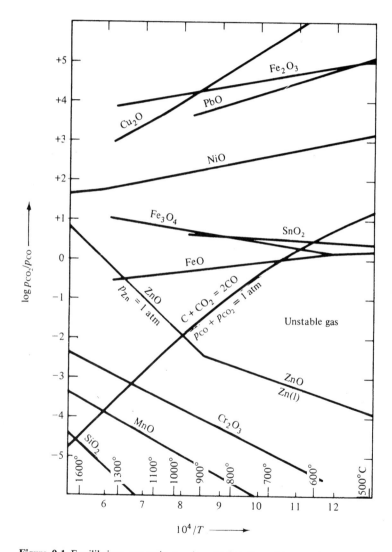

Figure 9-1 Equilibrium gas ratio p_{CO_2}/p_{CO} as function of inverse temperature for the reduction of various oxides.

The reduction of ZnO represents an exceptional case. At low temperatures ZnO is reduced to form liquid Zn, and the gas ratio is a function of temperature only. At higher temperatures zinc vapor is formed, $ZnO + CO = Zn(g) + CO_2$. In this case the gas ratio is given by $p_{CO_2}/p_{CO} = K/p_{Zn}$ where K is the equilibrium constant and p_{Zn} the partial pressure of zinc vapor. Of these, the equilibrium constant is a function of temperature only, whereas p_{Zn} may have any value. In Fig. 9-1 the gas ratio p_{CO_2}/p_{CO} is plotted for $p_{Zn} = 1$ atm. This curve intersects the curve for liquid zinc at the boiling point of zinc, 907°C, that is, liquid and gaseous zinc are in equilibrium with the same atmosphere, and consequently in equilibrium with each other. For a partial zinc pressure of 0.1 atm the gas ratio is displaced upward by one logarithmic unit, and the intersection with the curve for liquid zinc will be at about 740°C, which is the dew point of zinc vapor at 0.1 atm. For a zinc pressure of 10 atm the curve is displaced downward by one logarithmic unit, and so on. The special consequences which this behavior has for the thermal reduction of ZnO will be discussed in Chap. 10.

In Fig. 9-1 is also shown the gas ratio p_{CO_2}/p_{CO} for the *Boudouard reaction* $C + CO_2 = 2CO$. In this case the gas ratio is given by $p_{CO_2}/p_{CO} = p_{CO}/K$, that is, it is a function of p_{CO} and consequently of the total pressure $p_{CO} + p_{CO_2}$. This is a result of the reaction resulting in an increase in the number of gas molecules. In Fig. 9-1 the gas ratio p_{CO_2}/p_{CO} is plotted for a total pressure $p_{CO} + p_{CO_2} = 1$ atm. For higher or lower values of the total pressure the curve will shift upward or downward in accordance with this equilibrium expression.

For the reduction of a metal oxide with CO in the absence of solid carbon we may disregard the curve for the Boudouard reaction. In this case reduction of the oxide takes place when the gas ratio in the atmosphere falls below the equilibrium line in question. Thus the oxides of copper, lead, and nickel will be reduced at gas ratios between 10^5 and 10^2, that is, at very small concentrations of CO in the gas. This means that if originally pure CO is used, practically all of it will be converted into CO_2 before the reaction stops. The reduction of the oxides of manganese and silicon requires a gas ratio which is virtually free of CO_2. This means that with originally pure CO the reaction will stop as soon as minute amounts of CO_2 are formed. In other words, reduction of Cr_2O_3, MnO, and SiO_2 with CO gas is, in practice, impossible.

If solid carbon is present in the reaction mixture the two reactions

$$MeO + CO = Me + CO_2 \tag{1}$$

and

$$C + CO_2 = 2CO \tag{2}$$

take place simultaneously. Simultaneous equilibrium between MeO, Me, and C will occur at the temperature where the curves for the two reactions intersect. Thus it is seen that SnO_2, Sn, and C may be in equilibrium with an atmosphere of $p_{CO} + p_{CO_2} = 1$ atm at about 610°C. This means that SnO_2 may be reduced by carbon at any temperature above 610°C. Similarly, Fe_3O_4 will be reduced to

FeO above 650°C, FeO to Fe above 700°C, and so on. MnO and SiO_2 may be reduced by carbon at temperatures above about 1400 to 1600°C, all assuming 1 atm total gas pressure.

If the reaction takes place at the equilibrium temperature, the resulting gas mixture will be given by the value which corresponds to the intersection between the two curves. Thus for the reduction of SnO_2 at 610°C the p_{CO_2}/p_{CO} ratio is about 3, and for the reduction of MnO at 1500°C it is about 10^{-4}. If the reaction occurs above the equilibrium temperature the resulting gas mixture will have a value intermediate between the values for reactions (1) and (2), closest to the value for the reaction which has the highest reaction rate. Thus it is not strictly correct to say that the reduction with carbon is given by the equation $MeO + C = Me + CO$, as often is done. This would be the case only for the reduction of the most stable oxides where the concentration of CO_2 in the resulting gas mixture is extremely low.

It will be noted from Fig. 9-1 that the curve for NiO and for the oxides of the more noble metals do not intersect the carbon curve even at low temperatures. In these cases the temperature of reduction and the composition of the reaction gas mixture is determined entirely by reaction kinetics. It is a general observation that reduction of the oxides requires at least a few hundred degrees Celsius, and the reaction on the carbon surface occurs only above red heat. In practice even higher temperatures may be required.

The enthalpy for the different reactions in Fig. 9-1 may be deduced from the slope of the curves by the van't Hoff relation, Eq. (3-8). We see that the reduction of the oxides of the relatively noble metals with CO is exothermic, whereas the reduction of the less noble metal oxides is endothermic. Also the reaction between C and CO_2 is strongly endothermic. Therefore, the reduction of practically all metal oxides with carbon is endothermic, and the enthalpy of reduction increases with increasing stability of the oxide. Thus the reduction of the stable oxides with carbon requires not only a high temperature, but also a large amount of heat at that temperature. In practice this means that the necessary fuel and energy requirements increase considerably with increasing stability of the oxide.

Reduction with Hydrogen

Hydrogen is of less industrial importance in the reduction of oxides than carbon and carbon monoxide, but may be used under certain special conditions. The gas ratio p_{H_2O}/p_{H_2} for the reduction of important metal oxides is shown in Fig. 9-2. The curves are similar to the ones for reduction with CO and their relative order is, of course, the same. Since the reaction $H_2 + \frac{1}{2}O_2 = H_2O(g)$ is less exothermic than the reaction $CO + \frac{1}{2}O_2 = CO_2$ the reduction of MeO with H_2 is less exothermic than the corresponding reaction with CO. This shows up in the slope of the curves, and hydrogen at high temperature is a better reducing agent than carbon monoxide, but is a poorer one at lower temperature.

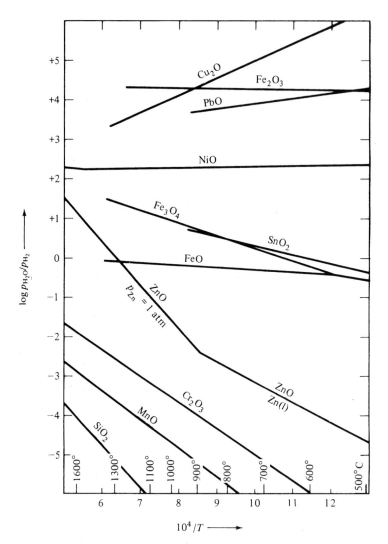

Figure 9-2 Equilibrium gas ratio p_{H_2O}/p_{H_2} as function of inverse temperature for the reduction of various oxides.

Very often the reduction is carried out with mixtures of CO and H_2. In such a case the reaction will occur so that the water–gas equilibrium $CO + H_2O = CO_2 + H_2$ is satisfied. This reaction shifts to the right at low temperatures, which again is another way of saying that carbon monoxide is a better reducing agent than hydrogen at lower temperatures.

One reducing agent of considerable importance is natural gas, which consists essentially of methane. If we calculate the equilibrium constant for a reaction such as $4MeO + CH_4 = 4Me + CO_2 + 2H_2O$ we will find that at high temperatures methane appears to be a better reducing agent than either CO or H_2. This is an example of the many pitfalls which one may fall into by careless application of thermodynamic data. As shown in Sec. 8-1 methane is unstable above 500°C and decomposes into carbon and hydrogen. The hydrogen will reduce the metal oxide according to its own equilibrium constant and the carbon according to its own. Therefore, when natural gas is used as a reducing agent, e.g., in the production of sponge iron, it is first subjected to a partial oxidation to give CO and H_2 as discussed in Sec. 8-1.

Another pitfall in the calculation of reduction equilibria is that the condensed phases which are actually formed may not be the ones which were assumed in the calculation. This is for example the case for the reduction of many oxides with carbon if the metal involved is a strong carbide-former. In that case the overall reaction will be $MeO + 2C = MeC + CO$ and the product will be a carbide rather than a metal. A calculation of the equilibrium to form metal is then unrealistic. In Chap. 14 some examples of this kind of reaction will be discussed.

Other Reducing Agents

In special cases a metal oxide may be reduced by means of another metal which has a higher oxygen affinity. It is seen from Fig. 9-1 that the equilibrium SiO_2–Si occurs at a lower p_{CO_2}/p_{CO} ratio, i.e., at a lower oxygen pressure, than the equilibrium FeO–Fe and Cr_2O_3–Cr. This means that silicon may reduce FeO and Cr_2O_3 by reactions of the form $Si + 2FeO = SiO_2 + 2Fe$ and $3Si + 2Cr_2O_3 = 3SiO_2 + 4Cr$. Other metals such as Al and Mg have even stronger oxygen affinities and may also be used as reducing agents.

Such *metallothermic reactions* are usually carried out between solid or molten reactants to give molten products. The entropy of reaction, $\Delta S°$, is in that case small, and as the Gibbs energy of reaction, $\Delta G°$, is negative this means that the enthalpy of reaction, $\Delta H°$, is also negative, i.e., the reaction is exothermic. The reaction is, therefore, favorable at low temperature, and is usually carried out slightly above the melting temperature of the products. In some cases, e.g., in the aluminothermic reduction of Cr_2O_3 or Fe_2O_3, the heat of reaction suffices to heat the reaction mixture from room temperature to above 2000°C, which is necessary to melt the products completely. In other cases some cold or inert material may have to be added to avoid overheating.

Metallothermic reactions are never complete. A final equilibrium will be established where some reducing metal remains unreacted in the metallic product and some of the metal oxide remains unreacted in the oxide slag. These amounts depend, of course, on the difference in oxygen activity between the reducing metal and the metal in question. Further discussion on metallothermic reactions will be given in Chap. 14.

9-2 REDUCTION OF IRON OXIDES

In Fig. 9-3 the p_{CO_2}/p_{CO} ratio for reduction of the various iron oxides, as well as the Boudouard reaction for $p_{CO} + p_{CO_2} = 1$ atm, are shown in greater detail. It follows that the wustite phase, $Fe_{1-x}O$, is unstable below about 570°C and that below this temperature magnetite, Fe_3O_4, reduces directly to metallic iron. It follows further that for a gas mixture with $p_{CO} + p_{CO_2} = 1$ atm the area to the right of the Boudouard line represents an unstable condition, the gas having a

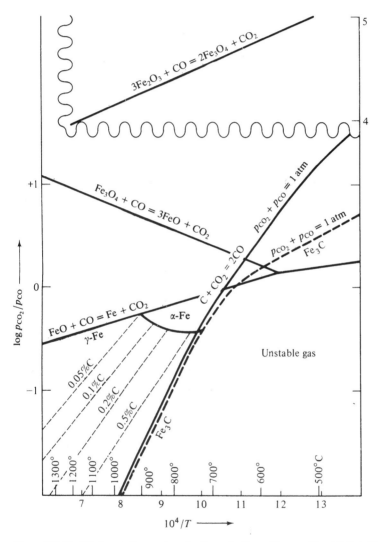

Figure 9-3 Equilibrium ratio p_{CO_2}/p_{CO} for the reduction of iron oxides. Metastable equilibria for the formation of Fe_3C as well as equilibrium carbon contents in austenite are given by dashed lines.

tendency to precipitate soot until the Boudouard line is reached. The sooting reaction is rather slow, however, and in practice gas mixtures with a high CO content may be preserved all the way to room temperature. Also, the equilibrium between Fe_3O_4, iron, and the gas mixture may be established in this region, even though the gas as such is unstable.

In the presence of a $CO–CO_2$ gas mixture the iron phase will pick up some carbon. This is particularly noticeable for the γ-modification of iron, which can dissolve up to almost 2 percent carbon, forming the solid solution austenite. In the austenite phase the chemical activity of carbon increases from zero for pure iron to unity for saturation with graphite. Thus between the gas and austenite the equilibrium $CO_2 + \underline{C}$ (diss) $= 2CO$ will be established and with the equilibrium constant $K = (p_{CO})^2/(p_{CO_2} \cdot a_C)$. For any given temperature and gas composition there will be a certain carbon activity, i.e., a certain carbon content in the austenite. This is shown in Fig. 9-3 which gives curves for constant carbon contents. These curves intersect the Boudouard line at austenite compositions which are saturated, i.e., in equilibrium with graphite.

As already mentioned the sooting or graphite formation is very slow, and gas mixtures may be established which are supersaturated with respect to graphite, corresponding to a carbon activity greater than unity. In this range the gas may react with iron to form cementite, Fe_3C. Cementite is thermodynamically unstable with respect to graphite, but is more readily formed. Also by the reduction of wustite and magnetite at temperatures below 700°C cementite may be formed by the reactions $3FeO + 5CO = Fe_3C + 4CO_2$ and $Fe_3O_4 + 6CO = Fe_3C + 5CO_2$. In Fig. 9-3 gas ratios are shown for these reactions which may be called metastable equilibria. See Sec. 5-7.

Finally, at high temperature and in the presence of carbon, iron will form a eutectic mixture between austenite and graphite. This mixture, which melts at about 1150°C and with about 4.2 percent C, is responsible for the fact that in the iron blast furnace low-melting pig iron rather than pure iron is formed.

Magnetizing Reduction

Magnetizing reduction, sometimes called magnetizing roast, is a process whereby low-grade hematite ore is reduced to magnetite, thus making it more suitable for subsequent magnetic concentration. It is seen from Fig. 9-1 that Fe_2O_3 may be reduced to Fe_3O_4 at a p_{CO_2}/p_{CO} ratio of about 10^5. Also the reaction is slightly exothermic. In order for the reduction to occur with sufficient velocity, however, temperatures around 700°C are required. In order to heat a low-grade ore with, say, 30 percent Fe to 700°C, appreciable quantities of heat are required, i.e., the amount of fuel exceeds greatly the quantity required for the reduction only.

For this reason magnetizing reduction has not found any wide industrial application. The economy of the process could be greatly improved if the heat content of the hot reduced ore could be recovered and returned to the process. This recovery is not as simple as, for example, in the firing of iron ore pellets where the hot pellets may be used to preheat the combustion air. If air were brought into contact with freshly reduced magnetite, this would oxidize immediately to form hematite, and the process would have defeated its own purpose. The only possible way to recover the heat content of the reduced ore would be through some intermediate material or a partition wall which prevents the ore from getting into direct contact with the air. Such a heat transfer is rather complicated, however, and has a low efficiency.

9-3 THE IRON BLAST FURNACE

The iron blast furnace of today is the result of development through many centuries, from the primitive hole in the ground used by the early iron age man through the first primitive blast furnace of late medieval times to the use of coke and hot blast which were introduced in the eighteenth and nineteenth centuries. Around the middle of the twentieth century another great leap forward was made by the introduction of better burden preparation, higher blast temperature and, in some cases, increased furnace pressure. Between 1940 and 1980 the smelting capacity of the blast furnace has been about doubled, without increase in furnace size, and the coke consumption has been reduced from about 1000 kg to below 500 kg per tonne of iron produced. These improvements have only been possible through a better understanding of the reactions which take place in the furnace and of the effect of these reactions on smelting capacity and coke economy.

The principle of the blast furnace is shown in Fig. 9-4. The furnace consists of a shaft about 20 to 30 m high, made of a steel shell lined with refractory bricks. The furnace is charged with *ore, coke,* and *fluxes.* The purpose of the fluxes is to make a slag of suitable composition, and they are usually lime or limestone, sometimes with dolomite added. In the bottom of the furnace *hot air blast* is introduced through the *tuyeres.* The furnace gases are taken out from the top of the furnace and molten *pig iron* (hot metal) and *molten slag* are withdrawn from the *hearth* in the bottom of the furnace. The inverted cone between the *stack* and the hearth is called the *bosh.*

The chemical reactions which take place in the furnace are shown to the right in Fig. 9-4. Also the CO_2/CO ratio of the gas as it passes through the furnace shaft is shown. At the tuyere level air reacts with coke to give the overall reaction $2C + O_2 = 2CO$. This represents the main source of heat and reducing gas. In the upper part of the stack the ore is reduced through the stages

$$3Fe_2O_3 + CO = 2Fe_3O_4 + CO_2$$

$$Fe_3O_4 + CO = 3FeO + CO_2$$

$$FeO + CO = Fe + CO_2$$

These reactions begin when the temperature has reached a few hundred degrees Celsius, but the main reactions occur in the range 700 to 1200°C.

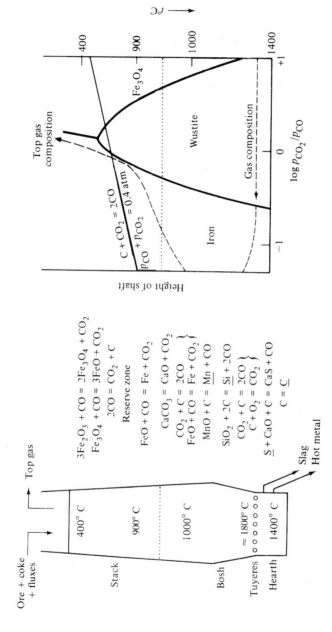

Figure 9-4 Schematic illustration of the iron blast furnace (left), chemical reactions (middle), equilibrium gas ratios, and the ratio of the gas as it moves up the furnace (right). Notice the slow increase in temperature from 900 to 1000°C. The dotted line at the 950°C level represents the border between the "indirect" and "direct" reduction zones.

In modern practice the reduction of Fe_2O_3 and Fe_3O_4 to FeO is completed before the reduction to metallic iron starts. This gives the most efficient ultilization of the CO content in the gas, and is achieved by the use of highly reactive ores of small lump size, for example, pellets. For coarser and less reactive ores the reactions are apt to overlap.

The last reduction of wustite to iron occurs only after the ore has reached the lower part of the stack where the temperature has increased to above 1000°C. In this region the reaction on the coke surface is sufficiently rapid to make the wustite reduction and the Boudouard reaction take place simultaneously. Thus CO_2 formed by the reduction of FeO may react with carbon to give the overall reaction

$$FeO + C = Fe + CO$$

This reaction is sometimes called "direct reduction" even though it actually occurs through the gas phase. The reaction between coke and CO_2 in this case is sometimes called "solution loss," the meaning being that some carbon reacts (dissolves) before it reaches the tuyeres. In comparison, the reaction between CO and iron ore in the upper part of the stack, where the temperature is too low for the coke to react, is called "indirect reduction."

In the lower part of the furnace, where the products are mainly molten, the following reactions occur to some extent:

$$MnO + C = \underline{Mn} + CO$$

$$SiO_2 + 2C = \underline{Si} + 2CO$$

These reactions may also be called direct reductions since they produce CO.

The composition of the furnace gas may be illustrated by the dashed line in Fig. 9-4. We assume that we follow a "pocket" of gas from the tuyeres to the top. Right inside the tuyeres oxygen in the air has reacted to give essentially CO_2; a little farther into the furnace this has reacted to form almost pure CO, and the temperature is of the order of 1800 to 2000°C. Rising through the shaft our gas pocket oscillates between the ore surface and the coke surface, and the composition of the gas will be intermediate between that for equilibrium with ore and that for equilibrium with coke. When the temperature has dropped to about 1000°C the reaction on the coke surface becomes sluggish, whereas reaction on the ore still occurs. The composition of the gas is now mainly determined by the equilibrium FeO–Fe. If the ore is very reactive a major part of the "indirect" reduction of FeO may take place below 900°C and give a CO_2/CO ratio around 0.5. For less reactive ores the reduction will need a higher temperature and give a lower CO_2/CO ratio. Moving toward the top of the stack the gas ratio will increase further due to the reduction of Fe_2O_3 to FeO. This last effect will be less pronounced if the Fe_2O_3 and FeO reductions overlap.

It is in this connection interesting to notice that a high CO_2/CO ratio is dependent on the ore being more reactive than the coke. An unreactive coke, therefore, is required for high furnace efficiency. Also, in order to carry the weight of the burden, the coke should be mechanically strong.

One reaction which may occur is the precipitation of soot around 500 to 600°C, that is, below the temperature where the Boudouard reaction reverses. This soot will mainly be precipitated on the ore, and will be carried with the ore to higher temperatures where it again is gasified. This closed-circuit circulation does not affect the overall stoichiometry of the blast furnace, and has little effect on the coke consumption. It may be beneficial in so far as it breaks open some dense ores. On the other hand it tends to ruin the brickwork. It also carries heat from the lower to the upper part of the furnace, and in general the sooting reaction is regarded as harmful.

Stoichiometry of the Blast Furnace

It was believed for a long time that the most efficient blast furnace was one where all the ore was reduced "indirectly" and all the coke reacted at the tuyeres (Grüner's theorem). A study of the chemistry of the blast furnace shows clearly that this is not so. This will be illustrated by the following case:

We will assume a burden which per mole of Fe_2O_3 contains 0.5 mol of SiO_2 and 0.1 mol of Al_2O_3. In order to give a suitable slag 0.5 mol of CaO has to be added, e.g., as limestone. We furthermore assume that 0.05 mol of SiO_2 is reduced to silicon, which dissolves in the iron, together with 0.4 mol of carbon to give a pig iron with 1.2 percent Si and 4.1 percent C. We assume now that all Fe_2O_3 is reduced to wustite before metallic iron is formed, and furthermore that 50 percent of the wustite is reduced "directly." The CO_2/CO ratio in the gas during the "indirect" reduction of FeO is assumed equal to 0.4, which corresponds to equilibrium at about 950°C.

Assuming counter-current plug flow and disregarding the nonstoichiometry of wustite we get the following stoichiometric scheme (basis one mole of Fe_2O_3):

Indirect reduction, 400 to 950°C

$$Fe_2O_3 + 1.5CO_2 + 3.75CO = 2FeO + 2.5CO_2 + 2.75CO$$

$$FeO + 0.5CO_2 + 4.75CO = Fe + 1.5CO_2 + 3.75CO$$

Decomposition of limestone, 950°C

$$0.5CaCO_3 + 4.75CO = 0.5CaO + 0.5CO_2 + 4.75CO$$

Direct reduction, etc., 950 to 1400°C

$$FeO + C = Fe + CO$$

$$0.05SiO_2 + 0.1C = 0.05\underline{Si}(diss.) + 0.1CO$$

$$0.4C = 0.4\underline{C}(diss.)$$

Tuyere reaction, $\simeq 1800°C$

$$3.65C + 1.825O_2 + 6.87N_2 = 3.65CO + 6.87N_2$$

We see that we will need a total of 5.15 mol of carbon or about 550 kg of carbon per tonne of iron in the ore, a value which would have been somewhat higher if the nonstoichiometry of wustite had been considered. If, on the other hand, all the reduction to iron had been "indirect" and maintaining the same CO_2/CO ratio at the FeO level, a total of 7.15 mol of carbon would have to be burned at the tuyeres, giving a total carbon consumption of 7.65 mol, corresponding to about 820 kg of carbon per tonne of iron. Conversely if a larger part of the FeO had been reduced directly even less carbon would be needed. Thus from a purely stoichiometric viewpoint a low carbon consumption is associated with a high degree of "direct" reduction. We shall later discuss the effect which "direct" reduction has on the heat balance for the blast-furnace process.

So far we have assumed the CO_2/CO ratio at the FeO level fixed to 0.4 corresponding to a given reducibility of the ore. For less reducible ores this ratio would be less, say 0.3. It is easily seen that in this case, assuming 50 percent "direct" reduction, an additional 1.25 mol of carbon would have to be burned at the tuyeres giving a total carbon requirement of 6.4 mol. If, on the other hand, the gas ratio for a more reducible ore were, say, 0.5 the total carbon requirement would be reduced to 4.4 mol, or about 480 kg carbon per tonne of iron in the ore.

We furthermore see that the carbon consumption increases with the presence of $CaCO_3$ in the burden. In order to maintain a CO_2/CO ratio of, say 0.4 at the FeO level an additional 2.5 mol of carbon has to be burned at the tuyeres for every mole of CO_2 expelled from the limestone. If CaO had been used instead of $CaCO_2$ and maintaining 50 percent "indirect" reduction the carbon consumption would be reduced from 5.15 to 3.9 mol, or to about 420 kg carbon per tonne of iron in the ore. This corresponds approximately to the very best coke rates so far reported.

It therefore follows that a low coke consumption is dependent on

1. A highly reducible ore giving a high CO_2/CO ratio
2. CaO instead of $CaCO_3$ in the burden
3. A high degree of direct reduction

The two first requirements are met by the use of pellets, or sinter where the necessary limestone has been sintered in. The third requirement is closely dependent on the amount of available heat in the furnace and on the blast temperature. This follows from the fact that the direct reduction reaction is strongly endothermic, whereas the combustion at the tuyeres as well as the indirect reduction are exothermic reactions. In order to illustrate this we shall work out a heat balance for the above reaction scheme.

Heat Balance of the Blast Furnace

We will assume the furnace is divided into two zones: (1) An upper zone which extends down to a burden temperature of 950°C. In this zone preheating of the burden and indirect reduction takes place. (2) A high-temperature zone from 950 to 1400°C. In this zone the direct reduction, and calcination of limestone, as well as the melting of iron and slag, take place. Also the combustion of coke at the tuyeres is considered as belonging to this zone. The temperature difference between the burden and the furnace gases at the junction between the two zones is assumed to be 50°C. We may then work out the heat balance for the two zones, disregarding for the time being the heat losses. See Table 9-1.

Comments on the heat balance (*a*) The enthalpy of decomposition of CO and formation of CO_2 in the stack are based on the values for graphite, whereas in

Table 9-1 Heat balance for iron blast furnace

1. Stack zone 25 to 950°C

Input		kJ	Output		kJ
$Fe_2O_3 + 0.5SiO_2$ $+ 0.1Al_2O_3$ $+ 0.5CaCO_3$ $+ 5.15C$	25°C	0	Fe $\rbrace$		36
			FeO		53
$4.75CO$ $\rbrace$		149	$0.5SiO_2$	H_{298}^{1223}	31
$0.5CO_2$	H_{298}^{1278}	24	$0.1Al_2O_3$		11
$6.87N_2$		213	$0.5CaCO_3$		52
$2(C + O_2 = CO_2)$ $\rbrace$	$-\Delta H_{298}$	787	$5.15C$		86
$Fe + \tfrac{1}{2}O_2 = FeO$		264	$Fe_2O_3 =$ $2Fe + \tfrac{3}{2}O_2$ $\rbrace$	ΔH_{298}	822
			$2CO = C + O_2$		221
Total input		1437			
Output		1312			
Surplus		125	Total output		1312

2. Bosh and hearth zone 950 to 1400°C

Input		kJ	Output		kJ
$Fe + FeO + 0.5SiO_2$ $+ 0.1Al_2O_3$ $+ 0.5CaCO_3 + 5.15C$ $\rbrace$	H_{298}^{1223}	269	$4.75CO + CO_2$ $+ 6.87N_2$ $\rbrace$	H_{298}^{1273}	386
$4.75(C + \tfrac{1}{2}O_2 = CO)$		586	118 g hot metal $\rbrace$	1400°C	149
$0.5CaO + 0.4SiO_2 =$	$-\Delta H_{298}$		52 g slag		94
$0.3CaSiO_3 + 0.1Ca_2SiO_4$ $\rbrace$		41	$0.5(CaCO_3 = CaO + CO_2)$		89
$0.05Si = 0.05\underline{Si}$		6	$FeO = Fe + \tfrac{1}{2}O_2$ $\rbrace$	ΔH_{298}	264
			$0.05(SiO_2 = Si + O_2)$		45
Total input		902	Total output		1027
			Input		902
			Deficiency		125

the bosh and hearth, where coke is actually consumed, the somewhat larger values which apply to amorphous carbon are used.

(b) The sooting reaction in the stack and the gasification of the soot in the bosh are disregarded. These reactions would tend to increase the heat surplus of the stack and the deficiency of the bosh.

(c) The heat content of hot metal and molten slag at 1400°C are based on the values 1260 kJ/kg for gray cast iron, and 1800 kJ/kg for the slag. These values are rather uncertain.

(d) Since in gray cast iron at room temperature silicon is present in solution, the exothermic heat of solution of silicon is included on the input side. Since carbon is present as graphite, its solution does not give any additional heat effect.

(e) The enthalpy of slag formation is taken as the sum for the formation of 0.3 mol of $CaSiO_3$ and 0.1 mol of Ca_2SiO_4. These are the major minerals in the solidified slag. No great error is introduced by neglecting the solution of Al_2O_3.

We see that the stack zone has a heat surplus of 125 kJ. This corresponds to a temperature of the top gas, $2.5\ CO_2 + 2.75CO + 6.87N_2$, of about 350°C. If heat losses were considered this temperature would be lower. The bosh and hearth zone have a deficiency of 125 kJ which is covered by preheating the air, $1.825O_2 + 6.87N_2$, to about 500°C. If heat losses were considered a higher blast temperature would be required.

If the burden had contained less coke, the heat deficiency of the bosh and hearth zone would increase considerably, due to less coke being burned at the tuyeres, as well as to the higher degree of direct reduction. Very soon the necessary blast preheat would be too high to be practical. With more coke in the burden the necessary blast preheat would be less.

We furthermore see that the temperature difference between the burden and the gas, which rather arbitrarily was assumed equal to 50°C at the 950°C level, increases toward the top of the furnace, i.e., the shaft has a heat surplus. If less coke were used, corresponding to a higher blast preheat, the gas volume per tonne of burden would decrease and we would eventually get to the point where the gas was unable to heat the burden to 950°C. Further increase in the blast temperature would then be of little value. If, on the other hand, more coke were used, corresponding to a lower blast preheat, the volume of the furnace gas and the top gas temperature would increase. Thus a high blast preheat gives a cold top gas and vice versa.

From the above discussion it follows that the temperature difference between the gas and the burden has its minimum at around 950°C. At higher levels in the furnace the gas has a higher heat capacity than the burden, at lower levels the burden has the higher heat capacity due to endothermic reactions such as direct reduction, limestone calcination, and the melting of the products. The 950°C level is called the *thermal pinch-point* of the furnace, and the temperature difference between the gas and the burden may in some cases be close to zero.

Graphical Representation of the Iron Blast Furnace

An alternative way of presenting the material and heat balances of the iron blast furnace was given by Rist.[1] This graphical representation is based on the general principles for counter-current reactors, as discussed in Sec. 6-7, and it is shown in Fig. 9-5 for the preceding case of iron-making and with a burden free of carbonates. The molar O/C ratio in the furnace gas and the molar O/Fe ratio in the burden are used, respectively, as abscissa and ordinate. In addition to oxygen combined with iron the ordinate gives oxygen removed by partial reduction of SiO_2, MnO, etc. as well as oxygen introduced through the blast,

---

[1] A. Rist: in W. K. Lu (ed.) "Blast Furnace Ironmaking," McMaster University, Hamilton, Ont., 1977, pp. 4–1 to 4–37. See also: *III Journ. Int. Siderurgie*, Luxemburg, 1962, p. 113.

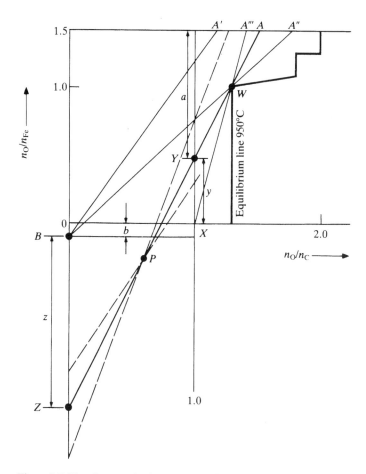

Figure 9-5 Rist diagrams for iron ore reduction.

all relative to the iron content of the burden. For practical reasons the zero point for the ordinate axis is chosen where all oxygen has been removed from the iron oxides. This does not imply that oxygen from SiO_2, MnO, etc., and from the blast have negative values. Thus in Fig. 9-5 the vertical distance $0–B = b$ gives oxygen removed from SiO_2, MnO, etc., and the distance $B–Z = z$ gives oxygen in the blast.

If the furnace gas for the time being is considered free of H_2 and H_2O an abscissa value of $O/C = 1$ corresponds to all carbon being present as CO, and an abscissa value of 2 corresponds to all being present as CO_2. The *equilibrium line* for reduction of the iron oxides at a constant temperature of, say, 950°C is shown in the upper right quadrangle of Fig. 9-5. Contrary to the lines shown in Sec. 6-7 for liquid extraction the equilibrium line is not smooth, but shows breaks at the composition of the various iron oxides. For O/Fe values between 1.5 (Fe_2O_3) and 1.33 (Fe_3O_4) the equilibrium gas is almost pure CO_2, corresponding to $O/C \simeq 2$. For O/Fe values between 1.33 and about 1.13, i.e., for equilibrium between magnetite and wustite, the O/C ratio at 950° is about 1.8, and between 1.05 and 0, corresponding to equilibrium between wustite and iron about 1.3. For lower temperatures, which could apply for highly reactive ores, the magnetite–wustite line would shift toward lower, and the wustite–iron line toward higher O/C values.

In order for the furnace gas to be reducing to the iron oxides the *operating line* must run to the left of the equilibrium line, and is in Fig. 9-5 shown to touch the equilibrium line at point W, which corresponds to initial reduction of wustite at 950°C. Point W may be called the *chemical pinch-point* of the furnace. In principle any operating line which touches the equilibrium line at that point, and which does not cross the equilibrium line, would satisfy the chemical requirements for a counter-current iron-making shaft furnace, and in Fig. 9-5 some alternative lines are drawn. In an iron blast furnace, however, the *thermal requirements* would also have to be satisfied. This is obtained by the burning of a certain amount of carbon at the tuyeres, corresponding to the operating line having also to run through point Z on the ordinate axis.

Actually the position of point Z has to be calculated from a heat balance for the lower half of the furnace, taking into account the temperature of the blast, the slag volume, and heat losses. Without going into detail it can be shown that for a given blast temperature and heat loss, and independent of whether the operating line touches point W, the line has to run through a certain point P, which is indicated in Fig. 9-5. The reason for this is as follows: On the ordinate axis the vertical distance $A–Y = a$ corresponds to the amount of oxygen removed by indirect reduction, and the distance $Y–X = y$ corresponds to that removed by direct reduction. The latter reaction, as well as the reduction of SiO_2, MnO, etc., are strongly endothermic. Thus the vertical distance $Y–B = y + b$ gives a measure of the endothermic load on the lower half of the furnace, whereas the distance $B–Z = z$ gives a measure of the exothermic contribution from the tuyere reaction. If, due to poor indirect reduction, the extent of direct reduction is increased, this has to be compensated by burning more carbon at the tuyeres, i.e., by increased amount of blast. Two alternative operating lines which would satisfy the thermal requirement of the furnace are indicated in Fig. 9-5, pivoting around point P. One of these lines would not satisfy the *chemical* requirements, however.

The *slope* of the operating line corresponds to the ratio C/Fe, that is, the number of moles of carbon needed per mole of iron, not including the carbon which dissolves in the hot metal. It follows from Fig. 9-5 that this slope decreases the closer the operating line runs to the point W, and the less carbon is burned at the tuyeres. The first requirement is met by using a highly reactive ore and closely counter-current plug flow. The second requirement is met by shifting the point P to the left, which is possible by the use of a high-blast preheat. The CO_2/CO ratio of the top gas is given by point A, and it is seen that this increases with decreasing slope of the operating line.

Most modern blast furnaces operate with some hydrogen and water vapor in the gas, which derive from steam or hydrocarbons added to the blast, as will be discussed later in this section. This requires some modification of the Rist diagram without affecting its principle.

The operating line of a given blast furnace may be calculated from the concentration of CO_2, CO, and N_2 in the top gas, the nitrogen giving a measure of the amount of blast used, and the amount of steam or hydrocarbons added to the blast. By comparing the operating line with the known equilibrium line it can be deduced how close it runs to point W, that is, how well the furnace is performing.

The principle of the Rist diagram may be used also for other iron-making processes, such as electric iron smelting and sponge iron processes, which will be discussed later in this chapter.

Furnace Capacity

So far we have not said anything about the factors which determine the smelting capacity of the blast furnace. This is determined primarily by the amount of blast which may be passed through the furnace without upsetting the above stoichiometric scheme and heat balance. With increasing gas velocity the retention times of the gas and the burden decrease, and the gas has less chance to come to thermal and chemical equilibrium with the ore in the shaft. This will lead to a lower CO_2/CO ratio in the top gas and a higher coke consumption. Also the temperature difference between the gas and the burden will increase. As far as one can judge, these factors do not seem to be limiting in modern blast furnace practice. To the contrary, it seems that a high degree of chemical and thermal equilibrium is established even at high production rates, and that some furnaces are unnecessarily tall. Thus in a modern blast furnace reduction of Fe_2O_3 to wustite is complete well before reduction of wustite to iron starts. As a result there will be a zone in the furnace where all the iron is present as wustite and where essentially no reaction takes place. This is called the *reserve* zone. If, occasionally, there is some irregularity in the operation of the furnace, this reserve zone will provide a reserve of freshly reduced wustite and will prevent the higher oxides from entering the iron-producing zone. As the reduction of wustite to iron is rather slow below about 900°C, the gas being very close to equilibrium, this will also be the temperature of the reserve zone. If the reactivity of the ore could be improved further some initial reduction of wustite should be possible below that temperature, which would further improve the efficiency of the furnace.

More serious is the fact that with increased gas velocity the distribution of the gas in the furnace becomes less uniform. Channels are formed with poor contact between gas and burden. Also the burden may become partly fluidized. This upsets the stoichiometric scheme which was based on an ideal counter-current flow. In recent years great efforts have therefore been made to ensure a controlled gas distribution. This is obtained by extensive burden preparation. By screening out the fines from the ore and by a uniform distribution of the burden a more uniform gas distribution is obtained together with less fluidization and dust losses.

Another improvement is to increase the gas pressure in the furnace by say 0.5 to 1 atmosphere. This causes the linear velocity of the gas to decrease, and as shown in Sec. 6-4, the linear velocity and not the mass velocity is responsible for the pressure drop and dust losses. Increased furnace pressure, therefore, makes it possible to increase the smelting capacity without increase in the degree of fluidization or in the dust losses. We may, therefore, add to the previous list of requirements for good furnace economy

4. Uniform burden distribution
5. Increased furnace gas pressure

Use of Oxygen, Steam, and Hydrocarbons in the Blast Furnace

In the previous discussion the heat deficiency of the bosh and hearth was covered by preheating the blast. Another way to balance the heat requirements would be to use oxygen-enriched blast, thus reducing the heat content of the gas leaving the bosh zone.

Oxygen-enriched blast differs from preheating in one respect: it does not supply any additional heat to the furnace. It only shifts heat from the stack to the lower part. This puts a limit on the amount of oxygen which may be added without disturbing the heat balance of the stack. Thus in the preceding example the heat deficiency of the hearth, 125 kJ, could be covered by reducing the nitrogen content from 6.87 to 2.83 mol corresponding to a blast with 39 percent oxygen. This would, however, decrease the heat supply to the stack by 125 kJ, and there would not be enough heat in the gas to heat the burden to the prescribed 950°C. Indirect reduction would be incomplete and there would be an increased heat load on the hearth, resulting in an increased coke requirement and a decreased CO_2/CO ratio in the top gas. For any given burden there is an optimal oxygen content of the blast, usually 25 to 30 percent, which gives a balanced heat distribution corresponding to a minimum coke requirement and a maximum CO_2/CO ratio.

Oxygen-enriched blast has not found any great application in the smelting of iron ore. Its greatest potentialities lie in the smelting of ferroalloys. The production of, e.g., ferromanganese, requires a very high hearth temperature, and most of the reduction is direct. In this case a balanced heat distribution may be obtained by a combination of oxygen-enriched blast and high blast preheat.

The addition of *steam* to the blast is another recent invention. Steam reacts with coke at the tuyeres to give H_2 and CO. This is a strongly endothermic reaction, and it is only possible if the blast is given an additional preheat. On the other hand, steam addition lowers the nitrogen content of the furnace gas, and the hydrogen produced appears to be beneficial for the ore reduction. Also, steam addition counteracts extreme local temperatures in the tuyere zone, temperatures that are harmful to the even descent of the burden.

In recent years considerable progress has been made with the addition of natural gas or oil to the blast. At the tuyeres these burn to form mixtures of CO and H_2. The heat of reaction is inadequate, however, to cover the increased heat capacity of the gas, and an increase in blast temperature is necessary. By the addition of hydrocarbons to the blast, furnace requirements for coke are reduced, and one may ask whether there is any limit to the amount of hydrocarbons that may be added, and if the blast furnace may run on hydrocarbons only. One practical limit is set by the degree of preheat which may be given to the blast, and which is limited to about $1000°C$. More important is the fact that a decrease in coke volume impairs the gas permeability of the burden. Thus, if for no other reason, a certain amount of coke seems to be necessary in order to give the burden a sufficient porosity.

A discussion of the chemistry of the blast furnace is not complete without mention of recent experiments with preheated and prereduced ore. From the heat balance shown it was clear that the stack had a heat surplus. Also, the gas was sufficiently rich in CO to prereduce Fe_2O_3 to Fe_3O_4 and FeO. Preheating and prereduction of the ore outside the furnace would in that case be of little value. If, on the other hand, a very high blast preheat or high oxygen addition were used, corresponding to a low coke requirement and low gas volume, we may get to the point where the stack has a heat and gas deficiency. In such a case preheating and prereduction of the ore may be beneficial. Whether it improves the overall economy of the blast furnace process is an open question, however.

9-4 ELECTRIC IRON SMELTING

In countries with cheap electricity and expensive coke, electric iron smelting may be an economic alternative to the blast furnace. So far electric smelting has found its widest application in the Scandinavian countries, in Italy, Yugoslavia,

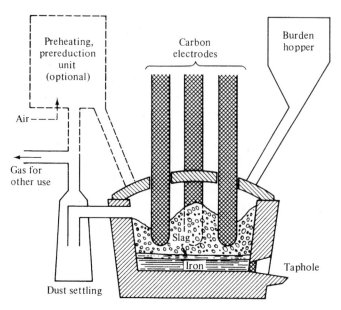

Figure 9-6 Electric pig iron furnace. The current passes mainly between the electrodes and the iron bath, and most of the heat is evolved under the electrodes.

Venezuela, and Japan. The furnace type is shown in Fig. 9-6. The main difference from the blast furnace is that heat is supplied by electricity instead of by the burning of coke at the tuyeres. The chemical reduction of the iron ore is still done by coke, however. Since no air is introduced the gas volume is much less than in the blast furnace, and is limited to what is produced by "direct" reduction, more or less according to the reaction

$$Fe_2O_3 + 3C = 2Fe + 3CO \qquad \Delta H_{298} = 482 \text{ kJ}$$

If we assume that the gas, 3CO, leaves the reduction-smelting zone with a temperature of 1200°C, it has a heat content of 113 kJ. If we disregard heat losses, the gas will be able to preheat the burden, $Fe_2O_3 + 3C +$ say 0.5CaO, to about 500°C—with heat losses to an even lower temperature. At these temperatures the "indirect" reduction is very slow, and only about 10 to 15 percent of the oxygen in the ore is removed by that reaction, corresponding to the top gas holding only 10 to 15 percent CO_2. The rest of the top gas is CO with the exception of about 5 percent N_2 which may have come from the coke or from air drawn in together with the burden. Because there is very little preheating of the ore and little indirect reduction, a shaft would serve no purpose, and the modern submerged arc furnaces have no shaft. Because of the low gas volume the gas is essentially cooled to room temperature before it leaves the furnace.

Referred to the Rist diagram, and including some reduction of SiO_2, MnO, etc., the operating line for the electric iron process would be given by line $B–A'$ in Fig. 9-5. We see that this line runs far to the left of the chemical pinch-point

W, and from a chemical viewpoint the electric iron process is not very elegant. The gas has a high calorific value, and in an integrated iron- and steelworks it may be used to heat soaking-pits and holding furnaces. In other cases the gas may have to be burned to waste. The coke consumption per tonne of pig iron amounts to 350 to 400 kg. This does not have to be of the same high quality as for the blast furnace, however. Coke from the gas works and up to 50 percent coke breeze may be used. The electric energy consumption is about 2000 kWh/tonne, depending somewhat on the desired quality of the iron.

In recent years considerable work has been done in an effort to lower the electric energy requirements. This is possible by preheating and prereduction of the burden. Preheating serves a dual purpose: (1) it reduces the electrical energy necessary to heat and smelt the burden, and (2) it brings the burden to a temperature where the "indirect" reduction by the furnace gases is faster, and thereby relieves the smelting zone of part of the endothermic direct reduction. If by preheating of the burden the gas could be brought into equilibrium with wustite at point W in Fig. 9-5 the top gas composition would be given by point A'' with correspondingly lower carbon and energy consumption. Preheating and prereduction may be carried out either in a special shaft or in a rotary kiln. In both cases the furnace top gas may be used as fuel. Thus in principle it is possible to design a unit wherein all the furnace gas is used up, and with a corresponding reduction in the electric energy requirements. By combination of preheating and prereduction of the burden electric energy requirements down to about 1000 kWh/tonne pig iron have been reported for pilot plant tests.

9-5 BEHAVIOR OF IMPURITY ELEMENTS DURING IRON-MAKING

In addition to iron oxide, carbon, and the normal slag-forming oxides, SiO_2, CaO, Al_2O_3, and MgO, the blast furnace burden may contain variable amounts of the elements Mn, Ti, S, P, Zn, Pb, Cu, and alkali metals.

The blast furnace slag should have a melting temperature of 1400°C or less. Also, in order to give good desulfurization (see below) the CaO/SiO_2 ratio should be high, 1.1 to 1.2. Further discussion on slags is given in Chap. 11.

Manganese and silicon oxides are partly reduced in the hearth zone of the furnace, and are partitioned between metal and slag. The reduction of both SiO_2 and MnO are strongly endothermic reactions. Therefore, the degree of reduction increases with increasing hearth temperature. Complete equilibrium is never obtained. As an example, if complete equilibrium were established at 1500°C between SiO_2 in the slag, solid carbon, and one atm CO gas, the iron would contain 5 to 20 percent Si, depending on the slag composition. In practice the iron contains between 0.5 and 2 percent Si, corresponding to a distribution ratio (percentage Si in slag)/[percentage Si in metal] of about ten. The reason for the nonattainment of equilibrium is believed to be the slowness of CO evolution during the reduction.

In front of the tuyeres the temperature may be sufficiently high for the reaction $SiO_2 + C = SiO(g) + CO$ to take place. Volatile silicon monoxide will follow the furnace gases to colder regions where the above reaction reverses causing accretions which impair the even descent of the burden. By addition of steam or hydrocarbons to the blast extreme temperatures are avoided and the formation of silicon monoxide is greatly reduced.

Manganese is reduced more easily than silicon, and the ratio (percentage Mn in slag)/[percentage Mn in metal] may be around unity. Thus, if the slag weight were half of the iron weight, the manganese in the burden would distribute itself with about one-third in the slag and about two-thirds in the metal.

Titanium occurs in some iron ores. On reduction in the blast furnace it is partly converted to lower oxides, such as Ti_2O_3, or it may react to form carbides and nitrides. These compounds all have high melting points and impair the fluidity of the slag. Only small amounts of titanium can, therefore, be tolerated in the blast furnace burden. In the electric furnace, however, less reducing conditions may be established, and titanium will remain as TiO_2 which dissolves easily in the slag. In special cases it is possible to smelt ores that give a TiO_2 content in the slag up to 80 percent, see Sec. 11-1. The titanium content of the hot metal will be less than 1 percent.

Sulfur derives mainly from the coke, which usually contains around 1 percent S. In the blast furnace shaft the sulfur is picked up by the iron. In the hearth it is transferred to the slag by the reaction

$$\underline{S} + CaO + C = CaS + CO$$

This reaction is strongly endothermic and is favored by high temperatures. Also it is favored by a slag with high CaO activity, i.e., a basic slag. Laboratory studies[1] have shown that the equilibrium distribution ratio (percentage S in slag)/[percentage S in metal] may be as high as 400 at 1500°C. In industrial practice ratios between 50 and 100 may be obtained. This shows that equilibrium is not established, in the same way as already shown for the silica reduction. It is interesting in this connection to observe that the reaction

$$2\underline{S} + 2CaO + \underline{Si} = 2CaS + SiO_2(slag)$$

appears to be very fast and is close to equilibrium. Thus there is a clear correlation between the silicon content of the iron and the degree of desulfurization; high silicon content gives low sulfur contents and vice versa.

A sulfur distribution of, say, 50 would mean that if the coke consumption were 750 kg/tonne of pig iron, the coke contained 1 percent sulfur, and the slag weight were half of the iron weight, the sulfur content of the slag would be 1.45 percent and of the iron 0.029 percent. This agrees approximately with industrial values. Further discussion on desulfurization will be given in Sec. 13-5.

[1] G. G. Hatch and J. Chipman: *Trans. AIME*, **185**: 274 (1949).

Phosphorus derives mainly from the iron ore, which may contain variable amounts of phosphates. Phosphorus reduces relatively easily and alloys with the iron. Under the strongly reducing conditions which prevail in the blast furnace hearth the reduction of phosphorus is almost 100 percent complete; the ratio (percentage P in slag)/[percentage P in metal] may be of the order of one-tenth. The phosphorus distribution cannot be significantly improved by changes in the operating conditions, and the phosphorus content of the iron follows directly from the phosphorus of the ore. Thus if phosphorus cannot be removed at the ore-dressing stage it will have to be removed in the steel-making process. See Sec. 13-3.

Zinc is found in certain iron ores, particularly in pyrite cinder. Zinc oxide is reduced in the lower part of the blast furnace stack according to the reaction $ZnO + CO = Zn(g) + CO_2$. This equilibrium is shifted to the right at high temperatures (see Fig. 9-1). Zinc vapor will follow the furnace gases into the colder regions of the furnace. Here the equilibrium shifts to the left and zinc vapor is reoxidized. The ZnO is precipitated on the burden and follows the burden downward into the hot zone where it again is reduced. In this way zinc will move in a closed circuit between the lower and upper part of the stack, and any zinc added to the burden will accumulate in this circuit. Eventually the zinc oxide content will form accretions in the upper part of the stack and will plug the passage of burden and gas. Thus zinc in the blast furnace burden is very harmful and should be avoided. In the electric iron-smelting furnace there is practically no shaft. The zinc oxide formed by reoxidation will therefore have less time to deposit and will be carried with the gas out of the furnace. Thus the electric furnace is able to treat iron ores with a higher zinc content than is the blast furnace.

Lead and copper are also present in pyrite cinders. They are both easily reduced and follow the iron to the hearth of the furnace. Lead is practically insoluble in liquid iron, however, and forms a separate metallic phase. Liquid lead is heavy and very fluid, and seeps into cracks in the hearth sole. In some cases it accumulates in such quantities that it may be tapped through holes drilled into the furnace sole. In other cases the lead is recovered during relining of the furnace.

Copper is soluble in liquid iron and goes completely into the pig iron. Since copper is more noble than iron it cannot be removed in the subsequent steel-making processes. Any copper added to the iron, either in the blast furnace or in the steel-making processes, will appear in the final steel. When this steel eventually returns as scrap to be remelted it forms a closed circuit where any addition of copper accumulates. Copper is therefore unwanted in the iron ore.

Alkali metals, sodium and potassium, enter the furnace with the coke as well as with the ore and lime. In the tuyere zone some of it will be reduced to metal vapor which follows the gas flow to colder regions. Here it is reoxidized and reacts with the gas to form alkali carbonates, which like zinc oxide and silica may form accretions and impair the burden flow. Part of the alkali metals may also react with carbon and nitrogen from the blast to form alkali cyanides.

Alkali metals are therefore unwanted, and in recent years great efforts have been made to reduce the alkali content of the burden, as well as to choose a slag composition and operating conditions which make the alkali oxides dissolve in and be discharged with the slag.

9-6 ALTERNATIVE REDUCTION PROCESSES FOR IRON

Smelting in the iron blast furnace is by far the most important process for the production of iron, and is likely to continue dominant in the future. Alternative methods may have potentialities under the following conditions:

1. Lack of good coking coals
2. Availability of other reducing agents such as cheap coal, oil or natural gas, or cheap electric power
3. Desirability of limited production capacity and great flexibility
4. Availability of high-grade iron ore concentrates

One example of alternative iron-making is the electric smelting process which has already been discussed. In addition a large number of other processes have been designed, and some of these are in industrial operation. Most of these have as their aim to reduce the ore at a temperature below the melting point of iron and produce an unmelted product, *sponge iron*. Based on a high-grade iron ore concentrate this sponge iron may be refined directly to steel in an electric steel furnace or even used directly in powder metallurgy. Sometimes the production of sponge iron is called "direct reduction," the meaning being that the blast furnace has been bypassed; it has nothing to do with the term direct reduction as previously applied to bosh reactions in the blast furnace.

Sponge iron processes may roughly be divided into two groups, those based on solid fuel and those based on gaseous reducing agents.

Solid Fuels

The fuel may be either coke breeze, cheap coals, or lignite. In all cases the main reducing agent is solid carbon. As shown in Figs. 9-1 and 9-3 solid carbon may reduce wustite at temperatures in excess of about 700°C. The reaction on the carbon surface is very slow below 950°C, however, and in practice temperatures of 1000 to 1200°C are used. The gas produced by the reduction may have a ratio CO_2/CO of about 0.2, and the overall reaction may therefore be described by the stoichiometric equation

$$\tfrac{7}{3}Fe_2O_3 + 6C(\text{amorph}) = \tfrac{14}{3}Fe + 5CO + CO_2$$

This reaction is strongly endothermic with an overall ΔH_{298} of 916 kJ. In addition there is the heat needed to raise the reaction products to 1000°C, about 372 kJ, giving a total heat requirement of 1288 kJ for 14/3Fe or 4930 kJ/kg of iron. The main problem in this type of process is, therefore, to cover the necessary heat requirements. One possibility would be by electricity, giving a theoretical energy requirement of 1.37 kWh/kg of iron, to which must be added energy to cover heat losses. A rotary kiln furnace, where the ore-coke mixture acted as an electric resistor, was developed in the twenties by Professor Bo Kalling in Sweden, but it has not found any industrial application.

In most other versions the energy is introduced by the burning of fuel. One source of fuel is the reaction gas itself. The burning of 5CO (1000°C) with air (25°C) to give $5CO_2 + 9.4N_2$ (1000°C) releases a total of 1050 kJ or 4020 kJ/kg of iron. The remaining heat has to be supplied by additional fuel. This has the drawback, however, that the combustion of fuel requires oxidizing conditions, whereas the reduction of the ore requires reducing conditions. The combustion and the reduction zones, therefore, have to be separated in such a way that heat may flow from one zone to the other, but without infiltration of the combustion gases.

In the Swedish *Höganäs process* high-grade iron ore and coke breeze are placed layerwise in refractory retorts (saggers), which are made either of fireclay or of silicon carbide. The saggers are heated either in a *ring kiln* (see Sec. 8-2) or in a *tunnel kiln*, and by means of additional fuel, e.g., available blast furnace or producer gas. In both types of kilns the hot saggers are cooled from the outside by means of the combustion air, which thereby is being preheated. In order to facilitate the flow of heat into the saggers, these are made rather narrow, with a diameter of about 0.3 m. Nevertheless the total treatment including heating and cooling takes several days. The product is used mainly for the production of iron powder. A continuous counterpart to the Höganäs process is the Italian *Kinglor–Metor* process.

Another version is the *rotary kiln* where the reaction gases together with additional fuel are burned in the space above the ore-coal mixture. By convection and radiation, and promoted by the rotating action of the kiln, the heat is transferred from the combustion zone into the reduction zone. One difficulty with this type of process is the control of temperature and atmosphere composition. If fuel and combustion air are introduced at the discharge end of the kiln, this is apt to become too hot and too oxidizing, resulting in sticking of the charge and incomplete reduction. In the *SL–RN process*, which has found some industrial application, the combustion air is introduced in controlled quantities along the entire length of the kiln. This makes it possible to maintain a constant temperature of about 1000°C throughout the kiln, and also an atmosphere which becomes increasingly reducing toward the discharge end.

In both the Höganäs and the SL–RN process the coke or coal contains some sulfur. In order to prevent this from being picked up by the freshly reduced iron, some burned lime is added to the reaction mixture, where it reacts with the sulfur to form calcium sulfide. After the products have been cooled in a protective atmosphere the sponge iron is separated from excess coke and from the lime-sulfide mixture. The coke may be returned to the process or used for other purposes within the plant, e.g., in the sintering plant.

Gaseous Reducing Agents

These may be derived either from natural gas or oil or may be produced from solid carbon, e.g., in a gas producer. In all cases the raw fuel is converted into mixtures of CO and H_2 by partial combustion or reforming as discussed in Sec. 8-1. In one version (the H-iron process) this gas is again converted into essentially pure hydrogen.

Gaseous reduction has the advantage that the heat of reaction is small; reduction with CO is even slightly exothermic. It also may occur at lower temperature than reduction with carbon. On the other hand the equilibria are less favorable. Thus at a temperature of, say, 850°C the equilibrium ratios CO_2/CO and H_2O/H_2 for the reduction of wustite are both around one-half, which means that only one-third of the gas may be utilized for the reduction. The resulting gas mixture may partly be used for prereduction $Fe_2O_3 \rightarrow FeO$ in a counter-current system, and by combustion with air, to supply the necessary calories to heat the ore to the reaction temperature. Still, there will be a considerable excess of reduction gas which cannot find immediate use in the process.

If the sponge iron plant is part˜of an integrated iron- and steelworks the excess gas may find use for the heating of holding furnaces, etc., but even in this case the supply may easily exceed the demand. In other cases the gas may be treated to remove CO_2 and H_2O and returned to the process, or it may be burned to waste. In some modern processes water is first taken out by condensation, whereupon the remaining CO_2–CO–H_2 mixture is used in the reforming of the raw fuel. In this way a good gas economy is obtained.

In the *Midrex process* iron ore or pellets are reduced in a shaft furnace, Fig. 9-7, which resembles the upper half of a blast furnace. Natural gas, which is essentially methane, is first reformed to a mixture of CO and H_2 by reaction with returned top gas from the furnace. The reforming is carried out at about 1000°C in pipes which contain a nickel catalyst. As the reforming reaction is endothermic the necessary heat is supplied by burning some natural gas as well as the remainder of the top gas in the chamber surrounding the pipes. The reformed gas, which contains about 50 percent of H_2 and 35 percent of CO as well as minor amounts of CO_2, CH_4, and N_2, is introduced into the shaft furnace at a temperature of 800 to 900°C, at which temperature the reduction of wustite to iron takes place. In the upper part of the shaft the hematite pellets are prereduced to wustite, and the temperature of the gas drops to 300 to 400°C. The top gas passes through a scrubber to remove dust and to condense some of the water vapor, whereupon the cleaned top gas which contains about equal amounts of CO and CO_2 together with appreciable amounts of hydrogen and some water vapor is used partly to reform the raw natural gas, partly to

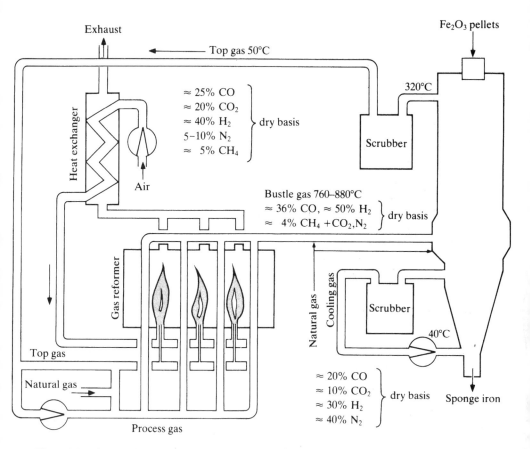

Figure 9-7 Principle of the Midrex sponge iron process. (*Reprinted with permission of the Korf-Stahl Aktiengesellschaft.*) Gas compositions are approximate, and vary with the quality of the natural gas. In modern plants the natural gas is also preheated.

supply heat for the reformer. In the lower part of the Midrex shaft the metallized product is cooled to essentially room temperature by means of a separate flow of cooling gas.

Referred to the Rist diagram, Fig. 9-5, the operating line for the gas in the Midrex shaft is given by the line $X-A'''$, and we see that if only CO and CO_2 are considered, the top gas will have a CO_2/CO ratio of about 0.67, which is more than enough to prereduce hematite to wustite. For lower temperature for the wustite reduction the CO_2/CO ratio in the top gas will be somewhat higher.

If the natural gas had contained sulfur this will accumulate as H_2S in the circulating gas, and will eventually lead to destruction of the nickel catalyst. In a modified version of the Midrex process that part of the top gas which is to be recirculated is, therefore, first used as cooling gas in the lower part of the shaft. Here sulfur is picked up by the freshly reduced iron thus being removed from the gas circuit. As the total amount of sulfur is small, this will not greatly affect the quality of the sponge iron product.

By the efficient utilization of the top gas for gas reforming and the hot exhaust for preheating of combustion air *and* natural gas the fuel energy consumption (net calorific power) is reported to be about 11×10^6 kJ/tonne of metallized product. Somewhat similar to the Midrex process is the *Purofer* process developed in Germany.

The *HyL process* developed in Monterrey in Mexico is in *principle* not much different from the Midrex process. The main difference is that the ore is maintained in a *stationary* position in a number (4 to 5) of *fixed beds*, and the gas is made to flow from one bed to the other. Thus the HyL process relates to the Midrex process in the same way as the ring kiln (Fig. 8-6) for limestone calcination relates to the shaft furnace (Fig. 8-5). As in the Midrex process natural gas is first converted to a mixture of CO and H_2 which at a temperature of 900 to 1000°C is passed through a bed which contains partially reduced ore, which is reduced to its final metallized state. The gas from this bed is now cooled to condense most of its water vapor, and is then reheated regeneratively and introduced into a bed which contains a less reduced ore. The same operation is repeated for the third bed which contains the fresh hematite ore, which thereby is heated to reaction temperature and prereduced. The remaining one or two beds are either being charged with fresh ore or are being cooled by cooling gas, and the metallized product removed. By keeping the ore and metallized product in a stationary position, difficulties are avoided which, in a shaft furnace, are encountered from sintering and sticking of the charge. Furthermore, the reaction may be carried out at increased pressure of a few atmospheres, with correspondingly higher reaction rates. With recent improvements the fuel consumption is reported to be about the same as for the Midrex process.

In the technical literature the HyL reactors are often called "retorts." As mentioned in Sec. 6-1 the term retort should be reserved for reactors with external heating. A better name would be *fixed-bed* reactors.

The *Wiberg process* is shown schematically in Fig. 9-8. High-grade sinter or pellets are charged to the shaft at the right. At the bottom of the shaft a hot (950°C) mixture of about two parts CO and one part H_2 is introduced. In the lower part of the shaft FeO is reduced to iron at a temperature of about 950°C, resulting in a gas mixture with $CO_2/CO \simeq H_2O/H_2 \simeq 0.5$. Half way up the shaft about $\frac{3}{4}$ of this gas is withdrawn for reforming and recirculation, whereas the remaining $\frac{1}{4}$ continues up the shaft where it prereduces Fe_2O_3 to FeO. Three-quarters of the way up the shaft air is admitted and the remaining gas burns to give enough heat to preheat the cold charge to about 1000°C. In order to avoid overheating of the charge a small part of the gas has to be bled out before combustion.

The circulation gas passes through the electrically heated *carburettor* where it reacts with a hot coke bed to give all CO and H_2. In order to compensate for the hydrogen removed in the stack gas some hydrocarbons, e.g., oil or propane, are introduced in the carburettor. The hot gas (about 1100°C) then

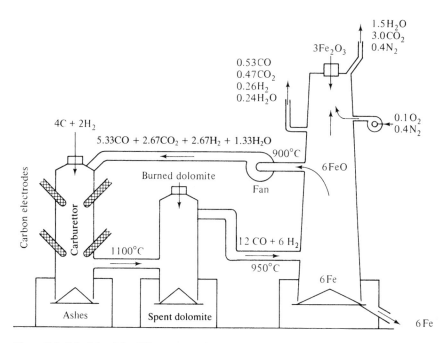

$1.5 H_2O$
$3.0 CO_2$
$0.4 N_2$

$3 Fe_2O_3$

$0.53 CO$
$0.47 CO_2$
$0.26 H_2$
$0.24 H_2O$

$0.1 O_2$
$0.4 N_2$

$4C + 2H_2$

$5.33 CO + 2.67 CO_2 + 2.67 H_2 + 1.33 H_2O$

900°C

6 FeO

Carbon electrodes

Carburettor

Burned dolomite

Fan

1100°C

12 CO + 6 H_2

950°C

6 Fe

Ashes

Spent dolomite

6 Fe

6 Fe

Figure 9-8 Principle of the Wiberg sponge iron process.

passes through a tower with burned dolomite, which absorbs any sulfur which may have come from the coke, and is finally introduced in the bottom of the shaft. The freshly reduced iron pellets are partly cooled in the furnace and then discharged into closed containers (in order to avoid air oxidation) and transferred to the electric steel furnace.

When the Wiberg process was first developed it operated with essentially pure CO. Since the reduction of FeO with CO is exothermic this resulted in overheating and sintering of the charge. With the introduction of hydrogen the heat of reduction was sufficient to cover only the heat losses, resulting in a constant temperature.

In the Wiberg furnace practically all the fuel is converted into CO_2 and H_2O. The process, therefore, has a very low energy consumption. Typical figures are: 111 kg coke, 63 kg oil, and 940 kWh/tonne of iron sponge (90 percent reduced). This corresponds to a total of about 9.2×10^6 kJ/tonne of sponge which is the lowest value reported for any iron-making process. The drawback of the process is that it is based on expensive energy (coke and electricity). Also, the process has a low production capacity resulting in high capital investments. It has found some application in Sweden, where it delivers the melting stock for the production of high-quality steel, but probably has limited potentialities elsewhere.

Several attempts have been made to reduce ore concentrates in a fluidized bed. The iron powder produced could then be briquetted for melting or used directly for powder metallurgical purposes. Unfortunately, it seems that it is difficult to maintain a fluidized bed of iron powder at high temperature. The freshly reduced particles have a tendency to weld together, destroying the fluidization. In the *H-iron process* the reduction was carried out at low temperature, about 500°C, and with hydrogen of about 30 atm as reducing agent. In that way sintering of the bed was avoided. Other processes, such as the *Esso–Fior process*, operate at higher temperatures, and some of these seem to have found industrial application.

One drawback of all sponge iron processes is that the product is subject to corrosion and rusting during storage and transportation. The processes should, therefore, preferably operate in direct connection with a steel plant.

In addition to the sponge iron processes some alternative iron-making processes have been designed to produce a molten product, semisteel. In these processes the ore is introduced together with fuel (coal dust, oil, or natural gas) and oxygen in a burner. This gives sufficient temperature to melt the iron. In order to reduce the ore the CO_2/CO ratio in the effluent gas has to be low—in the order of one-tenth. The effluent gas also being very hot gives the process a low energy efficiency and high fuel consumption, and, unless this gas can be used intelligently, the processes do not seem to have great potentialities.

9-7 PROCESS EFFICIENCY

By the *efficiency* of a reduction process we usually mean the ratio between theoretical and practical energy requirements. For iron-making neither of these is easy to evaluate, however. Starting from essentially pure Fe_2O_3 at 25°C and producing essentially pure iron (steel) of the same temperature we may say that the theoretical energy requirement is that for the reaction

$$Fe_2O_3 = 2Fe + \tfrac{3}{2}O_2 \qquad \Delta H_{298} = 821 \text{ kJ}$$

or about 7.33 million kilojoules per tonne of iron. This, therefore, represents the minimum energy required for a combined iron and steel works. In practice we allow other terms to be included in the theoretical requirements. Thus, looking only at the production of pig iron, we allow also the reduction of small amounts of silicon and manganese oxides to be included. This brings the total chemical requirement up to 7.5×10^6 kJ/tonne of pig iron. Then the pig iron has to be melted and brought to 1400°C, which requires an additional 1.3×10^6 kJ. To this we may add the heat necessary to decompose the limestone and melt the slag, altogether about 2.5×10^6 kJ, giving a total $\underline{11.3 \times 10^6 \text{ kJ/tonne}}$ of pig iron.

We said *may add* because melting of the slag is not an inherent necessity of iron-making provided the gangue minerals could be removed in some other way, but it is necessary in order for the furnace to work properly. On the other hand, one could argue that the physical heat of the furnace gases as well as the heat lost in cooling water, etc., should also be included, since the furnace cannot operate unless the top gas has a certain temperature and the tuyeres and hearth are properly cooled. If this line of thought were continued, one would soon get to the point where "theoretical" energy requirements became equal to the practical ones, and the efficiency became equal to 100 percent. Therefore, in the following, only the energies needed to reduce the oxides and melt the iron and slag are included in the theoretical requirements.

The practical energy consumption is essentially given by the coke consumption. If this is, say, 650 kg coke at 30 000 kJ/kg we get a total input of 19.5×10^6 kJ/tonne of hot metal. This would give an apparent efficiency of $11.3/19.5 = 58$ percent. But we have then forgotten the heating value of the blast furnace gas. This is a total of 6.6×10^6 kJ, using the blast furnace example of Sec. 9-3. Of this about one-third or about 2.2×10^6 kJ is used to preheat the blast, the remaining 4.4 being available for other uses in the works or for sale. In working out the efficiency of the blast furnace process, therefore, credit should be given for the heating value of the excess gas which gives an efficiency of $11.3/(19.5 - 4.4) = 75$ percent. It is clear, however, that the economic value of a gas calorie is less than of a coke calorie, and that the above efficiency does not give a true picture of the economy of the process.

Similarly it is difficult to compare the efficiency of a blast furnace to that of an electric furnace. If an electric furnace uses 2000 kWh and 400 kg coke per tonne of iron, this corresponds to a total of about 19×10^6 kJ. After subtraction of the heating value of the gas, about 6×10^6 kJ, the efficiency of the electric furnace becomes $11.3/13 = 87$ percent. It is clear, however, that this comparison is not fair to the blast furnace. In most places in the world the electric calorie is much more expensive than the coke calorie, which would be in disfavor of the electric furnace. Also, in some places there is no need for the furnace gas, or it is paid for very poorly. This would make the economy of the electric furnace less attractive.

As a whole the expression *efficiency*, even though it may give some approximate ideas, is not a good measure for the economy of a process. This has to be worked out in each individual case, taking into consideration the local cost of coke and electricity, as well as the possible market for and the price obtained for the by-product gases. Also the capital investments must be considered.

9-8 REDUCTION OF OTHER OXIDE ORES

Reduction of most other oxide ores follows the same principles as outlined for iron-making. Owing to differences in stability of the various oxides details of the reduction will differ, however. Oxides of metals more noble than iron such as lead, tin, copper, and nickel are easily reduced by smelting with carbon, for example in a blast furnace. Metals less noble than iron, such as manganese, chromium, and silicon are more easily obtained by electric smelting.

The standard method for production of *lead* is first to roast or sinter the sulfide ore to give lead oxide. This is smelted with carbon in the *lead blast furnace.* Because of the lower temperature and the higher CO_2/CO ratio in the furnace gas, the coke requirements are considerably less than in the iron blast furnace. Also preheating of the blast is less important. Under the slightly reducing conditions which exist in the furnace, any iron oxide present will not be reduced, but will enter the slag together with silica and other stable oxides. If the ore does not contain enough iron oxide to form a suitable slag, scrap iron may be added. In that case iron also acts as a reducing agent, $PbO + Fe = Pb + FeO$, thus lowering the coke requirements and increasing the smelting capacity of the furnace.

The slag from the lead furnace has a melting point around 1200°C. This, therefore, is the minimum temperature of the hearth; the lead itself melts at 327°C. The slag is very corrosive and would attack all ordinary refractories. The lower part of the furnace shaft is, therefore, made of water-cooled steel plates (water jackets). On these, a layer of slag solidifies and forms a lining. The hearth or *crucible* is usually lined with high quality magnesia brick, see Sec. 11-6.

The lead blast furnace, usually, has a rectangular cross section with the tuyeres located along the long sides. This makes it possible to increase the size of the furnace by increasing the length of the rectangle, without increase in the distance between the tuyeres.

The products from the lead blast furnace are first of all the *crude lead* (lead bullion), which contains impurities such as antimony, tin, arsenic, and copper. Also any noble metals, silver, or gold, would collect in the lead phase. The *slag* contains essentially iron silicate with some lime. If the charge contains any zinc oxide, this will also be found in the slag phase. If the charge contains sulfur, this will react with some of the iron to form a *sulfide melt* or *matte* (Chap. 12). Most of the copper in the charge will enter the matte. If the charge is rich in antimony or arsenic these may react with iron to form a separate liquid phase, the *speiss*, in which any cobalt or nickel are enriched. The matte and the speiss have densities between those of the lead and the slag phases. When the furnace is tapped it is, therefore, possible to separate as many as four liquid phases and treat them further, to refine the lead and to recover valuable by-products. These treatments will be discussed in Chap. 13.

Tin ores may also be reduced in the blast furnace but are more commonly smelted in a reverberatory furnace. Tin is only slightly more noble than iron. This makes it impossible to separate tin and iron in one smelting operation. Usually the ore is first reduced at a moderate temperature to give a tin of low iron content. The ferrous slag then contains considerable quantities of tin. This slag is reduced further in a second step to give a tin–iron alloy "hardhead" and a slag of low tin content, which is discarded. The tin–iron alloy is returned to the first step, where the iron, acting as a reducing agent, is slagged.

Alternatively[1] the tin ore could be reduced almost completely in the first

---

[1] An interesting discussion on the thermodynamics and technology of tin smelting is given by T. R. A. Davey and J. M. Floyd: *Proc. Australasian Inst. Min. Metall.,* **219**: 1 (1966); **223**: 75 (1967); *Australian Min.,* **61**: 62 (1969).

Table 9-2 Comparison of blast furnace and electric furnace

	Pig iron		Ferromanganese 80%	
	Blast furnace	Electric furnace	Blast furnace	Electric furnace
Coke kg/tonne	550	380	1500	400
kWh/tonne		2000		2600
kg coke/kWh	0.085		0.42	

step giving the tin–iron alloy and a low-tin slag. The alloy is then smelted with additional concentrate to give essentially pure tin and a slag which is returned to the first step. In this case the second step is in principle not much different from the fire-refining process which will be discussed in Chap. 13.

Copper could be produced from oxide ores or dead roasted sulfide concentrate by smelting with carbon. This type of smelting, *black copper smelting*, is not very common. The resulting copper is contaminated with other elements such as iron, and needs further refining. Today most copper oxide ores are treated by hydrometallurgical methods (see Chap. 15). A major part of the copper in the world is, however, produced directly from sulfide ores by *matte smelting* which will be discussed in Chap 12. Likewise *nickel* is usually produced by matte smelting, but is also produced by smelting of oxide ores with carbon. Since the ore usually contains iron, the product will be a *ferronickel*. This may be used directly as an additive in steel-making.

The less noble metals *manganese, chromium,* and *silicon* may all be produced by smelting in the electric furnace. Since they are all mostly used as additives in steel-making the smelting is usually carried out to give *ferroalloys*. In the case of manganese and chromium, the ores always contain some iron oxides, and iron will enter the metallic phase. Ferromanganese usually contains about 80 percent Mn, and ferrochromium about 70 percent Cr. In the case of silicon, scrap iron or iron ore is usually added to the charge, giving ferrosilicons with 45 to 90 percent Si. The purpose of the iron addition is, in this case, to facilitate the reduction of silicon by lowering its chemical activity. The furnaces used are similar to the one shown for electric iron smelting (Fig. 9-6), but sometimes the roof is omitted, and the furnace gases burn above the charge. Because of the higher reduction temperature the electric furnace becomes increasingly competitive relative to the blast furnace with increasing stability of the oxide. This is shown in Table 9-2, where some figures for the smelting of pig iron and ferromanganese in the blast furnace and in the electric furnace are compared.

For the same coke price the ferromanganese plant is able to pay five times as much per kilowatthour as the pig iron plant. Further discussion on the production of ferroalloys will be given in Chap. 14.

PROBLEMS

9-1 On the basis of Fig. 9-1 and for a pressure $p_{CO_2} + p_{CO} = 1$ atm:

(a) Give the temperature at which carbon may reduce PbO, SnO_2, FeO, and SiO_2.

(b) Is it possible for Si to reduce Cr_2O_3 and if so at what temperature?

(c) Give the temperature at which carbon may reduce ZnO if the pressures of Zn(g) and (CO_2 + CO) are both one atmosphere. Give the approximate composition and total gas pressure of the resulting gas mixture.

9-2 Low-grade hematite ore is to be reduced to magnetite for subsequent magnetic separation. The ore contains 33 percent Fe, all as Fe_2O_3, the balance being SiO_2. The process is carried out in a shaft furnace, and the fuel is producer gas with 25 percent CO, 10 percent H_2, 5 percent CO_2 and 2 percent H_2O, balance N_2. The ore is heated to the necessary reduction temperature, 650°C, by burning the furnace gas with air (25°C). The reduced ore is cooled in counter-current with the producer gas, which is initially at 25°C. Make a heat balance for the process, separately for the preheating, reduction, and cooling zones. Calculate the necessary amounts of producer gas and its calorific power per tonne of ore, and calculate the temperature of the cooled ore when the top gas temperature is 100°C. Heat losses are disregarded.

9-3 For the case given in Prob. 1-4 calculate the weight of carbon consumed by "direct" reduction and calculate the corresponding weight of oxide oxygen that is removed by "indirect" reduction.

9-4 In a blast furnace the coke requirement per tonne of iron is decreased by an increase in the blast temperature. How does this change affect:

(a) The temperature of the top gas.

(b) The volume of the top gas per tonne of iron.

(c) The CO_2/CO ratio in the top gas? The temperature and composition of the hot metal and the quality of the ore are assumed unchanged.

9-5 In a blast furnace which operates satisfactorily and produces hot metal with 4 percent C and 1 percent Si, the blast temperature is increased without changes in the burden. How will this affect the operation of the furnace and the quality of iron produced?

9-6 An iron ore containing 10 percent SiO_2 and 3 percent Al_2O_3, balance Fe_2O_3, is smelted in an electric furnace with coke and limestone. The ashes in the coke are disregarded. The produced iron contains 4 percent C and 1 percent Si. The slag has a ratio (wt % CaO)/(wt % SiO_2) = 1.1.

(a) On the basis of one kilogram of hot metal calculate the weight and composition of the slag.

(b) The furnace gas has a ratio $CO_2/CO = 0.2$. Calculate the necessary amounts of carbon and calculate the weight and volume (STP) of the gas.

(c) Make a heat balance for the process and calculate the electrical energy per kilogram of iron when the raw materials are introduced at 25°C, the slag and iron are withdrawn at 1400°C, and the gases at 100°C. The heat content, relative to 25°C, of the hot metal is 1260 kJ/kg and of the molten slag 1800 kJ/kg. The solid slag may be regarded as a mixture of $CaSiO_3$, Ca_2SiO_4, and Al_2O_3. The enthalpy of solution of Si in iron at 25°C is -125 kJ/mol. Heat losses are disregarded.

BIBLIOGRAPHY

Bogdandy, L. von, and H.-J. Engell: "Die Reduktion der Eisenerze," Springer Verlag, Berlin, 1967, "The Reduction of Iron Ores," Spinger Verlag, Berlin, 1971.

Kitaev, B. I., Yu. G. Yaroshenko, and V. D. Suchkov: "Heat Exchange in Shaft Furnaces," Pergamon Press, London, 1967.

Peacey, J. G., and W. G. Davenport: "The Iron Blast Furnace," Pergamon Press, Oxford, 1979.

"Direct Reduction of Iron Ore, a bibliographical Survey," The Metals Society, London, 1979.

TEN

VOLATILE METALS

In the previous chapter we assumed the vapor pressure of the metal to be negligible during the reduction process. This would be the case for metals such as iron or copper. For other metals the vapor pressure may be appreciable, and in the pyrometallurgical production of zinc and mercury the metal is obtained as a gas rather than as a liquid. In other cases, e.g., in the production of ferrosilicon (Sec. 14-1), volatile metal compounds may be formed, and these may greatly affect the reduction process. Finally the volatility of metals or their compounds may be utilized in refining processes. Typical examples are the refining of impure zinc by distillation, which will be discussed in this chapter, and the removal of volatile impurities by vacuum treatment of metals, which will be discussed in Sec. 13-9.

10-1 VAPOR PRESSURE OF PURE METALS

Most metals evaporate to form monatomic vapor according to the reaction $Me(s \text{ or } l) = Me(g)$. The vapor pressure of the pure metal is, in that case, given by the relation $RT \ln p^\circ_{Me} = -\Delta G^\circ_v = -\Delta H^\circ_v + T\Delta S^\circ_v$. As long as there are no phase transformations in the condensed phase ΔH°_v and ΔS°_v are practically independent of temperature, and the vapor pressure varies with temperature according to the relation $\log p^\circ_{Me} = A/T + B$. This is illustrated by Fig. 10-1 which gives the vapor pressure of a number of metals as function of temperature.

For the vaporization of most substances the entropy of vaporization is between 85 and 100 J/K per mole of gas. This gives *Trouton's rule:* $\Delta H_v^\circ = 92\ T_b$. Here T_b is the normal boiling point of the substance, i.e., the temperature where its vapor pressure is one atmosphere. Compared with Richard's rule (Sec. 4-1) we see that the entropy of vaporization of metals is about ten times their entropy of fusion.

Whereas for most normal metals the vapor is monatomic, the vapors of metalloids and semimetals are often polymerized. Thus bismuth vapor contains Bi and Bi_2 gas molecules. Arsenic gives As, As_2, and As_4, etc. Even for copper, silver, and gold a small amount of dimer gas molecules have been observed. Also for the formation of polymers Trouton's rule applies

$$Me(s,l) = Me(g) \qquad \Delta S_1^\circ \simeq 92 \text{ J/K} \qquad (1)$$

$$2Me(s,l) = Me_2(g) \qquad \Delta S_2^\circ \simeq 92 \text{ J/K} \qquad (2)$$

which gives, for the dissociation,

$$Me_2(g) = 2Me(g) \qquad \Delta S_3^\circ \simeq 92 \text{ J/K} \qquad (3)$$

The dissociation enthalpy ΔH_3° is usually smaller than ΔH_1°, which makes ΔH_2° larger than ΔH_1°. Since $\Delta S_2^\circ \simeq \Delta S_1^\circ$ this means that $\Delta G_2^\circ > \Delta G_1^\circ$ and $p_2^\circ < p_1^\circ$.

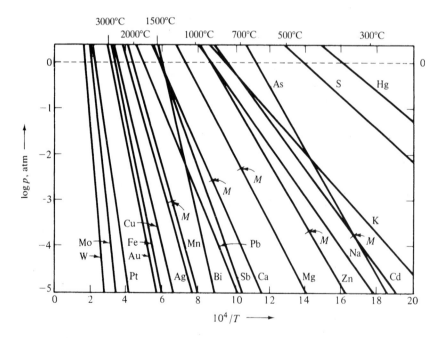

Figure 10-1 Vapor pressure of common metals, $M =$ melting point. (*After R. Schuhmann: "Metallurgical Engineering,"* vol. I, © 1952, Addison-Wesley, Reading, MA, Fig. 9-2. Reprinted with permission.)

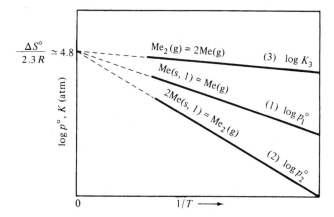

Figure 10-2 Log $p°$ and log K, vs. inverse temperature for vaporization, and dissociation of gaseous dimers.

This is shown in Fig. 10-2, and we see that the ratio of dimer to monomer in the saturated vapor is less than one, but increases with increasing temperature. If, on the other hand, the dissociation enthalpy, $\Delta H_3°$ had been larger than $\Delta H_1°$, the corresponding ratio would be larger than one, and would decrease with increasing temperature. The latter is the case for, e.g., sulfur where the average molecular weight of the saturated vapor decreases slightly with increasing temperature.

For unsaturated vapor, which is governed by reaction (3), we see that $RT \ln (p_{Me})^2/p_{Me_2} = -\Delta H_3° + T\Delta S_3°$ where $\Delta H_3°$ and $\Delta S_3°$ both are positive. This means that for constant total pressure the content of the dimer decreases with increasing temperature, and for constant temperature it increases with increasing pressure. Similar considerations may be made for the formation of higher gas polymers.

10-2 VAPORIZATION OF ALLOYS AND COMPOUNDS

For alloys of two normal metals A and B the vapor consists of A and B atoms. This is shown in Fig. 10-3(a), which illustrates the partial pressure of the two components as well as the total pressure for an ideal liquid or solid solution at two different temperatures. With the arbitrarily chosen values for the vapor pressure of the pure components we see that for T_1 and a total pressure of 1 atm the melt contains 58 atom percent and the vapor 87 atom percent of component B, the atomic fractions being equal to the partial pressure fractions. For the higher temperature T_2 the melt contains 18 atom percent and the vapor 50 atom percent of component B at $p_{tot} = 1$ atm. The composition of the melt and of the vapor for a constant pressure of 1 atm are shown in Fig. 10-3(b), as

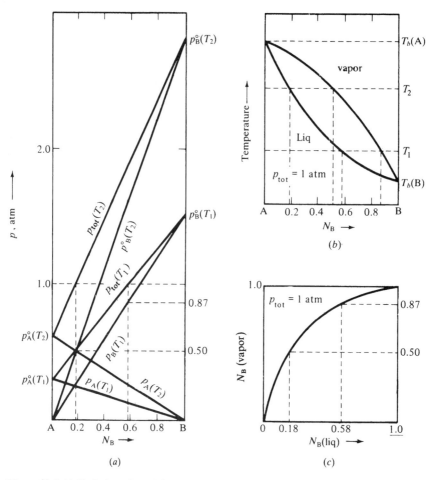

Figure 10-3 (a) Variation of partial and total vapor pressures with composition for an ideal solution A–B at two different temperatures. (b) Equilibrium between liquid and vapor phases as function of temperature at 1 atm pressure for an ideal solution A–B. (c) Same, in a N_B(vapor) vs. N_B(liq) plot T_b = boiling points of the pure components.

function of temperature. We see that at the normal boiling points of the two components the compositions of the melt and vapor coexist, whereas at intermediate temperatures the vapor phase is enriched in the most volatile component, B. Another way of illustration is given in Fig. 10-3(c), which shows the composition of the vapor directly as function of the composition of the liquid, also at constant pressure.

The fact that the most volatile component is enriched in the vapor phase makes possible separation of the two metals by *distillation*. It is clear, however, that a 100 percent separation is impossible. Even if the vapor phase is condensed and evaporated in several consecutive steps, the last condensate will always contain small amounts of the less volatile component, even though this amount may be made as small as desired.

For moderate deviations from ideality the behavior of Fig. 10-3 still applies, i.e., the volatile component will be enriched in the vapor phase. For larger deviations this may not be the case, however. This is illustrated in Fig. 10-4(a) and (b) for strong positive and negative deviations from ideality. We see that in both cases there will be a point where, for constant total pressure, the compositions of liquid and vapor are equal. We call these compositions constant boiling or *azeotropic mixtures*. Notice that for negative deviations from ideality we get an increased boiling point (maximum azeotrop), whereas for positive deviations the boiling point is lowered (minimum azeotrop).

One well-known example of a minimum azeotrop is in the system $H_2O-C_2H_5OH$ where the azeotropic composition at atmospheric pressure corresponds to 96 percent alcohol. An example of a maximum azeotrop is in the system H_2O-HCl where the constant boiling mixture contains 20.2 percent HCl. A metallurgical example of a maximum azeotrop is the system Mg–Hg where the liquid alloy Mg_2Hg vaporizes without change in composition.

A different kind of anomaly occurs if the two metals react to form volatile molecules. One example are the alkali amalgams, where the vapor, in addition to alkali and mercury atoms, contains molecules like HgNa or HgK. In this

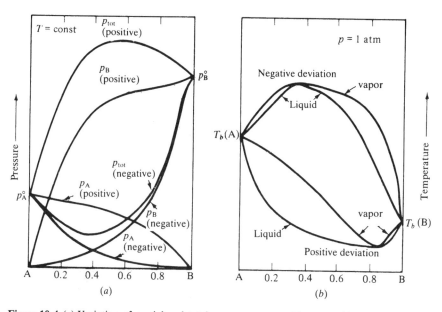

Figure 10-4 (a) Variation of partial and total vapor pressures with composition for strongly positive and strongly negative deviations from liquid ideality at constant temperature. (b) Equilibrium between liquid and vapor phases as function of temperature at constant pressure for strongly positive and negative deviations from liquid ideality (azeotropic solutions).

case the total vapor pressure will be more than the sum of the partial pressures of the two components.

For azeotropic mixtures it is not possible to separate the two components by simple distillation. By changing the total pressure, however, the azeotropic composition may be changed, and the system $H_2O–C_2H_5OH$ becomes normal, i.e., nonazeotropic, at pressures below about 0.1 atm. Another way to affect the vapor pressure curves is to add a third component which has a strong affinity for one of the components involved. Further discussion on the separation of metals by distillation will be given in Sec. 10-6.

Some metal *compounds* are volatile. The volatility may be of different kinds: (1) the compound evaporates as a compound molecule, or (2) the compound evaporates by dissociation. Examples of the first kind are As_2O_3 and SnS. An example of the second kind is ZnS, which evaporates according to the reaction[1]

$$ZnS(s) = Zn(g) + \tfrac{1}{2}S_2(g)$$

In principle a compound which evaporates by dissociation is analogous to an azeotropic mixture. On cooling, the above reaction reverses, and solid ZnS condenses. Thus the fact that a compound evaporates, and even condenses to form the same compound, does not prove that the vapor consists of molecules of the same composition as the compound.

A special type of distillation is possible if a metal reacts with an added component to form a volatile compound in one temperature range, and if this compound again dissociates to liberate the metal in another temperature range. Examples of such "catalytic distillation" or "transport reactions" are

$$2Al(l) + AlCl_3(g) \rightleftharpoons 3AlCl(g)$$

$$Ti(s) + 2I_2(g) \rightleftharpoons TiI_4(g)$$

The application of these reactions for metal refining will be discussed further in Sec. 14-7.

10-3 PRODUCTION OF VOLATILE METALS

Volatile metals may be extracted from their ores by reactions which are similar to those which apply for other metals. Thus mercury oxide may be decomposed by heating to above 450°C. Since this is above the normal boiling point of mercury, the vapor is formed: $2HgO(s) = 2Hg(g) + O_2$. Similarly HgS may be decomposed by roasting: $HgS + O_2 = Hg(g) + SO_2$. Both of these reactions shift to the left on cooling, and if equilibrium were established the reactions would reverse completely during condensation of the product. Fortunately the back-reaction is very slow at the temperatures in question, and this makes the

[1] C. L. McCabe: *Trans. AIME*, **200**: 969 (1954).

extraction of mercury a very simple process which in one operation produces the metal and separates it from the gangue and from other nonvolatile metals.

The decomposition of mercury sulfide may also be carried out by means of other reactants

$$HgS + Fe = Hg(g) + FeS$$

$$4HgS + 4CaO = 4Hg(g) + 3CaS + CaSO_4$$

For these cases the vapor contains only mercury, and there is no danger of back-reaction during cooling and condensation.

Mercury is a noble metal which is easy to produce. For the less noble metals the production is dependent on the following two conditions: (1) the necessary reaction equilibria must be satisfied, and (2) back-reactions during cooling and condensation must be avoided. This will be illustrated in connection with thermal production of zinc and magnesium.

10-4 THEORY OF ZINC OXIDE REDUCTION

About half of the world's production of zinc is made by reduction of sinter-roasted zinc ore with carbon. The thermodynamic condition for reduction of ZnO was shown in Fig. 9-1 and is given in greater detail in Fig. 10-5, which gives the gas ratio p_{CO_2}/p_{CO} for the reaction

$$ZnO + CO = Zn(g) + CO_2 \qquad p_{CO_2}/p_{CO} = K_1/p_{Zn} \qquad (1)$$

Since the gas ratio is also a function of the zinc pressure, it is given for $p_{Zn} = 0.1, 0.5, 1,$ and 10 atm. Also, the curve for reduction of ZnO to liquid zinc is shown. This intersects the curves for the various zinc vapor pressures at the temperatures where zinc vapor is in equilibrium with liquid zinc, i.e., at the dew points. The Boudouard reaction

$$CO_2 + C = 2CO \qquad p_{CO_2}/p_{CO} = p_{CO}/K_2 \qquad (2)$$

is also shown in Fig. 10-5, and with the gas ratio given for $p_{CO} = 0.1, 0.5, 1,$ and 10 atm.

In order for carbon to reduce ZnO continuously, both of the two equilibria (1) and (2) must be satisfied, i.e., the reduction will take place at the intersection of the above two sets of curves. Furthermore, for the reduction of pure ZnO with carbon in a retort the numbers of zinc and oxygen atoms are equal. This makes the pressures of Zn, CO, and CO_2 interrelated by the stoichiometric relation $p_{Zn} = p_{CO} + 2p_{CO_2}$. Referred to our discussion on the phase rule (Sec. 3-5) we have in the system ZnO–C three components Zn, O, and C and one restriction ($n_{Zn} = n_O$). With three phases (ZnO, C, and gas) we get one degree of freedom, i.e., the system is fixed at given temperature or given pressure. We assume first that under the conditions of reduction p_{CO_2} is small compared to p_{CO}, and we get $p_{Zn} \simeq p_{CO} \simeq \frac{1}{2}p_{tot}$. Thus for a total reaction pressure of 1 atm

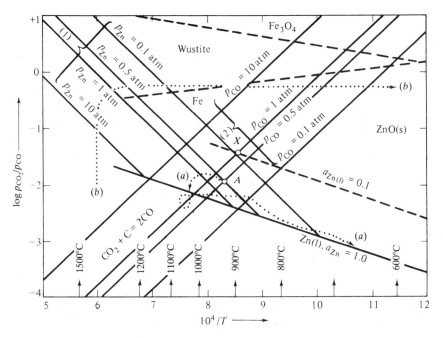

Figure 10-5 Equilibrium gas ratios for reduction of ZnO(s) to Zn(l) ($a_{Zn(l)} = 1.0$ and 0.1) as well as to Zn(g) ($p_{Zn} = 0.1$, 0.5, 1.0, and 10 atm), and for the Boudouard reaction for $p_{CO} = 0.1$, 0.5, 1.0, and 10 atm. Continuous reduction of ZnO with carbon at $p_{tot} = 1$ atm starts at point A. By further heating in the retort process and subsequent cooling and condensation the gas ratio follows the path (a)–(a). For the Imperial Smelting Processes blast furnace the gas ratio follows the path (b)–(b). Reduction of ZnO(s) to brass with $a_{Zn(l)} = 0.1$ and $p_{CO} = 1$ atm takes place at point X.

the curves for p_{Zn} and $p_{CO} = 0.5$ atm apply. We see that these curves intersect at about 920°C (point A). This, therefore, is the lowest temperature at which solid carbon can reduce zinc oxide at 1 atm total pressure. We furthermore see that at the point of intersection the CO_2/CO ratio is about 1.2×10^{-2}, that is, the CO_2 pressure is about 0.6×10^{-2} atm, in accordance with our assumptions of p_{CO_2} being small compared to p_{CO}.

An alternative way to determine the temperature of reduction is as follows: If we combine the equilibrium constants for Eqs. (1) and (2) with the stoichiometric relation $p_{Zn} = p_{CO} + 2p_{CO_2}$ and assume p_{CO_2} to be small compared to p_{CO}, we get

$$K_1 = \frac{(p_{CO} + 2p_{CO_2})p_{CO_2}}{p_{CO}} \simeq p_{CO_2}$$

$$K_2 = \frac{(p_{CO})^2}{p_{CO_2}} \simeq \frac{(p_{CO})^2}{K_1}$$

Inserting the known values for K_1 and K_2 at different temperatures we calculate p_{CO_2} and p_{CO} as well as p_{Zn} and p_{tot} as functions of temperature. This is shown in Fig. 10-6, and we see once more that the total pressure reaches 1 atm at about 920°C.

In Fig. 10-6 the vapor pressure for pure liquid zinc, p_{Zn}°, is also shown as function of temperature. We see that at 920°C this is higher than the partial pressure of zinc in the reacting mixture. Thus by the reduction zinc is formed as an unsaturated vapor. We see, however, that if we go to higher temperatures and pressures the Zn pressure from the reduction increases more rapidly than the saturation pressure, and the two curves intersect at about 1110°C corresponding to $p_{Zn} \simeq 5$ atm and $p_{tot} \simeq 10$ atm. Thus at high temperatures and pressure it should be possible to reduce ZnO directly to liquid zinc. To the knowledge of the author no attempts have been made to verify this in practice, however. It should be borne in mind that even though liquid zinc may be formed, the escaping gas will still contain appreciable amounts of zinc vapor which would have to be collected in order to make pressure reduction of zinc oxide an economic process.

If the reduction of ZnO with carbon was carried out in the presence of a nonvolatile metal which dissolved zinc, such as copper, a liquid alloy could be formed at atmospheric pressure. Thus in Fig. 10-5 the gas ratio for reduction of ZnO to an alloy with a Zn activity of 0.1 is shown as a dashed line, and we see that this intersects the Boudouard line for $p_{CO} = 1$ atm at about 890°C (point

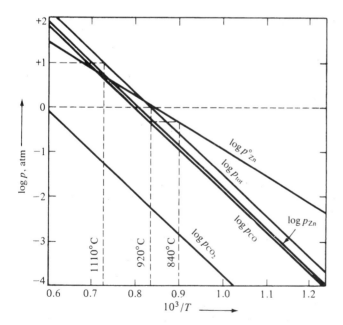

Figure 10-6 Equilibrium pressures of Zn, CO, and CO_2, as well as total pressure for retort reduction of ZnO with carbon as function of temperature. Also given is log p_{Zn}°, the saturation pressure over pure liquid zinc.

X). This is the basis for the direct production of brass from mixed Cu and Zn ores, a process which was known to man long before he was able to make pure zinc. We also see that in addition to CO, the gas will contain about 0.03 atm of CO_2 and about 0.1 atm of Zn vapor. We see, however, that if the temperature is increased much above 900°C the Zn pressure in the gas will increase rapidly, with a corresponding increase in Zn losses.

Reduction of Zinc Compounds

Roasted zinc ores contain various impurities and other zinc compounds such as ferrite, silicate, sulfide, and sulfate. Since iron is more noble than zinc, zinc ferrite will react below 900°C according to the reaction

$$ZnFe_2O_4 + 3CO = ZnO + 2Fe + 3CO_2$$

This liberates ZnO, and zinc ferrite is not particularly harmful in the carbothermic production of zinc.

Since silica is more stable than zinc oxide, zinc silicate will be reduced by the reaction

$$Zn_2SiO_4 + 2CO = 2Zn(g) + SiO_2 + 2CO_2$$

The Gibbs energy of formation of Zn_2SiO_4 from the oxides is about -29 kJ. For a given zinc pressure this gives a CO_2/CO ratio which is about half of that corresponding to the reduction of pure ZnO; this again means that in order to reduce zinc silicate with carbon at $p_{tot} = 1$ atm at higher temperature, about 1000°C, is required. Zinc silicate, therefore, is undesirable, and should, if possible, be decomposed by addition of lime during the preceding sintering process.

If the calcine contains zinc sulfide this is not decomposed during reduction with carbon at the prevailing temperature. Furthermore, *zinc sulfate* is reduced to sulfide: $ZnSO_4 + 4CO = ZnS + 4CO_2$. Thus both sulfide and sulfate in the calcine result in an equivalent loss of zinc, and should be avoided. Zinc sulfide may be decomposed by means of iron according to the reaction $ZnS + Fe = FeS + Zn(g)$. For this reaction a temperature of about 1200°C is required, and this is not obtained in the normal retort processes.

Kinetics of Reduction

The reduction of ZnO with carbon consists of: (1) reduction with CO on the ZnO surface; (2) the Boudouard reaction on the carbon surface; and (3) diffusion of the gases between the two surfaces. These three steps are coupled in series, and the slowest step will be rate-controlling. Measurements of the overall reaction indicate that this is slower than both the intrinsic reaction on the ZnO surface and on the carbon surface.[1] This should mean that diffusion

[1] D. W. Hopkins and A. G. Adington: *Trans. Inst. Min. Metall.*, **60**: 101 (1950–51).

between the two surfaces is rate-controlling. In principle it should therefore be possible to increase the reaction rate by increasing the gas velocity, for example by blowing CO through a fixed or fluidized bed of ZnO and carbon. To the knowledge of the author such a process has not been tried. Another way to increase the overall rate is to increase the surface area and decrease the distance between the two surfaces. This is obtained by intimate mixing of finely crushed oxide sinter and coke.

At the equilibrium temperature for the reduction the driving force and consequently the reaction rate are zero. Overheating to 1000 or 1100°C is necessary to get an acceptable reduction rate. At such a temperature the gas composition on the ZnO surface will approach equilibrium with ZnO, and the gas composition on the carbon surface will approach equilibrium with carbon. The evolving gas will have a CO_2/CO ratio between those which apply for the two surfaces, and closest to the one which has the highest reaction rate. The intrinsic reaction rate per unit area is larger for the ZnO reduction than for the Boudouard reaction. By using an excess of coke, however, the total area of the coke surface may be many times larger than that of the ZnO surface, and the resulting gas mixture may have a CO_2/CO ratio close to the value for equilibrium with carbon. In practice a low CO_2 content in the gas is desired, and this is accomplished by using 2 to 3 times the stoichiometric amount of carbon and the highest possible temperature, i.e., high driving rate.

Reoxidation and Condensation

We shall now discuss what happens to the Zn–CO–CO_2 gas mixture when this is cooled from the reaction temperature. If complete chemical equilibrium were established at all temperatures the reactions ZnO + CO = Zn(g) + CO_2 (1) and C + CO_2 = 2CO (2) would on cooling reverse to give essentially ZnO + C, that is, we would be back where we started. Fortunately reversion of reaction (2) is very slow, and the amount of soot formed on cooling of the gas is usually small.

Reversion of reaction (1) is much faster, and unless special precautions are taken most of the CO_2 in the reaction gas will on cooling react with an equivalent amount of zinc vapor. Thus if the gas had contained about 50 percent each of Zn and CO and 1 percent of CO_2, then 2 percent of its zinc content would reoxidize on cooling. If, by means of a carbon surplus and high temperature, the CO_2 concentration had been lowered to 0.1 percent, only 0.2 percent of the zinc would reoxidize.

Even though the amount of ZnO formed is small, it may have an adverse effect on the process in so far as it acts as nuclei for the condensation of zinc vapor. The droplets formed will be covered with a thin layer of zinc oxide, and this will prevent them from coalescing. In every zinc process a certain amount of so-called "blue powder" will be formed, which is zinc droplets covered with oxide, and the amount of "blue powder" usually increases with increasing CO_2 content in the gas.

In Fig. 10-5 the gas ratio in a retort process is illustrated schematically by the dashed line (a)–(a). The reaction starts at point A. As the temperature is increased the ratio will settle for some value intermediate between equilibrium with the ZnO and the C surfaces. On cooling of the gas reoxidation starts after the equilibrium line for $p_{Zn} = 0.5$ atm has been crossed, and the gas ratio decreases further during cooling and condensation of the gas.

From an equilibrium point of view condensation of zinc will start when the saturation pressure of liquid zinc is equal to the zinc pressure in the gas. For a zinc pressure of 0.5 atm this occurs at a temperature of about 840°C, see Figs. 10-5 and 10-6. In order to obtain 99 percent condensation, and maintaining a total pressure of 1 atm, the saturation pressure of zinc should be 0.01 atm, which occurs at about 600°C. In practice considerable supercooling occurs, and in industrial processes the zinc condenser is usually operated at about 500°C. Seeding with metallic zinc is also used.

In connection with the discussion of industrial zinc processes (Sec. 10-5) different condensers will be discussed, and it will also be shown how by very rapid cooling the back-reaction with CO_2 may be suppressed.

Heats of Reaction and Condensation

The reduction of zinc oxide with carbon is a strongly endothermic process. Thus for $ZnO + C = Zn + CO$, $\Delta H_{298} = 238$ kJ. To this must be added the heat content of the reaction products at 1000°C, including the heat of vaporization of zinc, a total of 138 kJ, giving a total heat requirement of 376 kJ/mol or about 5750 kJ/kg of zinc. One of the main problems in zinc production will be to supply this heat to the reaction mixture. In the retort processes this is achieved by using an outer combustion chamber and transferring heat by conduction through the retort wall and into the charge. This necessarily puts a limit on the size of the retorts and on the rate of reduction. In other processes the heat is evolved inside the charge, which allows larger reactors and higher reduction rates.

Cooling and condensation of the reaction gases releases large quantities of heat, about 1700 kJ/kg of zinc. In the condenser this heat has to be removed, and this again creates problems in the design of the condenser.

10-5 INDUSTRIAL ZINC PROCESSES

Until recently most carbothermic zinc production was either by the horizontal or by the vertical retort process. The main principle of both is the same, and they are shown in Fig. 10-7(a) and (b). The *horizontal retort* is usually made of fireclay and has a diameter of about 0.3 m and length of about 1.5 m. In the same way as previously discussed for the coking of coal (Sec. 8-1) and the Höganäs sponge iron process (Sec. 9-6) the small cross section is necessary in order for the heat to flow into the retort within a reasonable time. A large

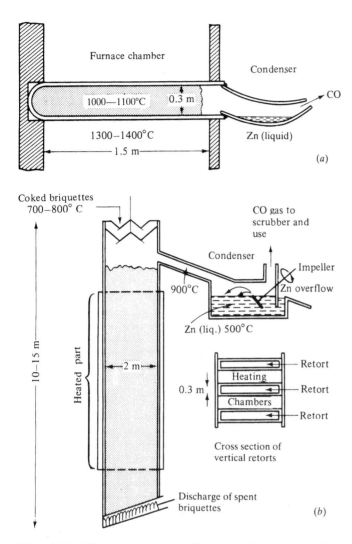

Figure 10-7 (*a*) Horizontal zinc retort, (*b*) Vertical (New Jersey) continuous zinc retort.

number of retorts are placed in a furnace chamber and heated by means of hot combustion gases. The escaping zinc vapor condenses in primitive air-cooled condensers, and the CO gas burns to waste at the mouth of the condensers. The operation, both charging and emptying of the retorts and draining of the condensers, is manual, and with the present high costs of labor and fuel the process is becoming obsolete.

The *vertical retort* developed by the New Jersey Zinc Co. is made of silicon carbide and has the shape of a rectangular shaft 10 to 15 m high and 2 m long, whereas the width is about 0.3 m, or about the same as the diameter of the horizontal retort. The retorts are heated from the outside, and a number of retorts separated by combustion chambers form a battery, in the same way as in a coking plant.

The retorts are charged with briquettes made of zinc oxide sinter and coal, prebaked and preheated to 700 to 800°C. On further heating to above 900°C the carbothermic reduction takes place, and by the time the briquettes have reached the bottom of the retort virtually all zinc is expelled, and the spent briquettes are discharged through a water lock. In the bottom of the retort a small amount of spent combustion gas is introduced in order to prevent the zinc from condensing in the lower part of the retort, and to sweep it upwards.

The gas is withdrawn near the top of the retort and passes into the condenser. This contains a pool of liquid zinc kept at a temperature of about 500°C. By means of a rotating impeller a shower of liquid zinc is formed, on which the vapor condenses. The action of the impeller also helps to break down the films of ZnO and to make the droplets coalesce. Also, to some extent, the rapid cooling helps to suppress reoxidation. In order to remove the heat of condensation the liquid zinc pool has to be continuously water-cooled to maintain its constant temperature. After condensation of the zinc the gas, which is essentially CO, is washed in a scrubber and is used as part of the fuel necessary to heat the retorts.

The heat may also be produced electrically. In the *Josephtown zinc furnace*, which is a shaft about 15 m high, electrical energy is introduced through a system of electrodes, and the coke in the charge acts as the resistor. Since in this case the heat is evolved *inside* the charge there is no limitation to the cross section of the shaft, and in the Josephtown furnace the cross section is 2 to 4 m with a correspondingly larger production capacity.

Both in the New Jersey and in the Josephtown furnaces the remaining ashes are discharged in solid form. They usually contain some lead and are shipped to a lead smelter.

In the *Sterling process* the mixture of zinc sinter and coke is heated in an electric arc furnace to a temperature at which a liquid slag is formed. Any lead will under these conditions be reduced to liquid metal, and any sulfur will react with iron and copper to form a liquid matte. Also some metallic iron is formed. The zinc vapor is withdrawn from the furnace and is condensed in the same way as described for the New Jersey vertical retort. The Sterling process has not found any wide industrial application: one reason may be the large electric energy consumption.

The most recent, and from the theoretical viewpoint most interesting, development in zinc smelting is the *ISP blast furnace process* developed by Imperial Smelting Processes Ltd., England, and this appears now to become the dominating carbothermic zinc process. The furnace, the principle of which is shown in Fig. 10-8, is water-jacketed and resembles a lead smelting furnace. The burden, zinc sinter and coke, is preheated to about 500°C before it is charged into the furnace through a double bell-and-hopper lock. Preheated air is introduced at the tuyeres and reacts with coke to give CO and a temperature above

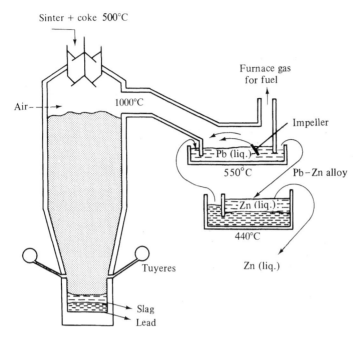

Sinter + coke 500°C

Furnace gas
for fuel

Air-

1000°C

Impeller

Pb (liq.)

550°C

Pb–Zn alloy

Zn (liq.)

440°C

Tuyeres

Zn (liq.)

Slag

Lead

Figure 10-8 Principle of the Imperial Smelting Processes blast furnace for smelting of mixed zinc and lead sinter.

1300°C. Rising through the shaft CO reacts with ZnO and with other oxides like PbO to give a top gas mixture with about 5 percent Zn, 10 percent CO_2, and about 20 percent CO, the remainder being nitrogen.

The composition of the gas in the furnace is illustrated in Fig. 10-5 by the dashed line (b)–(b), and we see that the top gas which has a CO_2/CO ratio of about 0.5 and $p_{Zn} \simeq 0.05$ is stable above 950 to 1000°C but will reoxidize below this temperature. In order to maintain a sufficiently high temperature a small amount of air is introduced in the top of the furnace, where it burns with CO under evolution of heat. It is interesting to notice that air is introduced to *prevent* reoxidation. The hot gas mixture is withdrawn at 1000°C and *shock-cooled* to about 550°C in a condenser where the gas is met by a shower of liquid lead, which causes the zinc to condense and dissolve in the lead without appreciable reoxidation.

Zinc is dissolved in the lead up to a concentration of about 2.5 weight percent. This melt is withdrawn and is cooled to about 440°C in the *lead-cooling launders*. At this temperature the melt separates into two liquids, one a zinc phase with about 1 percent lead, the other a lead phase with about 2.25 percent zinc. The zinc phase is withdrawn for further processing, whereas the lead phase is returned to the condenser. Thus the zinc content of the lead phase changes only by about 0.25 percent, and about 400 tonnes of lead need to be circulated between the condenser and the cooling launders for every tonne of zinc produced.

The zinc obtained from the ISP furnace contains about 1 percent of lead as well as minor amounts of cadmium, arsenic, and antimony which derived from the ore. This corresponds to the so-called "prime western" quality, which may be used for galvanizing purposes as well as for the production of some brass qualities. For the production of special zinc alloys, however, considerably higher-purity zinc is needed, and in order to meet these requirements the metal has to be refined further by *distillation* which will be discussed in the next section. Alternative to cooling and separation into two liquid phases the lead–zinc melt from the condenser can be subjected to *vacuum distillation*, whereby an intermediate grade of zinc can be obtained directly.

As is seen from Fig. 10-5 iron oxide in the charge will not be reduced under the prevailing conditions, and enters the slag together with the gangue minerals. Any lead oxide in the charge is reduced to metal which is withdrawn from the furnace hearth. Any sulfur in the charge combines with iron and copper to form a liquid matte. Thus the process is in the position to treat a complex zinc–lead ore in one operation and produce the two metals separately. Since in this process the heat is evolved *inside* the furnace, there is no limitation in cross section, and the furnace can be built to large production capacity.

10-6 REFINING OF ZINC BY DISTILLATION

The raw zinc from the retort or blast furnace processes contain as major impurities lead and cadmium. Of these cadmium is more volatile, and lead is less volatile than zinc. In the horizontal retort process, therefore, the first fraction of zinc distilled over contains most of the cadmium in the charge, whereas the last fraction contains 1 to 2 percent of lead. Of similar quality is the zinc from the ISP process. In order to produce a zinc grade which meets the requirements of the zinc alloy industry this metal has to be refined further by distillation.

The distillation is usually carried out in a system of two refluxing units, the principle of which is shown in Fig. 10-9. The columns are built of superimposed trays or plates of silicon carbide, which also is the material in the column walls. The lower half of the column is heated externally, whereas the upper half is well insulated. The impure zinc is introduced in the middle of the first column. During its flow through the lower part of the column zinc and cadmium evaporate, and the melt becomes enriched in lead. The vapor ascends to the upper part where further enrichment takes place by condensation of lead. As pointed out previously, however, a 100 percent elimination of lead from the vapor phase is not possible. The vapor from the first column is condensed and passed on to the second column. Here the temperature is somewhat lower, and zinc remains in the liquid state, whereas cadmium evaporates and is recovered by con-

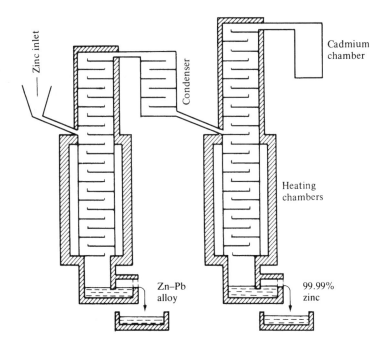

Figure 10-9 Double distillation column for purification of zinc.

densation of the overflowing vapor. Zinc with a purity of about 99.99 percent is withdrawn from the bottom of the second column, whereas from the first column a melt enriched in lead is obtained. Usually only about half of the zinc in the feed is driven off in the first column. Starting for example with a feed with one percent of lead the underflow from the first column will contain about two percent of lead. This impure zinc may either be marketed as such, or it may, by cooling to just above the melting point of zinc, be segregated into a lead melt with about two percent of zinc and a zinc melt with about one percent of lead, of which the latter may be returned to the distillation column.

Refluxing of zinc represents an application of the *counter-current* principle. Inside the column the descending liquid flows counter-current to the ascending gas phase. In the ideal distillation column equilibrium between the liquid and the gas is assumed to take place on each plate or tray, and the procedure outlined in Sec. 6-7 for solvent extraction may in principle also be applied to the distillation column. The distribution curve between gas and liquid phase, as shown in Fig. 10-3(*c*), is equivalent to the *equilibrium* line as shown in Fig. 6-12. As the liquid feed is introduced in the middle of the column the flow of liquid in the lower half of the column will be larger than in the upper. As a consequence the *operating line* will have a larger slope than in the upper half of the column. From a knowledge of the equilibrium line and the liquid flows in the two halves of the column the number of theoretical steps (theoretical plates) for a desired degree of separation may be derived in the same way as shown in Sec. 6-7 for solvent extraction. Detailed procedures for calculation of distillation columns are given in most textbooks on chemical engineering.

Refining of zinc by distillation is a heat-consuming process. It may be economic in localities with cheap fuel and with a demand for high-purity zinc. On the whole, however, the market for high-purity zinc is covered by the electrolytic process, which will be described in Chaps. 15 and 16.

10-7 ZINC FUMING

Zinc fume is finely divided zinc oxide, formed by oxidation of zinc vapor with air. Zinc fuming is used to produce either zinc oxide for the pigment industry or a high-grade oxide from a low-grade raw material. Zinc oxide for the pigment industry may be produced from zinc oxide ore (sinter), but the best quality is made from metallic zink scrap. In the latter case the zinc is brought to boiling, the vapor is oxidized with air, and the fumes are collected in bag filters.

For the production of zinc pigment or industrial zinc oxide from sinter or other low-grade raw materials various methods exist, which all are based on the same principle. In the *Waelz process* a rotary kiln is employed. Low-grade roasted zinc ore, or other zinc oxide material, is mixed with coal and heated to the point where zinc vapor is formed. The vapor burns immediately with air above the charge, and the finely divided zinc fume is carried with the furnace gases and is collected in bag filters. The heat of oxidation radiates back to the charge and helps to supply the heat needed for the reduction.

In the *half-shaft furnace* a complex oxide is smelted with coke and air. Zinc vapor is formed and is oxidized with additional air in the space above the charge, and is carried off with the furnace gases. Other metals in the charge may be tapped from the hearth either as metal (lead) or as a matte (iron and copper), whereas the stable oxides form a slag. Some lead is also evaporated as PbS, which oxidizes to give $PbSO_4$ which mixes with the zinc fume.

Zinc fuming may also be applied to *molten slag* from lead smelting, a slag which may contain as much as 20 percent ZnO. In this case the slag is transferred to a water-jacketed vessel, and a mixture of fuel and combustion air is blown through the melt whereby zinc oxide is reduced. Again the vapor is oxidized with secondary air above the melt, the heat of combustion radiates back to the melt, and the zinc oxide is carried off with the furnace gases. Like in the half-shaft furnace lead in the slag will be volatilized, mainly as PbS which oxidizes to $PbSO_4$.

Owing to the heat which radiates back from the combustion of the vapor, the heat requirement for zinc fuming is much less than for the reduction to metal. Going from essentially zinc oxide raw material to essentially zinc oxide product the chemical enthalpy change is small, and the major part of the applied fuel energy is found as physical heat in the furnace gases, which may have a temperature of more than 1300°C. This heat is usually recovered in steam boilers, whereby the gas is cooled to 100 to 200°C before it passes on to the bag filters, and the amount of steam produced is almost the same as if the fuel had been used directly. Thus we may say that the zinc fuming processes only "borrow" the fuel.

The zinc oxide product is very finely divided. If it is of sufficiently high purity it may be used as a pigment. If less pure it is usually calcined in a rotary kiln above 1000°C to produce clinker oxide which again may serve as raw material for the production of zinc metal.

It should finally be mentioned that the principle of slag fuming may also be applied to *tin slags*. As mentioned in Sec. 9-8 separation of tin from iron by simple reduction is difficult due to their similar oxygen affinities. In the slag fuming process, fossil fuel and iron sulfide are blown through the tin slag. This gives volatile SnS, which again may be oxidized to SnO_2 in the gas above the slag bath.

10-8 MAGNESIUM

The major part of the world's magnesium production is obtained by the electrolysis of fused magnesium chloride (Sec. 16-7). This process requires large capital investments, and when, during the second world war, the demand for magnesium increased rapidly, a number of thermal magnesium plants were built, particularly in the United States. A few of these are still in use and some new ones are being built.

Thermal magnesium processes are of two types: (1) metallothermic, and (2) carbothermic. The reducing agent used in the metallothermic processes has, in all industrial processes, been high-grade (75 to 90 percent) ferrosilicon. In laboratories other metals have been tried.

In the *silicothermic* magnesium process burned dolomite is reacted with ferrosilicon according to the equation

$$2CaO + 2MgO + Si = 2Mg(g) + Ca_2SiO_4(s)$$

From thermodynamic data the equilibrium Mg pressure at, for example, 1200°C is calculated as 16 mmHg (2.1 kPa). Experimental studies[1] have given Mg pressures of about 30 mmHg (4 kPa) at 1200°C, in reasonably good agreement with the calculations. In order to obtain a Mg pressure of one atmosphere, temperatures in the order of 1700°C would be needed. Although this is

[1] L. M. Pidgeon and J. A. King: *Disc. Far. Soc.*, **4**: 197 (1948).

not impossible to obtain in practice, all industrial processes so far are carried out at lower temperatures and under reduced pressure.

The reaction takes place between three solid phases, and, in addition to the gaseous product, a fourth solid phase is formed. One would therefore expect this process to be slow. Experiments have shown that the rate of reaction may be increased by briquetting the reactants with the addition of a few percent of fluorspar. Fluorspar does not participate in the reaction, but acts as a catalyst, probably by forming a low-melting eutectic through which the reactants may diffuse.

In the silicothermic magnesium process, magnesium vapor is the only gaseous product. Thus there is no danger of back-reaction during cooling and condensation of the gas. This is one of the advantages of the process.

In the few industrial installations which are operating at present the reaction is carried out in horizontal retorts made of heat resistant (30 percent Cr + 20 percent Ni) steel, and under a pressure of less than 1 mmHg (0.1 kPa). A large number of retorts are placed in a furnace chamber and heated from the outside either by hot combustion gases or by electric heating elements. The temperature inside the retorts is about 1170°C. As for zinc reduction a small retort cross section is needed.

Magnesium vapor condenses in the cooled part of the retort, which extends outside the furnace chamber. Since the working pressure in the retort is less than the vapor pressure of magnesium at its melting point, the metal condenses to form a solid crust. After completion of the reduction the vacuum is broken, the magnesium crust and the spent charge are removed, and the retort once more filled.

The energy requirements of the process may be calculated as follows: The enthalpy of the reaction at 25°C is 165 kJ per mole of Si. In addition 530 kJ are needed to heat the products, $2Mg + Ca_2SiO_4$, to 1170°C, giving a total of 695 kJ or 14 500 kJ = 4 kWh/kg of magnesium. To this must be added heat losses and energy needed to heat unreacted mixture, and losses due to incomplete magnesium recovery. These may result in a doubling of the above figure. Furthermore, consideration must be given to the energy required to produce the necessary ferrosilicon, about 15 kWh/kg Si, or about 10 kWh/kg Mg. In total this gives a little less than 20 kWh/kg Mg or about the same as for the electrolytic process. The advantage of the silicothermic process is lower capital investment and greater flexibility.

The retort process is discontinuous, and each retort has a small capacity. During the years several attempts have been made to make the process continuous and to adapt it to larger units. In the *Knapsack–Griesheim* process the reduction is carried out inside a chamber which is heated internally by means of graphite resistors. The reaction mixture is fed continuously into the chamber, where it is heated by radiation from the graphite resistors, and the spent charge is withdrawn continuously from the bottom of the chamber. The process is said to operate at 20 to 30 mm pressure (3 to 4 kPa), part of which is Mg vapor, the remainder being hydrogen and argon which is introduced together with the charge. The condensation takes place at 675 to 700°C, that is, liquid metal is obtained.

In the *Magnetherm* process alumina is added to the reaction mixture. This reacts with the oxide product to give a slag with melting point below 1500°C, thus making it possible to carry out the reduction in the molten state. The furnace is carbon lined and heated by means of a graphite electrode which dips into the slag, this again acting as the electrical resistor. The reaction temperature is kept at about 1500°C and the pressure at 5 to 10 mmHg ($\simeq$ 1 kPa). The resulting magnesium is condensed below its melting point. Occasionally the temperature of the condenser is raised above the melting point, and liquid metal is withdrawn. It is important that the temperature of the carbon electrode should not be too high, in which case some CO would have been formed and caused reoxidation of the metal during condensation.

At regular intervals the slag and residual ferrosilicon are tapped from the furnace, and the process is continued. One of the drawbacks of the process seems to be the large consumption of alumina, about 1 kilogram per kilogram of Mg, which again results in large quantities of slag. The economy of the process seems to depend on whether this slag can be used, e.g., as an admixture to cement. The residual 40 to 45 percent ferrosilicon could possibly find some use in iron foundries.

In addition to ferrosilicon various other alloys or compounds have been tried for the reduction of magnesium oxide. These may be Fe–Al, Fe–Al–Si, Ca–Si, and CaC_2. Also scrap aluminum has been suggested. Of these Fe–Al–Si may be the most promising. It can be made by reduction of bauxite with carbon in a submerged arc furnace. The reaction with burned dolomite will be mainly by Al according to the equation

$$12CaO + 21MgO + 14Al = Ca_{12}Al_{14}O_{33} + 21Mg(g)$$

If the reaction could be carried out around 1500°C it would give molten slag and residual alloy, and Mg vapor of atmospheric pressure. The vapor could be condensed above the melting point to give molten magnesium.

Carbothermic Reduction

Carbothermic reduction of MgO takes place according to the reaction: $MgO + C = Mg(g) + CO$. The Gibbs energy of reaction becomes zero around 1850°C corresponding to $p_{Mg} = p_{CO} = 1$ atm. Unfortunately the reaction reverses on cooling and, unless the cooling is made very rapidly, complete reversion to MgO and carbon will occur.

In the industrial process which operated in Permanente in California during the second world war the reduction was carried out in a carbon-lined furnace. Briquettes of MgO and carbon were introduced, and heat was generated by an electric arc between a graphite electrode and a coke bed to give a temperature above 1850°C inside the entire furnace chamber. The gas which was withdrawn from the furnace was shock-cooled to below 250°C by means of an inert gas. This could be hydrogen, but in Permanente natural gas was used. For one mole of Mg about 50 moles of hydrogen or somewhat less of natural gas is needed. Magnesium condenses in the form of solid particles covered with oxide and soot which resulted from partial reoxidation. This dust was collected in bag filters for further processing. The gas could be recirculated to the process, but would in that case have to be treated to remove the CO and prevent this from accumulating. In Permanente the used gas was passed on to a nearby cement factory where it served as fuel, and where the additional CO did no harm.

The Mg dust from the bag filters is highly *pyrophoric*, i.e., it would catch fire if exposed to air. Also, because of its oxide coating it would not coalesce to bulk liquid when melted. Therefore, it had to be processed by *resublimation*. The powder was first briquetted under inert atmosphere, and the briquettes were transferred to a retort of heat-resistant steel where they were sublimed under vacuum. Owing to the low heat conductivity of the briquettes the time required for sublimation was long, with a corresponding increase in capital investment.

The theoretical energy requirement for carbothermic magnesium amounts to 754 kJ per mole, or 31 500 kJ = 8.72 kWh/kg of magnesium. Since in practice the Mg recovery was only about 70 percent and since the heat losses amounted to about 30 percent, the total energy for the reduction became about 17 kWh/kg Mg. In addition there is the energy required for vacuum sublimation, about 6 kWh, giving a total of 23 kWh/kg Mg. This is somewhat more than for electrolytic production (about 20 kWh/kg). A large part of the energy and operating expenses is connected with the condensation and resublimation of the metal. If these steps could be improved for example along the lines used in the ISP zinc blast furnace, it might well be that carbothermic magnesium could have its renaissance.

PROBLEMS

10-1 At 600°C the vapor pressure of pure liquid Zn is 10 mmHg and of pure liquid Cd 100 mmHg.

(a) Assuming that liquid Zn–Cd alloy forms an ideal solution, calculate the composition and total pressure of the vapor over an alloy with 70 atomic percent Zn.

(b) Actually the alloy shows positive deviation from ideality. In which direction will this affect the values calculated under (a)?

10-2 (a) From data in App. C calculate the vapor pressure of pure zinc at 900, 1000, and 1100°C.

(b) Liquid aluminum at the three temperatures is exposed to Zn vapor of 0.5 atm. Calculate the equilibrium zinc content in the aluminum if the system Zn–Al is assumed to be ideal.

(c) In which direction would the zinc contents calculated under (b) shift if the Zn–Al system showed negative, alternatively positive, deviation from ideal mixing?

10-3 Figure 10-10 shows part of the equilibrium diagram Fe–Zn.

(a) Assuming Raoult's law to apply for the liquid and Henry's law for the solid, draw the activity of zinc, relative to the pure liquid, at 900, 1000, and 1100°C.

(b) Calculate the vapor pressure of pure zinc at the same temperatures, and draw the composition line for the condensed phases at $p_{Zn} = 1$ atm as function of temperature.

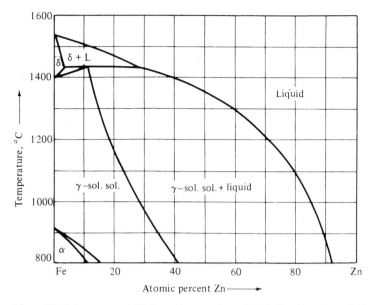

Figure 10-10 Iron–zinc equilibrium diagram above 800°C. (Partly estimated). Vapor phase omitted.

10-4 A mixture of ZnO and metallic Cu is reduced with carbon to produce brass with 30 percent Zn, for which the activity, referred to Zn(l) is 0.1.

(a) By means of data in App. C calculate the temperature where the reaction CO pressure is one atmosphere, as well as the equilibrium pressures of CO_2 and Zn at that temperature.

(b) Calculate the theoretical Zn recovery if the reduction is carried out at the temperature obtained under (a).

(c) How would you suggest the Zn losses in the vapor could be decreased?

10-5 The gas from a retort for carbothermic reduction of ZnO contains 50.3 percent Zn, 49.0 percent CO, and 0.7 percent CO_2.

(a) Calculate the percentage of zinc that will reoxidize by reaction with CO_2.

(b) From data in App. C, calculate the temperature of initial condensation, as well as the temperature where condensation is 99 percent complete. The total pressure is constant = 1 atm, and reoxidation is assumed completed before condensation starts.

10-6 In the Imperial Smelting Processes blast furnace shown in Fig. 10-8 the sinter consists of 50 percent ZnO, 20 percent PbO, 20 percent FeO, and 10 percent SiO_2. The coke is regarded as pure amorphous carbon. All ZnO is reduced to vapor and PbO to liquid lead, whereas FeO and SiO_2 form a molten slag. The gas after reduction contains 7 vol % Zn, and the CO_2/CO ratio is 0.5.

(a) Make a mass balance for the process and calculate the weight of carbon and volume of air per kilogram sinter as well as the volume and composition of the top gas.

(b) Make an overall heat balance for the process, when the raw materials are introduced at 500°C, the liquid products are withdrawn at 1300°C, and the top gas temperature is 1000°C. Introduction of air in the furnace top is disregarded. Calculate the necessary blast temperature if heat losses are disregarded.

10-7 Burned dolomite may be reduced by aluminum according to the reaction

$$21MgO(s) + 14Al(l) + 12CaO(s) = 21Mg(g) + Ca_{12}Al_{14}O_{33}(s)$$

From data given in App. C calculate the Mg pressure at 1200 and 1400°C and estimate the temperature where the pressure is one atmosphere.

BIBLIOGRAPHY

Bunce, E. H., and E. C. Handwerk: New Jersey Zinc Company Vertical Retort Process, *Trans. AIME*, **121**: 427 (1936).

Emley, E. F.: "Principles of Magnesium Technology," Chapter III, Pergamon Press, London, 1966.

Handwerk, E. C., G. T. Mahler, and L. D. Fetterolf: The Sterling Process, *J. Met.*, **4**: 581 (1952).

Lewis, G. N., and M. Randall: "Thermodynamics," rev. ed. by K. S. Pitzer and L. Brewer, chap. 33, McGraw-Hill Book Co. Inc., New York, 1961.

Morgan, S. W. K.: Production of Zinc in a Blast Furnace, *Bull. Inst. Min. Metall.*, no. 609, Aug. 1957; see also: *Min. World*, **19**, no. 11: 58 (1957).

Pidgeon, L. M., and W. A. Alexander: Thermal Production of Magnesium—Pilot-plant Studies on the Retort Ferrosilicon Process, *Trans. AIME*, **159**: 315 (1944).

Weaton, G. F., H. K. Najarian, and C. C. Long: Production of Electrothermic Zinc at Josephtown Smelter, *Trans. AIME*, **159**: 141 (1944).

ELEVEN

SLAGS AND REFRACTORIES

By slags we understand molten mixtures of metal oxides and silicates, sometimes with phosphates or borates. Also, some slags may contain sulfides, carbides, or halides. Slags are formed in the smelting of ores or in the refining of crude metals, and contain in general those elements which are not reduced in the reduction processes or which are oxidized in the refining processes. Thus the slag collects some of the unwanted components in the ore, and by virtue of its immiscibility with the metallic melt a separation of these components from the wanted metal is obtained. In order to give the slag the desired melting point, viscosity, density, or chemical properties, other components, "fluxes," are sometimes added. Common fluxes are lime and magnesia in iron- and steelmaking, fluorspar in steel refining and silica in copper smelting.

By refractories we understand the heat-resistant materials which are used in furnace linings, in roofs and gas ducts, as checker work in regenerators, etc. High-quality refractories are also used for tubes and crucibles in the laboratory. Like slags refractories are usually composed of oxides of the less noble metals: silica, alumina, magnesia, lime, or chromium oxide, but the composition is such as to give the highest possible melting points. Certain nonoxides, such as carbon and silicon carbide, are also used as refractories.

The melting relations both for slags and refractories may be understood from the phase diagrams for the metal oxide systems in question. In this chapter the most important oxide systems will be discussed with emphasis on melting relations, thermodynamics, and physical properties. Since it is impossible in one diagram to illustrate all possible mixtures of slag-forming oxides, the discussion will be based on the two ternary systems SiO_2–CaO–Al_2O_3 and

SiO_2–CaO–FeO, corresponding to the blast furnace and steel-making slags respectively, and the effect of other components on these systems will be considered. Furthermore, various physical properties of slags and refractories will be discussed.

11-1 BLAST FURNACE TYPE SLAGS

The system SiO_2–CaO–Al_2O_3 is of particular interest for the understanding of iron blast furnace slags, but is also the basis for alumina–silicate refractories, glassware, and portland cement. Liquidus isotherms for some selected temperatures are shown in Fig. 11-1. We see that the melting points for the three pure oxides are all high, between 1700 and 2600°C, but are lowered by addition of the other components. Melting points below 1400°C are found in two regions: for mixtures with about 40 to 70 percent SiO_2 and 10 to 20 percent Al_2O_3; and

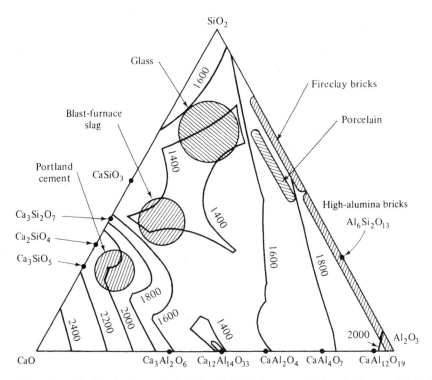

Figure 11-1 Liquidus isotherms in the SiO_2–CaO–Al_2O_3 system. (*After A. Muan and E. F. Osborn: "Phase Equilibria among Oxides in Steelmaking,"* © *1965 Addison-Wesley, Reading, MA, Fig. 80. Reprinted with permission.*) Approximate areas for portland cement, iron blast furnace slag, glass, porcelain, and alumina–silicate bricks are indicated. In glass and porcelain most of the lime is replaced by alkali oxides.

for mixtures of about equal amounts of lime and alumina and with up to 10 percent SiO_2. Most iron blast furnace slags fall within the first region, whereas the second region corresponds to the so-called lime aluminate slags which are formed in the smelting of certain high-alumina ores with lime.

In the silica–alumina system the melting points barely fall below 1600°C and for most of the system the liquidus temperature is above 1800°C. This forms the basis for the alumina–silicate refractories: fireclay bricks, mullite, and high-alumina bricks. Laboratory refractories such as mullite and alumina also belong to this system, whereas porcelain is alumina–silicates to which some alkali oxides are added.

The composition areas for portland cement and for glassware are also indicated in Fig. 11-1. The former contains somewhat more lime than blast furnace slag, and melts at a higher temperature; the latter contains more silica. Also, in most glasses a major part of the lime is replaced by alkali oxides.

Thermodynamic Properties

The Gibbs energy of mixing and the chemical activities in the SiO_2–CaO–Al_2O_3 system have been studied by several investigators, and a fairly consistent picture is obtained. The curves given in Fig. 11-2 show the chemical activities referred to the pure, solid oxides at 1600°C as standard state.[1]

We see that the activity of silica decreases rapidly when the lime content exceeds that of the metasilicate, $CaSiO_3$, and is very low for highly basic slags. In consequence the activity of lime decreases from rather high values for basic slag to very low values for acid slag. The activity of alumina decreases less rapidly from unity for alumina-saturated slags to zero for pure lime-silicate melts. Applying Eq. (4-5) we may conclude that the integral Gibbs energy of mixing has a pronounced minimum around the calcium orthosilicate composition, Ca_2SiO_4, whereas the minimum is less for lime–alumina slags and even less for alumina silicates. This, of course, is the result of the very large negative Gibbs energy of formation for Ca_2SiO_4 from the oxides, about -125 kJ, as compared to about -17 kJ for $CaAl_2O_4$ and about -4 kJ for Al_2SiO_5 (all at $T \simeq 1800$ K).

Effect of other Components

In addition to the three main components, iron blast furnace slags may contain variable amounts of magnesia, manganese oxide, and titanium oxides. *Magnesia* and *manganese oxide* show behavior similar to that of lime, although their

---

[1] F. C. Langenberg and J. Chipman: *Trans. AIME*, **215**: 958 (1959), see also J. F. Elliott et al.: "Thermochemistry for Steelmaking," vol. II, Addison-Wesley Publ. Co. Inc., Reading, Mass., 1963.

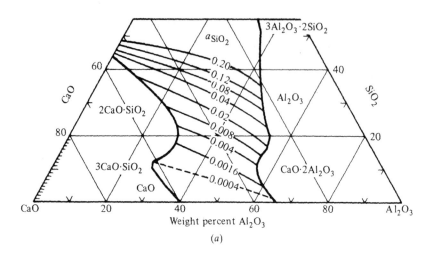

(a)

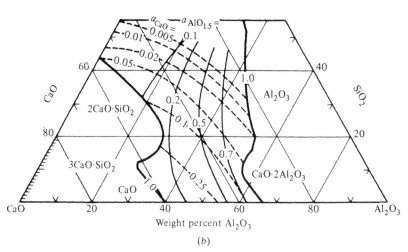

(b)

Figure 11-2 Isoactivity curves in the system SiO_2–CaO–Al_2O_3 at 1600°C. (a) SiO_2 activities, (b) CaO and $AlO_{1.5}$ activities. (*After Elliott, Gleiser, and Ramakrishna: "Thermochemistry for Steelmaking," vol. II, © 1963 Addison-Wesley, Reading, MA, pp. 590–591. Reprinted with permission.*) Note: $a_{Al_2O_3} = (a_{AlO_{1.5}})^2$.

affinity toward silica and alumina is somewhat less. Also, the silicates have relatively low melting points around the compositions $MgSiO_3$ and $MnSiO_3$.

If, in a slag of high SiO_2 content, some of the lime is replaced by equal amounts of magnesia or manganese oxide, this will not affect the melting temperature greatly. In a slag of lower SiO_2 content replacement of CaO by MgO or MnO will lead to a decrease in melting temperature. Thus the melting point of Ca_2SiO_4 is depressed by several hundred degrees when around 10 percent CaO is replaced by the same amount of MgO. At even higher MgO contents the melting point again increases (assuming 0 to 30 percent Al_2O_3 in every case). The ability of MgO to lower the melting temperature in the orthosilicate range is utilized in blast furnace practice where dolomite is sometimes added to give a slag with up to 10 percent MgO. The melting point in the orthosilicate range is also lowered by manganese oxide, but the effect is less than for magnesia.

The chemical activity of MnO in blast furnace type slags has been studied by Richardson and coworkers.[1] For a given MnO content the activity is low in slags of high SiO_2 content, corresponding to a low MnO activity coefficient, whereas for low SiO_2 contents the MnO activity is considerably higher. If lime is added to a $MnO-SiO_2$ melt, this causes a pronounced increase in the MnO activity. Conversely, but to a less degree, we should expect the addition of MnO to $CaO-SiO_2$ melts to increase the lime activity. We may say that both CaO and MnO are strongly bound by high-silica slags, whereas in low-silica slags they tend to "salt" each other out. A similar behavior is believed to apply for magnesia. Thus the addition of magnesia to blast furnace slags not only lowers the melting point but will also increase the lime activity.

Addition of *titanium oxide* to slags shows an entirely different behavior. TiO_2, which melts at 1842°C, is partially immiscible with molten SiO_2, whereas with CaO it forms the stable compound $CaTiO_3$. If TiO_2 is added to slags of $CaSiO_3$ composition the melting point is lowered, and an area of low-melting slags extends all the way to a eutectic composition with 80 percent TiO_2 and 20 percent CaO, which melts at 1460°C. See Fig. 11-3.

Although few thermodynamic data exist, we may conclude that the activity of a given TiO_2 content is less in a low-silica than in a high-silica slag. This is evident from the shape of the liquidus isotherms for primary solidification of TiO_2 in Fig. 11-3. These curves correspond to unit activity of TiO_2, and we see that this occurs at higher TiO_2 concentration in low-silica slags. Conversely, the addition of TiO_2 should lower the activity of CaO, but may even raise the activity of SiO_2 in the slag.

Titania slags are formed in the smelting of ilmenite-containing ores. In that case iron is reduced, whereas the less noble titanium is retained in the slag. Under favorable conditions slags with up to 80 percent TiO_2 may be produced, and may serve as a raw material for the production of pure TiO_2 or titanium compounds.

Slag Basicity

The concept of slag basicity has a firm position in the language of any metallurgist. Slags of high lime content are called basic, those of high silica content are called acid. This is analogous to the terminology used for aqueous solutions,

[1] K. P. Abraham et al.: *J. Iron Steel Inst.*, **196**: 82 (1960). S. R. Mehta and F. D. Richardson: ibid., **203**: 524 (1965).

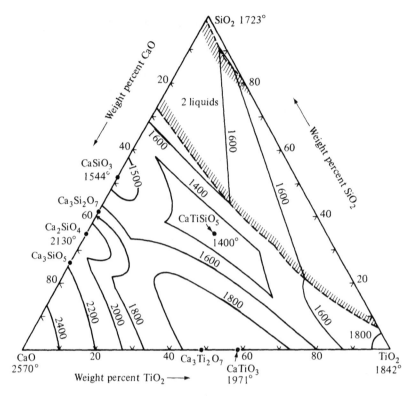

Figure 11-3 Liquidus isotherms in the SiO_2–CaO–TiO_2 system. (*After A. Muan and E. F. Osborn: "Phase Equilibria among Oxides in Steelmaking,"* © *1965 Addison-Wesley, Reading, MA, Fig. 82. Reprinted with permission.*)

and indeed if a solidified and finely ground slag is extracted with water the solution will show a higher pH for lime-rich than for silica-rich slags. But what is slag basicity, and is it possible to express this on an absolute scale? Is it possible to compare the basicity of say MgO and Al_2O_3 with that of CaO or SiO_2?

In aqueous solutions basicity or acidity is expressed by the pH value, which is the negative logarithm of the hydrogen ion activity. As will be shown in Chap. 15 the hydrogen ion activity can only be rigorously defined at infinite dilution. At higher concentrations the hydrogen ion activity cannot be measured independently of the cation activity, and the only activity which can be determined experimentally is that of the neutral acid or base. Thus, strictly speaking, it is not possible to compare the pH of, say, a 1 molal HCl and 1 molal HNO_3 solution.

It has been suggested that for liquid slags the oxygen ion activity should be used as a measure of slag basicity. The oxygen ion activity cannot be measured alone, however. We can only measure the activity of the neutral oxides CaO,

SiO_2, Al_2O_3, etc. Thus an absolute measure of slag basicity does not exist, and we cannot compare the basicity of one oxide with that of another on an absolute scale. What we can do is to compare the affinity of the various oxides for, e.g., silica, as expressed by the Gibbs energy of silicate formation or by the oxide activity in silicate slags. In that case we will find that these affinities decrease in the order CaO, MgO, FeO, Al_2O_3, TiO_2. If instead we measure the affinity of the same oxides for phosphorus pentoxide or boric acid, we will find approximately the same sequence, but the ratios between the affinities may be different. Conversely, if we want to compare the basicity of two slags on the basis of their ability to absorb P_2O_5, this may give a different result than if we compare their ability to absorb SO_3 or CO_2.

In industrial practice it is common to express the basicity of a slag by some basicity number, often called *B* value. Different expressions have been suggested, viz,

$$\frac{CaO}{SiO_2} \qquad \frac{CaO - 4P_2O_5}{SiO_2} \qquad or \qquad \frac{CaO + MgO}{SiO_2 + Al_2O_3}$$

The thing that these ratios have in common is that they only approximately express the ability of the slag for, e.g., desulfurization or dephosphorization. As long as the composition of the slag does not change drastically, however, any one of these numbers may serve the purpose of describing the relative "basicity" of a slag. Since the *B* values have no absolute significance, conversion of the concentrations from weight percent to mole percent serves no purpose.

Effect of Oxygen Potential on Slag Systems

Slags may be studied and used under greatly different conditions, from that of atmospheric air to the strongly reducing conditions of the iron blast furnace or electric smelting furnace. For slags consisting only of the "white" oxides SiO_2, Al_2O_3, CaO, and MgO these changes in oxygen potential do not noticeably affect the composition of the slag. For slags containing manganese, chromium, titanium, and iron oxides, however, considerable changes may occur. These metals may occur with different valencies, and the slag system which is studied under a given oxygen potential may not necessarily apply to other conditions. Thus if manganese slags are studied in air, some of the MnO is oxidized to Mn_2O_3. This is particularly the case for low-silica slags; we may say that Mn_2O_3 is more acid than MnO and is therefore stabilized by basic slags. Similar behavior is found for chromium slags; in air and for highly basic slags, some of the Cr_2O_3 is oxidized to CrO_3. On the other hand, under reducing conditions and particularly in acid slags, Cr_2O_3 may be partly reduced to CrO.[1]

[1] F. Körber and W. Oelsen: *Stahl Eisen*, **56**: 310 (1936); K. Tesche: *Arch. Eisenhüttenw.*, **32**: 437 (1961); G. W. Healy and J. C. Schottmiller: *Trans. AIME*, **230**: 420 (1964).

On cooling and crystallization of such a slag it decomposes according to the reaction $3CrO \rightarrow Cr_2O_3 + Cr$, and metallic chromium precipitates. It is interesting to note that the slag at high temperature in this case contains an oxide which does not exist as a stable compound at room temperature.

Another example is titanium slags, where TiO_2 may convert to lower oxides: Ti_3O_5, Ti_2O_3, and even to TiO. Thus the phase diagram which was shown in Fig. 11-3, and which was determined in air, may be greatly changed under reducing conditions. It is known that titania-rich slags on heating under strongly reducing conditions will thicken with the precipitation of solid phases. Chemical analyses by the author have shown that such reduced slags have an oxygen content corresponding to titanium being present approximately as Ti_3O_5. In order to smelt successfully titanium ores and produce a slag of high titanium content it is therefore necessary that the smelting is carried out under not too reducing conditions.

The effect of oxygen potential on ferrous slags will be discussed in Sec. 11-3.

11-2 NONOXIDE COMPONENTS IN SLAGS

Sulfur may dissolve in molten slags either as a sulfide or as a sulfate ion. Under the reducing conditions which exist in most metallurgical processes sulfide is the dominating form, whereas sulfate may be formed under oxidizing conditions.

The ability of a liquid slag to absorb sulfide, e.g., in the iron blast furnace, may be expressed by the so-called *sulfide capacity:* Sulfur dissolved in liquid iron has a certain sulfur activity or sulfur vapor pressure. The reaction with the slag may be expressed by the equations

$$\tfrac{1}{2}S_2 + CaO = CaS + \tfrac{1}{2}O_2$$

or

$$\tfrac{1}{2}S_2 + O^{2-} = S^{2-} + \tfrac{1}{2}O_2$$

The sulfide content of the slag may then be given by

$$(\%S) = C_s \sqrt{\frac{p_{S_2}}{p_{O_2}}}$$

Here C_S is called the *sulfide capacity* of the slag, and it is easily seen that it increases with increasing CaO activity and with decreasing activity coefficient for CaS in the slag. Indeed, the determination of sulfide capacities, combined with CaS solubility data, has been one of the methods used to determine the activity of CaO in molten slags.

The sulfide capacity of a large number of slags has been measured by

Richardson and coworkers.[1] It increases strongly with increasing contents of " basic " oxides: CaO, MnO, and FeO in the slag and is higher for ferrous and manganous slags than for corresponding lime slags, and the lowest for magnesia and alumina silicate melts. Sulfide capacities for various slags are shown in Fig. 11-4.

Under oxidizing conditions sulfur reacts with the slag according to the equation

$$\tfrac{1}{2}S_2 + \tfrac{3}{2}O_2 + O^{2-} = SO_4^{2-}$$

with the corresponding *sulfate capacity*

$$C_{SO_4} = (\%S)/\sqrt{p_{S_2} \cdot p_{O_2}^3}$$

where (%S) is the weight percentage of sulfate sulfur in the slag. The sulfate capacity is found to vary with slag composition in a way similar to that of the sulfide capacity. Thus the ratio C_{SO_4}/C_S appears to be rather independent of slag composition. On this basis Wagner[2] has suggested that any one of these may be used as a conventional expression for the " basicity " of the slag.

Carbide slags Under strongly reducing conditions, e.g., in electric reduction furnaces or in the electric steel-making furnace, calcium carbide may be formed by the reaction

$$CaO + 3C = CaC_2 + CO$$

Calcium carbide, which melts around 2300°C, may dissolve appreciable amounts of oxide whereby its melting point is lowered. Thus CaC_2 and CaO form a eutectic mixture with about 50 percent CaO which melts at about 1800°C. Furthermore Shanahan and Cooke[3] report that a mixture with 50 percent CaC_2 and about 10 percent Al_2O_3 melts at 1620°C. The same authors found that at 1400 to 1500°C a slag with 50 percent CaO and 50 percent Al_2O_3 is able to dissolve only about 0.6 percent CaC_2, whereas the solubility of CaC_2 in $CaO-Al_2O_3-SiO_2$ melts is insignificant. What usually is called a carbide slag is then probably an emulsion of a silicate slag and a carbide melt, the latter with considerable amounts of dissolved oxides.

Fluoride slags The most common halide added to metallurgical slags is fluorspar which melts at 1386°C, but when mixed with silicate slags may give melting points below 1200°C. The most important fluoride addition is to the reducing slag used in electric steel-making. In order to obtain a good desulfurization a highly basic slag is used, with a CaO/SiO_2 ratio between 2 and 3. Such a

[1] F. D. Richardson and C. J. B. Fincham: *J. Iron Steel Inst.*, **178**: 4 (1954). K. P. Abraham and F. D. Richardson: ibid., **196**: 309 (1960). R. A. Sharma and F. D. Richardson: ibid., **198**: 386 (1961); **200**: 373 (1962); **233**: 1586 (1965).
[2] C. Wagner: *Met. Trans. B*, **6B**: 405 (1975)
[3] C. E. A. Shanahan and F. Cooke: *J. Appl. Chem.*, **4**: 602 (1954).

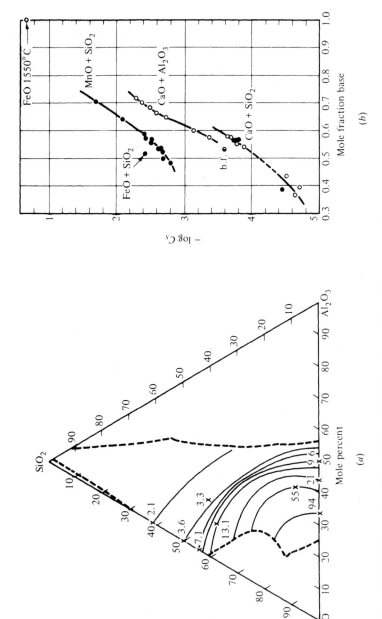

Figure 11-4 Sulfide capacity of slags. (a) $C_S \cdot 10^4$ in SiO_2–CaO–Al_2O_3 melts at 1650°C. (*From C. J. B. Fincham and F. D. Richardson: Proc. R. Soc. vol. A 223, p. 40 (1954)*). (b) CaO–Al_2O_3, CaO–SiO_2, and MnO–SiO_2 melts at 1500°C, b.f. = blast furnace slags. (*From K. P. Abraham and F. D. Richardson: J. Iron Steel Inst., vol. 169, p. 316 (1960)*).)

slag has a melting point above 2000°C, however. On addition of about 20 percent of CaF_2 the melting point is depressed to below 1500°C, with 50 percent CaF_2 to about 1230°C.[1] An even greater melting-point depression is obtained by a combination of fluorspar and alumina additions.

Addition of fluorspar to the slag has the character of a dilution, and does not seem to affect much the activity of the other components. Thus a combination of a high lime activity, with correspondingly high sulfide capacity,[2] and low melting point may be obtained.

Gases in slags The most common gases soluble in slags are carbon dioxide and water vapor. *Carbon* dioxide most probably dissolves according to the reaction

$$CO_2 + O^{2-} = CO_3^{2-}$$

In alkali-oxide systems carbon dioxide at atmospheric pressure dissolves to give an essentially stoichiometric carbonate melt. In lime-silicate slags the solubility is very small, and few measurements exist. For a given slag composition the solubility increases with increasing CO_2 pressure, and decreases with increasing temperature. At CO_2 pressures above 120 atm complete liquid miscibility between Ca_2SiO_4 and $CaCO_3$ (mp 1290°C) has been observed.[3]

Water is soluble both in basic and in acid slags, whereas the solubility shows a minimum for so-called neutral slags with $CaO/SiO_2 \simeq 1$. The solubility is of the order of 50 ppm of hydrogen in the slag. Most probably water dissolves according to the reaction

$$H_2O(g) + O^{2-} = 2OH^-$$

In basic slags it may exist as free OH^- ions, whereas in acid slags the reaction may be

$$H_2O(g) + \overset{|}{\underset{|}{-Si}}-O-\overset{|}{\underset{|}{Si}}- = -\overset{|}{\underset{|}{Si}}-OH + HO-\overset{|}{\underset{|}{Si}}-$$

In the latter case solution is accompanied by the breaking of an oxygen bridge in the silica anion network. These equations are supported by the fact that the solubility is proportional to the square root of the water vapor pressure.[4]

Elemental hydrogen is not soluble in "white" $CaO-Al_2O_3-SiO_2$ slags, but dissolves in ferrous and manganous slags. In these cases the reaction is probably

$$H_2 + 2Me^{3+} + 2O^{2-} = 2OH^- + 2Me^{2+}$$

---

[1] E. M. Levin, C. R. Robbins, and H. F. McMurdie: "Phase Diagrams for Ceramists," American Ceramic Society, Columbus, Ohio, 1964.

[2] G. J. Kor and F. D. Richardson: *Trans. AIME,* **245**: 319 (1969).

[3] W. Eitel: *Neues Jahrb. Mineral., Geol. Palaeontol.,* Beilageband, **48**: 63 (1924). See also W. Eitel: "Physical Chemistry of Silicates," University of Chicago Press, Chicago, Ill., 1954, p. 898.

[4] J. Chipman et al.: *Trans. AIME,* **206**: 1568 (1956).

Here Me is a metal which may occur with variable valency. We may say that hydrogen has been oxidized, and dissolved as water vapor by the reduction of the metal ion from the higher to the lower valency.

The presence of hydroxide ions in steel-making slags may be a vehicle by which hydrogen is transferred from the hot combustion gases to the molten steel bath. Apart from this, gas solubility does not seem to be of great importance for metallurgical processes. In geological processes, however, where high gas pressures exist, solubility of both carbon dioxide and water vapor plays an important role, and affects greatly the melting point and physical properties of the silicate melt.

11-3 FERROUS SLAGS

Ferrous slags are formed in steel-making as well as in copper- and lead-smelting. In these processes the conditions are much more oxidizing than in the iron blast furnace, and iron is partly retained in the slag phase. The major components in these slags are FeO, CaO, and SiO_2, but some of the iron is also present as Fe_2O_3. In addition the slag may contain variable amounts of MgO, MnO, Al_2O_3, P_2O_5, and ZnO. For a discussion we start with the ternary system SiO_2–CaO–FeO and shall discuss the effect of oxygen potential and other components on this system. This is illustrated in Fig. 11-5 together with the ternary systems SiO_2–FeO–Fe_2O_3 and CaO–FeO–Fe_2O_3.

When ferrous slags are melted in equilibrium with metallic iron, the iron is mostly in the divalent state, but some trivalent iron is also present. Thus a pure wustite melt in equilibrium with iron contains about 10 percent Fe_2O_3. When silica is added the Fe_2O_3 content decreases, and is insignificant at silica contents beyond Fe_2SiO_4. On the other hand, when lime is added to the wustite melt the Fe_2O_3 content for equilibrium with metallic iron increases to about 20 percent. Thus we may say that silica stabilizes the divalent, whereas lime stabilizes the trivalent iron oxide.

We see that at steel-making temperature of 1600°C a range of molten slags extends from pure wustite melt across the diagram to compositions around $CaSiO_3$, and is bordered on the acid side by saturation with solid silica, on the basic side by saturation with solid lime and solid Ca_2SiO_4. The lowest melting points are found for slags with about equal amounts of silica and iron oxide and 10 to 20 percent lime; these melt below 1100°C. Two areas are of particular interest in steel-making: (A) Slags practically saturated with silica and with very little lime, and (B) Slags practically saturated with solid lime or solid Ca_2SiO_4. The former covers the acid steel-making slags which may be melted in furnaces with silica lining, the latter covers the basic slags which may be melted in contact with magnesite or dolomite linings.

Copper- and lead-smelting slags resemble the acid steel-making slags, but contain more Fe_2O_3, that is, they are formed under more oxidizing conditions and are not in equilibrium with metallic iron. Also, in order to increase their

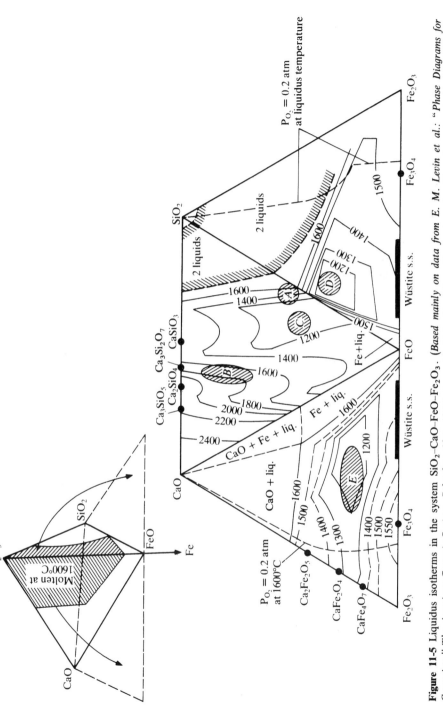

Figure 11-5 Liquidus isotherms in the system SiO_2–CaO–FeO–Fe_2O_3. (*Based mainly on data from E. M. Levin et al.: "Phase Diagrams for Ceramists," The American Ceramic Society, Columbus, Ohio, 1964.*) *A* = acid steelmaking slags, *B* = basic steelmaking slags, *C* = Cu and Pb smelting slags projected into SiO_2–CaO–FeO system, *D* = same, projected into the SiO_2–FeO–Fe_2O_3 system, *E* = lime ferrite slags.

fluidity some lime is usually added. In Fig. 11-5 the composition range for these slags is indicated by their projections into the FeO–CaO–SiO$_2$ and FeO–Fe$_2$O$_3$–SiO$_2$ systems.

Whereas in the past most slags used in copper smelting and refining have been iron silicates with only small amounts of lime and other oxides the *lime ferrite* slags have recently come up as an interesting alternative. As seen from Fig. 11-5 the ternary system FeO–Fe$_2$O$_3$–CaO shows melting temperatures below and around 1200°C for slags with about 20 weight percent CaO.[1] These slags are particularly well suited for the Bessemer (converting) and refining stages in copper smelting, as the melting temperature remains low even at high oxygen potentials. This is in contrast to the iron-silicate slags, where solid magnetite is formed under oxidizing conditions. It is important that the silica content in lime-ferrite slags should be low, as silica leads to formation of high-melting Ca$_2$SiO$_4$ with corresponding increase in the melting temperature.

The thermodynamics of ferrous slags has been studied rather extensively by several investigators. The activity of FeO in slags in equilibrium with iron was studied by Taylor and Chipman.[2] Their diagram for 1600°C is shown in Fig. 11-6. A characteristic feature of this diagram is the rather high activity values for compositions along the pseudobinary line FeO–Ca$_2$SiO$_4$. In comparison,

[1] Y. Takeda, S. Nakazawa, and A. Yazawa: *Can. Metall. Q.* **19**: 297 (1980).
[2] C. R. Taylor and J. Chipman: *Trans. AIME*, **154**: 228 (1943).

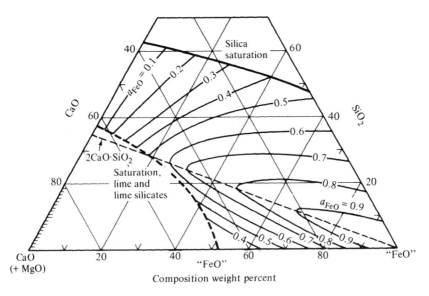

Figure 11-6 Activity of FeO in SiO$_2$–CaO–FeO slags in equilibrium with iron at 1600°C. (*Data by Taylor and Chipman as reviewed by Elliott, Gleiser, Ramakrishna: "Thermochemistry for Steelmaking," vol. II, © 1963 Addison-Wesley, Reading, MA, p. 587. Reprinted with permission.*)

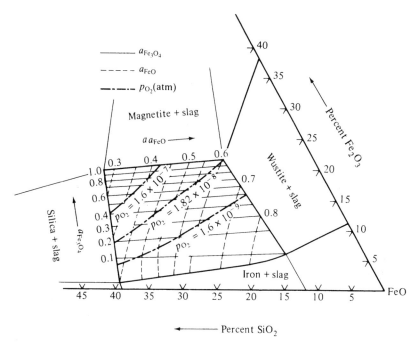

Figure 11-7 Activities and oxygen potential in SiO_2–FeO–Fe_2O_3 system at 1300°. (*Data by Schuhmann et al., and Muan and Osborn as reviewed by N. Korakas: Trans. Inst. Min. Metall., vol. 72, p. 35 (1962–1963).*)

the FeO activity both for the FeO–SiO_2 and for the FeO–CaO systems is roughly proportional to the FeO concentrations. This behavior may be attributed to the much higher stability of calcium than of iron orthosilicate. On the basis of the ionic theory for molten slags, and making use of the *Flood equation* given in Sec. 4-7, it has been possible to calculate the activity of FeO in the slag from the relative stability of Ca_2SiO_4 and Fe_2SiO_4, and results in fairly good agreement with the experimental data have been obtained, at least for the more basic slags.

The thermodynamics of the FeO–Fe_2O_3–SiO_2 system has been studied by Schuhmann and coworkers[1] and by Muan.[2] Their results are in fairly good agreement and are summarized in Fig. 11-7 which gives the activity of FeO and Fe_3O_4 as well as the oxygen potential at 1300°C as function of slag composition. Similarly the thermodynamics of the lime-ferrite slags at 1300°C is shown in Fig. 11-8 which gives the oxygen potential as well as the activities of CaO and FeO relative to respectively solid CaO and liquid iron-saturated FeO. Again we see that slags with around 20 percent CaO are molten at 1300°C all

[1] R. Schuhmann and P. J. Ensio: *Trans. AIME*, **191**: 401 (1951)
[2] A. Muan: *Trans. AIME*, **203**: 965 (1955).

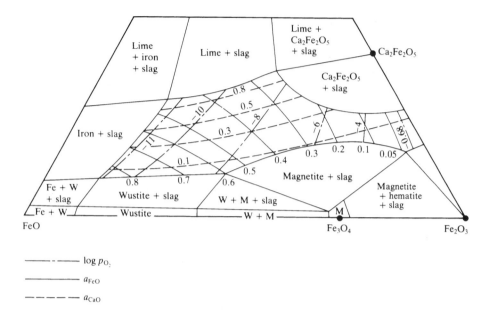

Figure 11-8 Oxygen potential and activities in lime ferrite slags at 1300°C. (*Data from Y. Takeda, S. Nakazawa, and A. Yazawa: Can. Metall. Q. vol. 19, p. 297 (1980)*). W = wustite, M = magnetite.

the way from equilibrium with solid metallic iron (log $p_{O_2} \simeq -11$) to equilibrium with air (log $p_{O_2} = -0.68$), and that the activity of FeO decreases from around 0.8 to less than 0.05 in the same range. In comparison the FeO activity in iron-silicate slags at 1300°C (Fig. 11-7) cannot be brought below 0.3, corresponding to log $p_{O_2} \simeq 10^{-7}$, before solid magnetite is precipitated. The importance of these facts for copper smelting will be discussed in Sec. 12-3.

Additional Components in Ferrous Slags

Acid steel-making slags often contain some manganese oxide; basic steel-making slags contain magnesium and phosphorus oxides. Copper and lead slags may contain some zinc oxide, and all slags may contain some alumina. Manganese, magnesium, and zinc oxides resemble lime in so far as they have high melting points and form rather stable silicates. Also, the activity coefficient for these oxides will be high in basic and low in acid slags.

Phosphorus pentoxide resembles silica in so far as it forms very stable compounds with lime and somewhat less stable compounds with ferrous oxide. As a consequence the activity coefficient for P_2O_5 is very low in basic, and high in acid, slags. In the CaO–FeO–P_2O_5 ternary system the high stability of the $Ca_3P_2O_8$ and $Ca_4P_2O_9$ compounds gives rise to a range of liquid immiscibility between a calcium phosphate melt and a melt of almost pure iron oxide.

Thus for a rough evaluation of the properties of steel-making, copper, and lead slags the MgO, MnO, and ZnO contents should be added to the CaO content, and P_2O_5 should be added to the SiO_2 content, and the resulting composition may be applied to the phase diagram and activity curves in Figs. 11-5 to 11-7. Alumina does not seem to be strongly bonded either in acid or in basic slags, and may be regarded as a neutral dilutant.

11-4 PHYSICAL PROPERTIES OF SLAGS

In addition to the melting point the viscosity plays an important role in the melting behavior of slags. For so-called "newtonian liquids" the viscosity μ at constant temperature and composition is independent of the rate of shear. In metric (SI) units the unit for viscosity is the *poise*, g/(cm·s), or the *dekapoise*, kg/(m·s) = Pa·s. The viscosity for a large number of silicate slags has been measured by various investigators as function of temperature as well as of slag composition, and is illustrated by the isoviscosity curves for the $CaO–SiO_2–Al_2O_3$ system shown in Fig. 11-9. We see that the viscosity at constant temperature decreases with increasing lime content, and increases with increasing content of silica and alumina. Between a basic and an acid slag the viscosity value may differ by a factor of one thousand or more. If lime is replaced by other

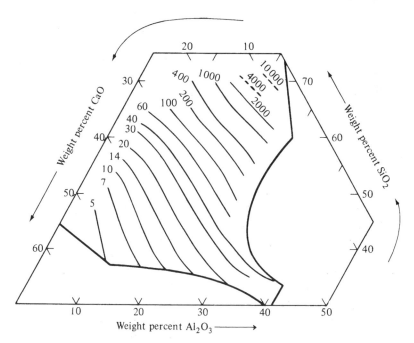

Figure 11-9 Viscosity in poises for $SiO_2–CaO–Al_2O_3$ slags at 1450°C. (*After J. S. Machin and T. B. Yee: J. Am. Ceram. Soc., vol. 3, p. 200 (1948).*). 1 poise = 0.1 Pa·s.

basic oxides, MgO, MnO, or FeO, the viscosity is not much affected; if anything there is a further decrease. Thus it seems that the viscosity decreases with increasing concentration of what are commonly called basic oxides. This again may be attributed to the breaking of oxygen bridges in the silicate anion network as discussed in Sec. 4-7.

An interesting exception is the viscosity of titania slags. As mentioned, in Sec. 11-1, TiO_2 in slags behaves like an acid, i.e., it forms stable compounds with lime and no compounds with silica. The viscosity of a lime-silicate slag is decreased by the addition of TiO_2, however, which shows that low viscosity and affinity toward silica do not always run parallel. Thus by addition of titania to siliceous slags it should be possible to obtain a combination of high SiO_2 activity and low viscosity. Conversely, Al_2O_3 increases the viscosity of blast furnace type slags almost as much as silica (Fig. 11-9), whereas chemically it appears to be rather "neutral."

The viscosity data mentioned above all refer to slags which are completely molten. If solid particles are present in the slag this will give a rapid increase in the viscosity. Thus the viscosity of titania slags increases on reduction and precipitation of solid phases of lower oxides. Similarly the viscosity of blast furnace type slags increases when the solubility limit for Ca_2SiO_4 is exceeded. The known effect of fluorspar in increasing the fluidity of steel-making slags is probably also a result of its fluxing of solid particles rather than of an effect on the true viscosity of the liquid phase.

The effect of temperature is greatly different for acid and for basic slags. For basic slags the apparent viscosity increases only slightly with decreasing temperature, and then rapidly as solid phases are precipitated. For high-silica slags the viscosity increases more gradually with decreasing temperature. Supercooling of the liquid is quite common, and the viscosity of the slag increases from that of the liquid to that of the glassy solid. Basic slags are often called *short* and acid slags are called *long*, illustrating the behavior of the viscosity on cooling of the slag. The effect of temperature on slag viscosity may be expressed by the *activation energy for viscous flow*. This is about 85 kJ for basic slags, increasing to more than 250 kJ for acid slags.[1]

Density

Relatively few measurements exist on slag density. The density seems to be more or less a linear function of slag composition, although deviations exist. In blast furnace type slags the density increases with increasing lime content, and in ferrous slags with increasing concentration of iron oxide.[2] In a ferrous slag the density is lowered when some of the iron oxide is replaced by lime, a fact which is utilized to obtain a good separation between the slag and the matte in copper smelting (Sec. 12-3).

[1] J. O. M. Bockris et al.: *Trans. Faraday Soc.*, **51**: 1734 (1955).
[2] J. Henderson et al.: *Trans. AIME*, **221**: 807 (1961).

Surface Tension

The surface tension is of importance for the wetting of metals and refractories by slags and for the rate at which various components dissolve in the slag. The surface tension of liquid slags is of the order of 0.3 to 0.5 J/m^2, and in general decreases with increasing silica content of the slag. Low surface tension is obtained by the addition of alkali oxides, boric acid, calcium sulfide, and phosphoric oxide, whereas lime, iron oxide, and alumina increase the surface tension of the slag. The surface tension decreases slightly with increasing temperature; over a range of 200°C the decrease in surface tension is about 10 percent.

Related to the surface tension is the tendency of certain slags to *foam*. Foaming of the slag may be harmful in open-hearth steel-making where the foam prevents the heat from being transferred from the flame to the bath. The nature of foaming is not fully understood and few systematic measurements exist.

Electrical conductivity of slags will be discussed in Sec. 16-1.

11-5 REFRACTORIES

In the same way as the metallurgist wants his slags to be low-melting and fluid, he wants his refractories to be high-melting and strong. A high melting point alone is not enough to give a good refractory, however, and below are listed the most important properties which are relevant to the value of refractory materials:

1. Melting point
2. Strength at high temperature
3. Thermal shock resistance
4. Resistance toward molten slags and metals
5. Resistance to oxidation and reduction
6. Storage stability
7. Cost

Refractories may be classified in the following groups:

A. Oxides

a. Silica
b. Fireclay (alumina–silicates)
c. High alumina
d. Chromite (mainly $(Fe,Mg)(Cr,Al)_2O_4$) and magnesia–chromite (mainly $MgCr_2O_4 + MgO$)
e. Burned magnesite (MgO) and dolomite (MgO + CaO)
f. Forsterite (Mg_2SiO_4)
g. Special oxides (ZrO_2, ThO_2, BeO)

B. Nonoxides

a. Carbon and graphite
b. Silicon carbide
c. Rare compounds (TiC, TiB$_2$, BN, sialons)
d. Metals (Mo, W, Fe).

Silica and fireclay are often called *acid* refractories, whereas magnesite, burned dolomite, and forsterite are called *basic*. Alumina and chromite are sometimes called neutral refractories. Some oxide refractories, such as stabilized zirconia and β-alumina, show high ionic conductivity. These are *solid electrolytes* and will be discussed in Chap. 16.

For industrial use refractories are mostly made from natural raw materials: silica sand, fireclay, silimanite rock, bauxite, chromite ore, magnesite and dolomite rocks, etc. For laboratory refractories specially purified raw materials are used. In order to decompose the carbonates and obtain a certain grain growth, the raw materials for basic refractories are usually precalcined at temperatures up to 1700°C, or they may be premelted and crushed to suitable grain-sizes. In molding the bricks a suitable binder is usually added. Thus silica bricks are normally made from silica sand with the addition of small amounts of lime or iron powder. Silicon carbide bricks are made from silicon carbide grains and clay. Fireclay bricks are made from a mixture of previously burned and crushed brick and raw fireclay, whereby excessive shrinkage during firing is avoided. The final firing of most bricks is made at temperatures between 1300 and 1700°C.

The porosity of the final product is determined by the grain distribution in the raw material, the amount of binder, and the temperature and duration of firing. By proper choice of the above variables a low porosity may be obtained; for laboratory refractories the porosity may be practically zero. Nonporous basic bricks may also be cast directly from the molten oxide. For most industrial refractories, however, the porosity is between 10 and 30 percent, the cheaper qualities usually having the highest porosity.

When refractory bricks are used, e.g., as a furnace lining, they are bonded together with a mortar which usually is of the same composition as the brick. An interesting exception is the use of sheet iron between magnesite bricks. When the furnace is heated the iron oxidizes to FeO which enters into solid solution in MgO (see Fig. 11-10(*b*)) and forms a monolithic structure.

Refractory materials may also be prepared as a plastic mix which is rammed directly into the furnace shell and is fired in place. This method is used particularly for silica and basic masses and for carbon. Another type is so-called *refractory concrete*, which is composed of fireclay or high-alumina material with some calcium–aluminum cement and water added. This concrete mix may be poured into place, where it hardens at room temperature.

For insulating purposes special porous bricks are used. These may be prepared by the addition of, for example, sawdust to the raw mixture, or by the addition of compounds which evolve gases during firing of the bricks. Insulating bricks are mechanically weaker than dense bricks and are less resistant toward slags. Therefore they are used in the outer layers of a furnace lining and are covered on the inside by denser and more refractory material. In recent years aluminum silicate *fibers*, containing from 40 to 85 percent Al_2O_3, have come up as an important insulating material, which can withstand temperatures up to 1700°C. Other insulating materials are asbestos and diatomaceous earth; the latter may be molded into bricks. These materials disintegrate readily above 1000°C and are not normally classified as refractories.

11-6 OXIDE REFRACTORIES

The highest melting points are obtained for the pure oxides, and increase in the order SiO_2, Al_2O_3, Cr_2O_3, CaO, MgO. The melting points are lowered by the addition of foreign oxides, but the effect differs greatly for different oxide systems. This is seen by comparing the diagrams SiO_2–FeO and MgO–FeO in Fig. 11-10. The addition of only a few percent of FeO to silica causes the formation of a eutectic phase which melts below 1200°C; at 60 percent FeO the mixture is completely molten at that temperature. In comparison, when FeO is added to magnesia it enters into solid solution. A mixture with 60 percent FeO is completely solid even at 1700°C, and liquid FeO is able to dissolve only about 10 percent MgO at 1600°C.

Addition of silica to alumina causes only a slight decrease in the melting point, and the rather high-melting mullite phase is formed, see Fig. 11-11. Similarly the addition of silica to magnesia causes the formation of the high-melting forsterite phase, Mg_2SiO_4, and the sesquioxides Al_2O_3 and Cr_2O_3 form rather high-melting phases (spinel) with MgO and FeO.

Industrial refractories are hardly ever pure. Therefore, they do not have sharp melting points but rather a softening region. This region is determined not only by the phase diagram for the main components but also by the presence and distribution of impurities. Most refractories consist of grains of rather pure and refractory materials embedded in a bonding matrix. The matrix may have a higher concentration of impurities and therefore a lower melting point. Thus fireclay bricks, which are mainly alumina silicates, already start softening between 1400 and 1500°C, depending on their alumina content and purity, whereas from the phase diagram softening should occur at the eutectic temperature of 1590°C. The softening temperature is dependent also on whether the brick is an equilibrated mixture of phases or not. A high-alumina brick with, say, 80 percent Al_2O_3 may be a stable combination of corundum and mullite, or it may be an unstable combination of corundum and silica. The latter will have a lower melting temperature.

The practical service temperature of a brick depends on whether the brick is loaded or not. Thus under load a fireclay brick may start softening up to 200°C lower than if it were not loaded. Silica brick, on the other hand, retains its strength practically all the way to its melting point.

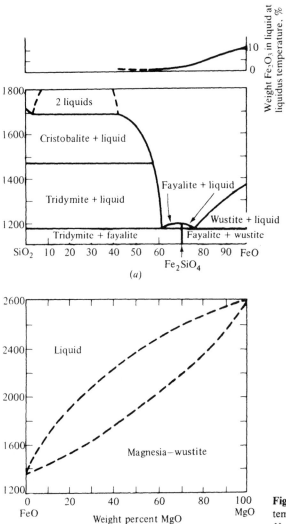

Figure 11-10 (a) Binary SiO_2–FeO system. (b) Binary FeO–MgO system. (*From N. L. Bower and J. F. Schairer: Am. J. Sci.*, vol. 29. p. 151 (1935).)

In order to test the softening and melting behavior of bricks certain standardized methods have been developed. For a discussion of these methods the reader is referred to the bibliography.

Resistance to Thermal Shock

This is first of all a function of the thermal expansion coefficient, mechanical strength, and modulus of elasticity, but also of the occurrence of polymorphic transformations. For refractories without polymorphic transformation the dimension increases rather linearly with increasing temperature, as illustrated by

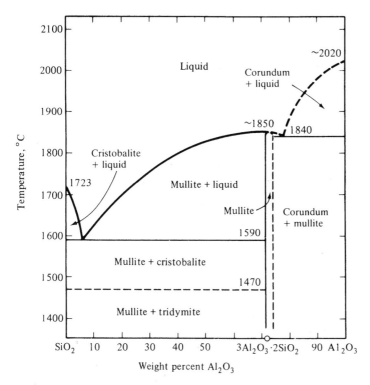

Figure 11-11 Binary SiO$_2$–Al$_2$O$_3$ system. (*From A. Muan and E. F. Osborn: "Phase Equilibria among Oxides in Steelmaking,"* © *1965 Addison-Wesley, Reading, MA, Fig. 19. Reprinted with permission.*)

the curves for fireclay, chrome, corundum, and magnesite in Fig. 11-12. Silica on the other hand undergoes crystallographic transformations on heating: at 573°C ($\alpha \rightarrow \beta$-quartz), at 870°C (β-quartz → tridymite) and at 1470°C (tridymite → cristobalite). The second transformation is very slow, however, and β-quartz may transform directly to cristobalite around 1250°C and further to tridymite. Furthermore, both tridymite and cristobalite may supercool to room temperature, but undergo $\beta \rightarrow \alpha$ inversions around 200°C. The transformations from α to β phases and from β-quartz to tridymite occur under volume expansion; particularly large is the $\alpha \rightarrow \beta$ transformation in cristobalite. As a result, the volume of silica brick, which usually is a mixture of all three modifications, increases rather strongly up to about 400°C but remains practically constant above that temperature. This gives silica a very good resistance toward thermal shock at high temperature, whereas serious cracks may occur if the brick is cooled below 400°C.

The low thermal expansion coefficient at high temperatures, and the fact that it retains its strength almost to its melting point, makes silica brick particularly suited for sprung (self-supporting) arches, and it was previously the dominating refractory for roofs in steel-making and reverberatory furnaces. Its slag

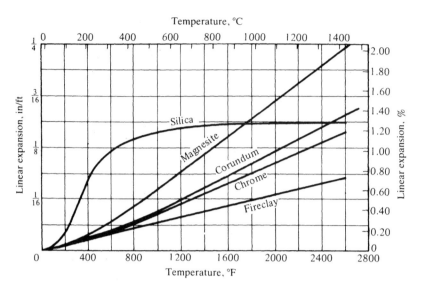

Figure 11-12 Reversible expansion curves for common refractory bricks. (*From R. Schuhmann Jr.: "Metallurgical Engineering," vol. I, © 1952 Addison-Wesley, Reading, MA, Fig. 10-2. Reprinted with permission.*)

resistance is poor, however, and its maximum service temperature is below 1700°C. In recent years, therefore, roofs of basic bricks, like magnesia, are becoming more common. Owing to their large thermal expansion coefficient these cannot be made self-supporting, however, and a steel hook embedded in each brick enables them to be suspended from a steel frame above the furnace.

An entirely different behavior is found for *vitreous silica*, so-called quartz-glass, which is used for laboratory ware. Vitreous silica has a very low thermal expansion coefficient over the entire temperature range and may be shock-cooled from white heat into ice-water without cracking.

Another interesting laboratory refractory is zirconia, ZrO_2, which is useful for its high melting point and chemical stability. Pure zirconia transforms around 1000°C under a considerable volume change, and requires very careful heating and cooling. Upon addition of about 10 percent lime, however, a third modification is obtained which is stable over the entire temperature range. Such *stabilized zirconia* is a very useful laboratory refractory. Another case is calcium orthosilicate, Ca_2SiO_4, which melts at 2130°C and might be considered a good and inexpensive refractory. Unfortunately it transforms on cooling below 725°C under a large volume expansion and disintegrates to powder. In comparison, forsterite, Mg_2SiO_4, which melts at 1890°C, shows no transformation, and has found some use as a refractory material.

Related to thermal cracking is the so-called *thermal spalling*. Spalling is the result of thermal stresses set up when the temperature at the surface of a wall changes more rapidly than in the interior. High spalling resistance is obtained for refractories which combine low thermal expansion coefficient with high heat conductivity and high mechanical strength. *Structural spalling* occurs when a slag reacts with a refractory and changes its volume or other physical properties.

Slag Resistance

The resistance of refractories toward slags is determined first of all by equilibrium relations. It is clear that a slag which is already saturated with a solid phase cannot further attack a refractory consisting of that solid phase. Thus a silica-saturated slag may be melted in a silica crucible, and a highly basic slag may be melted in furnaces with magnesite or dolomite lining. The relative temperature of the slag and the lining is of importance here. In a Bessemer converter or in an electric arc furnace the lining is colder than the slag, and is not easily attacked, whereas in a laboratory tube furnace the crucible is hotter than the slag, with a correspondingly greater danger of slag attack.

Another important factor is the viscosity of the slag. In a silica- or fireclay-lined container the slag near the lining will dissolve the refractory and become very viscous. Further attack by the slag is then only possible by diffusion through the viscous slag layer. As a result such refractories are fairly resistant even toward slags which are not completely saturated with the refractory oxides. In comparison basic refractories give a very fluid melt and are more easily attacked.

Slag resistance is also affected by the physical structure of the brick. A porous refractory is easily eaten up by the slag, which soaks into the pores, whereas a dense single-phase refractory is only slowly attacked. The binding matrix is more rapidly attacked than the solid grains, causing grains to loosen and become dispersed in the slag without actually dissolving.

Finally, two bricks which in themselves have high melting points, may form a low-melting-point slag on contact. Thus if silica or fireclay bricks come into contact with magnesite or dolomite bricks a low-melting-point blast furnace type slag is formed. Certain bricks, such as chromite, do not slag either with acid or basic bricks, and may be used as an intermediate layer.

Resistance Toward Oxidation and Reduction

Oxide refractories may be affected by oxidation as well as by reduction. As already mentioned silica refractories are easily attacked by ferrous oxide under reducing conditions. In air, however, the trivalent ferric oxide is formed, and is much less aggressive toward silica. Thus whereas in equilibrium with metallic iron eutectic melting occurs at 1178°C and about 40 percent silica, these values change to 1450°C and about 18 percent SiO_2 for melting in air. Similarly, the resistance of chromite bricks toward silica and acid slags is higher under oxidizing than under reducing conditions.

Under strongly reducing conditions oxide refractories may be partly reduced. This is particularly the case if the refractory is in contact with highly reactive metals. Thus if aluminum is melted in a quartz crucible, reduction will take place

$$4Al + 3SiO_2 = 2Al_2O_3 + 3Si$$

Also volatile products may be formed: SiO from silica refractories and magnesium vapor by reduction of magnesia refractories, see Sec. 13-9.

In the melting of the least noble metals such as aluminum, titanium, niobium, etc., the choice of refractories is quite a problem. Aluminum may be melted in alumina crucibles, but at very high temperatures volatile Al_2O may be formed. For some reactive metals zirconia, thoria, or beryllium oxide crucibles may be used. Titanium and niobium cannot be melted in any known oxide refractories, however, without some reaction and solution of oxide by the metallic phase. The special problems associated with the the melting of reactive metals will be discussed in Chap. 14.

Storage Stability

Basic refractories may, on storage in air, absorb water vapor and carbon dioxide. This is particularly the case for burned dolomite which contains free lime. One way to avoid this reaction is to mix the dolomite with tar, which acts as a waterproof coating, the tar-dolomite mixture being used as a ramming mix for furnace lining. Another method is to fire the dolomite at high temperature with the addition of some siliceous material. Magnesium silicate is a convenient addition, and the reaction will be of the type

$$3CaO + Mg_2SiO_4 = Ca_3SiO_5 + 2MgO$$

The product is called *stabilized (burned) dolomite*. It is important that the silica addition is limited to give only the tricalcium silicate, which does not disintegrate, and not the orthosilicate which disintegrates on cooling.

11-7 NONOXIDE REFRACTORIES

Carbon is the most common nonoxide refractory, and is used either as amorphous carbon or as graphite. Carbon remains solid up to temperatures above 3000°C. Also it has a low thermal expansion coefficient, it is strong and very resistant toward thermal shock. Its main limitation is that it burns readily in air, and can only be used under reducing conditions, in vacuum, or in a protective atmosphere. Carbon and graphite are not attacked by "white" blast furnace type slags below 1700 to 1800°C, but react readily with ferrous slags. Some metals such as copper and aluminum may be melted in carbon crucibles, whereas molten iron, ferroalloys, and nickel will dissolve several percent of carbon. Because of its high resistance to slag attack, carbon, either as blocks or as a monolithic lining, is frequently used in the iron blast furnace hearth where the metal already is carbon-saturated.

A mixture of carbon or graphite with fireclay gives a refractory which combines the best properties of both materials: good resistance to oxidation and slag attack and good thermal shock resistance. This material is often used for crucibles in nonferrous melting and casting.

Silicon carbide is used in boilers, around tapping holes, and in places where strength at high temperature and shock and slag resistance are required. Because of its high thermal conductivity it is also used for retorts in the carbothermic production and distillation of zinc. Silicon carbide decomposes without melting above 2200°C. It is fairly resistant toward oxidation below 1500°C, being protected by a layer of silica. At higher temperatures it oxidizes slowly to silica and carbon dioxide. The protective silica layer is attacked by basic slags and particularly by alkali oxides, whereas it is fairly resistant toward acid slags. Silicon carbide is readily attacked and dissolved by ferrous metals.

Silicon carbide is also used for electrical heating elements either directly ("Globar," "Silit"), or indirectly as material for channels filled with anthracite or coke, these materials acting as the electrical resistor.

Rare compounds These are particularly used for laboratory crucibles, etc. Some compounds which have been used are TiC, TiB_2, BN, and CeS. *Titanium boride* is very resistant toward fluoride melts. Being a good electrical conductor it has been suggested as a suitable lead to the molten aluminum cathode in aluminum electrolysis. *Boron nitride* shows similar chemical resistance, but is an electrical insulator, and has been tried for thermocouple protection tubes in the aluminum industry. *Cerium sulfide* combines a high melting point with low volatility, but is readily oxidized in air. The main limitation in the use of these rare refractories is their high cost.

In recent years ceramics composed of silicon, aluminum, oxygen, and nitrogen have received considerable attention. These *sialons* may be considered as derived from silicon nitride, Si_3N_4, where some of the silicon and nitrogen atoms are replaced by aluminum and oxygen. Sialons retain their strength to high temperatures, and, as they are also oxidation resistant, they have been suggested as material for high-temperature turbines, etc., but have so far found no application in extractive metallurgy.

Metals have limited applications as refractories, being readily oxidized in air. Exceptions are the platinum group metals which are used for laboratory crucibles, etc., but which are very expensive. Under reducing conditions or protective atmospheres crucibles of molybdenum or tungsten have been used for laboratory studies on molten slags. Similarly, crucibles of iron or nickel may be used below their melting points. Molybdenum and tungsten may also be used as electrical heating elements, provided the atmosphere is free of oxygen.

An interesting recent development is heating elements of *molybdenum silicide*. These are resistant in air up to 1700°C, being protected against oxidation by a glassy layer of silica. If this layer is broken it heals itself by oxidation at high temperature.

Finally, the use of *water-cooled metal* which is used under conditions where no other refractories are feasible, should be mentioned. Thus in the copper or lead blast furnace the slag is highly corrosive and would attack all known refractories. The furnace is here lined with water-cooled steel plates, *water jackets*. A crust of slag solidifies on the steel sheet, and a thermal balance is established between the heat conducted from the furnace and that removed by the cooling water. In that way a slag layer of constant thickness is maintained, the temperature of which does not exceed the melting point of the slag.

A similar principle is employed in the melting of highly reactive metals such as titanium or niobium. Here the crucible is made of copper, which is either water-cooled or cooled with a liquid alkali alloy. The heat is generated by an electric arc between an electrode and the molten metal bath, and only a thin solidified metal crust separates the melt from the crucible, see Sec. 14-6.

PROBLEMS

11-1 In the phase diagram shown in Fig. 11-3 all phases are assumed stoichiometric and with insignificant solid solution.

(a) Draw the pseudobinary diagram $CaSiO_3$–TiO_2 as function of temperature and composition.

(b) Which phases exist in equilibrium for a slag with 40 percent CaO, 30 percent SiO_2, and 30 percent TiO_2 at 1600, 1400, and 1200°C?

11-2 For a slag with 40 percent SiO_2, 40 percent CaO, and 20 percent Al_2O_3 the following activities are read from Fig. 11-2: $a_{SiO_2} = 0.08$, $a_{CaO} = 0.035$, $a_{Al_2O_3} = (a_{AlO_{1.5}})^2 = 0.015$. By means of data in App. C calculate the chemical activity of Si and Al as well as the partial pressure of SiO(g) and Ca(g) for equilibrium between the slag, graphite, and 1 atm of CO, all at 1600°C.

11-3 From data in App. C calculate the activity of FeO in equilibrium with metallic iron, graphite, and 1 atm of CO at 1500°C, and, by means of Fig. 11-6, which is assumed to apply also at 1500°C, estimate the corresponding equilibrium concentration of FeO in a slag with $(\%CaO)/(\%SiO_2) = 1$.

11-4 By means of Stokes' law (Eq. 6-11) and viscosity data given in Fig. 11-9 calculate the settling velocity of a spherical pig iron droplet, 1 mm diameter, in a slag with 60 percent SiO_2, 30 percent CaO and 10 percent Al_2O_3 at 1450°C. The densities of iron and slag are taken to be 7.15 and 2.45 g/cm^3 respectively. Remember that 1 poise = 1 g/(s · cm). Repeat the calculation for 0.1 mm droplets.

11-5 To melt a slag with 50 percent SiO_2, 30 percent CaO, and 20 percent Al_2O_3 at 1400 to 1500°C you have the choice between crucibles of silica, mullite, magnesia, and graphite. Which of these would be your first and which your second choice?

11-6 Which inexpensive crucible would you use to melt a slag of composition Fe_2SiO_4 at 1200 to 1400°C?

11-7 A mullite thermocouple tube is used in contact with graphite in a flowing argon atmosphere at 1500°C. A possible reaction would be $Al_6Si_2O_{13} + 2C = 3Al_2O_3(s) + 2SiO(g) + 2CO(g)$. Calculate the reaction pressure $p_{SiO} + p_{CO}$ and comment on the usefulness of the mullite tube.

BIBLIOGRAPHY

Alper, A. M. (ed.): "High Temperature Oxides" parts I–IV, Academic Press, New York, 1970–1971.
Eitel, W.: "The Physical Chemistry of Silicates," University of Chicago Press, Chicago, Ill., 1954; "Silicate Science" I–V, Academic Press, New York, 1964-66.

Elliott, J. F., M. Gleiser, and V. Ramakrishna: "Thermochemistry for Steelmaking," vol. II, Addison-Wesley Publ. Co. Inc., Reading, Mass., 1963.

Harders, F., and S. Kienow: "Feuerfestkunde," Springer Verlag, Berlin, 1960.

Levin, E. M., C. R. Robbins, and H. F. McMurdie: "Phase Diagrams for Ceramists," The American Ceramic Society, Columbus, Ohio, 1964. Supplement volume 1969.

Muan, A., and E. F. Osborn: "Phase Equilibria among Oxides in Steelmaking," Addison-Wesley Publ. Co. Inc., Reading, Mass., 1965.

Norton, F. H.: "Refractories," 3d ed., McGraw-Hill Book Co. Inc., New York, 1949.

Ryshkewitch, E.: "Oxide Ceramics," Academic Press, New York and London, 1960.

Schuhmann, R. Jr.: "Metallurgical Engineering," vol. I, Addison-Wesley Publ. Co. Inc., Reading, Mass., 1952.

CHAPTER
TWELVE

MATTE SMELTING

By mattes we understand mixed, molten sulfides of heavy metals, often containing some oxides. Of the greatest industrial importance is iron–copper matte, which is an intermediate product in the extraction of copper from sulfide ores. Next comes the copper–nickel matte. One advantage of matte smelting is that some mattes may be processed to give the desired metal directly. This is, for example, the case for iron–copper mattes which, on oxidation and slagging of the iron component, may be converted to metallic copper by the so-called *roast-reaction*. In this way the fuel requirements will be less than for the alternative *roast-reduction*, i.e., to roast the sulfide ore and reduce the resulting oxide with carbon. Another advantage is the relatively low melting temperatures of the matte. Thus a matte with equal amounts of iron and copper sulfides melts below 1000°C, whereas an alloy of the same metals would require more than 1400°C.

12-1 IRON–COPPER MATTES

Iron–copper mattes are sometimes described as binary mixtures of FeS and Cu_2S. This is only approximately correct. First of all the sulfur content, i.e., the mole ratio: $n_S/(n_{Fe} + \frac{1}{2}n_{Cu})$, may vary noticeably with the sulfur pressure above the melt. This is seen from Fig. 12-1(a) and (b) which show part of the binary systems Fe–S and Cu–S. In the iron–sulfur system complete liquid miscibility exists between iron and iron sulfide. In the copper–sulfur system a region of immiscibility exists, but the molten sulfide phase contains variable amounts of sulfur. In both systems the sulfur pressure at, say, 1200°C increases from values

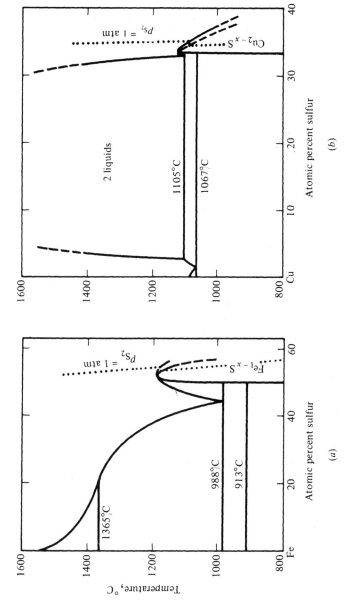

Figure 12-1 (a) Part of the iron–sulfur system. (b) Part of the copper–sulfur system. (*After M. Hansen: "Constitution of Binary Alloys," McGraw-Hill Book Co. Inc., New York, 1958.*)

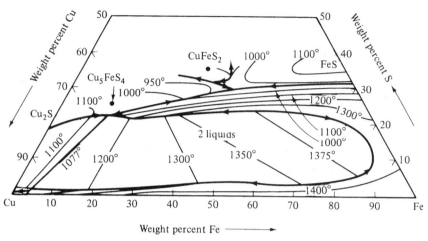

Figure 12-2 Liquidus surface of the iron–copper–sulfur system. (*Mainly from H. Schlegel and A. Schüller: Z. Metallkunde, vol. 43, p. 421 (1952).*)

of about 10^{-6} atm for equilibrium with the metallic phase to 1 atm, which is reached at the approximate compositions $Fe_{0.9}S$ and $Cu_{1.85}S$.

Figure 12-2 shows part of the ternary iron–copper–sulfur system. We see that the field of liquid immiscibility extends from the Cu–S side practically across to the Fe–S side. Also here the sulfur pressure increases from about 10^{-6} atm for equilibrium with the metallic phases to 1 atm along the pseudobinary line $Fe_{0.9}S$–$Cu_{1.85}S$ (all at 1200°C). The eutectic temperatures decrease to a minimum near the pseudobinary line. This is also seen from Fig. 12-3 which

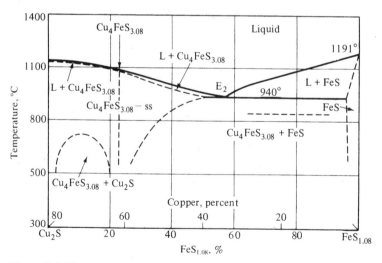

Figure 12-3 Binary Cu_2S–$FeS_{1.08}$ system. (*From H. Schlegel and A. Schüller: Z. Metallkunde, vol. 43, p. 421 (1952).*)

gives the phase diagram for the pseudobinary system $Cu_2S-FeS_{1.08}$. It is clear, however, that the shape of the pseudobinary diagram will change with changing sulfur pressure. This, probably, is the reason why different authors give somewhat different diagrams for the pseudobinary system.

In addition to iron, copper, and sulfur the mattes may dissolve variable amounts of oxygen. This is apparent from the fact that for industrial mattes the sum Fe + Cu + S rarely exceeds 95 weight percent. The oxygen content in the matte depends on the iron-to-copper ratio, the prevailing oxygen and SO_2 pressures, and whether a liquid slag is present or not. Most industrial matte smelting is carried out under conditions where the SO_2 pressure is between 0.1 and 1 atm; under these conditions, and in the absence of a liquid slag, the melt will, with increasing oxygen pressure, dissolve oxygen until a saturation limit where solid magnetite is formed. The solubility of oxygen for $p_{SO_2} = 1$ atm and magnetite saturation has been studied by Johannsen and Knahl[1] among others. Their results are shown in Fig. 12-4(a) and (b), and we see that the oxygen solubility increases strongly with increasing iron-to-copper ratio as well as with increasing temperature. Also the sulfur content has a maximum for about 25

[1] F. Johannsen and H. Knahl: *Erzmetall*, **16**: 611 (1963).

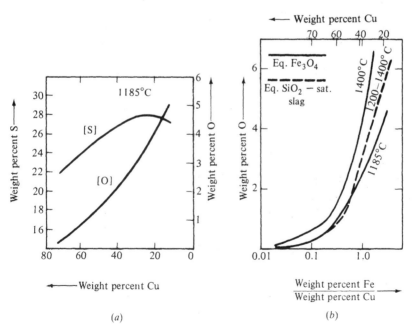

Figure 12-4 (a) Sulfur and oxygen contents in mattes saturated with magnetite and 1 atm SO_2. (b) Oxygen contents of mattes in equilibrium with magnetite (solid lines) and silica saturated iron silica slag (dashed line). (*After F. Johannsen and H. Knahl: Erzmetall, vol. 16, p. 611 (1963), and U. Kuxmann and F. Y. Bor: Erzmetall, vol. 18, p. 441 (1965).*)

percent copper in the matte. These values are believed to be approximately correct also for $p_{SO_2} = 0.1$ atm.

In the presence of slag-forming oxides, such as silica and/or lime, a liquid slag will be formed and will affect the oxygen content of the matte. This is shown in Fig. 12-4(*b*), where the dashed curve gives the oxygen content of the matte in equilibrium with silica-saturated iron–silicate slag and under a SO_2 pressure of 1 atm.[1] It is interesting to notice that in this case the oxygen content is practically independent of temperature between 1200 and 1400°C, whereas for saturation with solid magnetite the oxygen content increases with increasing temperature.

12-2 THERMODYNAMICS OF COPPER SMELTING

Matte smelting and converting of an iron–copper sulfide concentrate is an *oxidation* process where iron and sulfur are gradually oxidized and removed in a slag and gas phase respectively. Under certain conditions solid magnetite may also be formed. The equilibria between the various liquid, solid, and gaseous phases will, in addition to temperature and oxygen potential, depend on the presence of other slag-forming oxides, i.e., on the activity of FeO in the slag, and on the prevailing SO_2 pressure. For a silica-saturated iron-silicate slag, where the FeO activity is about 0.3, and for 1 atm of SO_2 the various phase equilibria can be calculated from existing thermodynamic data as function of temperature and oxygen potential as shown in Fig. 12-5.[2]

At low oxygen potential an iron–copper matte with a_{FeS} between 1 and 0.1 will coexist with the slag. As the oxygen potential is raised the activity of FeS will decrease due to oxidation and slagging

$$FeS(matte) + \tfrac{3}{2}O_2 = FeO(slag) + SO_2 \tag{1}$$

Below a certain temperature, which increases with increasing oxygen potential, solid magnetite will be formed

$$3FeS(matte) + 5O_2 = Fe_3O_4(s) + 3SO_2 \tag{2}$$

Coexistence between molten slag and magnetite is governed by the equilibrium

$$3Fe_3O_4(s) + FeS(matte) = 10FeO(slag) + SO_2 \tag{3}$$

As the oxygen potential is increased further the activity of FeS will decrease to values of less than 10^{-3}, and the matte will be almost pure Cu_2S (white metal). On further oxidation this will convert to metallic copper (blister copper) by the roast-reaction

$$Cu_2S(l) + O_2 = 2Cu(l) + SO_2 \tag{4}$$

---

[1] U. Kuxmann and F. Y. Bor: *Erzmetall*, **18**: 441 (1965).

[2] E.-B. Johansen et al.: *J. Met.*, **22**: 39 (September 1970), *see also:* T. Rosenqvist: *Met. Trans.* **9B**: 337 (1978).

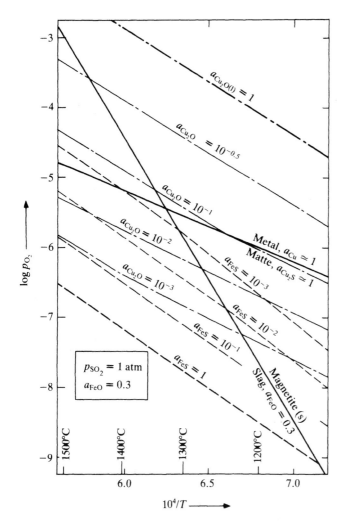

Figure 12-5 Oxygen potential for copper smelting equilibria as function of temperature, $p_{SO_2} = 1$ atm, $a_{FeO} = 0.3$ relative to FeO(l).

As is seen from Fig. 12-5 metallic copper will be formed before precipitation of magnetite at temperatures above about 1300°C, whereas at lower temperatures magnetite will be formed before metallic copper. For coexistence with white metal the blister copper contains about 1 weight percent sulfur, see Fig. 12-1(b), a value which will decrease on further increase in the oxygen potential.

During the increase in the oxygen potential there will be an increase in the activity of cuprous oxide according to the equilibria

$$Cu_2S(l) + \tfrac{3}{2}O_2 = Cu_2O + SO_2 \tag{5}$$

$$2Cu(l) + \tfrac{1}{2}O_2 = Cu_2O \tag{6}$$

For mattes with FeS activity of 0.1 or more the activity of Cu_2O will be less than 10^{-3}, whereas for coexistence between white metal and blister copper it will be of the order of 0.1. Although this means that a pure Cu_2O phase will not be formed it is evident that copper losses in the slag will increase strongly with increasing oxygen potential, as will be discussed later.

Figure 12-5 refers to conditions where the FeO activity is 0.3 and the SO_2 pressure 1 atm. For other values of these parameters the various lines will shift to the left or to the right according to the equilibria (1) to (6). Thus the line for coexistence between liquid slag and solid magnetite, equilibrium (3), will shift to the left, i.e., to higher temperatures with increasing FeO activity in the slag, and to lower temperatures with decreasing FeO activity and decreasing SO_2 pressure. Thus magnetite is more easily slagged at low FeO activity and low SO_2 pressures.

In Fig. 12-5 only the *activities* of FeS and Cu_2O are given. In order to calculate the corresponding *concentrations* in the matte and slag phases it will be necessary to know something about the thermodynamics of these phases.

Thermodynamics of Iron–Copper Mattes

Unfortunately very few measurements exist for the relation between activity and concentration in iron–copper mattes. On the basis of measurements by Krivsky and Schuhmann[1] on oxygen-free mattes and some unpublished measurements by the author for oxygen-saturated mattes we may, to a first approximation, conclude that the mattes can be described by the *Temkin model* (Sec. 4-7). Thus a mixture of Cu_2S, FeS, and FeO does not contain the three molecules as such, but rather a mixture of Cu^+ and Fe^{2+} cations embedded in a network of S^{2-} and O^{2-} anions. The chemical activities of Cu_2S and FeS will then be given by

$$a_{Cu_2S} = \left(\frac{n_{Cu}}{n_{Cu} + n_{Fe}}\right)^2 \frac{n_S}{n_S + n_O} \quad \text{and} \quad a_{FeS} = \frac{n_{Fe}}{n_{Cu} + n_{Fe}} \cdot \frac{n_S}{n_S + n_O}$$

These expressions are expected to apply also for moderate deviations from stoichiometry, i.e., for metal surplus or deficiency relative to a mixture of the stoichiometric compounds.

At high oxygen concentrations in the matte, deviations from the Temkin model is to be expected due to strong interaction between the iron and the oxygen ions in accordance with the exchange reaction

$$FeS + Cu_2O = FeO + Cu_2S$$

having a large negative standard Gibbs energy change. Thus in accordance with Eq. (4-25) one should, for high oxygen contents in the matte, expect the activity

[1] W. A. Krivsky and R. Schuhmann, Jr.: *Trans. AIME,* **209**: 981 (1957).

of FeS to be less and the activity of Cu_2S to be more than calculated from the Temkin model. As this applies to low-copper mattes these deviations are of minor importance to industrial copper smelting.

Copper Losses in Slag

Copper losses may be of two kinds: (*a*) dispersed matte droplets, and (*b*) chemically slagged copper. As far as one can judge about half of the losses in industrial copper-smelting are in the form of dispersed matte droplets. This part may be decreased by prolonged settling time, and by decreased density and viscosity of the slag. The chemically dissolved copper is most likely present as cuprous ions, Cu^+, although some cupric ions and neutral copper atoms cannot be excluded.[1]

Such a slag will have a certain activity of Cu_2O, given by the equilibrium $2Cu^+ + O^{2-} = Cu_2O$ whereby, according to the Temkin model, $a_{Cu_2O} = N_{Cu}^2 N_O$. For a slag of essentially constant composition we would expect N_O to be essentially constant, and N_{Cu} to be proportional to the copper percentage in the slag, that is, $a_{Cu_2O} = K(\% \ Cu)^2$. Measurements by Toguri and Santander[2] among others have shown that for sulfur-free and silica-saturated iron-silicate slag the constant K, which may be called the "activity coefficient" for Cu_2O, is about 10^{-3}. This makes $(\% \ Cu) \simeq 30(a_{Cu_2O})^{1/2}$. Thus for $a_{Cu_2O} \simeq 0.05$, which corresponds to coexistence between white metal and blister copper, the copper concentration in the slag should be about 7 weight percent. For $a_{Cu_2O} = 10^{-4}$, which applies for a medium copper matte, the copper concentration should be about 0.3 percent. These values agree with the better industrial data.

The "activity coefficient" for Cu_2O is expected to be a function of slag composition, however. Thus matte smelting slags will contain up to 1 percent sulfur, present as S^{2-} ions. Due to interaction between the copper and sulfide ions we should expect sulfur to lower the "activity coefficient" for Cu_2O, that is, give a higher copper percentage. This does not imply that some copper in the slag is present as oxide and some as sulfide. As already mentioned, copper is most probably present as ions, from which an oxide activity as well as a sulfide activity may be derived from the Temkins model.

The copper-oxide activity coefficient will change also with the overall composition of the slag. Thus there is evidence[3] that for a given Cu_2O activity the copper concentration of a *lime ferrite* slag is about 60 percent of that for a silica-saturated slag, corresponding to the activity coefficient of Cu_2O and the N_O value being about 3 times larger than for the siliceous slag.

On cooling of the slag, copper crystallizes either as a sulfide or as a metallic phase, copper oxide and silicate being relatively unstable. For slags of high

[1] F. D. Richardson and J. C. Billington: *Trans. Inst. Min. Metall.*, **65**: 273 (1956).

[2] J. M. Toguri, and N. H. Santander: *Can. Metall. Q.*, **8**: 167 (1969).

[3] U. Kuxmann, and H. Bussmann: *Erzmetall*, **27**: 353 (1974); *see also* A. Yazawa: *Erzmetall*, **33**: 377 (1980).

sulfur content, as encountered in the matte-smelting stage, the copper precipitates as an iron–copper sulfide phase. Its presence in the solidified slag, therefore, does not necessarily indicate a matte inclusion. For slags of low sulfur content, as encountered in the Bessemer stage, copper often comes out as metallic droplets, formed by the reaction $Cu^+ + Fe^{2+} = Cu + Fe^{3+}$. This reaction shifts to the right as solid magnetite is precipitated during cooling of the slag.

12-3 INDUSTRIAL COPPER SMELTING

Copper ores are usually sulfides, of which chalcopyrite, $FeCuS_2$, and bornite, Cu_5FeS_4, are the most important, but also other iron–copper sulfides as well as pure copper sulfides exist. In nature these minerals are usually associated with pyrite, FeS_2, or pyrrhotite, $Fe_{1-x}S$. A flow sheet for the production of copper from such ores was given in Fig. 1-1. After mineral dressing the concentrate may be subjected to a partial roast, and is then smelted to give a matte which contains almost all the copper and part of the iron, whereas the remaining iron is oxidized and discarded in a slag together with gangue minerals and fluxes which are added during the smelting process.

The matte may contain between 30 and 70 percent Cu. The grade of the matte depends on ore composition and on economic considerations, and may differ considerably from one plant to another. Thus if a low-grade matte is produced, a relatively large amount of iron will have to be removed in the subsequent Bessemer stage, with corresponding increase in process expenses. On the other hand, for a high-grade matte the copper losses in the slag will increase, and for every ore an optimum matte grade is chosen.

The matte grade may be regulated by means of the preliminary roast; the more extensive the roast the higher will be the matte grade. If the concentrate contains more than 25 percent Cu the roasting stage is often omitted; a small amount of iron will always oxidize during the smelting process.

In most cases the concentrate does not contain enough silica to react with iron oxide and give a suitable slag. Therefore some quartz is usually added. In some plants lime or limestone is also added to give a slag with 10 to 20 percent CaO. As mentioned in Sec. 11-3, this is done in order to lower the melting point and increase the fluidity of the slag, thus giving a better phase separation.

The smelting may be carried out in different types of furnaces. Formerly water-jacketed *shaft furnaces*, similar to the type used for lead smelting, were used. A shaft furnace requires lump ore, however, and either a high-grade lump ore is required, or the concentrate has to be agglomerated by sintering or briquetting. In the shaft furnace the crude or partly roasted ore is smelted with coke as a fuel. If the ratio of coke to combustion air is adjusted to give essentially stoichiometric combustion to CO_2, very little reaction occurs between the ore and the combustion gases, and the process is mainly a melting process. If, on the other hand, more air than needed for the combustion of the coke is used, the remaining oxygen will oxidize the iron sulfide, thus increasing the grade of the matte. In the extreme case the shaft furnace may be run without coke, in which case all the heat is provided by the oxidation of iron sulfide. This type of smelting, *pyrite smelting*, is only possible for low-grade copper ores, high in pyrite or pyrrhotite.

Since copper ores today are mostly obtained as high-grade concentrates of fine grain size, the shaft furnace is mostly replaced by the *reverberatory furnace* (from reverberate = to reflect). The principle of the reverberatory furnace is shown in Fig. 12-6. It is a long hearth furnace, covered with a roof and heated by the combustion of oil or powdered coal. The furnace is usually lined with magnesite (MgO) bricks, but acid (silica) linings are also used. The roof may be of arched silica bricks or of suspended magnesite bricks. The furnace is charged with hot preroasted calcine or with wet flotation concentrate. Furthermore, slag from the Bessemer converter as well as silica and limestone for slag formation are charged. The raw materials are charged in the part of the furnace which faces the burner. The molten matte is tapped from tapholes along the middle of the furnace, whereas the slag is tapped from the flue end of the furnace. This is done in order to give matte particles, which are suspended in the slag, a chance to settle.

The temperature needed to melt the slag and the matte is about 1200°C, which means that the hot combustion gases must leave the furnace at or above that temperature, and with a correspondingly high heat content. In order to utilize this heat, the gases are usually passed through a steam boiler before they go to gas cleaning and discharge. Alternatively, the hot combustion gases could be used to preheat the combustion air, thus bringing the heat back to the furnace.

Matte smelting in the reverberatory furnace is essentially a melting process. The one chemical reaction which is of particular interest is the slagging of magnetite from the converter slag and similar slagging of hematite from roasted calcine. As it was shown in Sec. 12-2, Fig. 12-5, the solution of magnetite is favored by low FeO activity and high temperature. If all the magnetite is not properly slagged, it may accumulate in the furnace as a crust along the bottom

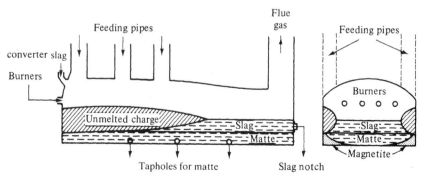

Figure 12-6 Reverberatory furnace for matte smelting (schematic). (*After O. Barth, in P. Queneau: "Extractive Metallurgy of Copper, Nickel, and Cobalt," Interscience Publishers, New York, 1961.*)

and side refractories. Also small magnetite crystals in the slag may prevent the settling of matte droplets, thus increasing the copper losses in the slag.

The copper content of the slag may be about 0.3 to 1.0 percent, increasing with increasing matte grade. A copper content of 0.5 percent may look small, but when one considers that the weight of the slag may be considerably more than the weight of the matte, it is clear that it may represent a noticeable loss and affect the overall economy of the process.

Whereas most copper ores today are smelted in reverberatory furnaces, two other furnace types are of importance: *electric smelting* and *flash smelting*. The electric furnace resembles the ones used for iron smelting, Fig. 9-6, but is often rectangular and with the electrodes in line. Because of the larger slag volume the slag acts as the main resistor, and the carbon electrodes dip into the slag. The electric furnace has the advantage compared to the reverberatory furnace that higher temperatures may be obtained, which facilitates the slagging of magnetite. Since the volume of the furnace gases is much less than for the reverberatory furnace, less heat is lost in the gases, and heat recovery is not necessary. Also, the gas is more easily cleaned. On the other hand since electrical energy is usually more expensive than fuel energy, electric smelting is likely to be confined to countries with cheap electric power.

Flash smelting was developed simultaneously by the International Nickel Company (INCO) in Canada and Outokumpu O/Y in Finland. In both cases oxidation of the iron sulfides gives the heat needed for the smelting process. Thus flash smelting employs the principle of pyrite smelting, adapted to fine-grained concentrate. If the oxidation takes place with cold air it is unable to give the necessary temperature of 1200 to 1300°C, however. In the INCO process the oxidation is carried out with essentially pure oxygen. This has the advantage that the furnace gases have a very high concentration of SO_2 which may be recovered and converted to pure liquid SO_2 by simple compression and condensation.

In the Outokumpu process, which is shown in Fig. 12-7 preheated air is used, sometimes with oxygen added, the hot flue gases being used for preheating. If there is still a heat deficiency some light fuel oil (naphtha) may be burned together with the concentrate. The reactants are introduced through a burner in the top of the combustion shaft, and the partially burned sulfides and fluxes collect in the hearth where they separate into molten matte and slag. The flexibility in the choice of oxidants makes the Outokumpu process very versatile, and a wide range of matte grades may be produced from a wide range of concentrates.

The hot flue gases leave the furnace through the uptake and are usually passed on for heat recovery, gas cleaning, and sulfuric-acid production. In one variation of the process fossil fuel, e.g., coal, is introduced into the uptake whereby SO_2 is reduced to elemental sulfur

$$SO_2 + C = \tfrac{1}{2}S_2 + CO_2$$

The sulfur vapor is condensed and represents a valuable by-product.

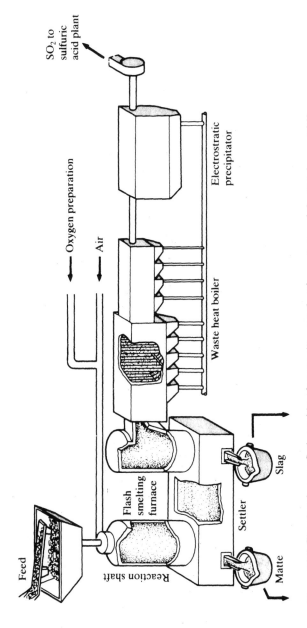

Figure 12-7 The Outokumpu flash smelting furnace. (*Reprinted with permission of the Outokumpu O/Y.*)

In the flash smelting processes a high-grade matte, 50 to 60 percent Cu, is usually produced. Because of poor settling of matte droplets the copper content of the slag is usually high, above 1 percent. In order to recover copper from the slag this may be subjected to various treatments. One possibility is to treat the molten slag with iron sulfide or low-grade concentrate in an electric furnace. When iron sulfide sinks through the slag dissolved copper may be converted to sulfide, viz.,

$$2Cu^+ + FeS = Cu_2S + Fe^{2+}$$

Also liquid iron sulfide helps to wash out fine matte droplets which are dispersed in the slag. Instead of iron sulfide other reducing agents like carbon or even scrap iron could be used.

Another possibility is to subject the slag to slow cooling. This makes the copper content crystallize as an iron–copper–sulfide phase. The solidified slag is crushed and ground, and the copper-rich phase is taken out by flotation. The treatment thus resembles that of ore-dressing of a low-grade copper ore.

Further Treatment of the Matte

The iron–copper matte from the matte smelting furnace is normally treated in the *Bessemer converter*. The most common type is the *Peirce–Smith converter*, the principle of which is shown in Fig. 12-8. The converter has the form of a horizontal cylindrical drum, and is usually lined with magnesite bricks which during the process become impregnated with magnetite.

The first stage of the converter process is the oxidation and slagging of iron sulfide. This is done by blowing air through the molten matte with simultaneous addition of silica. The oxidation of iron sulfide gives more than enough heat to maintain the temperature at about 1300°C, and sometimes cold material may have to be added to prevent overheating. The slag produced usually has a lower silica content than the matte-smelting slags. As the copper content of the remaining matte increases, a point is reached where solid magnetite is formed. Thus the slag from the Bessemer converter always contains considerable amounts of dispersed magnetite crystals. Its copper content is also high, at about 4 percent. As previously mentioned, this slag is usually returned to the matte-smelting furnace, but it could also be treated separately.

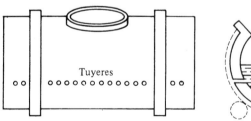

Figure 12-8 Peirce-Smith converter (schematic).

After practically all the iron has been slagged and raked off, more matte is added to the converter and the process is repeated until a sufficiently large volume of essentially pure Cu_2S (white metal) is obtained. After removal of the last slag, the blowing is continued, and copper sulfide is oxidized to give metallic copper:

$$Cu_2S + O_2 = Cu + SO_2$$

Also this reaction is sufficiently exothermic to maintain the necessary temperature (see heat balance Sec. 2-5). The Bessemer process produces a metal, "blister copper," with about 2 percent impurities, mostly iron, sulfur, and oxygen. The refining of this impure metal will be discussed in Chaps. 13 and 16.

Continuous Copper Smelting

In recent years considerable work has been made in developing processes where the smelting of the concentrate, the slagging of the iron, and the production of blister copper are carried out in a single furnace.[1]

As shown in Fig. 12-5 the conversion of Cu_2S to $Cu + SO_2$ requires a rather high oxygen potential, between 10^{-6} and 10^{-5} atm. Under these conditions the activity of Cu_2O in the system is of the order of 0.05 and the equilibrium copper content of the slag will be between 7 and 8 percent.[2]

In the *WORCRA process* (named after the inventor H. K. Worner and Conzinc Riotinto of Australia Ltd) this high-copper slag is treated with low-grade matte and concentrate in a counter-current system, whereby the copper content is reported to be reduced to between 0.5 and 1 percent. In the Canadian *Noranda process* the copper-rich slag is allowed to cool and crystallize, and is subjected to ore-dressing, whereby most of the copper is recovered.[3]

The most recent of the continuous copper processes is the *Mitsubishi process* which is shown in Fig. 12-9. In this process three reactors or furnaces are used, which operate in series. In the first furnace the concentrate and siliceous flux are reacted with oxygen-enriched air to give a matte with 60 to 70 percent Cu and an iron silicate slag. The matte and the slag, which has a copper content around 1 percent, are passed on to the second furnace. This is electrically heated and some reducing agent is added to bring the copper content of the slag below 0.5 percent. This slag is discarded. The matte from the second furnace is passed on to the third reactor, the converter, which is blown with air. Here the remaining iron is slagged, and Cu_2S is blown to blister copper. In this reactor *limestone* is used as a flux, whereby *calcium ferrite slag* is formed. As it was shown in Figs. 11-5 and 11-8 lime ferrite slags are low melting even at very

[1] H. K. Worner: "Continuous Smelting and Refining by the WORCRA Process," *Symp. on Advances in Extractive Metallurgy*, Inst. Min. Metall., 1967, Elsevier Publ. Co., 1968. N. J. Themelis et al.: *J. Metals*, **24**: 25 (April 1972).

[2] A. Geveci and T. Rosenqvist: *Trans. Inst. Min. Metall.*, **82**: C193 (1973).

[3] K. N. Subramanian, and N. J. Themelis: *J. Met.*, **24**: 33 (April 1972).

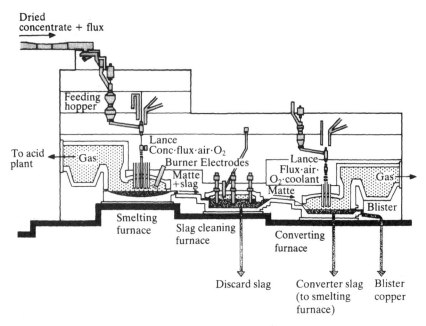

Figure 12-9 The Mitsubishi copper smelting process. (*Reprinted with permission of the Mitsubishi Metal Corporation.*)

high oxygen potentials, without danger of magnetite precipitation. Furthermore, as mentioned in Sec. 12-2, its Cu_2O activity coefficient is higher than for siliceous slags, which gives smaller copper losses. This makes it possible to carry out the conversion well beyond the point of coexistence with white metal, giving a blister copper of low sulfur content. The copper content in the resulting slag is in that case about 15 percent. This slag is returned to the first smelting furnace where its lime content helps to lower the melting temperature of the siliceous slag.

Impurity Elements in Copper Smelting

In addition to iron, sulfur, and gangue minerals, copper concentrates may contain a number of impurity elements. Most welcome are the noble metals, *silver* and *gold*. These will follow the copper and can eventually be recovered from the anode mud, which is formed during the electrolytic refining process (Sec. 16-6). Likewise most of the *nickel* will follow the copper and can be recovered from the refining electrolyte, whereas most of the *cobalt* in the concentrate will be lost in the smelter slag. The major part of the *zinc* will also go into the slag; the remainder will be found in the flue dust together with most of the *lead*. More troublesome are the metalloids *arsenic, antimony,* and *bismuth,* as well as *selenium* and *tellurium*. Some of these form volatile sulfides, and are partly volatilized during the roasting and smelting stages, and can be recovered from the flue dust, but some will eventually end up in the blister copper. As the activity coefficient of most of these elements in liquid copper is very low, and as they are rather noble, volatilization and alternatively oxidation from the metal is rather poor. An exception is bismuth, which has an activity coefficient larger than unity, and which to some extent may be volatilized during the converting stage. During the electrolytic refining most of the antimony, selenium, and tellurium and some arsenic will collect in the anode mud as copper compounds, whereas most of the arsenic and bismuth will accumulate in the electrolyte which, therefore, has to be withdrawn regularly for purification. In general the metalloid elements are unwanted, as they add to the refining cost.

12-4 NICKEL SMELTING

Nickel occurs in nature mainly as pentlandite, $(Ni,Fe)_9S_8$. It is usually associated with copper and iron sulfides, and often contains minor amounts of the platinum metals. Because of intimate intergrowth of the various minerals the nickel concentrate will be of lower grade than corresponding copper concentrates. The first part of the treatment is similar to that of copper smelting. The concentrate is first smelted to give a low-grade matte with about 10 percent nickel and, say, 5 percent copper. This is blown in a converter to remove most of the iron sulfides and to give a Bessemer matte with about 50 percent nickel and, say, 25 percent copper as sulfides. Contrary to the procedure for copper this matte cannot be blown directly to produce the metal, since for the roast-reaction $4NiO + Ni_3S_2 = 7Ni + 2SO_2$, the standard Gibbs energy change at 1200°C is positive, about $+20$ kJ. Therefore the Bessemer process is usually interrupted after the iron has been slagged, and the remaining copper–nickel matte is treated further by special methods, to separate nickel and copper and to obtain the metallic products.

One possible treatment is to roast the sulfide completely to oxide which is then reduced with carbon. This gives the so-called *Monel metal*, an alloy of about $\frac{2}{3}$ Ni and $\frac{1}{3}$ Cu, which has some of the properties of stainless steel and has found some direct application.

Another treatment of the Bessemer matte is that previously used by the Falconbridge nickel refinery in Norway. Here the matte is roasted, and the mixed oxide is leached with dilute sulfuric acid. This dissolves most of the copper and leaves a residue of essentially nickel oxide. This is reduced with coke in an electric smelting furnace, and the metal obtained is cast into anodes for electrolytic refining, while electrolytic copper is produced from the leach solution (see Chap. 16).

In a more recent Falconbridge process[1] the Bessemer matte is leached with HCl, whereby nickel together with iron and other impurities are dissolved. The

---

[1] P. G. Thornhill et al.: *J. Met.* **23**: 13 (July 1971).

solution is purified by solvent extraction (Sec. 15-6) to give pure $NiCl_2$, which subsequently is converted to metal.

A method which formerly was used by the International Nickel Company (INCO) is the *Orford process*. Although this today is of mainly historical interest it contains some interesting features. In the Orford process the copper–nickel matte is melted together with *sodium sulfide*. As is seen from Fig. 12-10, Cu_2S is completely miscible both with Ni_3S_2 and Na_2S. The presence of an intermediate compound indicates that the Cu_2S–Na_2S system shows negative deviation from ideality. In the system Ni_3S_2–Na_2S, however, there is a field of liquid immiscibility, and this field extends into the ternary system. From the tie-lines, which connect melts in mutual equilibrium, we see that Cu_2S is enriched in the Na_2S melt, whereas the copper content of the coexisting Ni_3S_2 melt is much less. By means of repeated melting with Na_2S two products were obtained, the one containing mostly nickel sulfide, the other most of the copper. These were processed further to produce the pure metals.

The main method used by the International Nickel Company is based on the phase diagram shown in Fig. 12-11. By very slow cooling of the Bessemer matte relatively large crystals of the various phases are obtained. In addition to essentially pure Ni_3S_2 and Cu_2S crystals, a small amount of a metallic phase is formed. In this phase the noble metals, mainly platinum metals, are enriched.

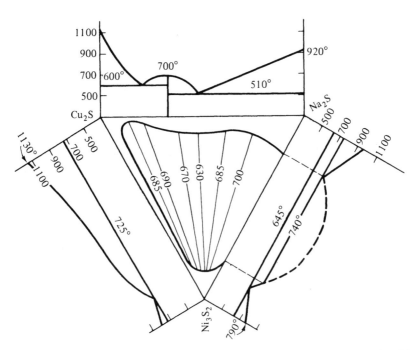

Figure 12-10 Liquid immiscibility in the Cu_2S–Ni_3S_2–Na_2S system. Notice the three binary border systems. (*After I. V. Zaitsev and G. I. Damskaya: ONTI, 1938.*)

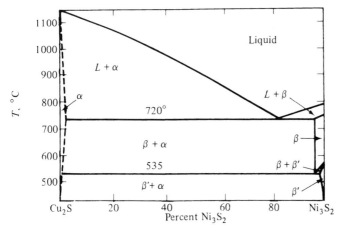

Figure 12-11 Pseudobinary Cu_2S–Ni_3S_2 system. (*From K. Sproule et al., in P. Queneau: "Extractive Metallurgy of Copper, Nickel, and Cobalt," Interscience Publishers, New York, 1961.*)

The solidified matte is treated further by ore-dressing methods: after free grinding, the metallic phase is removed by magnetic separation, and the nickel and copper sulfides are separated by selective flotation. This gives a nickel sulfide concentrate with about 70 percent Ni and 1 to 2 percent Cu and a copper sulfide concentrate with 73 percent Cu and about 5 percent Ni. These concentrates are treated further for production of the pure metals (see Chaps. 13 and 16). The metallic phase is treated for recovery of the platinum metals.

As already mentioned nickel sulfide cannot be blown to metal by the roast-reaction at temperatures around 1200°C. At very high temperatures, however, the Gibbs energy change of the roast-reaction becomes negative. Furthermore, by reducing the partial pressure of SO_2, either by vacuum or by dilution with an inert gas, the roast-reaction is enhanced. Thus, according to an INCO patent, nickel sulfide may be oxidized to give essentially pure metal by a combination of very high temperature (> 1600°C) and low SO_2 pressure (< 0.01 atm).

Refining of nickel by the *Mond process* will be discussed in Sec. 14-7, and by *solvent extraction* in Sec. 15-6.

12-5 OTHER MATTE-SMELTING PROCESSES

Lead occurs in nature mainly as galena, PbS. As mentioned in Sec. 9-8 the standard method for lead production is roast-reduction, i.e., the sulfide is converted to oxide, which again is reduced, e.g., by means of coke. An alternative method is the roast-reaction

$$2PbO + PbS = 3Pb + SO_2$$

The standard Gibbs energy change for this reaction is negative above 900°C, but is less so than the corresponding copper reaction. Furthermore PbS, PbO, and Pb are rather volatile, which sets an upper limit to the temperature which can be used without excessive vapor losses. The method is probably the oldest of all metallurgical processes: by controlled oxidation of lead sulfide ore on a hearth at temperatures around 900°C the metal was obtained and could be drained away. As the Gibbs energy change of the roast-reaction is only slightly negative, however, the chemical activity of PbO for coexistence between PbS and Pb will be appreciable, and lead losses in the slag, which is formed by gangue material in the ore, will be high.

With present-day availability of concentrates with low gangue contents the roast-reaction for lead has received renewed interest, and several new processes have been developed. In the Boliden process[1] PbS concentrate is melted and oxidized in an electric furnace. The oxidation is brought to the point where the crude lead still contains about 3 percent sulfur and has to be refined further. At the same time lead losses in the slag are appreciable. The lowest lead contents, about 4 percent, are obtained for a basic slag with about 34 percent CaO. In recent years further processes for roast-reaction of PbS have been suggested.

Lead sulfide can also be decomposed by means of metallic iron

$$PbS + Fe = Pb + FeS \qquad \Delta G^{\circ}_{1273} \simeq -25 \text{ kJ}$$

The reaction is facilitated by the fact that iron is practically insoluble in liquid lead; thus the metal produced will be quite pure. In order to obtain a good recovery an excess of iron is needed, whereby an iron-sulfide matte with a small lead content is formed.

Decomposition with iron can also be used on *lead sulfate* whereby, in addition to an iron-sulfide matte, a ferrous-oxide slag is obtained. Smelting with metallic iron is, therefore, particularly suitable for the processing of old lead battery scrap.

Antimony may be precipitated by the analogous reaction

$$Sb_2S_3 + 3Fe = 2Sb + 3FeS \qquad \Delta G^{\circ}_{1273} \simeq -200 \text{ kJ}$$

In order to lower the melting point of the iron sulfide some sodium salts are added. Since iron and antimony are miscible in the liquid state an alloy is obtained, which requires further refining. One way to refine this alloy is to treat it again with additional antimony sulfide, whereby iron, by the same reaction, is converted into sulfide. This second sulfide, which contains some antimony, is returned to the first smelting stage.

12-6 SPEISSES

Speisses are alloys of heavy metals like iron, cobalt, and nickel with arsenic and antimony, occasionally also with tin, which lower the melting temperature to around 1000°C. Speisses are only partly miscible with mattes, and if there is

---

[1] H. I. Elvander: in "Pyrometallurgical Processes in nonferrous metallurgy," AIME, New York, 1967, p. 225.

Table 12-1 $-\Delta H_{298}$ (kJ) **for formation of various sulfides, arsenides and antimonides.**†

FeS	100	CoS	96	NiS	94	Cu_2S	80
FeAs	—	CoAs	57	NiAs	72	Cu_3As	—
FeSb	9.2	CoSb	42	NiSb	84	Cu_2Sb	11.5

† Mainly from O. Kubaschewski and C. B. Alcock: "Metallurgical Thermochemistry" 5th ed., Pergamon Press, London, 1979.

enough arsenic or antimony in a copper charge a separate speiss melt may be formed. Because of their complexity and limited industrial importance the thermodynamics of speisses and their equilibria with other phases like matte and slag have not been much studied. An idea of their stability may be obtained from Table 12-1 which lists the negative enthalpies of formation of different sulfides, arsenides, and antimonides. Whereas the enthalpy of formation of the sulfides is approximately constant in the sequence iron to copper, the values for the antimonides show a pronounced exothermic maximum for cobalt and nickel. A similar variation should also apply for the Gibbs energy of formation.

Speisses also show high affinities for the platinum metals and gold. Thus the distribution ratio $(Pt)_{speiss}/(Pt)_{matte} \simeq 10^3$ and $(Au)_{speiss}/(Au)_{matte} \simeq 10^2$, where the parentheses give the weight percentages. In comparison the distribution ratio for silver is about unity.[1]

Speisses are also immiscible with liquid lead, and in the smelting of complex ores in the lead blast furnace (Sec. 9-8) one may get as many as four different molten products: (1) a metallic phase with lead and silver; (2) a matte with copper and iron; (3) a speiss with iron, cobalt and nickel; and (4) a slag with iron oxide, silica, lime, etc.

Further treatment of the speisses to recover the valuable metals is rather complicated and goes beyond the scope of the present text.

PROBLEMS

12-1 A copper concentrate contains 20 percent Cu and 40 percent S. The ore minerals are $CuFeS_2$ and FeS_2, the remaining gangue is SiO_2.

(a) Calculate the number of moles of Fe, Cu, S, and SiO_2 in one kilogram of concentrate.

(b) The concentrate (25°C) is flash smelted with oxygen enriched air (25°C) to give a matte with 40 percent Cu. All the oxygen in the blast is consumed. The matte is assumed to be a mixture of stoichiometric Cu_2S and FeS. The slag is stoichiometric Fe_2SiO_4. Calculate the number of moles of Cu_2S and FeS in the matte and of FeO and SiO_2 in the slag, and calculate the necessary addition of SiO_2 as well as the total oxygen in the blast, all per kilogram of concentrate.

[1] J. Gerlach, U. Hennig, and H.-S. Park: *Erzmetall*, **25**: 69 (1972).

(c) Matte and slag are tapped at 1250°C, the gases ($SO_2 + N_2$) are withdrawn at 1300°C. Make a heat balance for the process and calculate the N_2 content of the furnace gas as well as the O_2 percentage in the enriched air. The heat losses from the furnace are taken as 500 kJ/kg concentrate. The enthalpies of the reactions $CuS(s) + FeS(s) = CuFeS_2(s)$ and $Cu_2S(l) + FeS(l) =$ liquid matte are assumed to be zero.

12-2 A liquid matte contains 36 percent Cu, 36 percent Fe, 25 percent S, and 3 percent O.

(a) Calculate the corresponding atomic percentages, and comment on the stoichiometry of the matte.

(b) The matte is assumed to obey the *Temkin equation* (Sec. 4-7) with Cu and Fe as cations, S and O as anions. Calculate the corresponding activities of Cu_2S and FeS relative to the pure liquid sulfides. (Fe^{3+} is counted as Fe^{2+}.)

12-3 (a) Using data in App. C calculate the equilibrium oxygen pressure for conversion of pure Cu_2S to pure Cu at 1300°C if $p_{SO_2} = 0.1$ atm.

(b) Calculate the corresponding activities of FeS in the matte and Fe in the metal if the FeO activity in the slag, relative to supercooled liquid FeO, is 0.4.

(c) Calculate the corresponding activities of $Fe_3O_4(s)$ and $Cu_2O(l)$.

(d) In which direction would the above values be affected by (1) higher FeO activity, (2) higher SO_2 pressure, (3) lower temperature?

12-4 *Hearth smelting* of lead is controlled by the equilibrium $PbS + 2PbO = 3Pb + SO_2$. (a) Calculate the temperature where the SO_2 pressure is 1 atm if PbS, PbO, and Pb are present in their standard state. (b) In practice the process is carried out at 900°C, and PbO forms a slag with gangue minerals. Calculate the activity of PbO in the slag for equilibrium with pure PbS and Pb and 1 atm of SO_2, at 900°C.

BIBLIOGRAPHY

Biswas, A. K., and W. G. Davenport: "Extractive Metallurgy of Copper," 2d ed., Pergamon Press, Oxford, 1980.

Boldt, J. R., Jr., and P. Queneau: "The Winning of Nickel," Longmans Canada Ltd., Toronto, 1967.

Butts, A. (ed.): "Copper," Reinhold Publishing Corporation, New York, 1954.

Queneau, P. (ed.): "Extractive Metallurgy of Copper Nickel and Cobalt," AIME, New York, 1961.

THIRTEEN

REFINING PROCESSES

Metals obtained from the primary extraction processes often contain impurities which originate from the ore, the fluxes, or the fuel. For further use of the metals it is usually necessary to subject them to one or several refining processes. The refining may have as its purpose to produce the metal as pure as possible; in some cases extreme purity is desired. In other cases as, e.g., in steel-making the refining is done in order to give a product with controlled amounts of desired impurities. Finally, some refining processes are done in order to recover impurities which in themselves are not harmful, but which have intrinsic value, e.g., the recovery of silver from lead.

Refining processes are always based on the principle that different elements distribute themselves differently in different phases, and that these phases may be separated physically. Broadly speaking we may divide the processes into three main groups, (1) metal–slag, (2) metal–metal, and (3) metal–gas processes, although certain mixed cases exist. Among the metal–slag processes the oxidation and slagging of less noble elements is by far the most important, and may be given the common name of *fire-refining*. Here we find the important steel-making processes as well as the fire-refining of copper and lead. Among the metal–metal processes we find liquation and zone-refining the latter being used to produce metals of the highest purity. An important metal–gas process is refining by distillation, which was discussed in Chap. 10. Another example is the removal of gases, e.g., by the use of vacuum, which will be discussed in this chapter. Refining by electrolysis will be treated in Chap. 16.

13-1 FIRE-REFINING

Fire-refining is based on the difference in affinity for oxygen for the different elements. However, it also depends on the chemical activities of the elements dissolved in the metallic phase and of the oxides dissolved in the slag. In fire-refining the metal is subjected to oxidation either by means of air or oxygen, or by means of some oxide of the metal which is to be refined. Thus pig iron may be oxidized with oxygen or with iron ore, impure lead with air or with lead oxide, etc. The refining of impure tin by a second smelting with tin ore, which was mentioned in Sec. 9-8, is also an example of fire-refining.

The equipment used in fire-refining will depend on the metal in question and on the means of oxidation. For the oxidation with oxygen or air the reactions are strongly exothermic, and in some cases refining may be carried out without additional fuel. This is the case for the treatment of liquid pig iron in the *Bessemer converter*, Fig. 13-1. Here air is blown through the melt; the evolved heat is sufficient, not only to cover heat losses, but also to raise the temperature from about 1300°C in the pig iron to about 1600°C which is needed to melt the steel. In the *L–D process* shown in Fig. 13-2 pure oxygen is top-jetted on to the molten iron bath. Because of the absence of nitrogen the heat content in the escaping gases is much less than for the air-blown Bessemer process, and a fair amount of cold scrap or some iron ore may be charged in addition to liquid pig iron.

Top-jetting in the L–D process is done in order to protect the bottom lining of the converter from the very high local temperatures obtained when oxygen reacts with the iron bath. In recent years bottom-blown oxygen converters have been developed where the jet of oxygen is surrounded by an annular flow of hydrocarbons or inert gas. In that way extreme local temperatures are avoided, and these converter types are gaining increasing use in steel-making and refining.

An even better heat economy is obtained in the oxygen blown *Kaldo process*, Fig. 13-3. Here the CO which evolves from the refining process burns to CO_2 *inside* the converter, giving additional heat. Owing to heavy wear on the

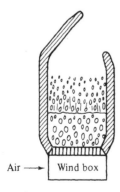

Air ⟶ | Wind box

Figure 13-1 Bessemer converter (schematic).

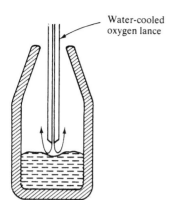

Figure 13-2 L–D converter (schematic).

refractories the Kaldo converter has not found much application in steel-making, however, but seems to find increasing use in the smelting and refining of nonferrous metals, where it is known as the "Top-Blown Rotary Converter," (TBRC).

For most cases of oxidation of impurities with metal oxides the heat of reaction is insufficient to maintain the necessary temperature, and the reactions may even be endothermic. In these cases a fuel-fired reverberatory furnace may be used as illustrated by the *open-hearth steel-making furnace* in Fig. 13-4. This is fired with oil or gas, and the hot furnace gases are passed through a regenerative system where they are used to preheat the combustion air. In that way the necessary temperature of about 1600°C is obtained. The charge for the open-hearth furnace usually consists of steel scrap and pig iron with some iron ore. The oxidation of impurities takes place partly by means of the ore and rust on the scrap, but also by oxygen in the furnace atmosphere which is picked up by the slag and transferred to the steel bath. Some oxygen may also be blown into the bath.

Similar fuel-fired furnaces are used for the refining of copper and lead. Because of the lower temperatures the combustion air does not need to be preheated for these metals. The oxidation may be carried out by blowing air into the molten metal bath or by means of oxide additions.

O₂-lance
(water-cooled)

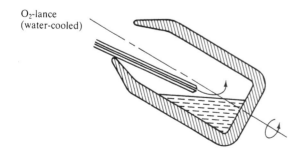

Figure 13-3 Kaldo converter (schematic).

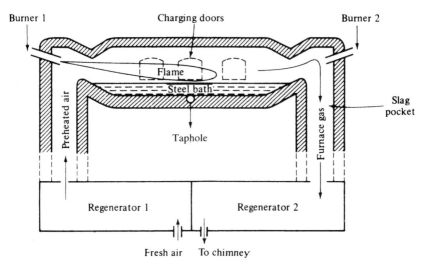

Figure 13-4 Open-hearth furnace (schematic).

The refining may also be carried out in an *electric arc furnace* as illustrated by the electric steel-making furnace, Fig. 13-5. Here the highest temperatures may be obtained. Because no fuel is used little heat is lost in the furnace gases. Also there is less oxidation from the atmosphere, and the process may be adjusted to oxidize only unwanted impurities, such as carbon, whereas valuable metals, such as chromium, may be retained in the steel. Also, slightly reducing conditions may be established.

Steel-making converters and furnaces may have *acid* or *basic* linings. In the first case the furnace lining is made of silica, and the slag obtained will be more or less saturated with silica. In basic furnaces the lining is of burned dolomite or burned magnesite. Sufficient lime is added to the slag to make it closely

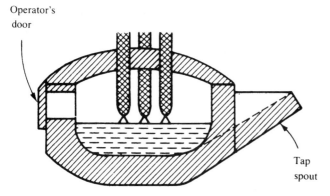

Figure 13-5 Electric arc furnace (schematic).

saturated with respect to the lining. Acid linings are cheaper, but the acid slag is unable to dissolve phosphorus oxide. Therefore, if the raw materials contain phosphorus, a basic process must be used. Basic open-hearth slag is also able to dissolve some sulfur, but for the best desulfurization the basic electric furnace is used. Most steel-making furnaces today have basic linings, acid steel-making being almost extinct.

13-2 THERMODYNAMICS OF LIQUID STEEL

A general discussion on the thermodynamics of metallic solutions was given in Chap. 4. Here the thermodynamics of common impurity elements in liquid steel will be discussed. Similar relations apply also for impurities in copper, lead, etc. As pointed out in Sec. 4-4, it is convenient when considering elements in diluted metallic solutions to use a standard state which makes the activity equal to the weight percentage at infinite dilution. For any other concentration the activity may be expressed as $a'_X = [\% \ X]f_X$ where f_X is an activity coefficient which becomes unity at infinite dilution, and which does not deviate strongly from unity for low concentrations of the dissolved element, X.

The value of the activity coefficient as function of composition as well as the Gibbs energy of the chosen standard state relative to that of the pure element or relative to some other standard state may be determined experimentally. This will be illustrated for some of the common elements in liquid steel.[1]

Oxygen in iron may be studied by the heterogeneous equilibrium:

$$\underline{O} + H_2 = H_2O \qquad K = \frac{p_{H_2O}}{p_{H_2}[O]f_O}$$

Here $\underline{O}$ denotes dissolved oxygen, [O] its concentration in weight percent, and f_O its activity coefficient.

The experimental gas ratio p_{H_2O}/p_{H_2}, which is proportional to the oxygen activity, is shown in Fig. 13-6(a) as a function of oxygen content for two different temperatures. We see that the oxygen activity increases with increasing oxygen content up to a limit which at 1600°C is about 0.23 percent. At this point liquid iron oxide is formed, and the oxygen activity remains constant inside the Fe–FeO two-phase field. Also, the solubility limit for oxygen in liquid iron increases with increasing temperature.

In Fig. 13-6(b) the expression $\log \{p_{H_2O}/(p_{H_2}[O])\} = \log K + \log f_O$ is shown as function of composition. With decreasing oxygen content, i.e., for $\log f_O \rightarrow 0$, this expression approaches $\log K$, which is found equal to 0.56 at

---

[1] The thermodynamic data given in this section are converted to SI units from data given by: G. Derge (ed.): "Basic Open Hearth Steelmaking," AIME, New York, 1964.

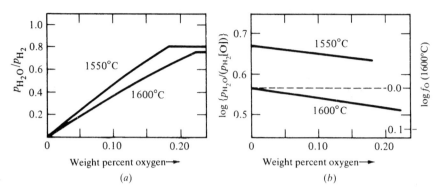

Figure 13-6 (a) Activity, (b) activity coefficient of oxygen in liquid steel. (*After T. P. Floridis and J. Chipman: Trans. AIME, vol. 212, p. 549 (1958)*.)

1600°C and equal to 0.67 at 1550°C. Combined with measurements at other temperatures we get for the above reaction

$$\Delta G° = -RT \ln K = -134\,700 + 61.21T \text{ J}$$

From other measurements we know that for the reaction

$$H_2 + \tfrac{1}{2}O_2(g) = H_2O \qquad \Delta G° = -251\,900 + 58.32T \text{ J}$$

Combination of the two gives

$$\tfrac{1}{2}O_2 = \underline{O} \qquad \Delta G° = -117\,200 - 2.89T \text{ J}$$

Here $\underline{O}$ denotes the standard state of an idealized one weight percent solution of oxygen in liquid iron. The fact that the melt cannot dissolve more than 0.2 to 0.3 percent oxygen does not prevent the use of such a standard state.

The value for $\log f_O$ at 1600°C may now be read directly from the right-hand scale of Fig. 13-6(b), and we find $\log f_O = -0.20$ [% O]. This expression does not seem to be greatly affected by changes in temperature.

Sulfur in iron may be studied by the similar reaction

$$\underline{S} + H_2 = H_2S \qquad K = \frac{p_{H_2S}}{p_{H_2}[S]f_S}$$

For this reaction the standard Gibbs energy is found to be

$$\Delta G° = 41\,170 + 27.36T \text{ J}$$

If this is combined with the Gibbs energy for the reaction

$$H_2 + \tfrac{1}{2}S_2(g) = H_2S \qquad \Delta G° = -90\,290 + 49.41T \text{ J}$$

we get for

$$\tfrac{1}{2}S_2 = \underline{S} \qquad \Delta G° = -131\,460 + 22.05T \text{ J}$$

Furthermore $\log f_S = -0.028$ [% S], rather independent of temperature.

Carbon in iron may be studied by the reaction

$$C + CO_2 = 2CO \qquad K = \frac{(p_{CO})^2}{p_{CO_2}[C]f_C}$$

When combined with the Gibbs energy for the reaction $C(graph) + CO_2 = 2CO$, we get for

$$C(graph) = \underline{C} \qquad \Delta G° = 21\,340 - 41.84T \text{ J}$$

Furthermore we find for the activity coefficient, $\log f_C = 0.22$ [% C].

Silicon in iron may be studied by the more complex reaction

$$\underline{Si} + 2H_2O = SiO_2(s) + 2H_2$$

If the measurements are carried out in a silica crucible, the activity of SiO_2 is unity, and

$$a'_{Si} = [Si]f_{Si} = \frac{1}{K}\left[\frac{p_{H_2}}{p_{H_2O}}\right]^2$$

From this we derive for the reaction

$$Si(l) = \underline{Si} \qquad \Delta G° = -119\,240 - 25.48T \text{ J}$$

and $\log f_{Si} = 0.32$ [% Si].

Similar measurements have been made to determine the Gibbs energy and activity coefficients for other metals dissolved in liquid iron.

Nitrogen and hydrogen dissolve in liquid iron by the reactions

$$N_2 = 2\underline{N} \qquad \Delta G° = 7200 + 47.78T \text{ J}$$

$$H_2 = 2\underline{H} \qquad \Delta G° = 63\,930 + 63.43T \text{ J}$$

Thus at 1600°C and 1 atm of N_2 or H_2 iron dissolves 0.045 percent of nitrogen and 0.0028 percent of hydrogen respectively, and the solubilities increase with increasing temperature. In accordance with *Sieverts' law* (Eq. 4-16) the solubility is proportional to the square root of the gas pressure. Owing to the small solubility, deviation from Sieverts' law has not been observed for these gases in liquid iron, i.e., the activity coefficient is unity. For solution in other metals, such as nitrogen in chromium, deviations have been observed at high N contents.

Multicomponent Solutions

The activities of carbon, sulfur, etc., in liquid steel are affected by the presence of other components. Thus the activity of carbon increases when silicon is added to the melt, but decreases on the addition of oxygen. The activity of sulfur is increased by carbon and silicon but is decreased by copper and manganese. In order to describe this phenomenon it is found convenient to maintain the same standard state as for the binary melts, and to introduce a second activity coefficient which describes the effect of the third component. Thus for the activity of X in the presence of Y

$$a'_X = [\% \ X] f_X f_X^Y \tag{13-1}$$

Here f_X^Y is a second activity coefficient, which is a function of $[\% \ Y]$, and which becomes unity when the concentration of Y goes to zero. For small concentrations of Y it is found to obey the relation

$$\log f_X^Y = e_X^Y [\% \ Y]$$

The coefficient e_X^Y is called the *interaction coefficient* of Y with respect to X. Similarly for the activity of Y, $a'_Y = [\% \ Y] f_Y f_Y^X$, where $\log f_Y^X = e_Y^X [\% \ X]$.

It has been found experimentally, and it can also be explained theoretically, that for low concentrations of X and Y the two interaction coefficients are connected by the relation[1]

$$e_X^Y = \frac{M_X}{M_Y} e_Y^X \tag{13-2}$$

Here M_X and M_Y are the atomic weights of the two elements. In recent years a large number of interaction coefficients have been determined experimentally as well as calculated from Eq. (13-2). A selection of these values is given in Table 13-1, together with similar coefficients for simple binary systems $e_X^X = \log f_X / [\% \ X]$.

The interaction coefficients differ greatly in value. Particularly large and negative is the effect of aluminum and silicon on dissolved oxygen. This must be attributed to the high stability of aluminum and silicon oxides, and to the persistence of the binding forces between the metal and oxygen atoms in the melt, although it is unlikely that actual molecules exist. Positive values are

---

[1] According to C. H. P. Lupis and J. F. Elliott: *Trans. AIME*, **233**:257 (1965), Eq. (13-2) should have the additional term: $+4.34 \times 10^{-3} (M_Y - M_X)/M_Y$. For most cases this term is insignificant.

Table 13-1 Interaction coefficients $e_X^Y \times 10^2$ in liquid steel†

Dissolved element X	Added element Y								
	Al	C	Cr	Cu	Mn	N	O	S	Si
C	4.8	22	−2.4	1.6		11.1	−9.7	9	10
H	1.3	6	−0.22	0.05	−0.14			0.8	2.7
N	0.3	13	−4.5	0.9	−2.0	0	5.0	1.3	4.7
O	−94	−13	−4.1	−0.9	0	5.7	−20	−9.1	−14
P						11.3	13.5	4.3	9.5
S	5.8	24	−2.2	−1.2	−2.5	3.0	−18	−2.8	6.6
Si	6.3	24		≃0	0	9.3	−25	5.7	32

† From: J. F. Elliott, et al.: "Thermochemistry for Steelmaking," vol. II, Addison-Wesley Publ. Co. Inc., Reading, Mass., 1963.

found for the effect of nitrogen and silicon on carbon and sulfur, and vice versa. This may be attributed to these elements having greater affinity for iron than for each other, and tending to displace each other from the solution.

For a *multicomponent solution* the total activity coefficient for each element will be determined by the effect of each component on that element. For diluted solution this is found to obey the relation

$$a'_X = [\% \ X] f_X f_X^Y \ldots f_X^Z = [\% \ X] f_X^*$$

Here f_X^* denotes the overall activity coefficient of component X, and is given by

$$\log f_X^* = e_X^X [\% \ X] + e_X^Y [\% \ Y] + \cdots + e_X^Z [\% \ Z] \tag{13-3}$$

This relation may be understood from statistical thermodynamics: The interaction coefficients e_X^Y, etc., give a measure of the bond energy between dissolved X and Y atoms, etc., relative to the X–Fe and Y–Fe bonds. For dilute solutions the energies of these bonds are independent of each other and additive, whereas for more concentrated solutions the Z atoms may affect the X–Y bonds, etc., and deviations from Eq. (13-3) are to be expected.

13-3 STEEL-MAKING REACTIONS

In steel-making, as well as in other fire-refining processes, oxygen has to be dissolved in the bath before it can react with dissolved impurities. In the absence of other slag-forming constituents liquid steel can dissolve oxygen up to the point where liquid FeO is formed, which at 1600°C occurs at 0.23 percent O. In the presence of slag-forming oxides the solubility is less, and is governed by the equilibrium $FeO(l) = Fe + \underline{O}$. Thus at 1600°C, $[\% \ O] = 0.23 a_{FeO}$, where a_{FeO} is the activity in the coexisting slag relative to pure liquid FeO. The chemical activity of FeO in $CaO-SiO_2-FeO$ slags was shown in Fig. 11-6, and we remember that it was quite high along the $FeO-Ca_2SiO_4$ line, and considerably lower for acid slags.

Carbon reactions

The most important steel-making reaction is the oxidation of carbon according to

$$\underline{C} + \underline{O} = CO \quad \text{with} \quad \Delta G° = -22\,380 - 39.66T \text{ J}$$

For a temperature of 1600°C this gives

$$\frac{p_{CO}}{a'_C a'_O} = 500 \quad \text{or} \quad [C] \times [O] = 0.002 p_{CO}/(f_C^* f_O^*)$$

The overall carbon activity coefficient increases and the overall oxygen activity coefficient decreases with increasing carbon content, the net effect being that their product is closely unity and may be disregarded.

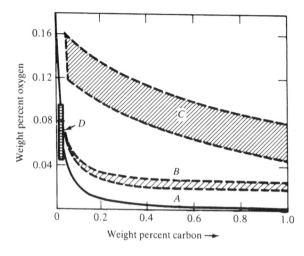

Figure 13-7 Carbon–oxygen relations in liquid steel. A = equilibrium with 1 atm $CO + CO_2$. B = observed values for basic open-heart. C = calculated from FeO activity of basic open-hearth (BOH) slag. D = observed values for basic Bessemer (Thomas) process. (*After "Basic Open-Hearth Steelmaking," AIME, New York, 1951.*)

In addition to CO the gas will contain some CO_2, controlled by the equilibrium $CO + \underline{O} = CO_2$. The concentration of CO_2 increases with decreasing carbon content, and may for low carbon steel amount to about 10 percent of the total gas pressure. Thus the curve A in Fig. 13-7, which relates the concentrations of carbon and oxygen for a total gas pressure of 1 atm at 1600°C deviates slightly from a perfect hyperbola.

In Fig. 13-7 are also shown, as cross-hatched areas, the compositions of the melt in actual steel-making practices, as well as the oxygen content which, in basic open-hearth practice, can be calculated from the FeO activity of the coexisting slag if equilibrium between slag and steel bath has been established. We see that for basic open-hearth practice the oxygen content of the bath is intermediate between what corresponds to equilibrium with its carbon content at $p_{tot} = 1$ atm, and what corresponds to equilibrium with the slag. This behavior is to be expected if we consider the kinetics of carbon oxidation: During the oxidation period, for which these data apply, oxygen is continuously being transferred from the slag to the bath, where it continuously reacts with carbon to give CO. It seems that the main resistance to this oxygen flow are at the slag–metal and at the metal–gas interfaces, whereas inside the steel bath the transfer of dissolved oxygen is very fast.

Figure 13-7 furthermore shows that in the air-blown basic Bessemer (Thomas) process the oxygen content even falls below the curve for $p_{CO} + p_{CO_2} = 1$ atm. In that process there is a very good contact between the steel bath and the gas, and because of nitrogen in the blast the $CO + CO_2$ pressure is less than 1 atm.

For the electric steel-making and oxygen-blown processes the oxygen content in the bath falls between the equilibrium curve and the curve for the basic open-hearth, indicating that equilibrium is nearly obtained in these processes.

Oxidation of Silicon

Oxidation of silicon is governed by the reaction

$$\underline{Si} + 2\underline{O} = SiO_2 \qquad \Delta G° = -594\,000 + 230.1T \text{ J}$$

For a temperature of 1600°C this gives

$$[Si] \times [O]^2 = 2.8 \times 10^{-5} \frac{a_{SiO_2}}{f_{Si}^*(f_O^*)^2}$$

The activity coefficient for oxygen is known to decrease and the activity coefficient of silicon to increase with increasing silicon content. Up to about 2 percent Si the two effects seem to cancel each other and may be disregarded.

For equilibrium with solid silica or with a silica-saturated slag, a_{SiO_2} is unity. This would be the case for the oxidation of silicon in iron as long as the oxygen content is less than 0.09 percent, corresponding to a silicon content of 3×10^{-3} percent and a FeO activity of 0.40. As shown in Sec. 11-3 this is the FeO activity of a silica-saturated iron silicate slag. For higher oxygen contents a silica-unsaturated slag will be formed until the melt is in equilibrium with pure liquid FeO at 0.23 percent O. The relation between silicon and oxygen content in the melt is shown in Fig. 13-8.

If basic oxides, for example, CaO, are added to the system a silicate slag with an activity of SiO_2 less than one will be formed. This will cause a decrease in the product $[Si] \times [O]^2$ compared to the value which applies for silica-saturated slag. This is shown in Fig. 13-8 for slags with a_{SiO_2} from 0.1 to 0.001. The latter value corresponds approximately to basic steel-making slags, and we see that the oxidation of silicon is much more favorable in basic than in acid slags.

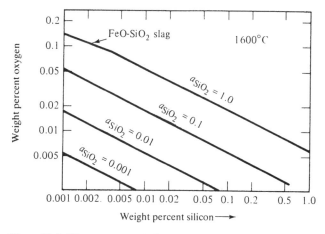

Figure 13-8 Silicon–oxygen relation as function of SiO_2 activity in slag.

The oxidation of silicon is a strongly exothermic process. This means that the equilibrium constant decreases with increasing temperature, and that the oxidation of silicon is more favorable at low temperature. In acid steel-making the oxidation occurs readily in the early part of the process, but a certain reversion of the reaction may occur at the highest temperatures which are achieved toward the end of the process.

Manganese Reactions

Manganese may, in the absence of other slag-forming components, be oxidized analogously to silicon. At high manganese and correspondingly low oxygen contents the reaction product is solid MnO

$$\underline{Mn} + \underline{O} = MnO(s) \qquad \Delta G^\circ = -291\,000 + 129.79T \text{ J}$$

This gives, for 1600°C, [Mn] × [O] = 0.047.

Above about 0.13 percent oxygen simultaneous oxidation of manganese and iron takes place, and the reaction product is a molten mixture of MnO and FeO. This reaction may be expressed by the two simultaneous equilibria

$$\underline{Mn} + \underline{O} = MnO(l) \qquad \qquad \Delta G^\circ = -244\,350 + 108.70T \text{ J}$$

$$\underline{Mn} + FeO(l) = Fe(l) + MnO(l) \qquad \Delta G^\circ = -123\,200 + 56.5T \text{ J}$$

The activity of MnO and FeO relative to the pure liquid oxides is found to be closely equal to their mole fractions. Thus $a_{MnO} + a_{FeO} \simeq 1$ and the equations may be solved to give the relation between manganese and oxygen in the metallic melt. Thus, at 1600°C, [Mn] × [O] + 0.33[O] = 0.074.

The relation between the manganese and oxygen contents of the liquid steel is shown in Fig. 13-9, and we see that with decreasing manganese content the

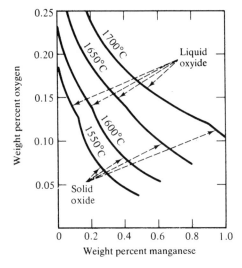

Figure 13-9 Manganese–oxygen equilibrium in liquid steel. (*From J. Chipman, J. B. Gero, and T. B. Winkler: Trans. AIME, vol. 188, p. 341 (1950).*)

oxygen content increases from the value which applies for equilibrium with solid MnO to the value which applies for liquid FeO. Also it is seen that the oxidation of manganese is favored by low temperature because it is an exothermic process.

For the composition of the oxide melt we have at 1600°C

$$K_{oxide} = \frac{(MnO)}{(FeO)[Mn]} = 3.1$$

where (MnO) and (FeO) denote mole percentages, i.e., approximately weight percentages, in the oxide melt. The same relation is found to apply also for *highly basic* slags, i.e., the ratio between the activities of MnO and FeO remains equal to the ratio between their concentrations provided the lime/silica ratio in the slag is larger than 2.5.

For *acid* slags this is no longer the case. Since MnO reacts more strongly with SiO_2 than does FeO the activity ratio a_{MnO}/a_{FeO} in silica-saturated slags is about one-eighth of the corresponding composition ratio. This gives, at 1600°C,

$$K_{acid} = \frac{(MnO)}{(FeO)[Mn]} \simeq 25$$

In other words manganese is more easily oxidized, and it is also a stronger deoxidant, in the presence of an acid slag, than alone.

For an acid slag there will also be a certain distribution of silicon between the slag and the metal, which may be expressed by the equilibrium

$$2\underline{Mn} + SiO_2 = 2MnO(l) + \underline{Si}$$

Also here the equilibrium ratio, expressed in concentrations rather than activities, will be a function of the composition of the slag. In the deoxidation of steel with manganese and silicon their ratio can be chosen to make these metals stronger deoxidants than either of them alone.

For silica-saturated slags the relation between slag and metal composition is shown in Fig. 13-10. The slag in this case contains about 50 percent SiO_2, the rest being FeO + MnO, and we see that with increasing Mn content of the metal the MnO content of the slag increases, whereas the oxygen content of the metal decreases and the silicon content increases.

Oxidation of Phosphorus

Phosphorus is oxidized less readily than manganese. The standard Gibbs energy change of the reaction

$$2\underline{P} + 5\underline{O} = P_2O_5 \text{ (g)}$$

has a large positive value. The oxidation is enhanced, however, by the presence of a basic slag in which the acid P_2O_5 is strongly bonded. A thermodynamic discussion of the reaction is hampered by the lack of basic data. Instead the

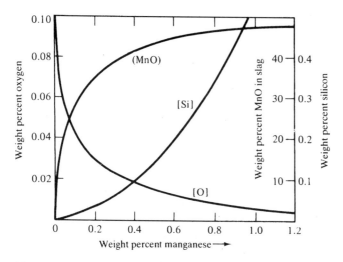

Figure 13-10 Mn–O–Si–MnO relations for SiO_2 saturated slag. (*From F. Körber and W. Oelsen; Mitt. Kaiser-Wilhelm-Inst. Eisenforsch. vol. 15, p. 271 (1933)*).)

empirical distribution ratio (% P_2O_5)/[% P]^2 is shown in Fig. 13-11 as a function of slag composition at 1600°C. We see that the distribution ratio increases strongly with increasing basicity and also with increasing FeO content of the slag. For acid slags the distribution ratio is practically zero.

The oxidation and slagging of phosphorus is a strongly exothermic process, which makes the distribution ratio increase with decreasing temperature. In basic open-hearth steel-making, therefore, considerable slagging of phosphorus takes place right after melt-down, even though the slag has not yet dissolved much lime. This first slag (flush slag) may be removed. On further heating of the bath, additional lime must be added in order to give a highly basic slag and prevent phosphorus from being reduced back into the steel.

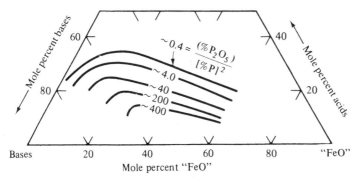

Figure 13-11 Phosphorus distribution as function of slag composition (*From Elliott, Gleiser, and Ramakrishna: "Thermochemistry for Steelmaking," vol. II, © 1963 Addison-Wesley, Reading, MA, p. 612. Reprinted with permission.*)

Other Elements

The oxidation equilibria for a large number of other elements dissolved in liquid iron have been studied experimentally. A summary of the results is shown in Fig. 13-12 which gives the oxygen *activity* of the metal as function of the alloy activity, all on a logarithmic scale. These curves refer to the oxidation of the element in the absence of other slag-forming oxides. The oxidation product may either be the pure oxide as in the case of SiO_2 and TiO_2 or it may be a mixed oxide like FeV_2O_4. Thus the curve for vanadium shows a break; at vanadium contents below 0.16 percent the coexisting phase is FeV_2O_4, whereas above that value V_2O_3 is formed. Similarly for manganese the break at 0.25 percent shows the transition from liquid FeO–MnO mixture to solid MnO.

In order to convert the data of Fig. 13-12 into weight percentages it is necessary to calculate the activity coefficients from Table 13-1. In some cases these deviate considerably from unity. Thus for high concentrations of chromium and titanium the oxygen activity coefficient is very small, and the oxygen concentration of the melt *increases* whereas the oxygen activity decreases. For aluminum rather large discrepancies have been observed between observed and calculated oxygen contents. This *may* be a result of the very large effect of aluminum on the activity coefficient of oxygen, but it is also possible that most experimental observations were not done under equilibrium conditions.

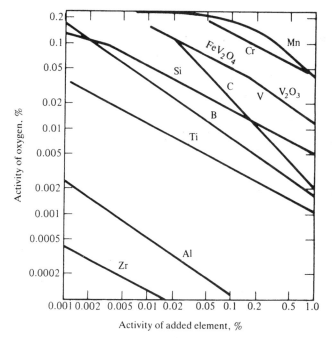

Figure 13-12 Comparison of deoxidizing power at 1600°C. (*From "Electric Furnace Steelmaking," vol. II, AIME, New York, 1963.*)

It should be pointed out that oxidation equilibria as given in Fig. 13-12 may be greatly altered in the presence of other slag-forming oxides or when two elements react together to form a mixed oxide product, as illustrated by the simultaneous oxidation of manganese and silicon.

All the oxidation reactions given in Fig. 13-12 are exothermic, making the solubility product of the corresponding oxide increase with increasing temperature. The effect is not equally large for all elements, however, and it is particularly small for carbon. This means that a reaction such as

$$Cr_2O_3 + 3\underline{C} = 2\underline{Cr} + 3CO$$

is endothermic, and will shift to the right at increasing temperature. This behavior is used in the refining of stainless steel. At normal steel-making temperature a large percentage of the chromium in the steel would necessarily be slagged simultaneously with the oxidation of carbon. By blowing oxygen into the bath, temperatures of the order of 1900°C are obtained locally. This causes the carbon to oxidize without appreciable chromium losses. Since the carbon is removed in the gas phase the reaction cannot reverse when the temperature later is decreased. A similar effect is obtained either by vacuum treatment or by purging the bath with an inert gas, such as argon. In this way stainless steel of low carbon content can be produced from medium-carbon raw materials (the argon oxygen decarburization, or AOD process).

The curves in Fig. 13-12 apply for the oxidation of impurity elements, but they also apply for the use of these elements as deoxidants. Deoxidation of steel and other metals will be discussed in Sec. 13-6.

13-4 FIRE-REFINING OF OTHER METALS

For metals such as copper and lead the activities of dissolved elements and the equilibrium constants for the fire-refining reactions are less well known than for steel. One known example is the oxidation of sulfur in copper according to

$$\underline{S} + 2\underline{O} = SO_2(g)$$

For sulfur contents above 0.1 percent and low SO_2 pressures a rather constant value of about 0.012 has been found at 1150°C for the product $[S] \times [O]^2/p_{SO_2}$.[1] For lower sulfur, and correspondingly higher oxygen contents, as well as for higher SO_2 pressures, the solubility product shows an increasing trend, corresponding to oxygen decreasing the activity coefficient of dissolved sulfur. The oxidation of sulfur in liquid copper is in many respects analogous to the oxidation of carbon in liquid steel. The reaction is slightly exothermic and is favored by low temperature.[2]

[1] F. Johannsen and U. Kuxmann: *Erzmetall,* **8**: 45 (1955).
[2] C. F. Floe and J. Chipman: *Trans. AIME,* **147**: 28 (1942).

For oxidation of other impurities in nonferrous metals we can as a general rule say that the less noble metals are oxidized most readily. Thus iron in copper is oxidized more readily than nickel. A contributing factor here is that iron in copper shows strong positive deviation from Raoult's law (Fig. 4-8), whereas the copper–nickel system is more nearly ideal. Antimony and tin in lead are removed by oxidation (lead softening) whereas the more noble bismuth is not attacked. Another contributing factor is the composition of the slag. Thus antimony and tin form acid oxides which are readily dissolved by the basic lead oxide slag. An even better slagging of these elements is obtained by the addition of alkali oxides.

Oxidation processes where condensed oxide products are formed are nearly always exothermic. This means that they are favored by low temperatures. Thus we may summarize the conditions for most favorable fire-refining:

1. The element should have a higher affinity for oxygen than the main metal
2. The element should have a high activity coefficient in the metallic melt
3. The oxide should have a low activity coefficient in the slag
4. The temperature should be low

13-5 DESULFURIZATION

In Sec. 13-4 it was shown that sulfur in copper may be removed by oxidation. A similar reaction for sulfur in iron is not possible, since iron would oxidize more readily than sulfur. Sulfur, both in pig iron and in steel, may be removed by *reduction* processes, i.e., by conversion to sulfides which dissolve in the slag or are formed as separate phases.

As mentioned in Sec. 11-2 molten slags are able to dissolve sulfur by the reaction

$$\tfrac{1}{2}S_2 + O^{2-} = S^{2-} + \tfrac{1}{2}O_2 \tag{1}$$

which, combined with

$$\tfrac{1}{2}S_2(g) = \underline{S} \tag{2}$$

gives

$$\underline{S} + O^{2-} = S^{2-} + \tfrac{1}{2}O_2 \tag{3}$$

Here the oxygen potential may be controlled by reaction with carbon, with some other reducing agent, or with iron itself. We may then write for the distribution ratio

$$\frac{(S)}{[S]} = \frac{C_S \, f_S^*}{K \sqrt{p_{O_2}}} \tag{13-4}$$

Here (S) and [S] denote the sulfur concentrations in weight percent in the slag and in the metal respectively, C_S is the *sulfide capacity* of the slag = $(S)\sqrt{p_{O_2}/p_{S_2}}$, and K is the equilibrium constant for reaction (2) above.

We see that the sulfur distribution ratio increases with increasing sulfide capacity and with increasing activity coefficient for sulfur in the metal as well as with decreasing oxygen potential. Since the activity coefficient of sulfur is increased by the presence of carbon and silicon, and because of the strongly reducing conditions in the iron blast furnace, desulfurization is much better in the blast furnace than in the steel-making furnaces. This is seen in Fig. 13-13 which gives the sulfur distribution ratio $(S)/a'_S$ against the FeO content of the slag, the latter being a measure of the oxygen potential. We see that for low FeO contents in the slag the distribution ratio is inversely proportional to the FeO content, as one should expect from Eq. (13-4). It also increases with increasing slag basicity. For molten pig iron the activity coefficient f_S^* is about 5, which gives an equilibrium distribution ratio (S)/[S] between 200 and 400 in the blast furnace. As mentioned in Sec. 9-5 the practical distribution ratio is much less, however.

With increasing FeO content the equilibrium distribution ratio decreases less rapidly and tends to level off at a constant value of 2 to 8 for slags with basicity ratio from 2 to 4 and with 5 to 20 percent FeO, that is, for basic open-hearth slags. This seems to be in contradiction to Eq. (13-4), but it must be remembered that in going from low to high FeO contents we have not only increased the oxygen potential but also changed the cationic composition of the slag. It seems that the increased oxygen potential is more or less compensated

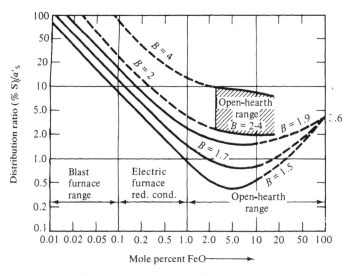

Figure 13-13 Sulfur distribution ratio as function of FeO content and slag basicity. (*From R. Rocha, N. J. Grant, and J. Chipman: Trans. AIME, vol. 191, p. 319 (1951).*) $B = (CaO + MgO)/(SiO_2 + Al_2O_3)$.)

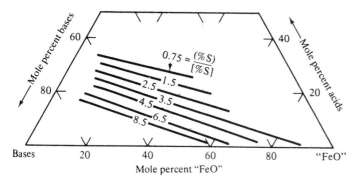

Figure 13-14 Sulfur distribution ratio as function of slag composition at 1600°C (*From Elliott Gleiser, and Ramakrishna; "Thermochemistry for Steelmaking," vol. II, © 1963, Addison-Wesley, Reading, MA, p. 606. Reprinted with permission.*)

by an increased sulfide capacity of the slag. Indeed, measurements have shown that the sulfide capacity for ferrous slags is considerably higher than for lime silicate slags. See Fig. 11-4(*b*).

The sulfur distribution ratio for steel-making slags may also be plotted directly as a function of slag composition as shown in Fig. 13-14, and again we see that it increases with increasing slag basicity and is rather independent of the FeO content.

The effect of temperature on the equilibrium sulfur distribution ratio is different in steel-making than in the blast furnace. The sulfide capacity C_S increases, and the constant K in Eq. (13-4) decreases with increasing temperature. In the blast furnace the oxygen potential tends to be defined by the reaction with carbon to give CO, and increases only slightly with increasing temperature, whereas in steel-making furnaces the oxygen potential is essentially determined by the FeO content of the slag, and increases strongly with temperature for a given slag. As a consequence the equilibrium sulfur distribution in the blast furnace, controlled by the reaction

$$\underline{S} + O^{2-} + C = S^{2-} + CO$$

increases strongly with increasing temperature. In the steel-making processes, where it is controlled by the reaction

$$\underline{S} + Fe = S^{2-} + Fe^{2+}$$

the sulfur distribution ratio, if anything, shows a slight decrease with increasing temperature.

The practical implications of these findings are that desulfurization should preferably be carried out in the blast furnace or on the hot metal before it is oxidized in the steel-making processes. In the basic open-hearth or O_2-blown converter the degree of desulfurization will necessarily be limited. With a sulfur distribution ratio of say 6 and with a slag weight of 10 percent of the steel, only about one-third of the sulfur in the charge will go into the slag. If better desulfurization is needed the slag volume must be increased, with corresponding increase in melting costs.

A better desulfurization is possible in the electric steel-furnace. After the end of the oxidation period the ferrous slag is removed, and a new, "white" slag is prepared from burned lime and fluorspar. By adding reducing agents such as carbon or silicon the FeO content of this slag can be kept below 1 percent. Under these conditions a sulfur distribution ratio up to 20 may be obtained, i.e., the desulfurization is almost as good as in the blast furnace. See Fig. 13-13.

Desulfurization outside the Furnace

Both in iron-making and in steel-making desulfurization may be done outside the furnace. In steel-making the so-called *Perrin process* may be used. Here a highly basic and iron-free slag is prepared in a separate electric furnace. This slag is poured into the ladle before the open-hearth furnace is tapped. When the molten steel falls through the slag a vigorous mixing takes place, and sulfur is transferred by the reaction $\underline{S} + Fe = S^{2-} + Fe^{2+}$, that is, iron itself acts as the reducing agent.

In iron-making a somewhat similar treatment is possible with *soda ash*. This forms a liquid slag and reacts with the molten pig iron, viz.,

$$\underline{S} + Na_2CO_3 + 2\underline{C} = Na_2S + 3CO$$

alternatively

$$\underline{S} + Na_2CO_3 + \underline{Si} = Na_2S + SiO_2 + CO$$

Desulfurization with soda ash has found particular application in the smelting of low-grade acid iron ores. For such ores it is more economical to run the blast furnace with an acid slag, and desulfurize the iron after the furnace, than to add limestone to the burden and make a basic slag.

Other desulfurizing processes for pig iron are based on the use of solid lime, calcium carbide, or calcium cyanamide. In all cases the reaction product is solid calcium sulfide. The main difficulty in these processes is to obtain an intimate mixing of the solids with the liquid iron. In the so-called *Kalling process* liquid pig iron is treated with burned lime in a rotary drum. Here carbon or silicon in the iron acts as the reducing agent, and solid calcium sulfide is formed. For a temperature of 1500°C one may calculate the equilibrium sulfur content in the iron as 0.0003 percent. In practice values of the order of 0.003 percent sulfur have been obtained, which means that equilibrium is not established completely.

In recent years considerable progress has been made with *injection* of various desulfurizers into the iron or steel bath. This can be CaC_2, Mg, or mixtures of lime and magnesium, which are injected through water-cooled lances by means of, e.g., nitrogen. As an example, by injection of CaO + Mg, sulfur values down to 0.005 percent have been reported.

Sulfur Oxidation from Slag

As already pointed out sulfur cannot be removed as SO_2 by oxidation of the steel bath. A different situation arises when there is a slag-layer between the steel and the atmosphere. At the interface between slag and atmosphere sulfur may be oxidized, viz.,

$$S^{2-} + \tfrac{3}{2}O_2 = SO_2 + O^{2-}$$

The possibility of sulfur removal through this reaction was pointed out by Larsen and Sordahl.[1] Thus we arrive at the interesting situation that a reaction which is not possible directly, is made possible by the intermediate action of the slag. We may say that the slag acts as a semipermeable wall which allows sulfur to diffuse out, but which limits the amount of oxygen which may diffuse into iron. In the extreme case a state of *partial equilibrium* (Sec. 5-7) may be established, iron and the atmosphere being in equilibrium with respect to sulfur but not with respect to oxygen. To what extent steel desulfurization actually takes place by the above reaction is not clear. One good example is in the oxygen-blown Kaldo process. Here a mass balance shows that the combined sulfur content of the steel and the slag is noticeably less than the sulfur introduced in the raw materials, and that some sulfur has been removed by the gas.

It should finally be pointed out that the above reaction may proceed also from right to left if the furnace atmosphere is rich in SO_2 and if the oxygen potential is low. This would be the case, e.g., for incomplete combustion of sulfur-containing fuel-oil. There is evidence that steel scrap may pick up sulfur from the atmosphere before melt-down, whereas the danger of sulfur pickup is negligible after the bath has been covered with a slag layer.

13-6 DEOXIDATION

In steel-making, after carbon, silicon, and phosphorus have been removed by oxidation, the bath contains up to 0.1 percent of oxygen. If such a steel were cast into ingots the oxygen would be enriched in the molten phase as the steel freezes, and very soon the solubility product of CO would be exceeded, and a vigorous gas evolution would occur. It is necessary, therefore, to remove a major part of the oxygen before casting. As deoxidants ferrosilicon and ferromanganese, sometimes silicomanganese, ferrotitanium, and aluminum are used. The same equilibria apply as given in Fig. 13-12. Usually most of the oxygen is removed with ferrosilicon and ferromanganese. This may be done in the furnace or in the ladle. If done in the furnace care must be taken that phosphorus is not

---

[1] B. M. Larsen and L. O. Sordahl: in "Physical Chemistry of Process Metallurgy," Interscience Publishers, New York, 1961, p. 1141.

reduced back from the slag, by first removing the slag. The final deoxidation may be done with aluminum or calcium–silicon either in the ladle or during the pouring of the steel into the ingot molds.

Deoxidation of liquid steel may be carried out to different oxygen levels dependent on the subsequent use of the steel ingot. With decreasing gas evolution during solidification the steels are classified as follows: "rimmed," "capped," "semikilled," and "killed" steel. For killed steel there is no gas evolution. Killed steel is, among other things, used for steel castings.

One problem in steel deoxidation is the removal of deoxidation products from the melt. If deoxidation is carried out with aluminum the resulting solid Al_2O_3 particles will be very small and will not easily separate out. As mentioned in Sec. 7-7 the separation is greatly enhanced by vigorous stirring of the melt whereby the particles are brought to adhere to the upper slag layer or to the ladle refractories. Another method is to inject lime together with aluminum, in which case liquid calcium aluminate, which separates out more readily, is formed. At the same time a good *desulfurization* is obtained.

Another way to obtain a steel with low content of inclusions is by *electro slag remelting*. In this method the first ingot is used as an electrode which dips into a molten slag pool, usually composed of $CaO + Al_2O_3 + CaF_2$, contained in a water-cooled ingot mold. By passing a heavy current from the ingot through the slag, the ingot melts inside the slag pool, the inclusions dissolve in the slag, whereas the molten steel collects and solidifies in the mold forming the final ingot. The interesting thing about this method is the use of a water-cooled ingot mold, which remains cold even though it contains a slag that is kept molten by means of the heavy electric current. Another application of the same principle is shown in Fig. 14-8 for remelting of reactive metals.

Deoxidation is done also in connection with fire-refining of *copper*. After oxidation of iron, sulfur, etc., the copper bath may contain about 1 percent oxygen. Since the oxygen potential in copper refining is much higher than in steel-making, weaker deoxidants may be used. Most common are carbon and hydrocarbons but ammonia has also been used. Common practice is to deoxidize by means of *wood*. After the slag has been removed, heavy tree trunks are thrust into the copper bath. The vigorous evolution of hydrocarbons stirs the bath, and the gases react with oxygen to form CO_2 and H_2O. In this way the oxygen content of the bath can be lowered to about 0.05 percent. At lower oxygen contents appreciable amounts of hydrogen will be picked up by the liquid copper.

For the production of so-called "oxygen-free" copper more powerful deoxidants, such as phosphorus or calcium–boron alloy, are used. It should here be pointed out that copper easily picks up oxygen from the air during pouring and casting, and for oxygen-free copper these operations must be done under a protective atmosphere.

For other nonferrous metals, such as lead, the solubility of oxygen is low, and deoxidation is not necessary.

13-7 REFINING BY METAL–METAL SEPARATION

In some cases impurities may be removed in a separate metallic phase. The separation of liquid zinc from lead was already mentioned in Sec. 10-5. Another example is the separation of *iron from tin*. Even after double smelting with tin ore the metal contains about 1 percent iron. Part of the phase diagram tin–iron is shown in Fig. 13-15, and we see that by cooling of the impure tin to above its melting point, 232°C, the compound $FeSn_2$ will separate out as a solid phase and give a liquid with less than 0.01 percent Fe. The process may be carried out by cooling the melt in a ladle, whereby the solid compound is allowed to settle, or impure solid tin may be heated slowly in a reverberatory furnace with a sloping hearth till above the melting point, whereby the almost pure liquid phase drains off. In the latter case the process is called *liquation*.

Somewhat similar to Fig. 13-15 is the phase diagram lead–copper. Raw lead from the blast furnace usually contains some copper. Part of this may be separated out by cooling the impure lead to above its melting point. The separation of copper is further enhanced by the presence of sulfur, arsenic, antimony, or tin in the raw lead. These elements form stable, solid compounds with copper, which separate out during cooling. Metallic copper as well as the solid compounds are lighter than lead, and rise to the top where they can be skimmed off.

Such solid phases which float on top of the metal are usually called *dross*. Drosses may be insoluble metal, but more often they are sulfides, arsenides, or oxides. The separation of drosses from the underlying metal is not as good as

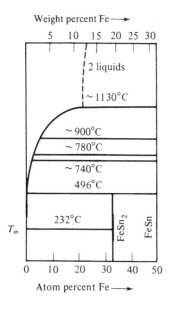

Figure 13-15 Part of tin–iron phase diagram. (*After M. Hansen: "Constitution of Binary Alloys,"* McGraw-Hill Book *Co. Inc., New York, 1958.*)

for liquid slags from metals; the drosses usually contain large amounts of entrained metal. By mixing the dross with sawdust or ammonium chloride the liquid phase coalesces and may partly be squeezed out. The remaining part may be recovered during the further treatment of the dross.

Further removal of copper from liquid lead is obtained by stirring in some elemental sulfur, which at the temperature in question will be below its boiling point. By reaction with elemental sulfur the copper content of the lead may be brought way below what can be calculated for the equilibrium

$$PbS(s) + 2Cu = Cu_2S(s) + Pb \tag{1}$$

This phenomenon is explained by liquid sulfur representing a much higher sulfur potential than is given by equilibrium (1), and that the two reactions

$$Pb + S(l) = PbS(s) \tag{2}$$

and

$$(2 - x)Cu + S(l) = Cu_{2-x}S(s) \tag{3}$$

are both progressing to the right. It appears that the *rate* of reaction (3) is much higher than that of reaction (2) and that some kind of partial equilibrium with elemental sulfur is approached for copper, but not for lead. To make the matter even more complicated the above observation refers to industrial conditions where the lead also contains impurities like tin and silver. In the absence of these impurities the phenomenon is not observed. It thus seems that tin and silver either catalyze reaction (3) or inhibit reaction (2). After all the sulfur has reacted or evaporated the copper content of the lead will again increase by reversion of reaction (1) until equilibrium with PbS is established.[1]

A type of reverse liquation takes place when the pure metal has a higher melting point than the impure melt. A typical example is given by the so-called *Pattison process*, which previously was used for desilvering of lead. Figure 13-16 shows the phase diagram lead–silver. Raw lead often contains around 0.2 percent of silver. If this is cooled slowly essentially pure lead will crystallize out and the liquid will be enriched in silver up to the eutectic composition of 2.5 percent Ag. In practice a system of repeated crystallizations and meltings was used, and a liquid phase with up to 2 percent Ag was obtained. Even though this still contained 98 percent Pb, it represented a ratio of concentration of about ten. Silver could now be recovered from the lead–silver alloy by fire-refining, *cupellation*, i.e., the 98 percent of lead was oxidized, and the 2 percent of the more noble silver remained as a metallic phase.

Desilvering of lead is today mostly done by the *Parkes process*. In this process some zinc is added to the impure lead. Zinc forms a number of stable, solid phases with silver, whereas, as can be deduced from the shape of the liquidus curve in Fig. 13-16, the lead–silver system shows positive deviation

---

[1] T. R. A. Davey: in "Lead–Zinc–Tin '80 Symposium," AIME, New York, 1979, pp. 477–507.

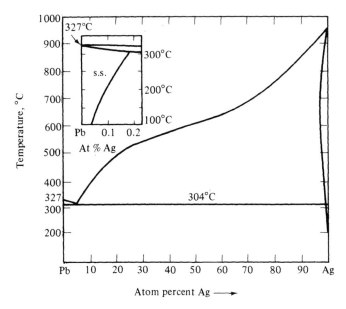

Figure 13-16 Lead–silver phase diagram. (*After M. Hansen: "Constitution of Binary Alloys," McGraw-Hill Book Co. Inc., New York, 1958.*)

from ideal mixing. On cooling of a lead–zinc–silver melt to just above the melting point of lead, solid zinc–silver phases will therefore precipitate, and may be skimmed off. In order to obtain a high silver recovery, double treatment with zinc is common.

The further treatment of the zinc–silver alloy, which contains appreciable amounts of entrained lead, is as follows: First the zinc is distilled off by heating the alloy in a retort to above the boiling point of zinc. The remaining lead–silver alloy, which may contain 50 percent of silver, is oxidized in a hearth furnace (cupellation) until all the lead has been oxidized. The remaining noble metal is refined by electrolysis to produce pure silver and to recover its gold content.

After removal of the zinc–silver compound the lead melt contains up to 1 percent zinc, which has to be removed. The dezincing of lead could be done by oxidation or by treatment with chlorine to form zinc chloride. In recent years dezincing has been done by vacuum treatment, whereby the zinc evaporates and may be recovered in metallic form for further use.

A method which resembles the Parkes process is sometimes used if the lead contains bismuth. In this case a Ca–Mg alloy is stirred into the melt. This forms a solid alloy with bismuth which may be skimmed off. More usually, however, bismuth is removed and recovered by electrolytic refining (see Chap. 16).

It should finally be mentioned that *arsenic* and *antimony* in the raw *zinc* from the ISP blast furnace (Sec. 10-5) may be removed by treatment with metallic sodium whereby solid NaAs and NaSb are precipitated out.

13-8 ZONE REFINING

Zone refining occupies a unique position among refining processes, although it is based on the same principle as the old-fashioned Pattison process. Zone refining is used to produce metals of extreme purity, e.g., transistor silicon. The principle of zone refining is shown in Fig. 13-17(a). The metal, in the form of a long bar, is moved slowly through a furnace with a narrow, hot zone, or through an induction coil. Passing through the hot zone or the coil a section of the metal melts in one end and freezes in the other, until after a complete passage the entire bar has been melted and solidified. As a result impurities that are enriched in the molten phase are transferred to the one end of the bar.

This is shown in Fig. 13-17(b), which illustrates a typical phase diagram near the melting point of the metal. If the initial concentration of the impurity X had been C_0, this concentration will not change when the metal melts, that is, $C_0 = C_L^o$. On slow freezing of the melt a solid with composition C_S^o will come out, and the melt will become enriched in the impurity element. As the liquid zone moves along the bar the concentration in the liquid phase will build up as shown in Fig. 13-17(c), the increase in impurity content of the melt being equal to the decrease in the solid as shown by the two cross-hatched areas. The composition of the two phases at the freezing front corresponds to the liquidus and solidus lines in the phase diagram. As the concentration in the liquid builds up near the freezing front, the impurities diffuse into the main liquid body. In that way the concentration in the liquid will increase as the liquid zone moves

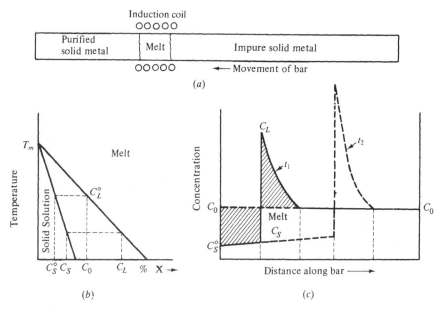

Figure 13-17 Principle of zone refining.

along the bar, see Fig. 13-17(c). Finally, when the molten zone reaches the end of the bar, it has taken up a substantial part of the impurity content in the original bar. The operation may now be repeated until, after several passages, most of the impurities are concentrated in the end of the bar, which subsequently is removed.

In the above case the impurity concentration was higher in the liquid than in the solid phase. For the opposite case the impurities will be concentrated in the main part of the bar, whereas the end will be purified. In both cases the higher the ratio between the two concentrations the higher will be the degree of purification for each passage. Also, the purification is enhanced by a high rate of diffusion in the liquid phase. The rate of movement of the bar must be adjusted to allow the impurity elements to diffuse through the liquid body, and to balance the supply and removal of heat with the heats of melting and solidification.

13-9 VACUUM REFINING

If the impurities in a metal have a pronounced vapor pressure, they may be distilled off, as in the removal of zinc from lead. Distillation may also be used if the main metal is more volatile than the impurities as illustrated in Sec. 10-6 for the refining of zinc. Finally, distillation and gas evolution may be used if non-volatile impurities can be converted into volatile compounds as illustrated by the oxidation of carbon in steel and of sulfur in copper. Refining processes in which gases or vapor are evolved are enhanced by decreased pressure. The use of reduced pressure and vacuum has made it possible to carry out many refining processes which could not be done at atmospheric pressure, or it has made it possible to carry out such processes at a lower temperature.

Among vacuum refining processes the deoxidation and degassing of liquid steel is of particular interest. As it was shown in Fig. 13-12 carbon is a rather poor deoxidant at atmospheric pressure. Its deoxidizing power is enhanced at lower pressure in accordance with the reaction $\underline{C} + \underline{O} = CO(g)$ shifting to the right. It can easily be shown that at a CO pressure of 10^{-3} atm, the solubility product $[C] \times [O]$ becomes equal to 2×10^{-6}, that is, carbon becomes a stronger deoxidant than aluminum. In practice the results are not so good, as illustrated by Fig. 13-18 which gives the product $[C] \times [O]$ in liquid steel as function of pressure. We see that the practical product is always higher than the theoretical one, the difference increasing with decreasing pressure. Below 1 mmHg the practical curve tends to level off at $[C] \times [O] = 10^{-4.2}$. The curve for nitrogen shows similar, but smaller, deviation, whereas for hydrogen the theoretical and practical curves agree rather closely above 1 mmHg pressure.

One reason for this discrepancy may be that the gases can only escape from the metal surface, and that the diffusivity of hydrogen is much higher than that of nitrogen, carbon, and oxygen. Another reason is the effect of refractory materials on the oxygen content. In the same way as carbon becomes a stronger

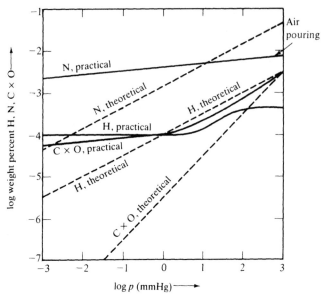

Figure 13-18 Effect of pressure on solubility of N, H, and C × O in liquid steel at 1600°C. (*After A. Tix et al.: J. Iron Steel Inst., vol. 191, p. 260 (1959)*.)

deoxidant at reduced pressure, common refractories become less stable. Thus silica may react with carbon or iron, viz.,

$$SiO_2 + 2\underline{C} = \underline{Si} + 2CO(g)$$

$$SiO_2 = SiO(g) + \underline{O}$$

Similarly for magnesia refractories

$$MgO(s) = Mg(g) + \underline{O}$$

The volatile products SiO and Mg being pumped off give an increase in the oxygen content of the bath.

In practical vacuum deoxidation and degassing one tries to give the steel melt as large a surface as possible. In one method the liquid steel is poured through a nozzle into an evacuated chamber, which contains either a second ladle or the ingot mold. When the steel flow enters the chamber it breaks into a shower of fine droplets, and a rapid degassing takes place. Another method is based on sucking the liquid steel into an evacuated pipette whereupon it again drains back into the ladle. Finally the degassing may be done by *remelting* the ingot in an evacuated electric arc furnace. For all processes pressures of the order of 1 mmHg are used, lower pressures giving no advantage.

Vacuum desulfurization is less well understood. Desulfurization may take place either by reaction with solid lime

$$\underline{S} + CaO + \underline{C} = CaS + CO(g)$$

or by the formation of volatile sulfides. For the above reaction the standard Gibbs energy at 1600°C is -97 kJ. This means that for a carbon content of 0.1 percent and a CO pressure of 10^{-3} atm the theoretical sulfur content should be 2×10^{-5} percent. In practice such values are not obtained, probably for the same reasons that make desulfurization with solid lime difficult also at atmospheric pressure.

In the absence of lime, sulfur may evaporate as atomic sulfur or as volatile sulfides. These may be CS_2 and CS for high-carbon steels and SiS for high-silicon steels. For most steels, however, the partial pressure of these compounds is too low to make vacuum desulfurization attractive.

Other metals which may be removed by vacuum treatment of steel are copper, zinc, and manganese. Particularly for copper vacuum treatment should be of interest, since copper is not easily removed from liquid steel by other methods. Measurements[1] have shown that for a pressure of 0.1 mmHg, copper may be removed down to 0.01 percent. It is interesting to notice that vacuum treatment did not remove arsenic, however, in spite of pure arsenic being much more volatile than pure copper. Obviously the activity coefficient of arsenic in steel is very low, whereas copper in steel shows positive deviation from Raoult's law.

Tin is another unwanted constituent in steel, and it cannot be removed by ordinary refining methods. Also, metallic tin has a low vapor pressure. In the presence of sulfur, however, volatile tin sulfide may be formed and there should be a possibility of removing this compound by vacuum treatment.

Vacuum refining of ferroalloys and rare metals and refining by transport reactions will be discussed in the subsequent chapter.

PROBLEMS

13-1 A steel melt contains 0.08 percent C, 0.90 percent Cr, and 0.02 percent S.

(a) By means of Eq. 13-3 and data in Table 13-1 calculate the activity, a'_C, of carbon in the melt.

(b) Calculate the corresponding activity of oxygen if the melt is in equilibrium with 1 atm of CO at 1600°C where the product $a'_C a'_O = 0.002$.

(c) Calculate the corresponding oxygen percentage.

13-2 A steel melt contains 0.90 percent C, 1.30 percent Si, and 0.10 percent S.

(a) Calculate the activity a'_S in the melt.

(b) Calculate the gas ratio p_{H_2S}/p_{H_2} at 1600°C as function of [% S], when for the reaction

$$\underline{S} + H_2 = H_2S \qquad \Delta G° = 41\,170 + 27.36T \text{ J}$$

(c) Pure hydrogen is bubbled through the melt, and the effluent gas is assumed to contain equilibrium amount of H_2S. Derive an expression for the rate of sulfur removal as function of residual sulfur percentage, and calculate the volume of hydrogen, in normal cubic meters per kilogram of steel, required to lower the sulfur content from 0.10 to 0.02 percent.

[1] H. Schenck and H.-H. Domalski: *Arch. Eisenhüttenw.*, **32**: 753 (1961); G. M. Gill et al.: *J. Iron Steel Inst.*, **191**: 172 (1959).

13-3 Calculate the partial pressures of S_2, CS_2, and SiS over a steel melt with 0.9 percent C, 1.3 percent Si, and 0.05 percent S at 1600°C, where for the reactions $C(s) + S_2 = CS_2(g)$, $\Delta G^\circ_{1873} = -33$ kJ; $2Si(l) + S_2 = 2SiS(g)$, $\Delta G^\circ_{1873} = -201$ kJ; $C(s) = \underline{C}(1\%)$, $\Delta G^\circ_{1873} = -57$ kJ; $Si(l) = \underline{Si}(1\%)$, $\Delta G^\circ_{1873} = -167$ kJ; $S_2(g) = 2\underline{S}(1\%)$, $\Delta G^\circ_{1873} = -180$ kJ.

13-4 For a given iron blast furnace slag at 1500°C the *sulfide capacity*, $C_S = (\% \ S)\sqrt{p_{O_2}/p_{S_2}} = 2 \times 10^{-4}$. Calculate the distribution ratio (% S in slag)/[% S in iron] for equilibrium between hot metal, slag, graphite, and 1 atm of CO, when for the reaction $\frac{1}{2}S_2 = \underline{S}$, $\Delta G^\circ = -131\,460 + 22.05T$ J, and $f_S^* = 5$. For the reaction $2C + O_2 = 2CO$, $\Delta G^\circ = -223\,400 - 173.3T$ J.

13-5 For the reaction $MgO(s) = Mg(g) + \underline{O}(\text{in Fe})$ $\Delta G^\circ_{1600°C} = 224$ kJ. Calculate the Mg pressure for equilibrium between a MgO crucible and liquid steel with 10^{-3} and 10^{-5} percent oxygen respectively. Activity coefficients are disregarded.

13-6 By means of data in App. C calculate the concentration of copper in liquid lead at 400°C for equilibrium with solid PbS and solid Cu_2S, when the solubility of solid copper in liquid lead is 0.2 percent and Henry's law is assumed valid.

BIBLIOGRAPHY

Belk, J. A.: "Vacuum Techniques in Metallurgy," Pergamon Press, London, 1963.

Bodsworth, C., and H. B. Bell: "The Physical Chemistry of Iron and Steel Manufacture," 2d ed. Longmans, London, 1972.

Bunshah, R. F. (ed.): "Vacuum Metallurgy," Reinhold Publ. Co., New York, 1958.

Derge, G. (ed.): "Basic Open Hearth Steel-making," AIME, New York, 1964.

Elliott, J. F., et al.: "Thermochemistry for Steel-making," vol. II, Addison-Wesley Publ. Co. Inc., Reading, Mass., 1963.

McGannon, H. E.: "The Making, Shaping and Treating of Steel," 9th ed. U.S. Steel Corp, Pittsburgh, 1971.

Parr, N. L.: "Zone Refining and Allied Techniques," George Newnes Ltd., London, 1960.

Peters, A. T.: "Ferrous Production Metallurgy," John Wiley and Sons, New York, 1982 (steel-making).

Ward, R. G.: "An Introduction to the Physical Chemistry of Iron and Steel-making," Edward Arnold (Publishers) Ltd., London, 1962.

Winkler, O., and R. Bakish (ed.): "Vacuum Metallurgy," Elsevier Publishing Co., Amsterdam, 1971.

FOURTEEN

RARE AND REACTIVE METALS, FERROALLOYS

The expression *rare*, or *less common metals* is not easy to define. It is used for some metals which only in recent years are being produced on an industrial scale, such as tungsten, molybdenum, beryllium, titanium, zirconium, and niobium. These metals are produced in relatively small quantities and are used, among other things, in atomic reactors, space ships, and jet engines. With the exception of tungsten and molybdenum, they all form extremely stable compounds with oxygen, nitrogen, and carbon, and very special methods are needed in order to obtain them in pure form.

Some of these metals may more easily be produced as *ferroalloys* to be used as additives in steelmaking. Other important ferroalloys are ferrochromium, ferromanganese, and ferrosilicon, the production of which was discussed briefly in Sec. 9-8. In this chapter first the production of ferroalloys and second the production of the more reactive metals in pure form will be discussed in more detail.

14-1 FERROALLOYS

Ferroalloys are relatively impure products, which, in addition to their valuable component, contain considerable amounts of iron, and which are used as additives in steel-making. The iron content may either be a result of iron in the ore, or it may have been added deliberately in order to facilitate the reduction process or to lower the melting point of the alloy. Examples of the first type are ferrochromium and ferromanganese. Chromium and manganese ores invariably contain some iron, and, since these metals are less noble than iron, the iron content of the alloy will be determined by the iron content of the ore. Examples of the second and third types are ferrosilicon where scrap iron is added in order to facilitate the reduction, and ferrotungsten or ferromolybdenum where iron additions lower the melting point.

In general ferrochromium and ferromanganese contain 70 to 80 percent of the valuable metal, ferrosilicon from 45 to 90 percent, and ferroalloys of the rare metals usually less. If the available ore is of low grade, it may sometimes be *enriched*. This could be done by a partial reduction, where some of the iron is reduced, and most of the valuable metal remains in the slag. This slag is then reduced further to produce a high-grade ferroalloy. Alternatively, the ore could be reduced completely to give a low-grade ferroalloy. This is again oxidized, e.g., in a Bessemer converter, to give a relatively pure iron and a slag rich in the valuable element, the latter again being reduced to give the ferroalloy. By this last method high-grade ferrovanadium is at present produced commercially from an iron ore which contains only a few tenths of one percent of vanadium.

Most ferroalloys are produced by reduction of oxide ores with carbon. Since the reduction takes places at temperatures well in excess of $1000°C$ the partial pressure of CO_2 is very low and the reduction may conveniently be expressed by the equation

$$MeO + C = Me + CO$$

If we examine the Gibbs energy curves given in App. C we find that the standard Gibbs energy for this reaction becomes negative above a certain temperature, which increases in the order Cr_2O_3, NbO_2, MnO, V_2O_3, SiO_2, TiO_2, MgO, and Al_2O_3, the reduction of the last oxide being favored above $2100°C$. Thus, if there were no side reactions, carbothermic production of the rare and reactive metals would be an easy task, $2000°C$ being easily obtained in an electric arc furnace. Unfortunately, with few exceptions, these metals also form very stable carbides. This is seen from the Gibbs energy curves which show that the stability of the carbides per mole of carbon increases in the order Mn_3C, NbC, Al_4C_3, CaC_2, SiC, $Cr_{23}C_6$, VC, TiC, and ZrC. Thus, if an oxide of one of the above metals is reduced with an excess of carbon, the product will be a carbide rather than the pure metal. If the reaction mixture had contained only enough carbon to give the pure metal, the product would be a mixture of metal carbide and unreacted metal oxide. These may react on further heating

$$MeC + MeO = 2Me + CO$$

At atmospheric pressure this type of reaction shifts to the right with increasing temperature. Conversely, at constant temperature it may be shifted to the right with decreasing pressure. Even though in this way it may be possible to decompose the carbide phase by heating it with an excess of metal oxide, the resulting metal still contains a considerable amount of dissolved carbon. This is, for example, the case for ferrochromium and ferromanganese, which in the molten state are able to dissolve up to 10 percent of carbon. Also, in order to reduce significantly the carbon content of the metal, the losses of oxide in the slag will be high. Nevertheless, by the combined use of vacuum and high temperatures it has been possible in recent years to produce low-carbon ferroalloys and also pure metals by carbothermic reduction. Some examples of this will be given in Sec. 14-3. In other cases low-carbon ferroalloys have been produced in the absence of carbon, and with silicon or aluminum as reducing agents. Some examples of this type of *metallothermic* reduction will be given in Sec. 14-2.

A further complication in the production of reactive metals at high temperatures is the formation of volatile compounds, particularly suboxides such as SiO and Al_2O. It is clear that if the partial pressure of the volatile oxide becomes significantly higher than the CO pressure, carbothermic production of the metal will be virtually impossible. An example of such difficulties will be given in connection with the discussion on ferrosilicon.

Ferrosilicon and Si-metal

Ferrosilicon is made by reducing quartz with coke in an electric low shaft furnace. Scrap iron or iron ore is added to give the desired grade. The most common qualities are 45 percent, 75 percent, and 90 percent Si. A product with 98 to 99 percent Si (Si-metal) may be produced by the same technique, but without addition of iron and with charcoal instead of coke.

In its simplest form the reduction of quartz with carbon may be given by the reaction

$$SiO_2 + 2C = Si + 2CO$$

As we already have seen, however, silicon will in the presence of carbon react to form silicon carbide. Also some volatile SiO will be formed. Even though carbothermic production of ferrosilicon and Si-metal has taken place on a commercial scale for decades, it is only in recent years that the thermodynamics and mechanism of the process have become relatively clear.

Applying the phase rule we find that in a system of three components, silicon, oxygen, and carbon, the sum $P + F$ (phases + degress of freedom) = 5. For an arbitrary temperature and pressure we can at the most have three phases, e.g., two condensed phases and a gas. Possible condensed phases are SiO_2, C, SiC, and Si, whereas the gas will be a mixture of essentially CO and SiO. Among these phases there will be a number of monovariant equilibria, viz.,

$$SiO_2 + C = SiO(g) + CO(g) \tag{1}$$

$$2SiO_2 + SiC = 3SiO(g) + CO(g) \tag{2}$$

$$SiO_2 + Si = 2SiO(g) \tag{3}$$

$$2Si + CO(g) = SiC + SiO(g) \tag{4}$$

$$SiC + CO(g) = 2C + SiO(g) \tag{5}$$

If, at a given temperature, the partial pressure of one gas component is given, the partial pressure of the other component is fixed for each of the above phase combinations, i.e., the gas composition is fixed.

For a total working pressure $p_{CO} + p_{SiO}$ of one atmosphere the partial pressure of SiO is shown in Fig. 14-1 as a function of temperature for each of the above monovariant equilibria, and we see that the set of curves intersect at the points A (1510°C) and B (1820°C), corresponding to the nonvariant phase combinations $SiO_2 + SiC + C$ and $SiO_2 + SiC + Si$ respectively. In this figure the solid curves refer to stable phase combinations, the dashed curves to unstable combinations. Thus, at point C, which corresponds to equilibrium between SiO_2, C, and Si, the system is unstable and would tend to react to give SiC. We also see that at points A and B the gas contains 0.4 and 67 percent SiO respectively.[1]

All curves so far refer to silicon being present as a pure phase. If iron is added the silicon activity will be lowered and the curves (3) and (4) will shift to lower partial pressure of SiO.

We may now outline the mechanism for the production of silicon and ferrosilicon in the electric low shaft furnace at atmospheric pressure: The reaction mixture which consists of quartz and coke, with or without scrap iron, descends in an atmosphere of CO and SiO. At temperatures above 1510°C coke may react with SiO, by reversion of reaction (5), to give SiC. A direct reaction between solid SiO_2 and C to SiC seems less likely. The mixture of SiO_2 and SiC now descends to the hottest part of the furnace. At temperatures above 1820°C

---

[1] The curves in Fig. 14-1 are based on thermodynamic data determined by K. Motzfeldt and coworkers at the Norwegian Institute of Technology. See also: A. Schei: *Tidsskr. Kjemi, Bergves., Metall.*, **27**: 152 (1967), and W. A. Krivsky and R. Schuhmann, Jr., *Trans. AIME*, **221**: 898 (1961).

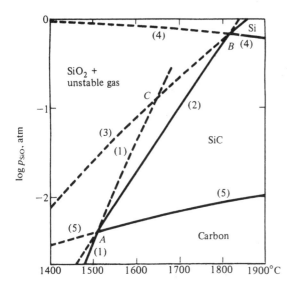

Figure 14-1 Predominance . fields and SiO pressure for $p_{CO} + p_{SiO} =$ 1 atm as function of temperature in the system Si–C–O. Solid lines correspond to stable, dashed lines to unstable phase equilibria. (*After A. Schei: Tidsskr. Kjemi. Bergves. Metall., vol. 27, p. 152 (1967)*.)

they react by a combination of reactions (2), (3), and (4) to give silicon, CO, and SiO(g).

We now shall consider the *stoichiometry* of the process. If we assume that the only reaction which occurs in the cooler parts of the furnace is the formation of SiC by reversion of reaction (5) we arrive at the following stoichiometric scheme (I), assuming counter-current plug flow:

$$> 1510°C: \quad 4C + 2SiO(g) = 2SiC + 2CO$$

$$> 1820°C: \quad 3SiO_2 + 2SiC = Si + 4SiO(g) + 2CO$$

The stoichiometry given for the second reaction is derived from the fact that the equilibrium gas mixture at point *B* contains about 67 percent SiO and only 33 percent CO. We see that according to this scheme the hot zone produces twice as much SiO as the carbon is able to absorb in the cooler zone. The remaining SiO would then have to escape from the top of the furnace, burning in the air to give a white smoke of SiO_2. At the same time only one-third of the quartz would be converted into metal with a correspondingly poor energy efficiency.

Fortunately for the silicon industry some other reactions seem to occur. First of all the gas from the high-temperature zone may contain somewhat less SiO than given for equilibrium at point *B*. In the same way as discussed for carbothermic reduction of ZnO (Sec. 10-4) the actual reaction temperature at high driving rates will be higher than the equilibrium temperature. If now the reaction *rate* on the SiC surface is higher than on the SiO_2 surface, i.e., if the reversion of reaction (4) is faster than reaction (3) this will result in a gas product with more CO and less SiO than given for point *B*.[1] To what extent this is actually the case is not known, however. Second, in addition to reacting with carbon, SiO may condense to a mixture of SiO_2 and Si in the cooler parts of the furnace by reversion of reaction (3) below, say, 1700°C. If we assume that all SiO which did not react with carbon, condenses in the furnace, we get the following stoichiometric scheme (II):

$$< 1700°C: \quad 2SiO(g) = Si + SiO_2$$

$$> 1510°C: \quad 4C + 2SiO(g) = 2SiC + 2CO$$

$$> 1820°C: \quad \begin{cases} Si + SiO_2 = 2SiO(g) \\ 2SiO_2 + 2SiC = 2Si + 2SiO(g) + 2CO \end{cases}$$

Here the third reaction is the reevaporation of the deposit formed in the first reaction. We see that in this case the furnace acts as a reflux condenser and with a certain amount of SiO_2 + Si circulating between the hot and the cooler zones, thus maintaining the equilibrium gas composition in the hot zone. As we see all the quartz is converted into metal, and no SiO is lost in the top gas. Most probably industrial furnaces operate somewhere between the schemes I

---

[1] Y. Otani et al.: Paper N. 112, 6th Int. Congr. on Electro-heat: Brighton, 1968.

and II, and considerable skill and experience is needed to catch the SiO gas and prevent it from escaping.

As seen from Fig. 14-1 reversion of reaction (3) does not correspond to stable equilibrium, however; reactions (2) and (1) having greater tendency to reverse. These reactions would result in a complete plugging of the furnace and would block the process. Fortunately they seem to occur only to a limited extent.

One drawback of the SiO condensation is that the deposit tends to plug up the furnace, preventing an even descent of the charge. The furnace mix therefore has to be poked regularly, an operation which is well known to every operator of silicon furnaces. Usually the need for poking makes it necessary to work without a roof causing the furnace gases and the smoke to escape into the atmosphere.

The addition of iron to the furnace greatly facilitates the process. Not only does iron lower the activity of silicon and therefore the SiO content of the gas at point *B*, but it also acts as an extra trap for SiO in the cooler part of the furnace by absorbing the silicon formed during condensation. The presence of scrap iron may also affect favorably the gas distribution in the furnace. It is well known that ferrosilicon is more easily produced than Si-metal, and with a higher recovery and lower power consumption. For the smelting of 45 percent FeSi the plugging of the charge is small, and such furnaces may be operated with a roof.

If pure quartz is used, the silicon process does not produce any slag. With impure raw materials a small amount of highly acid and viscous slag is formed. Some of the aluminum and calcium in the raw material will also be reduced and enter the metallic product, and may affect its quality. Any sulfur in the charge will react with silicon to give volatile silicon sulfide which escapes with the furnace gases, whereas phosphorus in the charge enters and contaminates the metal.

Other Ferroalloys

Compared to the ferrosilicon process carbothermic production of other ferro-alloys is relatively simple. High-carbon *ferromanganese* and *ferrochromium* may be produced by methods which resemble electric pig iron smelting (Sec. 9.4). Ferromanganese may even be produced in the blast furnace. The ore always contains some gangue, and in order to obtain a high metal recovery and good desulfurization enough lime is added to give a basic slag. The metallic products usually contain 7 to 9 percent carbon, and may be used as a deoxidant or alloy addition for low-alloy steels. For high-alloy steels, however, a lower carbon content is required. Some reduction in the carbon content may be obtained by remelting the alloy with an excess of ore, but only down to about 1 percent carbon. For lower carbon contents, which in particular is required for high-chromium steels, silicothermic reduction or vacuum treatment, which will be discussed later, may be used.

Ferromolybdenum and *ferrotungsten* may also be produced by carbothermic reduction. Owing to the high melting points of the alloys, the furnaces are not tapped, and the metal is allowed to cool inside the furnace. After a sufficient amount has been produced the furnace is cooled and later broken up to recover the lump of solidified alloy.

Silicon carbide is produced by reduction of quartz sand with an excess of coke in rectangular furnaces and is recovered as a solid product after cooling of the furnace. *Calcium carbide* is produced by reduction of lime with coke. Pure CaC_2 melts at about 2300°C, but its melting point is depressed by impurities, mainly unreacted lime, and the melt may easily be tapped from the furnace.

Phosphate rocks may also be reduced in the electric furnace. In order to facilitate the reduction an acid slag is used. Most of the phosphorus escapes in elemental form with the furnace gases and may be recovered by condensation or is oxidized to pure P_2O_5. The remaining part combines with iron to give molten *ferrophosphorus* with about 25 percent P.

In recent years several attempts have been made to produce *aluminum* by carbothermic reduction of the oxide by reactions analogous to those used for silicon. If successful, such a process would give smaller energy losses and higher production capacity per furnace than the present-day electrolytic method. So far carbothermic production of pure aluminum has been unsuccessful, however. More promising seems to be simultaneous reduction of alumina and silica to give an aluminum–silicon alloy with 50 to 60 percent Al. By cooling such an alloy to just above the eutectic temperature of 577°C solid silicon precipitates and leaves a melt with almost 90 percent Al. The latter may be used as raw material for Al–Si foundry alloys.

If impure raw materials, such as bauxite, are used, carbothermic reduction will give an alloy which, in addition to aluminum and silicon, contains iron, titanium, etc. Such an alloy may be used as a steel deoxidizer, or as a reducing agent for the production of metallothermic magnesium (Sec. 10-8), and has also been suggested as being suitable for refinement to pure aluminum by the subchloride process (Sec. 14-7).

14-2 METALLOTHERMIC REDUCTION

Silicon and aluminum, occasionally also magnesium, may be used as reduction agents in order to produce metals of low carbon content. Thus silicon may be used in the production of low-carbon ferrochromium, usually in the form of a ferro–chromium–silicon alloy with about 40 percent each of chromium and silicon. This alloy may be produced by carbothermic reduction of a mixture of chromium ore and quartz analogously to the production of ferrosilicon. In this alloy the solubility of carbon is small, as shown in Fig. 14-2, which gives the carbon solubility in iron, ferromanganese, and ferrochromium as function of the silicon content. The further process may differ between plants. One procedure is illustrated in Fig. 14-3. Here ferro–chromium–silicon is produced in one furnace.

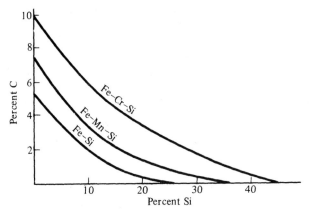

Figure 14-2 Carbon solubilities in ferroalloys (approximate).

In another furnace chromium ore is melted with lime to give a slag rich in Cr_2O_3. This again is reacted with the Fe–Cr–Si alloy in a system of two ladles in counter-current, whereby an effort is made to produce a final alloy low in silicon and a waste slag low in chromium.

The silicothermic reduction is exothermic, the heat of reaction being more than enough to cover heat losses from the ladle. The reduction $2CrO + Si =$

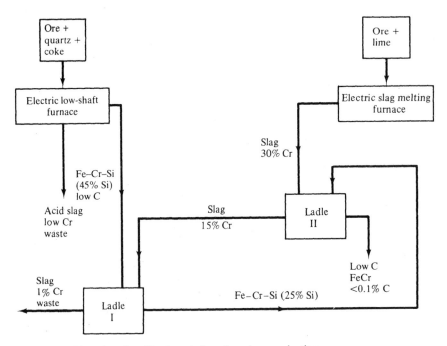

Figure 14-3 Flow sheet for silicothermic ferrochromium production.

$SiO_2 + 2Cr$ is favored by low temperature, and some cold metal may be added to keep the temperature just above the melting point of the alloy. Also it is favored by a basic slag. A similar process may be used for the production of low-carbon ferromanganese.

Ferrotitanium, ferrovanadium, and ferroniobium and also 99 percent chromium metal may be produced by *aluminothermic* reduction. With aluminum the reduction is even more exothermic than with silicon, and under favorable conditions the reaction may be done autogeneously, i.e., by igniting a mixture of cold oxide and aluminum powder. If the heat of reaction is not sufficient to give a molten product, the raw materials may be preheated to, for example, 500°C before they are ignited, or some compound is added to the mixture which increases the heat of reaction. As such a "booster" some higher oxide, for example CrO_3 or some oxygen-rich salt, may be used. Sometimes lime is added to lower the melting point of the slag.

Compared to carbothermic reduction where gaseous CO is formed the slag volume in metallothermic processes is large. As a result the total loss of valuable metal in the slag may be large even though its concentration is relatively small. Also it is not possible to produce a metal completely free of silicon or aluminum.

14-3 VACUUM REDUCTION

As already mentioned, the reaction between carbide and oxide is favored by low pressure, and in recent years considerable progress has been made on vacuum reduction and refining of reactive metals. Even after elimination of the carbide phase some carbon may remain dissolved in the metal. Since the solubility is less in solid than in liquid metals, the reaction is most favorable just below the melting point of the metal. One example is the conversion of high-carbon ferrochromium into a low-carbon product. Here the high-carbon alloy is produced by conventional methods. It is crushed and mixed with a suitable oxidant which may be silica sand, or it is oxidized slightly on the surface to give Cr_2O_3. The powder is now briquetted and heated to about 1375°C under a vacuum of about 1 mmHg. The reactions occurring are

$$Cr_{23}C_6 + 6SiO_2 = 23Cr + 6SiO + 6CO$$

alternatively

$$Cr_{23}C_6 + 2Cr_2O_3 = 27Cr + 6CO$$

In both cases a product with less than 0.01 percent carbon is obtained. For the first case the product contains about 6.5 percent silicon, whereas for the second case the silicon content is about 1.5 percent.[1]

---

[1] C. G. Chadwick: *J. Met.*, **13**: 806 (1961).

A similar process is employed for the production of pure *niobium*. Here pure oxide is reduced with carbon to give niobium carbide. This again is mixed with stoichiometric amounts of niobium oxide and heated in high vacuum to temperatures above 2000°C to give a product of low carbon content.

Vacuum reduction may also be carried out in one stage, i.e., by heating a mixture of oxide and carbon in vacuum. In this way almost pure vanadium has been produced by reaction between a mixture of V_2O_3 and carbon at 1700°C and under a vacuum of 10^{-4} mmHg. Carbon reduction gives a rather large gas volume, however, which requires a high pumping capacity. An alternative reducing agent would be silicon which would react according to the scheme

$$MeO + Si = Me + SiO(g)$$

Volatile SiO evaporates from the hot zone, but condenses in a cold trap before it reaches the vacuum pump. In this way a high vacuum may be maintained with a relatively small pumping capacity. It is probable that some pure niobium is being produced commercially by this method.

14-4 PRODUCTION OF PURE METALS

It is almost as difficult to define "pure" as it is to define "rare" metals. Some metals are sufficiently pure by normal production methods. For the more reactive metals small amounts of impurities such as carbon, nitrogen, and oxygen may affect the properties and usefulness of the metal. For the lack of a better definition we can say that a pure metal is one where the harmful effects of the impurities have been virtually eliminated. Whereas the more noble metals such as iron, copper, etc., may be obtained relatively pure by refining of crude metals, this cannot be done for the more reactive metals, the impurities usually being less reactive than the metal itself. Instead it is necessary to start with a pure raw material. This may be obtained by various hydrometallurgical methods. The ore is leached, the solution is purified, and a pure metal compound is precipitated by methods which will be discussed in Chap. 15. In other cases an impure ferroalloy may serve as the raw material for the production of a pure metal compound. In both cases the pure metal compound is again reduced to give the pure metal. Figure 14-4 illustrates a number of paths which may be followed in the production of relatively pure reactive metals from impure raw materials.

As already mentioned pure *oxides* may be reduced with carbon or silicon in vacuum, or with aluminum, magnesium, or calcium at atmospheric pressure. Oxides of the more noble metals such as molybdenum and tungsten may also be reduced with hydrogen in a tube furnace. A special complication in the reduction of oxides is the solubility of oxygen in the metal. This is illustrated by Fig. 14-5, which shows the relative partial Gibbs energy of oxygen in the binary

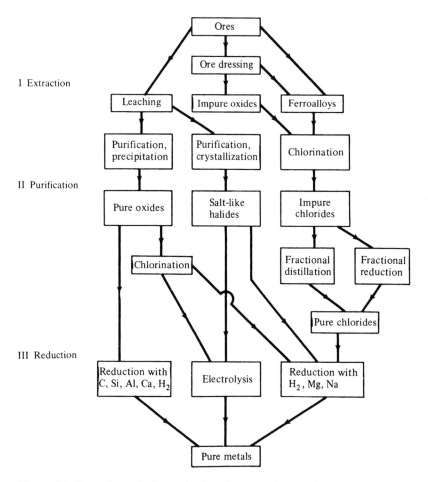

Figure 14-4 Alternative paths for production of pure reactive metals.

titanium–oxygen and zirconium–oxygen systems at about 1000°C. On the left-hand scale is given the oxygen potential in kilojoules, on the right-hand scale the corresponding gas ratio p_{H_2O}/p_{H_2} as well as the oxygen potentials for reduction with Ba, Al, Mg, and Ca. We see that the oxide phases as such may be reduced at oxygen potentials around -800 kJ, but the resulting metal will contain considerable amounts of oxygen in solid solution, the last traces of oxygen having potentials which are even less than that of the Ca–CaO couple. The physical properties, e.g., the ductility, of rare metals depend greatly on a low oxygen content, and in some cases it is impossible to obtain a sufficiently pure product by reduction of the oxide. For such cases the metal may more easily be obtained from its halides.

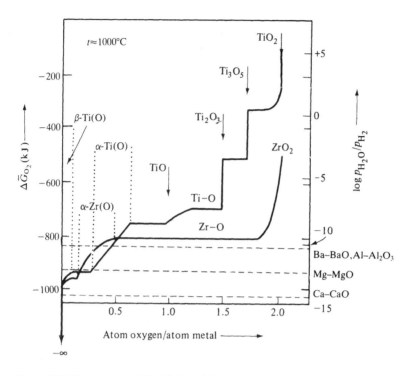

Figure 14-5 Oxygen potential in Ti–O and Zr–O systems around 1000°C. Dashed lines = potentials for reduction with Ba, Al, Mg, and Ca. Dotten lines = composition ranges for solid solutions of oxygen in α-Zr and in α- and β-Ti. (*Data from O. Kubaschewski: Bull. Inst. Min. Met., vol. 1 (1956); O. Kubaschewski and W. A. Dench.: J. Inst. Met., vol. 84, p. 440 (1955–1956)*.)

14-5 HALIDE METALLURGY

Halogens, in particular chlorine, are used in many metallurgical processes. In Sec. 8-3 the use of chlorine for the purification of certain ores was discussed. Chlorine is also used in the refining of some metals where it reacts with impurities to form condensed or volatile chlorides. Furthermore, some metals such as magnesium and sodium are prepared by electrolysis of a chloride melt. In this section the preparation of rare metals by *chemical* reduction of their halides will be discussed.

The reasons for the use of halides in the production of reactive metals may be listed as follows:

1. High vapor pressure makes possible purification by distillation
2. Low solid solubility in metal makes possible production of pure metals

In addition comes for some metals:

3. Aqueous solubility makes possible purification by crystallization
4. Low melting point and high electrical conductivity makes possible electro-lytic reduction

From the thermodynamic curves given in App. C we see that the Gibbs energy of formation of chlorides more or less follows the same sequence that applies for the oxides, but shows breaks which result from the melting and boiling points of the chlorides. An exception is carbon tetrachloride which is much less stable than CO and CO_2. Carbon, therefore, cannot be used as a reducing agent for chlorides. Also we see that, whereas the oxides of aluminum and calcium are considerably more stable than those of the alkali metals, the opposite is the case for the chlorides. Hydrogen chloride becomes increasingly stable with increasing temperature, and hydrogen will reduce many metal chlorides at high temperature.

The properties of the other halides resemble those of the chlorides, but they are less known. The stability decreases in the order: fluoride, chloride, bromide, iodide. Metal iodides are relatively unstable, and several of them decompose completely at high temperature.

In the presence of water or water vapor many halides are easily hydrolyzed

$$MeCl_2 + H_2O = MeO + 2HCl$$

Thus the tetrachlorides of tin and titanium hydrolyze to give a white oxide smoke on contact with moist air. Other chlorides, like those of iron and nickel, may be converted to oxides by reaction with steam at elevated temperature. Most resistant toward hydrolysis are the alkali halides which may be crystal-lized from aqueous solutions and heated in the presence of water vapor without reaction.

Some halides may also react with oxygen

$$MeCl_2 + O_2 = MeO + Cl_2$$

This reaction is more sluggish than the hydrolysis, and most halides are resis-tant toward dry air at room temperature.

Preparation of Halides

Anhydrous halides may be prepared by (1) crystallization from aqueous solu-tions, (2) halogenation of oxides, and (3) halogenation of ferroalloys.

Crystallization from aqueous solutions is mainly used for alkali halides. Most other halides are partly hydrolyzed in water. Thus $MgCl_2 \cdot 6H_2O$ may easily be crystallized from an aqueous solution. On heating, the crystal water may be expelled down to a composition $MgCl_2 \cdot 1.5H_2O$. On further heating this hydrolyzes, viz.,

$$MgCl_2 + H_2O = MgO + 2HCl$$

Thus it is not possible to obtain anhydrous magnesium chloride by simple heating of the hydrated salt. If the heating is done in a flow of HCl gas the anhydrous chloride may be obtained, however.

Certain complex rare-metal salts may be obtained directly from aqueous solutions: K_2TiF_6, K_2NbF_7, and $(NH_4)_2BeF_4$. The first two salts have been suggested as possible raw materials for the production of titanium and niobium. The third salt will, on heating, give off volatile NH_4F and produce anhydrous BeF_2.

Halogenation of oxides is the most common method used for preparation of anhydrous halides, in particular chlorides. In order to obtain a favorable equilibrium the oxide is mixed with a suitable reducing agent, usually carbon, but sulfur or sulfides may also be used. Examples:

$$2MgO + 2Cl_2 + C = 2MgCl_2(l) + CO_2$$

$$TiO_2 + 2Cl_2 + C = TiCl_4(g) + CO_2$$

The above reactions are exothermic, and once started they may take place autogeneously. Also the equilibrium is most favorable at low temperature, and a temperature of 700 to 1000°C, which is necessary to obtain a good reaction rate, is used.

The reactions may be carried out in a shaft furnace (Fig. 14-6). Briquettes of oxide and carbon are charged to the top of the furnace, and chlorine is introduced at the bottom. Heat to start the reaction and to maintain a suitable temperature is generated electrically in a coke bed which acts as an electrical resistor. Of the two examples given above $MgCl_2$ is formed as a liquid, and is tapped from the bottom of the furnace, whereas $TiCl_4$ is obtained as a gas which follows the furnace gases and may be recovered by condensation. Chlorination of TiO_2 is usually carried out in a *fluidized-bed* reactor.

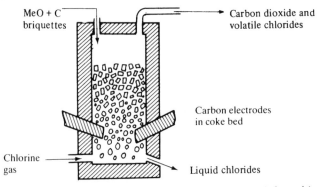

Figure 14-6 Shaft furnace for chlorination of metal oxides (schematic).

It is important that the raw materials are completely dry. If not, water will react to give HCl, causing a loss of chlorine. Also water vapor may cause hydrolysis of the metal chlorides.

One complication in the chlorination of oxides is the possible formation of oxychlorides, such as $WOCl_4$, $VOCl_3$, and $NbOCl_3$. Oxychlorides are often volatile and they are not easily separated from the chlorides. If such an oxygen-contaminated chloride is reduced to metal, oxygen will enter and contaminate the product.

Halogenation of ferroalloys or carbides would make it possible to produce a halide virtually free of oxyhalide. The reaction between a ferroalloy and, e.g., chlorine is strongly exothermic, and once started the reaction will proceed without additional heating. The resulting chloride will be impure and will need further purification to remove iron, etc.

Purification of Halides

Water soluble halides may be purified by recrystallization or reprecipitation by methods which will be discussed in Chap. 15. Also solvent extraction or ion exchangers may be used.

Volatile chlorides are most easily purified by *fractional distillation* by methods which were discussed in Chap. 10. As the boiling points are low, the distillation may be carried out in glass columns. Even chlorides as similar as those of niobium and tantalum may be separated. As an example[1] a chloride mixture with 67.2 percent $NbCl_5$ (b.p. 254°C), 6.0 per cent $TaCl_5$ (b.p. 239°C), 26.4 percent $FeCl_3$ (b.p. 315°C), and 0.4 percent $WOCl_4$ (b.p. 228°C) was separated by distillation in two columns into a niobium fraction with 99.98 percent $NbCl_5$ and 0.01 percent $TaCl_5$, a tantalum fraction with 93.2 percent $TaCl_5$, 0.6 percent $NbCl_5$ and 6.2 percent $WOCl_4$, and an iron fraction with 98.7 percent $FeCl_3$, 1.2 percent $NbCl_5$, and 0.06 percent $TaCl_5$.

Chlorides may also be purified by *fractional reduction*. By careful reduction of an impure niobium chloride gas with hydrogen at 350°C, $FeCl_3$ is reduced to $FeCl_2$ which condenses as a solid. By further reduction with hydrogen at 500°C, $NbCl_5$ is reduced to solid $NbCl_3$, whereas $TaCl_5$ is unreduced and is carried away with the gas flow.[2]

Reduction of Halides

Hydrogen reduction Hydrogen may be used to reduce the solid and less stable chlorides, such as VCl_3 or $CrCl_3$. The reduction may be carried out in an ordinary tube furnace. One complication is that the chlorides may either be

[1] B. R. Steele and D. Geldart: in "Extraction and Refining of the Rarer Metals," The Instn. of Min. and Met., London, 1957, p. 287.

[2] A. B. McIntosh and J. S. Broadley: in "Extraction and Refining of the Rarer Metals," The Instn. of Min. and Met., London, 1957, p. 272.

partly volatile or they may disproportionate to form volatile products. The volatiles will follow the gas flow, and have to be condensed and returned to the process.

At higher temperatures hydrogen may reduce also the more stable chlorides. As an example, it has been suggested to produce titanium by passing a mixture of H_2 and $TiCl_4$ into an electric arc. Titanium would be formed as a molten product, and the gas could be treated to remove HCl and returned to the process. This method has not found any industrial application.

Metallothermic reduction This is the most common method used for reduction of halides of the reactive metals. As reducing agent metallic magnesium, aluminum, or calcium is used. If the halide is nonvolatile the reduction may be carried out like a metallothermic reduction of oxides. Example:

$$BeF_2 + Mg = Be + MgF_2$$

The reduction is carried out in a graphite crucible with the addition of some calcium chloride which acts as a flux. The reaction is started at 650°C, and the temperature is gradually increased to 1300°C, which is necessary to melt and separate the products.

Reduction of the more volatile halides is done in a closed reactor, and in an atmosphere of helium or argon. Figure 14-7 shows the principle of the *Kroll reactor*, which is used for commercial production of titanium and zirconium. The reactor is made of steel, and is heated externally until the charge of magnesium ingots melts. The halide is then introduced. Solid halides such as VCl_3 or $ZrCl_4$ are introduced through a feeding mechanism, whereas liquid halides

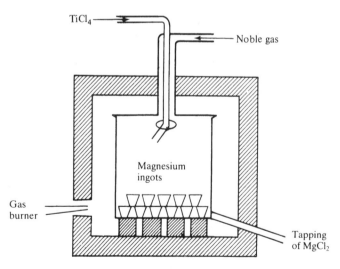

Figure 14-7 Kroll reactor (schematic). (*After H. C. Fuller, et al.: U.S. Bur. of Mines Rep., Inv. No. 4879, U.S. Department of the Interior, (1952)*.)

such as $TiCl_4$ are fed in through a pipe. Instead of magnesium, sodium may be used as reducing agent.

The reduction, for example, $TiCl_4 + 2Mg = Ti + 2MgCl_2$, is exothermic and produces enough heat to maintain a temperature of 800 to 1000°C. Higher temperatures are harmful, as they may lead to reaction with the steel container as well as to incomplete reduction. Solid titanium is deposited as a sponge on the reactor walls and is impregnated with molten magnesium chloride. After completion of the reaction a tap-hole is opened, and most of the chloride melt may be drained off. After cooling of the reactor the solid content is removed in a lathe, and remaining magnesium metal and chloride may be removed from the Ti sponge by vacuum distillation.

Theoretically the process should produce anhydrous magnesium chloride equivalent to the amount of magnesium metal consumed, and this chloride may be reelectrolyzed to produce magnesium and chlorine, of which the latter again may be used to produce more titanium tetrachloride. Thus we may say that magnesium acts as an intermediate carrier of energy from the electrolytic cell to the Kroll reactor. In practice there will be losses both of magnesium and chlorine.

The batchwise procedure and the elaborate separation of the products make the Kroll process relatively expensive. During the years a number of modifications have been suggested to make the process continuous. To the knowledge of the author none of these have been applied industrially, and the Kroll process seems still to be without serious competitors.

Electrolytic reduction of halides will be discussed in Chap. 16.

14-6 MELTING OF METAL SPONGE

In the Kroll process and some other reduction processes the high-melting metal is obtained as a sponge. In some cases this may be rolled to give a compact metal. In most cases, however, we want to melt the metal and cast it into ingots. This is particularly the case if we also want to add some alloying elements. Since the metal is very reactive, it must not come into contact with oxygen, nitrogen or carbon during the melting process. Also, ceramic refractories may not be used in most cases. Melting may then be done in a cooled crucible, usually of copper, the principle of which is shown in Fig. 14-8.

In earlier operations a *water-cooled* crucible was used. This had the disadvantage that if the molten metal should break through the copper wall there would be an explosive, exothermic reaction

$$Ti + 4H_2O = Ti(OH)_4 + 2H_2$$

Furthermore, the hydrogen released would burn in air. In modern practice, therefore, a *sodium-potassium* eutectic alloy, which is liquid at room temperature, is used as the cooling medium. It is interesting to notice that alkali

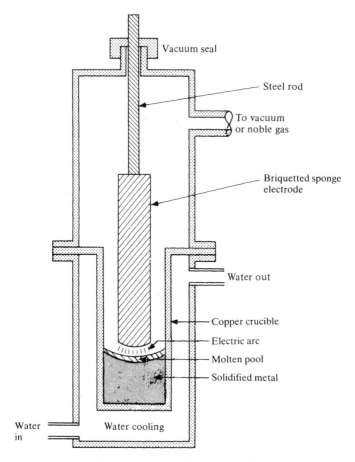

Figure 14-8 Principle of vacuum or noble gas arc melting of reactive metals.

metals, which normally are regarded as hazardous, here are used to avoid explosion and fire hazards.

The furnace is either filled with a noble gas, or it is evacuated. In the latter case we also get a degassing of the melt. The electrodes is usually made of compressed metal sponge, which is consumed continuously during melting, or nonconsumable tungsten electrodes may be used. In the latter case briquetted sponge, with alloying elements added, is fed continuously into the furnace. As the molten pool builds up, the electrode is raised slowly, the melt freezes in the bottom, and finally a solid ingot is obtained. In this way ingots of several tonnes may be produced.

In order to produce a homogeneous product this first ingot is usually made electrode to a second crucible where it is remelted to give a second ingot. This second ingot is cleaned for the half-melted surface layer, and is forged or rolled into products of various shapes.

In recent years *electron beam melting* has come up as an alternative to arc melting. An electron beam furnace is based on the same principle as an x-ray tube. The furnace is a closed chamber and is evacuated to 10^{-4} mmHg or better. A glowing tungsten filament acts as the cathode, emitting electrons. The anode is a compressed or sintered bar of metal sponge as well as the pool of molten metal, the latter being contained in a water-cooled copper crucible. When a dc source of 10 to 50 kV is applied to the furnace the electrons which are emitted from the cathode accelerate and hit the anode with great energy, and temperatures in excess of 3000°C may be obtained. In this way it has been possible to melt, among other metals, niobium, which has a melting point around 2500°C. The combination of high temperature and low pressure also makes possible an extensive purification of the metal. Thus a niobium sponge which originally contained 207 ppm of carbon, 2750 ppm of oxygen, and 580 ppm of nitrogen, gave after electron beam melting a product with 35 ppm carbon, 106 ppm oxygen, and 73 ppm nitrogen.[1] Notice that the decrease in oxygen is larger than the amount removed stoichiometrically by the reaction $\underline{C} + \underline{O} = CO(g)$, and that some oxygen probably evaporates as a volatile niobium suboxide. Also metallic impurities, such as iron, may evaporate in the electron beam furnace.

Electron beam furnaces have been built with power up to 1000 kW and there seems no reason why they cannot be built larger.

14-7 TRANSPORT REACTIONS

Whereas nonvolatile metals cannot be separated by ordinary distillation methods such separation may in some cases be possible by the use of a third element or compound which forms volatile compounds with the metal in question. Typical examples of such *transport reactions* or *catalytic distillations* are

$$Ni(s) + 4CO = Ni(CO)_4(g) \tag{1}$$

$$2Al(l) + AlCl_3(g) = 3AlCl(g) \tag{2}$$

$$Ti(s) + 2I_2 = TiI_4(g) \tag{3}$$

The standard Gibbs energies for these reactions are shown in Fig. 14-9 as function of temperature. For reactions (1) and (3) the temperature coefficient of the Gibbs energy is seen to be positive, corresponding to a negative entropy of reaction. This is a result of a *decrease* in the number of gas molecules during the reaction. For reaction (2) there is an *increase* in the number of gas molecules, corresponding to a positive entropy of reaction. The equilibrium constants for these reactions will, therefore, change strongly with temperature, which makes them suitable for metal refining purposes.

[1] J. A. Belk: "Vacuum Techniques in Metallurgy," Pergamon Press, London, 1963, p. 151.

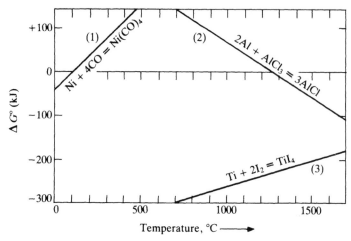

Figure 14-9 Standard Gibbs energies of some typical transport reactions.

In the *Mond process* for refining of nickel, impure nickel is reacted with CO gas at about 50°C. This causes volatile nickel carbonyl to be formed, whereas the impurity metals do not form carbonyl under these conditions. The formation of nickel carbonyl is exothermic, and once the reaction has been started the temperature of the reactor must be controlled by means of water-cooling. The gas is passed on to another reactor which is kept at about 230°C, and which contains metallic nickel seeds. Here the carbonyl dissociates, solid nickel is precipitated on the seeds, and the gas is recycled to the first reactor.

In the above case the process was carried out at atmospheric pressure. By increasing the pressure nickel carbonyl will be formed also at higher temperatures. Thus at 100°C the equilibrium constant for reaction (1) is about unity, which for a total pressure of, say, 18 atm gives 16 atm of $Ni(CO)_4$ and 2 atm of CO, that is, a 97 percent conversion of CO to $Ni(CO)_4$. In addition, the rate of reaction increases with increasing temperature. High pressures are, therefore, used in modern versions of the carbonyl process.

The *subchloride process* for aluminum, reaction (2), was explored as a possible way to produce pure aluminum from an impure aluminum–silicon–iron alloy formed by carbothermic reduction of bauxite (Sec. 14-1).[1] In this process the impure alloy is reacted with $AlCl_3(g)$ at about 1200°C, whereby $AlCl(g)$ is formed. On cooling of the subchloride to about 700°C the reaction reverses, pure aluminum condenses, and volatile $AlCl_3$ is re-formed and is returned to the process. The process has been tested on a pilot scale, but has not yet found any industrial application.

Reaction (3) above forms the basis for the *van Arkel–de Boer method* for production of high-purity titanium. The same principle has also been used for

[1] P. Gross et al.: "Catalytic distillation of aluminium" in "The Refining of Non-Ferrous Metals," Instn. of Min. and Met., London, 1950.

zirconium and chromium. As is seen from Fig. 14-9 the Gibbs energy of reaction (3) is negative in the entire temperature range of interest. This does not mean that the reaction cannot be reversed at high temperatures. In the same way as the formation of nickel carbonyl is favored by high pressures the dissociation of titanium tetraiodide is favored by low pressure in the reaction vessel. Furthermore, some of the I_2 molecules dissociate to monatomic iodine at high temperature, which enhances the reaction.

The principle of the method is illustrated in Fig. 14-10. Impure metal is placed along the walls of the glass container. In the middle of the container a filament of thin tungsten wire or of the metal in question is suspended, and may be heated electrically. A small amount of iodine is introduced, and the container is evacuated and sealed. On heating to a few hundred degrees centigrade iodine reacts with the crude metal, forming volatile iodides. When these hit the filament, which is heated to 1000 to 1500°C, they are again decomposed and the metal precipitates on the filament, whereas the iodine is liberated and may once more react with the crude metal.

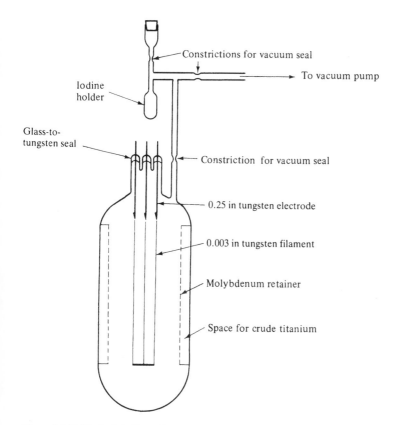

Figure 14-10 Typical de Boer deposition bulb. (*From I. E. Campell et al.: J. Electrochem. Soc., vol. 93, p. 271 (1948)*.)

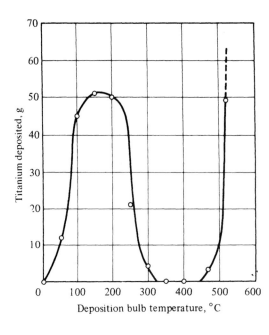

Figure 14-11 Titanium deposition as function of bulb temperature. (*From J. D. Fast: Z. anorg. Chem., vol. 241, p. 42 (1939)*.)

In the refining of impure titanium, first the gaseous TiI_4 is formed. Above 200°C this reacts with additional metal to give (probably) solid TiI_3. This is less volatile, but disproportionates to Ti and TiI_4 above 400°C. The rate of reaction therefore varies with the temperature of the crude metal as shown in Fig. 14-11. In order to have a high reaction rate the crude metal should have a temperature either of about 200°C or above 500°C. The higher temperature is mostly used. The method has found application for the production of high-purity metals for research purposes, but is too expensive for industrial use.

It will be noted that in the Mond and the van Arkel–de Boer processes the catalytic distillation takes place from low to high temperature. The same principle is used to improve the lifetime of electric light bulbs. Light bulbs usually burn out because of a hot spot formed on the tungsten wire. From this hot spot some tungsten will evaporate, which makes the wire thinner at that point. This again results in an increased electrical resistance, and an even higher temperature, which eventually will cause the wire to break. By the addition of small amounts of iodine and oxygen to the argon atmosphere of the bulb some volatile WO_2I_2 will be formed. As the formation of this compound is exothermic it is formed at low temperature but decomposes at high temperature. If a hot spot should develop on the tungsten wire, this will immediately lead to deposition of some tungsten on that spot until uniform wire thickness and temperature again are established.[1] As a consequence a higher temperature of the tungsten filament may be used, with a correspondingly higher yield of light, than for iodine-free bulbs.

---

[1] J. W. Hastie: "High Temperature Vapors," Academic Press, New York, 1975, pp. 204–206.

PROBLEMS

14-1 Calcium carbide is produced by the reaction $CaO + 3C = CaC_2 + CO$.

(a) By means of data in App. C calculate the temperature where the CO pressure is 1 atm, assuming that CaO, C, and CaC_2 are all present in their standard states.

(b) Calculate the corresponding pressure of Ca(g).

(c) Actually a molten eutectic mixture of CaC_2 and CaO is formed. Discuss the effect of temperature on the composition of this mixture as well as on the Ca losses by evaporation, assuming equilibrium with graphite and $p_{CO} = 1$ atm.

14-2 Vacuum refining of ferrochromium may be carried out by the reaction $Cr_{23}C_6 + 2Cr_2O_3 = 27Cr + 6CO$. Calculate the CO pressure at 1375°C if $Cr_{23}C_6$ and Cr_2O_3 occur in their standard states and the chromium activity in the alloy is 0.7.

14-3 On the basis of Fig. 14-5 calculate the activity of titanium as function of composition in the Ti–O system, and calculate the standard Gibbs energy of formation of the stoichiometric compounds TiO, Ti_2O_3, Ti_3O_5, and TiO_2. The activity of titanium at low oxygen contents is assumed to follow Raoult's law. *Note:* $d\Delta\bar{G}_{O_2} = 2\, d\Delta\bar{G}_O$.

14-4 From data given in the appendix discuss the "subhalide process" for aluminum refining. Estimate the composition of the equilibrium gas phase at 1200 and 700°C for $p_{tot} = 1$ atm, and estimate the theoretical heat requirements of the process in kilowatthours per tonne of aluminum refined.

BIBLIOGRAPHY

Dennis, W. H.: "Metallurgy of the Non-Ferrous Metals," Sir Isaac Pitman and Sons Ltd., London, 1961.
Durrer, R., and G. Volkert: "Metallurgie der Ferrolegierungen," 2d ed., Springer Verlag, Berlin, 1972.
"Extraction and Refining of the Rarer Metals," The Instn. of Min. and Metal., London, 1957.
The Instn. of Metallurgists, London: "Purer Metals," Iliffe Books Ltd., London, 1962.
Kroll, W. J.: The Pyrometallurgy of Halides, *Metall. Rev.*, **1** (8): 291–337 (1956).
McQuillan, A. D., and M. K. McQuillan: "Titanium," Butterworths, London, 1956.
Miller, G. L.: "Zirconium"; "Tantalum and Niobium," Butterworths, London, 1957; 1959.
Rolsten, R. F.: "Iodide Metals and Metal Iodides," John Wiley and Sons, Inc., New York, 1961.
"The Refining of Non-Ferrous Metals," The Instn. of Min. and Metal., London, 1950.
Zelikman, A. N., O. E. Krein, and G. V. Samsonov: "Metallurgy of Rare Metals," Transl. from Russian, Israel Program for Scientific Translations, Jerusalem, 1966.

FIFTEEN

HYDROMETALLURGY

By hydrometallurgy we understand processes for the isolation and winning of metals by the use of water or aqueous solutions. This covers a large variety of processes ranging from the leaching of ores or roasted sulfides through the purification of the solutions to the winning of the metals or their compounds by chemical or electrochemical precipitation. Electrolytic refining of impure metals may sometimes be classified as hydrometallurgy. In the present chapter only *chemical processes* will be considered, whereas electrolytic processes, be it winning (reduction) or refining, will be treated in the subsequent chapter on electrometallurgy.

In recent years hydrometallurgy has been supplemented with the use of other types of solvents, such as organic liquids, as well as with the use of solid organic resins (ion-exchangers). A short survey of the use and potentialities of such organic compounds will be given toward the end of this chapter.

15-1 THERMODYNAMICS OF AQUEOUS SOLUTIONS

The same laws of thermodynamics apply to aqueous solutions as to the types of solutions previously discussed: molten metals and slags. Certain features are characteristic for aqueous solutions, however, and will be discussed briefly.

For aqueous solutions of organic compounds, such as cane sugar, or of simple inorganic gases, such as oxygen and nitrogen, the solute is present as *molecules*. In these cases the solute will on increasing dilution follow Henry's law and the solvent will follow Raoult's law. For the large majority of inorganic substances, the so-called electrolytes (acids, bases, and salts), Henry's law is not obeyed, however. This will best be illustrated by an example:

The gas HCl is soluble in water up to 41 weight percent at 25°C, that is, the partial pressure of HCl reaches one atmosphere at that concentration. If we measure the HCl pressure, i.e., the chemical activity of HCl relative to the gas of one atmosphere, we will find that it falls off very rapidly with decreasing concentration. At low concentration we would find that it approaches the equation

$$p_{HCl} = K(HCl)^2$$

Here the parentheses give the concentration of HCl in some convenient unit. This relation is analogous to Sieverts' law which applies for the solution of diatomic gases in metals. This leads to the concept of *electrolytic dissociation*, $HCl(g) = H^+ + Cl^-$, where H^+ is the positively charged cation and Cl^- the negatively charged anion. For the formation of HCl from these ions we define an equilibrium constant

$$K = \frac{p_{HCl}}{a_{H^+} \cdot a_{Cl^-}}$$

where a_{H^+} and a_{Cl^-} are called the activities of the ions relative to some suitable standard state. At infinite dilution these activities become proportional to the corresponding ionic concentrations, and, since for the solution of HCl in pure water the concentrations of H^+ and Cl^- ions are equal and equal to the concentration of HCl, the Sieverts' law relation is obtained. Similarly for an aqueous solution of H_2SO_4 the chemical activity of the acid is at infinite dilution found to be proportional to the cube of the concentration, and so on for other electrolytes.

Traditionally the unit of concentration used for electrolytes is the *molality*, i.e., the number of moles per kilogram of water. In the same way as we established a standard state for dilute solutions in liquid steel according to which the activity becomes equal to the weight percentage at infinite dilution, we define the standard state for the ions so that their activity is equal to their molality at infinite dilution. For higher concentrations deviations occur which may be described by activity coefficients

$$a_{H^+} = m_{H^+} \gamma_{H^+} \qquad a_{Cl^-} = m_{Cl^-} \gamma_{Cl^-}$$

Here m denotes the molality of the ions and γ their activity coefficient, the latter being analogous to the activity coefficient, f, used in liquid steel. For a solution of HCl in pure water $m_{H^+} = m_{Cl^-} = m_{HCl}$, and we get

$$p_{HCl} = K(m_{HCl})^2 \gamma_{H^+} \gamma_{Cl^-}$$

We have here one equation with two unknowns, the two activity coefficients. It is in principle impossible to evaluate one of these independently of the other. Therefore, by *convention*, it is agreed to use the mean activity coefficient, defined by $\gamma_{\pm} = \sqrt{\gamma_+ \gamma_-}$, which gives $p_{HCl} = K(m_{HCl})^2(\gamma_{\pm})^2$.

400 PRINCIPLES OF EXTRACTIVE METALLURGY

This means that we cannot determine the true activity of the individual ionic species, but only that of the neutral compound. The product $m_{H^+}\gamma_\pm$ expresses a purely conventional activity.

By a second convention we agree to define the *chemical activity of the neutral compound* in solution equal to the product of the ionic activities

$$a_{HCl} = a_{H^+}a_{Cl^-} = m_{H^+}m_{Cl^-}(\gamma_\pm)^2$$

where for solution in pure water $m_{H^+} = m_{Cl^-} = m_{HCl}$ and $\lim \gamma_\pm = 1$ for $m_{HCl} \to 0$. This convention does not imply that the solution contains some dissociated and some undissociated HCl molecules. On the contrary, it is most likely that the dissociation of HCl and of other so-called *strong electrolytes* is almost 100 percent complete.

The relation between electrolytic concentrations, activities, and activity coefficients may conveniently be illustrated by Fig. 15-1 which shows the "activity" of H^+ and Cl^-, called $a_\pm$, as well as of HCl and also the mean activity coefficient as functions of HCl concentration. We see that for infinite dilution $a_\pm = m_{HCl}$ and $a_{HCl} = (m_{HCl})^2$, whereas at higher concentrations deviations occur. For a 1 molal solution the activities a_{H^+} and a_{Cl^-} are equal to 0.809, corresponding to $\gamma_\pm = 0.809$, and the activity of HCl = 0.655. An activity of unity for H^+, Cl^-, and HCl is obtained for a real solution of about 1.2 molal.

On a separate scale is shown the partial pressure of HCl gas above the solution. This decreases from 1 atm for a 19.2 molal (41 weight percent) solution to less than 10^{-12} atm for a 10^{-3} molal solution.

Thus we may summarize this survey of the thermodynamics of aqueous solutions of strong electrolytes:

1. The activity of the ions is taken equal to their molality multiplied by an activity coefficient, the latter going to unity when the concentration goes to zero
2. The activity coefficient of the positive and the negative ions are by convention taken equal, and equal to the mean activity coefficient
3. The activity of the neutral compound is expressed as the product of the activities of the ions, each raised to the power which corresponds to their number in the compound

For most electrolytes of moderate concentration $\gamma_\pm$ is between 1.0 and 0.1. This means that the *concentration* of a given ion may be up to ten times the corresponding activity. At higher concentrations $\gamma_\pm$ increases beyond unity (Fig. 15-1).

So far we have only discussed solutions of one single electrolyte, for example, HCl in water. When, in addition, there are other electrolytes, for example, $NaCl$ or HNO_3, the total Cl^- or H^+ concentration is given by the sum of the ions in the various electrolytes. In such a case the activities of H^+ and Cl^- are no longer equal, but the relation $a_{HCl} = a_{H^+}a_{Cl^-}$ still applies. Thus we see that by addition of either $NaCl$ or HNO_3 to a given HCl solution the activity of HCl will increase. Conversely, the activity of NaCl increases when either HCl or

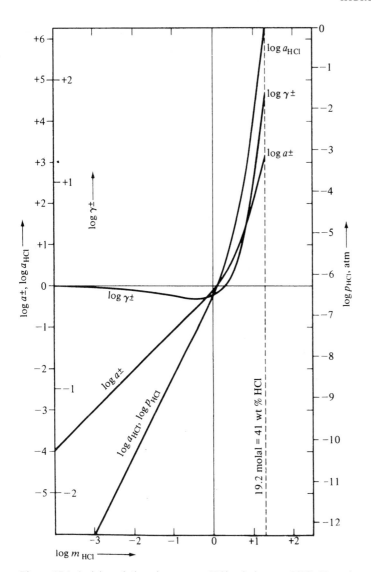

Figure 15-1 Activity relations in aqueous HCl solutions at 25°C. (*Data from G. N. Lewis and M. Randall: "Thermodynamics," revised edition by K. S. Pitzer and L. Brewer, McGraw-Hill Book Co. Inc. New York, 1961.*) Notice that the scale of log $\gamma_\pm$ is 2.5 times larger than for the other quantities.

a sodium salt, such as $NaNO_3$, is added to the solution. Under certain conditions this increase may be so large that the solution becomes saturated with respect to NaCl.

We mentioned that for all strong electrolytes the concentration of undissociated compound is practically zero. This is not the case, however, for the so-called *weak electrolytes*, such as carbonic and acetic acids. Here the concentration of the undissociated compound may be appreciable, although also in these cases the degree of dissociation becomes infinite at infinite dilution.

The weak electrolyte which is of utmost interest is water itself. Water dissociates according to the reaction $H_2O = H^+ + OH^-$. (The fact that H^+ is tied to water to give the H_3O^+ ion is of no consequence for this discussion.)

The dissociation constant is very small, about 10^{-14} at 25°C. This means that for pure water the concentration or activity of both H^+ and OH^- ions is 10^{-7} molal. If HCl or some other acid is added to water the activity of H^+ increases and the activity of OH^- decreases. The opposite effect is obtained by the addition of some base, for example, NaOH. Since the activity of water as such in all cases is essentially unity, the relation between the H^+ and OH^- activities is given by $a_{H^+} a_{OH^-} = 10^{-14}$. Thus the activity of H^+ will vary between unity for an acid of unit activity and 10^{-14} for a base of unit activity, and may exceed these limits for more strongly acid or basic solutions.

The hydrogen ion activity is conveniently expressed by its negative logarithm, which is known as the pH of the solution. Thus $pH = -\log a_{H^+}$. The pH of an aqueous solution is one of its most important properties and is of great importance for its behavior in hydrometallurgy. The fact that the true pH of a solution, like any other single-ion activity, cannot be measured, but has to be approximated by some standardized measuring technique need not worry us at this stage.

15-2 IONIC REACTIONS

In an aqueous system the ions may undergo reactions of various kinds. These may be classified in the following groups:

1. Reactions due to changes in pH (hydrolysis)
2. Reactions due to changes in oxygen potential (red-ox)
3. Complex formation
4. Precipitation of compounds

The two first groups of reactions apply to solutions where different ions do not interfere with each other, and will be discussed first.

Hydrolysis

By hydrolysis we understand reactions which involve water or its ions. Typical examples are

$$Al^{3+} + 3H_2O = Al(OH)_3(s) + 3H^+$$

$$Al(OH)_3(s) = AlO_2^- + H_2O + H^+$$

We see that on increasing pH the aluminum cation will first react to precipitate the hydroxide, which again will go into solution as an anion. Thus aluminum can occur as cation as well as anion. In principle we could imagine that all metal cations by increasing pH could form anions of various types. However, since the maximum pH obtainable in a saturated caustic solution is about 15, anions may not be formed in appreciable amounts for many metals.

For each of the above reactions there is a definite change in standard Gibbs energy, and an equilibrium constant which correlates the activity of the various ions with the pH of the solution.[1] Thus at 25°C Al^{3+} of unit activity coexists with $Al(OH)_3$ at pH = 3.3 and $Al(OH)_3$ coexists with AlO_2^- of unit activity at pH = 12.4.

Similarly certain anions may, on decreasing pH, undergo reactions

$$SO_4^{2-} + H^+ = HSO_4^-$$

$$S^{2-} + H^+ = HS^-$$

$$HS^- + H^+ = H_2S(g)$$

In a sulfate solution the HSO_4^- ion will be dominating at pH < 2, whereas in a sulfide solution the HS^- ion will dominate at pH < 14 and H_2S gas will be evolved at pH < 8.

Reduction-Oxidation (Red-Ox) Reactions

In addition to the pH the other important property of the solution is the so-called *oxidation potential*. In the present chapter this will be expressed in terms of the corresponding oxygen pressure, in the same way as was done for pyrometallurgical processes. An alternative expression in terms of the electrochemical potential will be given in Chap. 16.

In the presence of oxygen many metals may be brought into aqueous solution

$$2Me + O_2 + 4H^+ = 2Me^{2+} + 2H_2O$$

This is the overall reaction in the corrosion of most metals under aqueous and aerated conditions. Also, in some cases, Me^{2+} may be oxidized to ions of higher valency

$$4Me^{2+} + O_2 + 4H^+ = 4Me^{3+} + 2H_2O$$

In acid solutions some metals may dissolve with evolution of hydrogen

$$Me + 2H^+ = Me^{2+} + H_2$$

Thus for the equilibrium between Fe(s), Fe^{2+}, and H^+, each of unit activity and at 25°C, the corresponding oxygen pressure is 10^{-113} atm and the hydrogen

---

[1] See, e.g., "Selected Values of Chemical Thermodynamic Properties," U.S. Government Printing Office, Washington, D.C., 1952.

pressure 10^{15} atm. For the equilibrium between Fe^{2+}, Fe^{3+}, and H^+, each of unit activity, the oxygen pressure is 10^{-31} atm and the hydrogen pressure 10^{-26} atm. Even though these pressures are far outside the range which can be established or measured directly, they nevertheless express well-defined chemical conditions which can be established and measured by indirect methods.

We shall now see how a typical metal, Me, may behave on changes in pH and oxygen potential. We assume that the metal may form the cations Me^{2+} and Me^{3+} as well as the anions MeO_2^{2-} and MeO_2^-, and the solid hydroxides $Me(OH)_2$ and $Me(OH)_3$. Among the metal, oxygen, the various ions, the hydroxides, and water, the following equilibria may be established:

$$2Me + 4H^+ + O_2 = 2Me^{2+} + 2H_2O \tag{1}$$

$$4Me^{2+} + 4H^+ + O_2 = 4Me^{3+} + 2H_2O \tag{2}$$

$$Me^{2+} + 2H_2O = Me(OH)_2 + 2H^+ \tag{3}$$

$$Me^{3+} + 3H_2O = Me(OH)_3 + 3H^+ \tag{4}$$

$$Me(OH)_2 = MeO_2^{2-} + 2H^+ \tag{5}$$

$$Me(OH)_3 = MeO_2^- + H^+ + H_2O \tag{6}$$

Each of the above six equations is described by an equilibrium constant which correlates the activities of the various ions, the pH, and the oxygen pressure:

$$2 \log a_{Me^{2+}} + 4pH = \log p_{O_2} + \log K_1$$

$$4 \log a_{Me^{3+}} + 4pH = 4 \log a_{Me^{2+}} + \log p_{O_2} + \log K_2$$

$$0 = \log a_{Me^{2+}} + 2pH + \log K_3$$

$$0 = \log a_{Me^{3+}} + 3pH + \log K_4$$

$$\log a_{MeO_2^{2-}} = 2pH + \log K_5$$

$$\log a_{MeO_2^-} = pH + \log K_6$$

The numerical value for the equilibrium constants may differ greatly for the various metals, as illustrated by the fact that for many metals some of the above ions will not be formed in noticeable amounts. The above six equilibrium constants suffice for a complete description of the system. Any other reaction between the species involved may be obtained by arithmetic addition or subtraction of the above equations. Thus by the addition Eq. (1) + 2 Eq. (3) we get

$$2Me + 2H_2O + O_2 = 2Me(OH)_2 \tag{7}$$

with

$$0 = \log p_{O_2} + \log K_1 + 2 \log K_3$$

and so on.

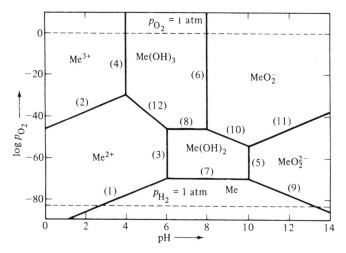

Figure 15-2 Predominance areas for the system Me–H_2O (schematic). The numbers in parentheses correspond to reaction equations in the text.

These relations are illustrated in Fig. 15-2 for a hypothetical metal and with arbitrarily chosen values for the various equilibrium constants. Notice that changes in the equilibrium constants only shift the location of the curves, whereas their slopes are fixed by the stoichiometry of the reactions.

The lines in Fig. 15-2 give the equilibria between the various solid phases and ions of *unit activity* as function of pH and oxygen pressure. The lines divide the figure into *predominance areas* for the various species in the same way as was previously shown in Fig. 8-7 for the roasting of sulfides at elevated temperatures. The main difference is that, whereas the lines for equilibria between solid phases are fixed and independent of the amounts of solids, the lines for equilibria with dissolved species depend on their concentration. Thus in Fig. 15-2 line (1) gives the equilibrium between solid metal and Me^{2+} of unit activity. For $a_{Me^{2+}} = 0.1$ the line would shift down by two logarithmic units. Similarly for equilibrium between $a_{Me^{2+}} = 0.1$ and $a_{Me^{3+}} = 1$, line (2) would shift up by four logarithmic units, and so on.

In Fig. 15-2 the curve for evolution of hydrogen of 1 atm pressure is drawn as a dashed line. At 25°C this curve corresponds to an oxygen pressure of 10^{-83} atm in accordance with the standard Gibbs energy for the reaction $2H_2O = 2H_2 + O_2$ being $+473$ kJ. Below this line hydrogen would be evolved under equilibrium conditions. Hydrogen evolution is very slow near the equilibrium line, however, and in practice this line may be greatly exceeded, and the equilibrium between Me and Me^{2+} may be established even at lower oxygen pressures.

Figure 15-2 shows that the metallic phase is stable only below the line (1)–(7)–(9). For a given pH an increase in the oxygen pressure will cause the metal to dissolve as cation or anion or it will convert into hydroxide. Conversely the metallic phase may be precipitated from the Me^{2+} or MeO_2^{2-} solutions by reduction, e.g., with hydrogen, between two limiting pH values.

On further increase in the oxygen pressure the trivalent ions Me^{3+} or MeO_2^- as well as the trivalent hydroxide may be formed. The highest oxidation potential is, theoretically, set by the line for $p_{O_2} = 1$ atm. In practice, however, this line may be greatly exceeded by means of chemical or electrochemical oxidation.

In Fig. 15-2 is shown, what frequently is the case, that the higher valency anions and hydroxides are more stable than those of the lower valency, and are therefore formed at lower pH. A typical example is iron where the ferric hydroxide precipitates at a pH around 2 to 3, whereas the ferrous hydroxide first precipitates around pH = 6 to 7. Also ferric hydroxide may be dissolved in strong caustic solution, forming the ferrite ion FeO_2^-, whereas a ferrous anion, FeO_2^{2-}, is less stable.

The example given in Fig. 15-2 was purely schematic. Thus it is not necessary that all metals form the species indicated. Instead of the simple hydroxides more complicated hydrated oxides, or even the anhydrous oxides, may be formed. Instead of the anions MeO_2^{2-} and MeO_2^- other anions of the types MeO_n^{2-2n} and MeO_n^{3-2n} as well as hydrated anions may be formed. Furthermore, some ions as well as solid phases may be *metastable*, but have a considerable lifetime.

Also for the different metals the location and size of the different predominance areas may differ greatly. Thus a relatively noble metal such as copper may exist in the metallic state under conditions where iron is present as Fe^{2+} ion. Iron oxidizes to the trivalent state under conditions where cobalt and nickel are still in their divalent state. Ferric hydroxide precipitates at a lower pH than, for example, $Zn(OH)_2$, and aluminum and silicon dissolve as anions under conditions where ferric hydroxide is still present. These differences in predominance areas of the various metals constitute one of the main principles used in hydrometallurgy.

Also anions may undergo oxidation and reduction. A typical example is sulfur. In acid solution H_2S will on oxidation give elemental sulfur

$$H_2S(g) + \tfrac{1}{2}O_2 = S(s) + H_2O$$

On further oxidation sulfate ions will be formed

$$S(s) + \tfrac{3}{2}O_2 + H_2O = SO_4^{2-} + 2H^+$$
$$S(s) + \tfrac{3}{2}O_2 + H_2O = HSO_4^- + H^+$$

Notice that sulfate formation produces H^+ ions, and thus becomes more favorable with increasing pH. In basic solutions elemental sulfur is no longer stable, and the sulfide ions oxidize directly to sulfate

$$HS^- + 2O_2 = SO_4^{2-} + H^+$$
$$S^{2-} + 2O_2 = SO_4^{2-}$$

In addition to sulfide and sulfate ions sulfur may form a number of inter-mediate sulfites and thiosulfates. From a thermodynamic viewpoint these are *metastable*, and will eventually disproportionate to give sulfate and elemental sulfur. Furthermore, at very high oxygen potential the persulfate ion, $S_2O_8^{2-}$, and the peroxymonosulfate ion, SO_5^{2-}, may be formed. These are very strong oxidants.

Similarly, the chloride ion will at high oxygen pressures oxidize to chlorine in acid solution

$$2Cl^- + \tfrac{1}{2}O_2 + 2H^+ = Cl_2(g) + H_2O$$

and to hypochlorite in basic solution

$$Cl^- + \tfrac{1}{2}O_2 = OCl^-$$

At very high oxygen potentials the chlorate, ClO_3^-, and the perchlorate, ClO_4^-, ions may be formed.

Complex Formation

Different ions may react with each other or with neutral molecules to give *complex ions*. Typical examples are

$$Ag^+ + 2CN^- = Ag(CN)_2^-$$
$$UO_2^{2+} + nSO_4^{2-} = UO_2(SO_4)_n^{2-2n}$$
$$Cu^{2+} + nNH_3 = Cu(NH_3)_n^{2+}$$

Notice that in the first case the silver ion changes from positive to negative charge. Thus, whereas silver in a nitrate solution is present as a cation, the addition of NaCN or KCN transforms it into an anion. The uranyl cation UO_2^{2+} will, on increasing sulfate ion concentration, form complexes where the number n increases from 1 to 3. For $n \geq 2$ the complex will have negative charge. Similarly Cu^{2+} may form ammonia complexes, *ammines*, where the number n increases with increasing ammonia concentration in the solution, but without affecting the charge of the ion.

Some complex ions may be very stable, corresponding to the equilibrium constant for their formation having a very large value. Thus, whereas silver nitrate easily reacts with HCl to precipitate AgCl, the chloride cannot be precipitated from a solution where silver is present as $Ag(CN)_2^-$ anion. On the contrary, solid AgCl will react with a NaCN solution to give the water-soluble silver cyanide ion. Similarly, whereas addition of NaOH to a Cu^{2+} solution would lead to precipitation of cupric hydroxide the addition of NH_3 gives the cupric ammine ion, which is stable at high pH values. Thus the pH–oxygen potential diagram which applies for the copper–water system, will be greatly modified when there is also ammonia present.

In addition to the well-known stable complexes there may be less stable complexes either between different ions or between ions and water. In the last case we talk about hydrated ions. The formation of weak complexes may be one of the reasons for deviations from ideal ionic solution which was discussed in Sec. 15-1.

Precipitation of Solid Compounds

In the same way as solid hydroxides may be precipitated by hydrolysis other ions may react to form solid precipitates

$$Na^+ + Cl^- = NaCl(s)$$

$$Ag^+ + Cl^- = AgCl(s)$$

$$2Ag^+ + S^{2-} = Ag_2S(s)$$

Each of these reactions is characterized by an equilibrium constant, the inverse of which is called the *solubility product* of the solid. Thus for NaCl at 25°C $a_{Na^+}a_{Cl^-} = 380$ in agreement with the known high solubility of NaCl. For AgCl the solubility product at 25°C is 1.7×10^{-10} and for Ag_2S it is 5.5×10^{-51}. The last value shows that Ag_2S may be precipitated even from a complex silver cyanide solution, since the equilibrium constant for formation of the $Ag(CN)_2^-$ ion is "only" 6×10^{18}.

In the same way as for complex formation the precipitation of solid phases may greatly alter the pH–oxygen potential diagrams which apply for the pure metal–water systems. Thus if copper and sulfur are present together there will be regions where the sulfides Cu_2S and CuS are the stable solid phases, and there will be regions where basic sulfates such as $CuSO_4 \cdot 2\ Cu(OH)_2$ and $CuSO_4 \cdot 3\ Cu(OH)_2$ are precipitated rather than the simple hydroxide.

Effect of Temperature on Ionic Equilibria

Like all other equilibria those between ions in solution and solid phases will be affected by temperature changes in accordance with the van't Hoff equation, i.e., endothermic reactions will shift to the right with increased temperature. Since the oxidation with oxygen is always exothermic this means that with increased temperature the different equilibria will shift to higher oxygen pressures, without necessarily any changes in the relative size and location of the different predominance areas. Likewise volatile components such as ammonia and carbon dioxide may be expelled by heating the solution.

Since the dissolution of solid compounds in water is usually endothermic their solubility will increase with increasing temperature. This is, for example, the case for $Al(OH)_3$ in alkaline solution, a fact which is utilized in the *Bayer process* for production of alumina (Sec. 15-4). On the other hand the solubility of $AlO(OH)$, $FeO(OH)$ and other iron hydroxides and basic sulfates in acid solution decreases with increasing temperature,[1] a fact which is utilized in the

---

[1] R. G. Robins: *J. Inorg. Nucl. Chem.* **29**: 431 (1967).

removal of iron from the acid leach solution in the hydrometallurgical zinc process. In other cases the effect of temperature is of greater importance for the kinetics of the process.

15-3 KINETICS OF LEACHING AND PRECIPITATION

The reactions between solid compounds and liquid phases are heterogeneous processes and are, in principle, not much different from the gasification of carbon or the reduction of metal oxides which were discussed in Sec. 5-5. The reaction rate may be controlled either by mass transfer to and from the solid surface or by the chemical reaction on the surface, as well as by combinations of the two steps. In the first case the rate of dissolution is increased by stirring of the liquid, in the second case the rate is independent of the motion of the liquid. In both cases the rate will increase with increasing concentration of the reacting components in the solvent. Also, mass transfer to and from the surface has a lower activation energy than the chemical reaction, and will be controlling at higher temperatures.

Of particular interest in hydrometallurgy is the leaching of valuable constituents from low-grade ores. Usually, in this case, the valuable mineral will be distributed throughout a matrix of worthless gangue. In order to make the mineral accessible for the solvent the ore has to be finely crushed. Even in this case the valuable mineral may only be approached by diffusion through very fine pores in the gangue. It is evident that in this case the leaching will not be affected much by the external stirring of the liquid.

A second phenomenon which may occur is the formation of insoluble intermediate products or by-products. One example is the formation of elemental sulfur in acid leaching of sulfide ores. Below the melting point of sulfur this layer is relatively porous and does not greatly hinder the diffusion of the solvent into the ore. Above 120°C, however, liquid sulfur is formed, covering the sulfide surface and greatly reducing the leaching rate.

From a chemical viewpoint leaching reactions may be of two types:

1. Simple dissolution without oxidation In this group we find the important Bayer process for leaching of bauxite as well as processes for leaching of copper oxide ores and of roasted sulfides. Leaching of oxides may have a high intrinsic rate, and are often mass-transfer controlled. The rate may differ greatly for different oxides, however. One example is the leaching of roasted copper–nickel matte with sulfuric acid (Sec. 12-4). From a thermodynamic viewpoint both copper and nickel oxides are soluble in sulfuric acid, but the copper oxide dissolves faster, thus making a separation possible.

2. Dissolution combined with oxidation In this group we find processes for leaching of sulfides of, e.g., copper and nickel, as well as of native metals. These reactions are often slow. Thus the dissolution of native silver in a cyanide solution may be described by the reaction

$$Ag + \tfrac{1}{4}O_2 + 2CN^- + H^+ = Ag(CN)_2^- + \tfrac{1}{2}H_2O$$

Measurements have shown that at low concentrations of cyanide ions, and relatively high oxygen pressures, the rate is controlled by, and is proportional to, the cyanide concentration. At higher cyanide concentrations the rate becomes proportional to the oxygen pressure and independent of the cyanide concentration.[1]

Similar behavior applies for the leaching of copper sulfides, and in most industrial processes the reaction between oxygen and the sulfide surface is the rate-controlling step. The rate of reaction can, in that case, be increased by increased oxygen pressure, a fact which is utilized in pressure leaching (Sec. 15-5). The reaction rate of oxygen may also be increased by the presence of iron in the solution. In contact with oxygen or air iron is oxidized to the trivalent state, the Fe^{3+} ion then acts as an oxidizing agent on the sulfide, itself being reduced to Fe^{2+}, which again is oxidized by the air. An interesting phenomenon is that the oxidation of the Fe^{2+} ion seems to be promoted by the presence of certain *bacteria*. Experiments have shown that the leaching rate of sulfides may be increased by a factor of ten to a hundred in the presence of such bacteria derived from mine waters. Different bacteria have been isolated and experimental work is being carried out to learn how to utilize these silent workers in hydrometallurgy.[2]

Precipitation from aqueous solutions may also be affected by kinetic phenomena. Precipitation may be of the following kinds: (1) precipitation of hydroxides by changes in pH without changes in oxygen potential; (2) precipitation of insoluble compounds or salts by addition of certain chemicals or by cooling; (3) precipitation by reduction or oxidation.

An example of the first kind is precipitation of $Mg(OH)_2$ from seawater by means of $Ca(OH)_2$. An example of the second kind is precipitation of CuS from Cu^{2+} solution and H_2S. Examples of the third kind are precipitation of FeO(OH) by oxidation of a Fe^{2+} solution and precipitation of Cu by reduction of Cu^{2+} solution with iron or hydrogen.

In each of the above groups we may have either *homogeneous* or *heterogeneous* precipitation, i.e., the reaction is either between dissolved species or between the solution and a second phase. In the first case nucleation of the precipitate may be the rate-controlling step. The rate may then be increased by the addition of *seeds* of the wanted product. A typical example is the precipitation of $Al(OH)_3$ in the Bayer process, which would not be possible without

---

[1] H. Y. Sohn and M. E. Wadsworth: "Rate Processes of Extractive Metallurgy," Plenum Press, New York, 1979, p. 172.

[2] J. D. A. Miller (Editor): "Microbiological Aspects of Metallurgy," Medical and Technical Publ. Co. Ltd., Aylesbury, 1971.

seeding. Seeding may also be used to control the grain size of the product. Thus in the reaction between seawater and quicklime $Mg(OH)_2$ comes out in a very finely divided form which is difficult to settle. If some seeds of $Mg(OH)_2$ are added, the precipitate becomes coarser and is easier to handle. Nucleation and grain growth is also affected by temperature. In general the precipitate becomes coarser by digestion at higher temperatures.

In the purification of *zinc leach solution* (Sec. 15-4) iron is precipitated as hydroxide by treatment with excess zinc oxide which acts as a base. An interesting phenomenon is here the ability of the precipitate to adsorb ions of other elements such as arsenic and antimony which would not precipitate alone under the prevailing conditions.

The precipitation of trivalent *iron* and *cobalt* hydroxides by oxidation of their divalent solutions bears a certain resemblance to the leaching by oxidation inasmuch as the oxidizing gas reacts very slowly. Cobalt cannot be oxidized to its trivalent state by means of air, even though this is thermodynamically possible. Instead the oxidation is carried out with chlorine, which is both a stronger and faster oxidizing agent.

Cobalt, nickel, and the more noble metals may be *reduced* from solutions of their salts by means of hydrogen. Hydrogen reacts very slowly at room temperature, however, and in practice temperatures between 150 and 200°C are used, as well as increased hydrogen pressure (Sec. 15-5). Also seeding with previously precipitated metal powder is used.

Of considerable industrial importance is the precipitation of metals by means of a less noble metal, which thereby goes into solution. A typical example is the *cementation* of copper by means of iron scrap. This reaction seems to be very fast and is probably controlled by the rate at which the ions diffuse to and from the iron surface. The rate may therefore be increased by stirring and by continuously rubbing off the metallic copper which deposits on the iron.

15-4 INDUSTRIAL APPLICATIONS

Industrial hydrometallurgical processes usually consist of the following steps:

1. Leaching of the ore or of an intermediate metallurgical product with acid, caustic, or a complex-forming solvent, often combined with oxidation
2. Purification of the solution by precipitation of insoluble compounds, cementation of unwanted metals, or by solvent extraction
3. Precipitation of wanted product, either as an insoluble compound or as a metal, either by chemical or electrochemical methods

Of these the second step may sometimes be omitted. On the other hand, if the ore contains several valuable metals, each of these may be precipitated or extracted in separate steps.

The *reactors* used in leaching depend on the nature of the ore, its grade, and particle size. For coarse- and medium-sized ore *fixed-bed* reactors (Fig. 6-1(a)), like heaps or vats, are used, and the leaching liquor is percolated through the bed. For finely ground ore *agitation leaching*, where the ore is kept in suspension in the liquor, is used. The agitation may either be by means of air (Pachuca tanks), in which case the reactor resembles a fluidized bed (Fig. 6-1(c)), or mechanical agitation is used. Leaching above atmospheric pressure is done in *autoclaves*, usually equipped with mechanical stirring. Recently high-pressure leaching has also been made in long tubes where the ore-and-liquor suspension moves through the tube, and where the leaching reaction is completed during the passage.

Production of Al_2O_3 by the Bayer Process

The dominating raw material for aluminum production is bauxite. This is an impure aluminum hydroxide with ferric hydroxide and silica as the main impurities. Most bauxites are treated by the *Bayer process* to produce pure Al_2O_3. In this process finely ground bauxite is treated with a strong *sodium hydroxide solution* (130 to 350 g Na_2O/l) in autoclaves at 5 to 30 atm pressure and at temperatures from 150 to 240°C. Aluminum goes into solution forming the AlO_2^- anion, whereas ferric hydroxide, TiO_2, etc., remain undissolved. Silica is also soluble in alkaline solution, but in the presence of sodium and aluminate ions it is again precipitated as a *sodium aluminum silicate*.

After the solution has become saturated with aluminum hydroxide the insoluble parts are removed by settling, washing, and filtration. The solution is cooled to room temperature and diluted with water. This decrease in temperature and pH brings the solution into the predominance area for $Al(OH)_3$. However, in order to precipitate the hydroxide, seeding with fresh $Al(OH)_3$ is necessary. When no further precipitation occurs the hydroxide is separated by thickening, washing, and filtration. The hydroxide is calcined at about 1200°C to give 99.5 percent Al_2O_3, while the solution is concentrated by evaporation and is returned to the leaching stage.

The hydroxides occurring in natural bauxite may either be the trihydrate, $Al(OH)_3$, "Gibbsite," or the monohydrate, $AlO(OH)$, "Böhmite." Of these the first is thermodynamically unstable with respect to the second, and therefore has the higher solubility. In order to dissolve the monohydrate higher temperature and also stronger caustic solution are needed than for the trihydrate. In both cases the metastable trihydrate is formed during the precipitation process.

If the bauxite had been high in silica the insoluble part from the leaching stage, *red mud*, would still contain considerable amounts of alumina. Special processes have been developed to recover this alumina. Thus the red mud may be calcined with lime and soda ash to give water-soluble sodium aluminate and insoluble calcium silicate, the former being leached out and treated as described above. The remaining red mud is of little value, but may find some use as an iron ore.

Leaching of Zinc Calcine

The most important zinc ore is the sulfide, which is the raw material both for carbothermic and electrolytic zinc production. Since ZnS dissolves very slowly in acid it is first roasted to give ZnO. The roasting is usually carried out so as to produce some zinc sulfate, thus compensating for the losses of sulfate ions during the process. Also the amount of unroasted sulfide must be kept to a minimum.

The leaching of the calcine and the treatment of the solution is illustrated in the flow-sheet given in Fig. 15-3, and may be described in the following way: The major part of the zinc calcine is treated in the *neutral leaching* stage. Here an excess of calcine is treated with the solution from the subsequent *acid leach* and with some return electrolyte. Because of the excess of ZnO the pH of the solution after leaching will be around 5. At this pH any iron and aluminum which had come from the acid leach will precipitate as hydroxides together with silica and adsorbed arsenic and antimony. In order to ensure complete precipitation of the iron this has to be present in its trivalent state. For this purpose air is blown through the solid–liquid suspension, and some MnO_2 is usually added. The solids are separated from the solution by settling and filtration. In addition to the precipitates they contain an excess of zinc calcine and are passed on to the acid leach.

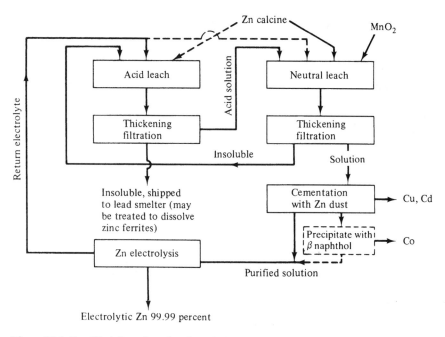

Figure 15-3 Simplified flow-sheet for electrolytic zinc production. Dashed lines represent optional routes. Instead of β naphtol a combintion of zinc dust and antimony may be used.

In the *acid leach* the solids are treated with an excess of return electrolyte. Some fresh calcine may also be added here. The zinc oxide is dissolved, whereas the major part of the iron remains undissolved, as oxide and ferrite, which are poorly soluble even at a pH of around 2. Thus all the iron, silica, etc., in the original ore is discharged in the solids from the acid leach, together with insoluble compounds of lead, silver, etc. If the calcine contains appreciable amounts of zinc ferrite the solid may be treated further by different methods to bring the zinc into soluble form (see below). After repeated leaching, settling, and filtration the undissolved solids may be shipped to a lead smelter.

The solution from the neutral leach contains, as main impurities, copper, cadmium, and sometimes cobalt. These are more noble than zinc and are removed by *cementation* on zinc dust, whereby an equivalent amount of zinc goes into solution. The cementated mixture may be treated further for recovery of these metals. Residual *cobalt* in the solution may be precipitated by means of *beta-naphthol* and *sodium nitride*, forming an insoluble cobalt compound, or it may be precipitated on zinc dust in the presence of some antimony. The latter practice is becoming more common.[1]

The neutral, purified $ZnSO_4$ solution is electrolyzed to give metallic zinc and sulfuric acid. During electrolysis only about half of the zinc in the solution is deposited. The remaining half is returned with the acid electrolyte to the leaching processes. Also, any *manganese* in the solution is oxidized on the *anode* to solid MnO_2, which is returned to the neutral leach. Further details about zinc electrolysis will be given in Chap. 16.

Leaching of zinc ferrite Most zinc calcines contain appreciable amounts of zinc ferrite, which do not dissolve during the standard acid leach at pH around 2, and therefore would represent an appreciable zinc loss. In order to bring zinc ferrite into solution a hot leach at pH of 1 or less is needed, but under these conditions iron oxide and hydroxide would also dissolve. If calcine were added to such a solution ferric hydroxide would come out as a slimy precipitate, which is difficult to filter. In recent years considerable work has, therefore, been made in improving the precipitation of iron from such solutions, thus making it possible to leach the ferrite. Without going into details these processes may be classified in two groups:

1. Goethite and hematite processes
2. Jarosite processes

Goethite, FeO(OH), and *hematite*, Fe_2O_3, are the thermodynamically stable precipitates for low concentration of other ions, and their solubility at a given pH decreases with increasing temperature. In order to obtain a precipitate which is easy to filter, the solution, after leaching of zinc ferrite, is reduced with,

[1] G. M. Meisner: *J. Met.*, **26**: 25 August 1974).

for example, SO_2 to bring iron into the ferrous state, and the pH is raised to 3 to 4 by addition of some calcine. The solution is now again oxidized with air or oxygen to form the ferric precipitate. If the oxidation is done below 100°C, that is, at atmospheric pressure, goethite is formed, whereas at temperatures above 150°C and correspondingly high pressure the precipitate is hematite.

In the *jarosite process* use is made of the fact that in the presence of sulfate and alkali ions a complex basic alkali iron sulfate (jarosite) will precipitate at a pH of about 1.5. Instead of alkali, ammonia is most commonly used, in which case the precipitate may be described by the formula $NH_4[Fe_3(SO_4)_2(OH)_6]$. Like for the goethite process the precipitation is carried out at temperatures somewhat below 100°C. The jarosite process has the advantage that it can more easily be incorporated into existing leaching plants. On the other hand the precipitate is bulky with a low iron content, whereas the goethite and hematite processes may possibly produce a product which can be used as an iron ore.

It should finally be mentioned that at the present tests are made to remove iron from the acid leach solution by *solvent extraction* (Sec. 15-6), without this yet having found any industrial application.

Leaching of Copper Ores

We distinguish between leaching of oxide ores or roasted sulfides, which require no further oxidation, and leaching of sulfide ores. Of these leaching of oxide ores is by far the most important, as such ores cannot easily be upgraded by ore-dressing methods. We furthermore distinguish between acid and ammoniacal leaching. Thus for oxide ores we have the two possibilities

$$CuO + 2H^+ = Cu^{2+} + H_2O$$

and

$$CuO + H_2O + 4NH_3 = Cu(NH_3)_4^{2+} + 2OH^-$$

Of these acid leaching with sulfuric acid is the most common, and the pregnant sulfate solution may be passed directly on to an electrowinning plant for deposition of metallic copper (Sec. 16-6). Acid leaching is not suitable for ores which contain carbonates such as limestone, however, as these would consume acid. In such cases ammoniacal leaching is used. As the ammoniacal solution cannot be electrolyzed directly it must either be boiled to precipitate $Cu(OH)_2$, or the copper value must be transferred to an acid sulfate solution by means of solvent extraction (Sec. 15-6). As already mentioned leaching of oxide ores is a relatively fast process, and a properly crushed ore may be leached within a few days.

In contrast, the leaching of sulfide ores is a slow process as it involves oxidation. Examples are the so-called *dump* and *heap* leaching processes, which have been used in Spain since 1752. In these processes low-grade sulfide ore—too low to be treated economically by ore-dressing methods—is exposed to oxidation by the atmosphere, and a dilute sulfuric acid solution, which itself is produced during the process, is percolated through the ore. In these cases the extraction of the copper values may take years. Copper in the pregnant solution can be recovered by cementation on scrap iron and the solution returned to the leaching process. The iron which enters the solution, either from the ore or from the scrap iron, will eventually hydrolyze and precipitate as hydroxide or basic sulfates in the leaching heap. As an alternative to cementation on scrap iron the copper values can be taken out of the leach solution by solvent extraction. As this gives a stripping solution which can be electrolyzed directly this practice has recently become more common.

Similar to heap and dump leaching is the so-called *in situ mining*. This is used on copper deposits which either are of too low a grade or are too deeply located to warrant mining by conventional methods. In these cases the ore body is first fractured by means of explosives, and dilute sulfuric acid and air are percolated through the fractured ore. Like in heap and dump leaching the copper values are recovered from the effluent pregnant solution by cementation or solvent extraction.

In recent years there has been a considerable interest in the leaching also of high-grade copper sulfide concentrates, and a large number of processes have been suggested and have been tried on a pilot plant scale. Of these only a few seem to be competitive to existing pyrometallurgical methods, however. The proposed processes are either acid or ammoniacal. Most attractive seem to be the acid processes as the sulfur in the ore may be oxidized to elemental sulfur, a product which is easy to store and easy to ship, and iron may be converted to oxide or hydroxide. As copper sulfide minerals, and in particular the most common chalcopyrite, oxidize very slowly in air most acid processes are based on the use of *ferric chloride* as oxidizing agent. During leaching this is reduced to ferrous chloride, which again must be oxidized to its ferric state by means of either air, oxygen, or chlorine.

It is interesting to notice that oxidation with ferric *chloride* is much faster than with ferric *sulfate*, even though the thermodynamics of the oxidation process is the same. Evidently the kinetics are more favorable in the presence of chloride anions. Another interesting phenomenon is that elemental *sulfur* and *ferric hydroxide* are found together, even though the oxidation potential needed to form ferric hydroxide should have oxidized elemental sulfur to sulfate. Evidently the latter reaction is rather slow.

Ammoniacal leaching of sulfides is somewhat faster than acid leaching. Processes have been suggested where such leaching is done with air or oxygen at atmospheric pressure and at temperatures close to 100°C, but by and large ammoniacal leaching is more often done at elevated pressure and at temperatures above 100°C, as will be discussed in the following section. In contrast to acid leaching ammoniacal leaching oxidizes all the sulfur in the ore to sulfate. This may either be converted to ammonium sulfate and sold as fertilizer or the solution may be reacted with lime to give calcium sulfate, which is discarded, and ammonia which is returned to the process. One drawback of ammoniacal leaching is that precious metals, such as gold and silver, are not dissolved, and will be lost in the tailings.

15-5 PRESSURE LEACHING AND REDUCTION

In recent years a number of processes have been developed to leach sulfide ores in autoclaves at temperatures above 100°C and at high oxygen pressures. The advantage of the elevated pressure and temperature is more favorable equilibria and, more important, a higher reaction rate. The leaching processes may be carried out either in acid or in ammoniacal solutions, the latter being particularly suited to bring cobalt, nickel, and copper into solution as ammonia complexes.

In the *Sheritt Gordon process* (Fig. 15-4) a Fe–Co–Ni–Cu sulfide ore is leached with ammonia at about 105°C and under an air pressure of about 8 atm.[1] This brings cobalt, nickel, and copper into solution, whereas iron is precipitated as hydroxide. Sulfur in the ore is mostly oxidized to sulfate, but some of it goes to thiosulfate and polythionate ions. After separation of the undissolved solids the pregnant solution is boiled at atmospheric pressure to expel the excess ammonia. This causes the metastable thiosulfate and polythionate ions to decompose and react with copper to precipitate copper sulfide, which is filtered off and shipped to a copper smelter.

The remaining solution now contains nickel and cobalt ammonia complexes, in a ratio of approximately forty to one. The solution is first treated with air at a pressure of 50 atm at about 200°C to oxidize all thiosulfates and is then reduced with hydrogen at about 15 atm and about 175°C. Under these conditions the nickel ammine ion is reduced to give metallic nickel

$$Ni(NH_3)_2^{2+} + H_2 = Ni + 2NH_4^+$$

To initiate the reaction it is necessary to add either metallic nickel powder seeds or a suitable catalyst. The hydrogen reduction is carried to the point where most of the nickel is precipitated but without affecting the cobalt in solution.

The solution now contains about equal amounts of cobalt and nickel in high dilution. In order to concentrate and separate the two metals the solution is first treated with hydrogen sulfide which precipitates a mixed sulfide. This is once more brought into solution by acid pressure oxidation. Once more ammonia is added and the solution is now oxidized to bring cobalt into its trivalent state, forming the *cobaltic ammine ion*, $Co(NH_3)_5^{3+}$, whereas nickel remains in its *divalent* state. On acidification of the solution the cobaltic ammine remains dissolved, whereas the nickel compound precipitates. The solution now contains cobalt and nickel in a ratio of about fifty to one. In order to remove the last traces of nickel some metallic cobalt is added which precipitates metallic nickel. The two nickel precipitates contain considerable amounts of cobalt and are returned to previous stages. The pure cobalt solution is finally reduced with hydrogen at about 175°C and 20 atm to give metallic cobalt powder.

---

[1] D. J. I. Evans: in "Advances in Extractive Metallurgy," Instn. of Min. and Met., London, 1968, p. 831.

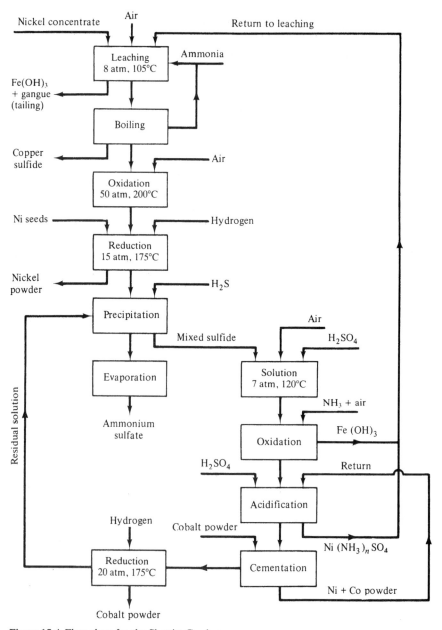

Figure 15-4 Flow-sheet for the Sherritt Gordon process.

The solutions after the nickel and cobalt reductions contain essentially ammonium sulfate, which is recovered by evaporation and is sold as fertilizer. Overall recoveries are 90 to 95 percent of nickel, 85 to 90 percent of copper, 75 percent of cobalt, and 60 to 75 percent of sulfur.

Pressure leaching at elevated temperature can also be done in *acid* solution. One example is pressure leaching of zinc concentrate,[1] which may become an alternative to the present roast-leach methods.

15-6 ION EXCHANGE AND SOLVENT EXTRACTION

Ion-exchangers are solids in which various ions are loosely bonded. The first known ion-exchangers were natural minerals, *zeolites*, but these have later been supplemented by organic resins, which are those most commonly used. When these resins are immersed in aqueous solutions an exchange may take place between ions in the solution and in the resin. Both cation and anion resins are known.

In metallurgy ion-exchangers found their first important use in the extraction of *uranium*. Uranium ores are not easily concentrated by ore-dressing methods, and the low-grade ore had to be treated by leaching. Both acid and alkaline leaching may be used, but acid leaching is most common. Under oxidizing conditions uranium dissolves in dilute sulfuric acid to form the *uranyl ion* UO_2^{2+}. This again is more or less complexed by the equilibrium

$$UO_2^{2+} + nSO_4^{2-} \rightleftharpoons UO_2(SO_4)_n^{2-2n}$$

As previously mentioned the complex will have a negative charge for $n \geq 2$. If such an anion is brought into contact with an *anion* exchanger we have an exchange reaction, viz.,

$$4R^+X^- + UO_2(SO_4)_3^{4-} \rightleftharpoons (R^+)_4UO_2(SO_4)_3^{4-} + 4X^-$$

Here R^+ stands for the resin cation and X^- for anions like Cl^- or NO_3^-. Since uranium is one of the few metals which can form anions in sulfuric acid solutions the exchange is very selective, other metals such as Ca, Fe, etc., being retained in the aqueous phase as cations.

The capacity of the resin may correspond to 3 to 4 milliequivalents per gram of dry resin, corresponding to 50 to100 g of U_3O_8 per liter of granulated resin, and is the largest for pH from 1.5 to 2. The contact between the leach solution and the ion-exchanger may be accomplished in different ways. Most common are perhaps fixed beds of granulated ion-exchanger through which the solution is percolated. By successively shifting the beds to solutions of higher concentration a counter-current extraction is obtained.

After one column of the ion-exchanger has been "loaded" with uranium it is first washed with water and then "stripped" by means of a strong (1 M) acidified chloride or nitrate solution. This brings the uranium back into an

---

[1] B. N. Doyle et al.: Proceedings of the XI Commonwealth Min. Met. Congr., Hong Kong, 1978, p. 645.

aqueous phase, whereas the resin is ready for a new loading cycle. The ratio of concentration between uranium in the stripping and the leach solution may be of the order of 20, and is combined with a high degree of separation from other metals. Finally uranium oxide is precipitated from the stripping solution by neutralization.

One weakness of ion-exchange resins is that they are susceptible to *poisoning* by certain ions which form strong and almost permanent bonds with the resins, thus ruining the ion-exchange properties. Particularly harmful are cyanides, thiosulfates, and polythionates, but also silica in the solution may precipitate and lower the ion-exchange efficiency. For this and for other reasons ion-exchangers have mostly been replaced by *solvent extraction* which will be discussed next. It should be mentioned, however, that as the poisoning problem has gradually been overcome, the ion exchange method seems recently to be getting a renaissance.[1]

Solvent Extraction

Some metallic compounds and ions show high solubility in certain organic solvents. Also the solubility may be highly selective, which makes possible a separation between different metals. The extraction may occur either by neutral ionic pairs being absorbed by the solvent and forming a loose chemical bond with the organic molecule, or by a cation or an anion being bonded to the molecule in exchange for one or more hydrogen cations or chloride or nitrate anions. In the latter case solvent extraction is similar to ion exchange, and the trade name *Lix* used for some of the first solvents stands for *liquid ion-exchanger*.

Like for solid ion-exchangers solvent extraction found its first important application in the extraction of uranium, but is today used within large parts of the nonferrous extractive metallurgy. Most important is perhaps its application to copper extraction. For this purpose *cation solvents*, known under trade names like *Lix* and *Kelex*, are used. The exchange reaction may in these cases be written as

$$2RH + Cu^{2+} = R_2Cu + 2H^+$$

Here R represents the anionic part of the organic molecule, and Cu^{2+} and H^+ represent ions in the aqueous phase. The R_2Cu compound is called a *chelate*. Similar exchange reactions may be written for other metal ions, but the value of the equilibrium or distribution constant differs for the different metals, making a separation possible. It follows also from the above equation that for a given metal the absorption of the metal in a chelating solvent becomes more favorable with increasing pH of the solution. For the Lix and Kelex solvents the

[1] A. R. Burkin: "Topics in Non-Ferrous Extractive metallurgy," Society of Chemical Industry, Oxford, 1980, p. 95.

distribution coefficient for copper is large, and copper is extracted even from acid solutions, whereas impurity elements like iron and zinc remain in the aqueous phase.

In the extraction process the organic compound is usually dissolved in a suitable solvent like kerosene. In order to obtain the highest possible extraction the aqueous and the organic phases are made to flow counter-current to each other. Most common is a system of *mixer-settlers* as was discussed in Sec. 6-7, and from a knowledge of the effective equilibrium distribution curve and the operating line for the system, the number of mixer-settlers needed to obtain a desired extraction can be derived.

The organic solvent is now passed on to the *stripping* system. This is another series of mixer-settlers where the solvent is treated with a sulfuric acid solution, e.g., a return electrolyte. This brings copper back into the aqueous phase, and the organic solvent is returned to the extraction system, whereas the aqueous phase is passed on to electrolysis. Like in ion-exchange processes a considerable *upgrading* may be obtained, the metallic concentration in the stripping solution being many times larger than in the original leaching liquor.

In addition to being highly selective for different metals, solvent extraction has the advantage that a metal like copper can be transferred from an ammoniacal leaching solution to a sulfuric acid solution, which is needed for electrodeposition. Solvent extraction can also be applied directly on an ore-leaching pulp. This makes it unnecessary to separate the leaching liquor from the undissolved gangue, and after extraction, the aqueous pulp is discarded. Processes are also under development where the metal values are precipitated, e.g., as hydroxides, or reduced to metal directly from the organic phase, thus combining the stripping and precipitation steps.

Treatment of Copper–Nickel Matte

An example where solvent extraction is used to remove impurities from a solution is given by the *Falconbridge* matte leach process.[1] Here the copper–nickel matte (Sec. 12-4) is leached with a strong (7.5 M) HCl solution. Nickel, cobalt, and iron go into solution with evolution of H_2S, whereas copper sulfide remains undissolved and is shipped to the copper smelter. After oxidation of Fe^{2+} to Fe^{3+} and of sulfide ions to elemental sulfur the solution is purified by solvent extraction in two stages. First iron is extracted by TBP (tri-butyl-phosphate), next cobalt and residual copper are extracted by TIOA (tri-iso-octyl-amine).

It is interesting to notice that with TBP iron is extracted as a *neutral* $FeCl_3$ molecule, whereas with TIOA cobalt and copper are extracted as *anions*, which are formed by complexing with the chloride ions.

The metals are now stripped from the respective solvents by means of acidified water. It is interesting in this connection to notice that cobalt and

[1] P. G. Thornhill et al.: *J. Met.* **23**: 13 (July 1971).

copper, which were present in the same solvent, are separated during the stripping operation; by changing the chloride ion concentration in the stripping solution a high degree of selectivity is achieved.

Nickel chloride is precipitated from the purified solution by the addition of HCl(g) whereby the solubility product of $NiCl_2 \cdot 4H_2O(s)$ is exceeded. The solid nickel chloride is reacted with steam at 850°C to give solid NiO, which is reduced to metal with hydrogen, whereas the remaining solution is returned to the leaching stage.

In a more recent Falconbridge process[1] copper–nickel matte is leached in a slightly acid solution and with chlorine as oxidant. This brings copper, nickel, cobalt, and iron into solution as mono- and divalent ions, and elemental sulfur is formed. The slurry is now *reduced* by means of some finely ground matte. This makes copper and metals like silver and bismuth precipitate as sulfides, whereas nickel, cobalt, and iron remain in solution. After removal of the solids the solution is again oxidized, and iron is precipitated as $Fe(OH)_3$, whereas cobalt is taken out by solvent extraction. The resulting, purified nickel chloride solution is electrolyzed with inert (titanium) anodes to give metallic nickel and chlorine, of which the latter is returned to the leaching stage.

PROBLEMS

15-1 We assume the following salts to behave ideally, that is, $\gamma_\pm = 1$, at 0.01 molal concentration in water: NaCl, Na_2SO_4, $CaCl_2$, and $MgSO_4$.

(a) Calculate the activities of each salt if they were alone at the above concentration.

(b) Calculate the activity of NaCl in a mixed solution of NaCl and $CaCl_2$, each at 0.01 molal concentration.

15-2 The mean ionic activity coefficient $\gamma_\pm$ in a 10 molal HCl solution is 10.5 and in a 0.1 molal solution 0.8.

(a) Calculate the pH and the activity of HCl at the two concentrations and calculate the HCl pressure at $m = 0.1$ when at $m = 10$ it is 5.5×10^{-3} atm.

(b) The HCl pressure becomes 1 atm at $m = 19.2$. Calculate $\gamma_\pm$ at this concentration.

15-3 The solubility product of Ag_2S in water at 25°C is 5.5×10^{-51} (activity is taken equal to molality).

(a) Calculate the concentration of Ag^+ and S^{2-} in a saturated solution of Ag_2S, assuming that these are the only ions formed.

(b) For the complexing reaction $Ag^+ + 2(CN)^- = Ag(CN)_2^-$ the equilibrium constant is 6×10^{18}. Calculate the concentration of $Ag(CN)_2^-$, Ag^+, and S^{2-} in a 0.1 molal NaCN solution saturated with $Ag_2S(s)$.

(c) If the above solutions are exposed to air S^{2-} is oxidized to SO_4^{2-}. How will this affect the solubility of Ag_2S, and what would be the difference between cases (a) and (b)?

15-4 $Al(OH)_3(s)$ may dissolve in water at 25°C by the two reactions (1) $Al(OH)_3 = Al^{3+} + 3(OH)^-$ with $K_1 = 5 \times 10^{-33}$ and (2) $Al(OH)_3 + OH^- = AlO_2^- + 2H_2O$ with $K_2 = 40$.

(a) For a solution saturated with $Al(OH)_3$ calculate and plot $\log a_{Al^{3+}}$ and $\log a_{AlO_2^-}$ for pH from 1 to 14.

[1] E.D. Stensholt et al.: *Trans. Instn. Min. Met.* **95** pp. C 10-16 (1986).

(*b*) Reaction (1) is exothermic, whereas reaction (2) is endothermic. How do you expect the solubility of $Al(OH)_3$ to be affected by temperature in the two cases?

15-5 In the neutral leaching of zinc calcine a 2 molal Zn^{2+} solution is approaching equilibrium with $Zn(OH)_2$ and $Fe(OH)_3$. Calculate the equilibrium Fe^{3+} concentration and pH of the solution when for the reactions $Zn(OH)_2 = Zn^{2+} + 2(OH)^-$ and $Fe(OH)_3 = Fe^{3+} + 3(OH)^-$ the equilibrium constants are 4.5×10^{-17} and 6×10^{-38} respectively. Activity coefficients are disregarded.

BIBLIOGRAPHY

"Advances in Extractive Metallurgy," The Instn. of Min. and Met., London, 1968.

Burkin, A. R.: "The Chemistry of Hydrometallurgical Processes," E. and F. N. Spon Ltd., London, 1966.

Burkin, A. R. (ed.): "Topics in Non-Ferrous Extractive metallurgy," Society of Chemical Industry, Oxford, 1980.

Burkin, A. R. (ed.): "Extraction Metallurgy of Nickel", Society of Chemical Industry, Chichester, 1987.

Fulda, W., and H. Ginsberg: "Tonerde und Aluminium," Walter de Gruyter & Co., Berlin, 1951.

Garrels, R. M., and C. L. Christ: "Solutions, Minerals and Equilibria," Harper and Row, New York, 1965.

Gerard, G., and P. T. Stroup (eds.): "Extractive Metallurgy of Aluminum, Vol. 1, Alumina," Interscience Publishers, New York, 1963.

Gittus, J. H.: "Uranium," Butterworths, London, 1963.

Habashi, F.: "Principles of Extractive Metallurgy," vol. II, Gordon and Breach, New York, 1970.

Latimer, W. M.: "Oxidation, Potentials," Prentice-Hall Inc., Englewood Cliffs, N.J., 1952.

"Progress Report on Advances in Extraction Metallurgy," Metallurgical Reviews 151, *Met. Mat.* **5**: (3) 47 (1971).

Queneau, P. (ed.): "Extractive Metallurgy of Copper, Nickel, and Cobalt," AIME, New York, 1961.

Ritcey, G. M., and A. W. Ashbrook: "Solvent Extraction" part II, Elsevier Sci. Publ. Co., Amsterdam, 1979.

CHAPTER
SIXTEEN

ELECTROMETALLURGY

In this chapter we shall discuss the application of electrolysis to the winning and refining of metals. Electrolytic *winning* is important for the very reactive light metals aluminum and magnesium, which almost exclusively are produced by electrolysis of fused salts. For other metals such as copper and zinc, electrolytic winning from aqueous solutions represents an alternative to pyrometallurgical processes. Finally, electrolytic *refining*, with aqueous electrolytes or fused salts, is important for the production of high-purity copper and aluminum, as well as for the recovery of valuable impurities, such as silver and gold from copper. Before we discuss the various metallurgical applications it will be necessary to discuss the properties of electrolytes and the theory of electrolysis.

16-1 CONDUCTANCE AND TRANSFERENCE

We distinguish between two main types of conductors:

1 Electronic conductors Here the current is carried by free electrons. This type is again divided into two groups, (*a*) metallic conductors; and (*b*) semiconductors. The conductivity of metallic conductors is many times higher than that of semiconductors, and decreases with increasing temperature, whereas semiconductivity usually increases with increasing temperature. Sulfides and oxides of heavy metals are often semiconductors, and the same is the case for semimetals such as silicon, arsenic, and carbon.

2 Ionic conductors In these the current is carried by positive and negative ions, i.e., it is associated with a transport of chemical matter. When the electric current flows across the boundary between an ionic and an electronic conductor, as at the metallic electrodes which dip into an electrolytic solution, the ions become discharged to give neutral elements or they may react with the electrode. Thus, at the electrode where the electrons enter the electrolyte we have *cathodic (reduction) reactions* such as

$$2H^+ + 2e^- = H_2(g)$$

$$Cu^{2+} + 2e^- = Cu(s)$$

At the electrode where the electrons leave the electrolyte we have *anodic (oxidation) reactions*

$$2OH^- = H_2O + \tfrac{1}{2}O_2 + 2e^-$$

$$Cu(s) = Cu^{2+} + 2e^-$$

These are examples of electrode reactions which form the basis for electrolysis.

In addition to the above two main types we may also have *mixed conductors*. In these the current is carried partly by electrons and partly by ions. Examples are molten iron silicates and molten mixtures of lead sulfide and lead chloride. Pure iron oxide and pure lead sulfide are almost exclusively electronic conductors and cannot be electrolyzed. By dilution with silica and lead chloride respectively, the electronic conductivity decreases, and the ionic conductivity becomes more dominating. Finally a point is reached where the melts are almost exclusively ionic conductors.

Ionic Conductance

The *specific conductance*, κ, of a substance is given by $\kappa = Is/VA$. Here I is the current, V the voltage drop, A the cross-sectional area in square centimeters, and s the length in centimeters. Thus the dimension of the specific conductance is ohm^{-1} cm^{-1}, often written mho/cm.

Whereas the specific conductance of metals is of the order of 10^5 ohm^{-1} cm^{-1}, the conductance of pure water is very low. Values down to $\kappa = 0.5 \times 10^{-7}$ ohm^{-1} cm^{-1} have been reported. The conductance increases strongly upon addition of ionic compounds (acids, bases, or salts), and within the range of dilute solutions it is proportional to the electrolytic concentration. This is seen from Table 16-1 which gives some values for KCl solutions.

If κ is divided by the number of gram equivalents per cubic centimeter we get the *equivalent conductance*, λ (a 1–1 salt is regarded as monovalent, a 2–1 salt as divalent, etc.). We see that the equivalent conductance of KCl approaches a limiting value λ_0 for infinite dilution. This is analogous to the activity coefficient approaching a limiting value.

Table 16-1 Specific conductance for KCl solutions

Normality	$\kappa(25°C)$	$\lambda = \dfrac{\kappa \cdot 1000}{normality}$
1	0.111	111
0.1	0.0129	129
0.01	0.00141	141
Infinite dilution	—	$150 = \lambda_0$

In Table 16-2 some values for the equivalent conductance for some electrolytes are given at 0.1 normal and infinite dilution.

Notice the very high conductance for HCl solution, which is the result of the high mobility of the hydrogen ion. In the electrolysis of metal salt solutions acid is often added in order to increase the conductance.

Salt melts The conductance of solid salts at room temperature is usually small, but increases rapidly with increasing temperature. This is caused partly by increased mobility of the ions and partly by an increased number of lattice defects over which the ions may move. Below the melting point the conductance is still smaller than for aqueous solutions of the same salt, but increases at the melting point by a factor of one thousand or more (for KCl by a factor of 3000). Above the melting point the conductance increases with increasing temperature, more or less according with the Arrhenius equation $\kappa = \kappa_0 \exp(-E_\kappa/RT)$, where the activation energy may be of the order of 40 to 80 kJ, that is, of the same order of magnitude as for diffusion in liquids.

In Table 16-3 the conductance at the melting point is given for some salts. We see that for the alkali halides the equivalent conductance is of the same order of magnitude as for aqueous solutions of the same salts. The highest conductance is found for LiF and LiCl which have the smallest cations. Notice also the great difference between Hg_2Cl_2 and $HgCl_2$. The latter is not an ionic compound.

Table 16-2 Equivalent conductance of aqueous electrolytes

	$\lambda(N/10)$	λ_0
KCl	129	150
NaCl	107	126
$AgNO_3$	109	133
KNO_3	120	145
HCl	391	426

Table 16-3 Conductance of fused salts at melting point†

	m.p.(°C)	κ(m.p.)	Density	mol·wt	λ(m.p.)
LiF	905	20.3	1.8	26.0	290
LiCl	780	7.59	1.5	42.4	214
NaCl	850	3.66	1.5	58.5	143
KCl	800	2.19	1.54	74.5	107
Hg_2Cl_2	529	1.00	5.9	472.1	40
$HgCl_2$	295	0.0052	5.1	271.5	0.01

† From P. Drossbach: "Elektrochemie geschmolzener Salze," Springer Verlag, Berlin, 1938.

Molten slags are also ionic conductors. In Table 16-4 the conductance κ is given for some silicate melts. Again in this case the conductance is highest for the smallest cations. The conductance also increases with decreasing SiO_2 content in accordance with increasing concentration of cations and decreasing size of the silicate anion network.

If the slags are cooled to form a glass they retain an appreciable conductance. Ordinary soda-glass has a noticeable conductance and the current is carried by the Na^+ ions.

Solid electrolytes Some solids have high ionic conductivity. One example is ZrO_2 with 5 to 10 percent CaO in solid solution. As Ca^{2+} replaces Zr^{4+} in the cation lattice there will be corresponding vacancies in the anion lattice through which the O^{2-} anions may easily move. Thus $ZrO_2 + CaO$ is an oxygen ion conductor. The specific conductance κ increases with increasing temperature, from about 10^{-4} ohm^{-1} cm^{-1} at 600°C to about 10^{-1} ohm^{-1} cm^{-1} at 1200°C for ZrO_2 with 15 mole percent = 7.5 weight percent CaO.[1]

[1] E. C. Subbarao: "Solid Electrolytes and their Applications," Plenum Press, New York, 1980, p. 40.

Table 16-4 Conductance for silicate melts†

Composition	κ(1750°C)	Composition	κ(1750°C)
$MgO \cdot 2SiO_2$	0.23	$MnO \cdot 2SiO_2$	0.55
$MgO \cdot SiO_2$	0.72	$MnO \cdot SiO_2$	1.8
$2MgO \cdot SiO_2$	2.15	$2MnO \cdot SiO_2$	6.3
		Composition	κ(1400°C)
$CaO \cdot 2SiO_2$	0.31		
$CaO \cdot SiO_2$	0.83	$FeO \cdot SiO_2$	1.5
$2CaO \cdot SiO_2$	1.15	$2FeO \cdot SiO_2$	3.25

† Mainly after J. O'M. Bockris, et al.: *Trans. Faraday Soc.*, **48**: 75 (1952).

Solid zirconia electrolytes have found wide application in metallurgical research and measurements. Thus by the use of a suitable reference electrode a measuring cell may be constructed to measure oxygen potentials down to 10^{-20} atm at 1000°C, and to about 10^{-10} atm at 1600°C. At lower oxygen potentials zirconia develops electronic conductance, and is no longer useful as an electrolyte. See "Electrochemical Measurements," Sec. 16-2.

Another solid electrolyte is so-called β alumina, which actually is a sodium aluminate with approximate composition $Na_2O \cdot 10Al_2O_3$. This contains cation vacancies and has a high conductivity for alkali ions. Beta alumina has among other things been used as electrolyte in sodium–sulfur and lithium–sulfur electrical storage cells.

Solid oxide electrolytes have also been used as heating elements in electrical resistance furnaces. In order to avoid electrolysis, alternating current must be used. In contrast to metallic heating elements they may be used in air up to around 2000°C. As their conductivity at room temperature is negligible, however, they must either continuously be kept warm, or they must be preheated by some auxiliary heating element.

Ionic Transport

An example of an electrolytic cell is shown in Fig. 16-1. Here two copper electrodes are immersed in a solution of $CuSO_4$. At the electrodes the following reactions take place:

$$Cu = Cu^{2+} + 2e^- \qquad \text{(anode)}$$

$$Cu^{2+} + 2e^- = Cu \qquad \text{(cathode)}$$

In the absence of other electrolytic reactions and in the absence of electronic conduction, a total of N (Avogadro's) electrons is transferred for every gram equivalent of copper. Measurements have shown that this corresponds to

$$F = 96\,487 \text{ coulombs} = 26.80 \text{ Ah} = 96.487 \text{ kJ/V}$$

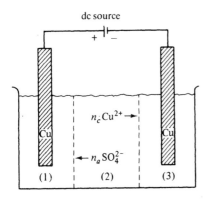

Figure 16-1 Principle of electrolytic transport: (1) anolyte; (2) main electrolyte; (3) catholyte.

all per gram equivalent of ions transferred. The quantity F is called one *Faraday*, and is the charge of N (Avogadro's) electrons.

In the electrolyte the dominating ions are Cu^{2+} and SO_4^{2-}. Under the influence of the electric field the positively charged *cations* move in the direction of the cathode, and the negatively charged *anions* in the direction of the anode. During the passage of one Faraday a total number of n_c gram equivalents of cation and n_a gram equivalents of anion will cross a plane perpendicular to the current flow. We call these numbers the cationic and anionic *transport numbers*.

In order to evaluate these numbers we imagine the electrolyte divided into three sections: (1) the space around the anode (the anolyte); (2) the main electrolyte; and (3) the space around the cathode (the catholyte). During the flow of one Faraday through the cell a total of one gram equivalent of copper goes into solution at the anode, and one gram equivalent is deposited on the cathode. From the anolyte a total of n_c gram equivalents of copper move into the main electrolyte and the same amount moves into the catholyte. From the catholyte a total of n_a gram equivalents of sulfate ions move into the main electrolyte and further into the anolyte. It is easily seen that, in order to maintain electroneutrality in the anolyte and catholyte, the sum of $n_c + n_a$ must be unity. Thus the transport numbers give the fraction of current which is carried by the two types of ions.

Since, by the passage of one Faraday, the total number of copper equivalents that are added to the anolyte or removed from the catholyte is $1 - n_c$, the transport numbers may be determined from the change in chemical composition of the anolyte or catholyte after the passage of one Faraday.[1] Examples of transport numbers in aqueous solutions are given in Table 16-5.

Notice the high cationic transport number for the acids, which is the result of the high mobility of the hydrogen ion. Likewise, the low value for NaOH shows a correspondingly high value for the OH ion mobility. Notice also that

---

[1] Since the ions usually are hydrated and carry a certain amount of water with them, thus affecting the two compositions by dilution or concentration, the above procedure will give only apparent transport numbers. For accurate work these will have to be corrected somewhat in order to give true transport numbers.

Table 16-5 Cationic transport numbers in 0.02 N solutions†

	n_c		n_c
NaCl	0.39	HNO_3	0.84
KCl	0.49	H_2SO_4	0.82
LiCl	0.33	HCl	0.83
$CuSO_4$	0.38	NaOH	0.18

† Partly from L. G. Longworth: *J. Am. Chem. Soc.*, **57**: 1185 (1935).

for KCl the cationic and anionic transport numbers are closely equal, a fact which is utilized in certain electrochemical measurements.

In Fig. 16-1 the only electrolytic process that had taken place was the transfer of copper from the anode to the cathode, whereas the electric current had been carried partly by the copper and partly by the sulfate ions. There are many cases where the elements that are dissolved from or deposited on the electrodes do not correspond to the ions which carry the current, or to any ion at all. Thus in the electrolysis of a NaOH solution, which essentially contains Na^+ and OH^- ions, oxygen is liberated on the anode and hydrogen on the cathode. The same occurs for a diluted H_2SO_4 solution. In the electrolysis of a $NaAg(CN)_2$ solution, which contains Na^+ and $Ag(CN)_2^-$ ions, silver is deposited on the cathode in spite of the fact that the silver-containing ion is negatively charged and moves toward the anode. Also in the electrolytic production of aluminum from a cryolite–Al_2O_3 melt the metal is deposited on the cathode in spite of the fact that it most probably is present in the melt as a complex anion.

If, in the last two examples, we tried to determine the apparent transport numbers for the Ag^+ and Al^{3+} cations, we would find them to have negative values. Thus there is no relation between the transport numbers and the kinds of elements that are dissolved from or deposited on the electrodes. These are determined by thermodynamic or kinetic reasons as will be discussed later.

One may ask how, in the above examples, an anion may be deposited on the cathode. Most probably one of the existing cations, for example, Na^+, is discharged to form a neutral atom. This again reacts chemically with the electrolyte to deposit the silver or aluminum with re-forming of the sodium ion. Similarly in the electrolysis of dilute H_2SO_4 solutions the SO_4^{2-} anion is discharged at the anode to give a neutral SO_4 radical, which again reacts with water to give oxygen and sulfuric acid. The stoichiometry and thermodynamics of the overall reactions are determined by the final products, however, and are independent of these intermediate steps.

Also for salt melts we have cationic and anionic transport numbers, although these are not easily determined experimentally. For silicate melts it appears that the current is carried almost exclusively by the cations. For mixed conductors where the current is carried partly by ions and partly by electrons we may talk about an *electronic transport number*, which is the part of the current carried by electrons. This part is not able to carry out electrolysis.

16-2 CELL TYPES AND POTENTIALS

We may classify electrolytic cells in two main groups:

1. Refining cells (transference cells)
2. Production cells (cells without transference)

The example which was shown in Fig. 16-1 illustrated the principle of a refining cell. If the anode and cathode reactions are added we get the total cell reaction $Cu = Cu$, that is, the reaction has been the transfer of one mole of copper from the anode to the cathode. In a transference cell the two electrodes need not be identical. Very often a metal is transferred from an impure anode or from an alloy or compound to a cathode of the pure metal.

An example of a production cell is given by

$$Pb,O_2/Cu^{2+}, SO_4^{2-}, H_2O/Cu$$

Here Pb,O_2 represents an insoluble lead anode on which oxygen is liberated. By convention the above notation denotes a cell where the positive current flows from left to right. Thus in the above case the anode reaction is

$$H_2O = 2H^+ + \tfrac{1}{2}O_2 + 2e^-$$

and the cathode reaction

$$Cu^{2+} + 2e^- = Cu$$

which give the total cell reaction

$$Cu^{2+} + H_2O = Cu + \tfrac{1}{2}O_2 + 2H^+$$

Thus, in a production cell the electrolyte is consumed, and the components are deposited on the electrodes. If the cell had been written

$$Cu/Cu^{2+}, SO_4^{2-}, H_2O/Pb,O_2$$

it would correspond to the cell reaction

$$Cu + \tfrac{1}{2}O_2 + 2H^+ = Cu^{2+} + H_2O$$

Whereas the former cell would require electric energy in order to react from left to right, the latter cell may react spontaneously and produce electric energy. It is therefore also called a *galvanic* cell. In principle any electrolytic cell, including transference cells, may be operated as a galvanic cell by reversing the electric current flow and by supplying the necessary reactants at the electrodes. In cases where all products of electrolysis remain on the electrodes the cell may be operated alternately as an electrolytic cell and as a galvanic cell, i.e., we will have a *storage cell* or *accumulator*.

Decomposition Potentials

We remember from the second law of thermodynamics (Sec. 3-2) that for reactions which occur at constant pressure and temperature, $dG \leq -\delta w'$. Here $\delta w'$ denotes work in excess of volume work, and the equality sign refers to reversible, the inequality sign to irreversible processes. In electrochemical processes w' is equal to the electrical energy which the cell can *produce*, i.e., it is the product of voltage and quantity of electricity. If an electrochemical reaction is carried out reversibly, i.e., at infinitely low current density, the cell voltage is denoted E and is called the *electromotive force* (emf) of the cell. We have in that case

$$\Delta G = -nFE \tag{16-1}$$

where ΔG is the change in Gibbs energy, n is the number of electrons transferred, and F is Faraday's number $= 96.5$ kJ/(V·eq). Notice that E is positive for negative values of ΔG, that is, for spontaneous reactions, and negative for positive ΔG values, i.e., for reactions which cannot occur unless work is applied from the outside. Thus, E is positive for galvanic and negative for electrolytic cells.

If in a cell reaction all components are present in their standard states, the Gibbs energy is denoted ΔG° and the emf is denoted E°. If for a reaction $AX = A + X$ the components are present with activities a_{AX}, a_A, and a_X, then from Eq. (3-5) we have

$$\Delta G = \Delta G^\circ + RT \ln \frac{a_A \cdot a_X}{a_{AX}}$$

and

$$E = E^\circ - \frac{RT}{nF} \ln \frac{a_A \cdot a_X}{a_{AX}} \qquad (16\text{-}2)$$

and similarly for other cell reactions.

Example 16-1 For the decomposition of water at 25°C we have:

$$2H_2O(l) = 2H_2(g) + O_2(g) \qquad n = 4$$

$$\Delta G^\circ_{298} = 473 \text{ kJ} \qquad E^\circ = -1.23 \text{ V}$$

This is the theoretical decomposition potential for water. In practice higher potentials, around 2 volts, are required in order to overcome irreversible resistance (overvoltage) on the electrodes.

Example 16-2 For the electrolytic production of magnesium at 750°C we have

$$MgCl_2(l) = Mg(l) + Cl_2(g) \qquad n = 2$$

$$\Delta G^\circ_{1023} = 477 \text{ kJ} \qquad E^\circ = -2.47 \text{ V}$$

Again, in practice higher voltages are used, mainly to overcome overvoltage and ohmic resistance in the electrolyte.

Example 16-3 For the electrolysis of Al_2O_3 at 1000°C we have

$$\tfrac{2}{3}Al_2O_3(s) = \tfrac{4}{3}Al(l) + O_2(g) \qquad n = 4$$

$$\Delta G^\circ_{1273} = 850 \text{ kJ} \qquad E^\circ = -2.20 \text{ V}$$

This would apply for the electrolysis of pure solid Al_2O_3, but applies also for a cryolite solution saturated, i.e., in equilibrium, with solid Al_2O_3. It applies for electrolysis with an *inert anode*, for example, Pt, whereby oxygen is liberated. In practice a carbon anode is used, and the overall reaction becomes

$$\tfrac{2}{3}Al_2O_3 + C = \tfrac{4}{3}Al + CO_2 \qquad n = 4$$

$$\Delta G^{\circ}_{1273} = 452 \text{ kJ} \qquad E^{\circ} = -1.17 \text{ V}$$

We see that the carbon anode has resulted in a lowering of the necessary potential, which is obtained at the expense of an equivalent carbon consumption. If the electrolyte had been unsaturated with respect to Al_2O_3 the potential would be higher. Thus for $a_{Al_2O_3} = 0.1$

$$E = E^{\circ} + \frac{R \cdot 1273}{4F} (\ln 0.1)^{2/3} = -1.21 \text{ V}$$

In order to overcome both resistance and overvoltage 4 to 5 volts are used in the industrial aluminum production.

Example 16-4 For the electrolytic refining of copper at 25°C we have

$$Cu = Cu \qquad \Delta G^{\circ} = 0 \qquad E^{\circ} = 0$$

Thus, for the transfer of copper from a pure anode to a pure cathode the theoretical potential is zero. If the anode had been an *alloy* with $a_{Cu} = 0.5$, and as $n = 2$, we would get

$$E_{298} = \frac{R \cdot 298}{2F} \ln 0.5 = -0.009 \text{ V}$$

i.e., a very small voltage. In addition there is the voltage to overcome ohmic resistance and possible overvoltages.

Electrochemical Measurements

Electrochemical cells may be used to measure Gibbs energies of reaction as well as chemical activities either in the electrolyte or in the metallic electrode. In general the thermodynamic quantities are derived from the measured reversible emf by means of Eqs. 16-1 and 16-2. If all reactants and products are present in their standard state, as for the decomposition of water or pure $MgCl_2$, the standard Gibbs energy is obtained directly. If the electrolyte is not in its standard state, its activity may be derived from the difference between E° and E.

For measurement of the activity of a component in one of the electrodes a transference cell, i.e., a cell where ΔG° and E° are zero, is usually used. This will be illustrated by the use of a solid $ZrO_2 + CaO$ electrolyte to measure the oxygen potential in metal–oxide systems or in liquid metals. In this case the cell may be written

$$Pt, p^{\circ}_{O_2}/ZrO_2 + CaO/p_{O_2}, Pt$$

with the cell reaction and charge number

$$O_2(p^\circ_{O_2}) = O_2(p_{O_2}) \qquad n = 4$$

Here Pt stands for the electrical leads, $p^\circ_{O_2}$ is the oxygen potential of the reference electrode and p_{O_2} is the oxygen potential to be measured. The charge number $n = 4$ derives from the fact that for each O_2 molecule $2O^{2-}$ ions have to be transferred. Applying Eqs. (3-3) and (16-1) this gives

$$\ln p_{O_2} = \ln p^\circ_{O_2} - 4FE/(RT)$$

As reference electrode either gaseous oxygen ($p^\circ_{O_2} = 1$ atm) or air ($p^\circ_{O_2} = 0.21$ atm) could be used, but a Me–MeO couple, like Ni–NiO, where the oxygen potential is well known, is more common. The cell may then be used to measure the oxygen potential of another Me–MeO couple or of a more complicated system. Solid electrolytes have found particular application in the measurement of oxygen potentials in liquid metals, like steel and copper, where they are used as a standard measuring device. It should be emphasized, however, that only the oxygen *activity* in the melt may be derived. In order to obtain the actual oxygen *concentration* it is necessary to know the oxygen activity coefficient as discussed in Sec. 13-2.

16-3 ELECTROCHEMISTRY OF AQUEOUS SOLUTIONS

The total electromotive force of a cell may be regarded as consisting of a cathodic and an anodic potential. Thus, for the cell

$$X/X^-, A^+/A$$

which corresponds to the cell reaction $A^+X^- = A + X$, we may write

$$E = E_{A^+/A} + E_{X/X^-}$$

Here the two *half-cell potentials* correspond to the reactions $A^+ + e^- = A$, and $X^- = X + e^-$, respectively. The first of these may be called a reduction, the second an oxidation. Alternatively, we could express the total emf as the difference between two reduction potentials

$$E = E_{A^+/A} - E_{X^-/X}$$

where $E_{X^-/X}$ corresponds to the reaction $X + e^- = X^-$. This latter convention is now internationally accepted.

The half-cell potentials may again be expressed in terms of standard half-cell potentials and activities in the electrolyte and in the electrode, viz.,

$$E_{A^+/A} = E^\circ_{A^+/A} - \frac{RT}{nF} \ln \frac{a_A}{a_{A^+}} \qquad (16\text{-}3)$$

where for monovalent ions $n = 1$. In this expression a_A and a_{A^+} give the activity of the element and ion respectively, and E° is the potential for the half-cell if all components occur in their standard states. As mentioned in Sec. 15-1 it is customary to apply a standard state for ions in aqueous solutions which makes their activity equal to their molality at infinite dilution. Thus, if the half-cell potential could be measured for decreasing ionic concentration, and with the electrode in its standard state, $E°$ as well as the ionic activity coefficient γ_+ could be calculated, the latter as function of concentration.

Unfortunately it is not possible to measure the emf of only one half-cell; we always have to work with a complete cell. Therefore, by *convention* it is agreed that the half-cell for the hydrogen reaction $2H^+(pH = 0)/H_2(1\ atm)$ shall be given the value of zero. The potential of any other half-cell may then be determined relative to the hydrogen potential. Similarly we cannot determine the individual ionic activity coefficients, only their product, from which the mean activity coefficient may be derived as shown in Sec. 15-1.

In Table 16-6 standard reduction potentials are given, all relative to the hydrogen potential. These form what we call the *electromotive series*, where negative values correspond to the most stable cations or unstable anions, and positive values correspond to the most noble metals or stable anions.

Electrolytic Separation

If a solution, which contains two cations A and B, where A is more noble than B, is electrolyzed under *closely reversible* conditions the more noble metal will be deposited on the cathode first. Assuming that the two metals do not form

Table 16-6 Reduction potentials in aqueous solutions at 25°C*

Half cell	$E°_{A^+/A}$ (volt)	Half cell	$E°_{X^-/X}$ (volt)
Ag^+/Ag	0.7991	Cl^-/Cl_2	1.3595
Cu^{2+}/Cu	0.337	$H_2O/H^+, O_2$	1.229
H^+/H_2	0.000	Br^-/Br_2	1.0652
Pb^{2+}/Pb	−0.126	I^-/I_2	0.5355
Sn^{2+}/Sn	−0.136	OH^-/O_2	0.401
Ni^{2+}/Ni	−0.250	S^{2-}/S	−0.48
Co^{2+}/Co	−0.277	Se^{2-}/Se	−0.92
Cd^{2+}/Cd	−0.403	Te^{2-}/Te	−1.14
Fe^{2+}/Fe	−0.440		
Cr^{3+}/Cr	−0.74		
Zn^{2+}/Zn	−0.763		
Mn^{2+}/Mn	−1.18		
Na^+/Na	−2.714		
K^+/K	−2.925		
Li^+/Li	−3.045		

* From W. A. Latimer: "Oxidation, Potentials," Prentice-Hall, Inc. 1951. *Note:* Latimer lists *oxidation* potentials, i.e., with opposite sign.

any alloy on the cathode, simultaneous deposition of the two metals will take place only when the potentials for their deposition become equal. Thus if their standard reduction potentials differ by an amount

$$\Delta E^\circ = E^\circ_{B^+/B} -- E^\circ_{A^+/A}$$

then, for simultaneous deposition[1]

$$\Delta E^\circ = \frac{RT}{nF}\left(\ln \frac{a_{A^+}}{a_{B^+}}\right) \tag{16-4}$$

where a_{A^+} and a_{B^+} denote the activities of the two ions in the solution. Similarly, if a solution of A^+ is exposed to metallic B a *local element* will set up and A will be precipitated by the reaction

$$A^+ + B = A + B^+$$

Again equilibrium will be established when Eq. (16-4) is satisfied.

Example 16-5 For a mixed solution of Cu^{2+} and Ni^{2+} at 25°C we have

$$\Delta E^\circ = E^\circ_{Ni^{2+}/Ni} - E^\circ_{Cu^{2+}/Cu} = -0.25 - 0.34 = -0.59 \text{ V}$$

This gives

$$\log \frac{a_{Cu^{2+}}}{a_{Ni^{2+}}} = \frac{-2 \times 96.5 \times 0.59}{19.144 \times 298} = -20$$

Taking the activities equal to the molalities this gives

$$m_{Cu^{2+}} = 10^{-20} m_{Ni^{2+}}$$

Thus a relatively small difference in E° makes possible a very high degree of electrolytic separation. In industrial processes the separation is usually poorer because deposition does not take place under reversible conditions, and because of specific overvoltage phenomena which will be discussed in Sec. 16-4.

Red-Ox Potentials

In the above discussion we considered only reactions between ions and neutral elements. This need not always be the case. Also reduction or oxidation from one valence state to another may be carried out electrochemically. A typical example is the half-cell

$$Fe^{2+}, Fe^{3+}/Pt$$

with the cell reaction

$$Fe^{3+} + e^- = Fe^{2+}$$

[1] This expression applies if the A and B ions both have the same charge n. If the A ion has the charge n and the B ion has the charge m, the activity of B^{m+} must be raised to the power of n/m.

Table 16-7 Red-Ox potentials at 25°C†

Electrode	$E°$ (volts)
Co^{3+}/Co^{2+}	1.82
MnO_4^-/Mn^{2+}	1.51
Fe^{3+}/Fe^{2+}	0.771
Cu^{2+}/Cu^+	0.153
Sn^{4+}/Sn^{2+}	0.15
Cr^{3+}/Cr^{2+}	−0.41

† From Latimer who lists *oxidation-reduction* potentials, i.e., with opposite sign.

In this case the Pt electrode merely acts as a receiver for the electrons. Such *reduction-oxidation (red-ox) potentials* may be measured relative to the hydrogen electrode, and some typical values are listed in Table 16-7.

For a metal Me we may determine standard half-cell potentials for reactions such as

$$Me^{n+} + ne^- = Me$$

$$Me^{m+} + (m-n)e^- = Me^{n+}$$

$$Me(OH)_n + nH^+ + ne^- = Me + nH_2O$$

all relative to the hydrogen electrode. We see that of these potentials the third one is also a function of pH, assuming unit activity for Me and $Me(OH)_n$. For a given metal these potentials may be plotted in a $E°$ vs. pH diagram. This is done in Fig. 16-2 for the same arbitrary metal Me which was illustrated in Fig. 15-2. Because the potential for the hydrogen reaction is dependent on pH the figure becomes distorted relative to Fig. 15-2. Diagrams like the one shown in Fig. 16-2 are called *Pourbaix diagrams*, and they are very useful for the discussion of electrochemical reactions.

In Fig. 16-2 are also shown the potentials for the evolution of hydrogen and oxygen, both at atmospheric pressure, and we see that they differ by a constant amount of 1.23 volts. This is in agreement with the emf of the water electrolysis being independent of the pH.

Liquid Junction Potentials

At this point a brief mention should be made of the potential in cells where the anolyte and catholyte have different compositions, as for example

$$H_2(1 \text{ atm})/H^+(a_1) \quad \vdots \quad H^+(a_2)/H_2(1 \text{ atm})$$

or

$$Zn/Zn^{2+} \quad \vdots \quad Cu^{2+}/Cu$$

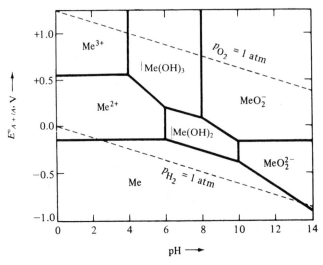

Figure 16-2 Predominance areas (Pourbaix diagram) for the system Me–H$_2$O as function of pH and reduction potential. Compare Fig. 15-2.

Here a_1 and a_2 denote different hydrogen ion activities and the dotted line the junction between the two compositions. In such cells the total emf is determined not only by the ɔlectrode reactions. In addition to the electrode potentials, which can easily be calculated, there will be a potential set up at the liquid junction. The latter is caused by the fact that the transport numbers of the cations and anions usually are different. Thus an electric *double layer* will be established at the liquid boundary, giving rise to an additional potential. This potential may be calculated from a knowledge of the transport numbers, or the experiment may be carried out in the presence of an excess of KCl. In the latter case most of the current will be carried by the K$^+$ and Cl$^-$ ions, and since their transport numbers are almost equal the liquid-junction potential may be virtually eliminated.

Liquid-junction potentials may affect the potential of some electrolytic processes, but they are usually small and are not likely to influence industrial electrolysis.

16-4 IRREVERSIBLE PHENOMENA

The electromotive force of a cell can be measured only if the cell reaction takes place under reversible conditions, i.e., at infinitely low current density. In practical electrolysis, which occurs at a finite rate, there will be additional potentials. One of these has already been mentioned: the voltage drop due to ohmic resistance in the electrolyte. A second voltage drop is due to ohmic resistance in the electric leads and in the electrodes. These terms are proportional to the current and to the resistance of the electrolyte and leads respectively. In addition there will be a voltage drop which we call the *overvoltage*, η, of the cell. Thus the total applied voltage may be expressed as:

$$V = -E + (R_1 + R_2)I + \eta$$

Here R_1 and R_2 are the resistances of electrolyte and leads, and I the current. The emf is given with negative sign, since it, by definition, is the voltage the cell can *produce*, whereas V is the voltage *applied* to the cell. Thus for *electrolytic* cells, where E is negative, the additional voltage drops cause an increase in the applied voltage, whereas for *galvanic* cells the additional voltage drops reduce the voltage from the cell.

The overvoltage, η, is caused either by the reactants not being supplied to the electrodes as fast as they are removed or by the reaction products not being removed as fast as they are supplied, and it may be divided into concentration overvoltage and activation overvoltage.

Concentration overvoltage is the result of macroscopic changes in the composition of the electrolyte in the vicinity of the electrodes. Thus in the electrolytic refining of copper, copper goes into solution at the anode and is deposited at the cathode. This creates a concentration cell, the emf of which works against the current and adds to the voltage needed for the electrolysis. Concentration overvoltage may be reduced or eliminated by stirring of the solution, and it usually does not greatly affect industrial electrolytic processes.

Activation overvoltage is a phenomenon which occurs on a molecular or atomic scale on the electrodes, and its nature is only partly understood. It may be measured as the potential between a reversible electrode, i.e., an electrode which does not carry any current, and a similar electrode with a certain current density. The overvoltage increases with increasing current density, and for many reactions and within a certain current range it is found to follow the *Tafel equation*

$$\eta = a + b \, \log \, (I/A)$$

Here η is the overvoltage, I/A the current density, and a and b are constants which are characteristic for the electrode reaction in question.

The magnitude of the activation overvoltage differs greatly for different electrode reactions and is also a function of the electrode material. It is generally large for the evolution of gases, particularly oxygen and hydrogen from aqueous solutions, but is known also for the evolution of CO_2 during aluminum electrolysis. In Table 16-8 some values for hydrogen and oxygen overvoltage are given for different electrode materials and for different current densities.

For the electrolysis of water where $E° = -1.23$ V the hydrogen and oxygen overvoltages on Pt electrodes at $I/A = 10^{-3}$ A/cm^2 give an additional 0.8 V or a total voltage of more than 2 V. On Ni electrodes the total voltage is about 1.9 V. The activation overvoltage may be reduced by giving the electrodes a *spongy surface*. It also decreases with increasing temperature, and industrial water electrolysis is usually carried out at 1.7 to 2.0 V.

**Table 16-8 Activation overvoltages for hydro-
gen and oxygen evolution**

Electrode	H_2 overvoltage at I/A (A/cm^2)		O_2 overvoltage at $I/A =$ 10^{-3} A/cm^2
	10^{-3}	1	
Al	0.58	0.78	0.5
Graphite	0.47	1.03	0.53
Cu	0.60	0.84	0.4
Fe	0.40	0.77	—
Hg	1.04	1.24	—
Ni	0.33	0.59	0.35
Pt solid	0.09	0.44	0.70
Pt sponge	0.01	0.07	0.40

For the deposition of metals, both from aqueous and molten salt electrolytes, the overvoltage is usually small, but may in certain cases be measurable.

We also have overvoltage in the dissolution of electrodes. Thus in the
electrolytic *combustion* of hydrogen with oxygen, which is the opposite of electrolytic decomposition of water, there is appreciable overvoltage. This makes
the voltage which can be obtained from such a *fuel cell* less than the reversible
emf of the cell. Also for anodic dissolution of certain metals there is a noticeable
overvoltage. This is the case, e.g., for nickel and lead, which may be used as
anodes in processes where oxygen is evolved in spite of the fact that their
reduction potentials are lower than for oxygen and that they therefore should
go into solution if equilibrium were established. Other examples are iron and
chromium which under anodic conditions become *passivated*. Particularly for
chromium this passivation is pronounced and is utilized, e.g., in stainless steel.
When a metal is passivated the overvoltage for its anodic dissolution makes its
reduction potential more positive, i.e., the metal appears to be more noble than
its reversible electrode potential shows.

Overvoltages may greatly alter the outcome of electrolytic processes. One
very important example is the electrolytic production of *zinc*. According to the
emf series zinc is less noble than hydrogen and we should expect only hydrogen
to be formed at the cathode. The hydrogen overvoltage is rather large, however,
and by proper operation the product of electrolysis is mainly zinc. Another
example is *anodic oxidation* processes: Working with anodes which give high
overvoltages for oxygen evolution strongly oxidizing conditions may be established and oxygen-rich compounds such as persulfates, chlorates, manganese
dioxide, and lead dioxide may be produced.

The relation between voltage and current density for electrolytic processes
is summarized in Fig. 16-3. The curve to the right illustrates a *production cell*
(electrowinning). Even at voltages below the reversible emf there is a small
current flowing, the *diffusion current*. This is caused by deposition of elements

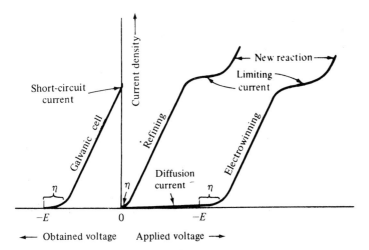

Figure 16-3 Current density vs. applied voltage for electrowinning and refining (right) and galvanic cell (left).
Note: for the galvanic cell the current direction is opposite to that of electrolysis. $-E$ = reversible emf with opposite sign.

which dissolve in the electrolyte and diffuse to the other electrode where they again become ionized. At the reversible emf there is no pronounced increase in current density, and the current is small until the activation overvoltage has been overcome. The current now increases more or less in proportion to the applied voltage in accordance with Ohm's law. Finally, at very high current densities the current may again level off as a result of increasing concentration overvoltage. If the voltage is increased further the current may again increase due to the deposition of other (less noble) reaction products.

For *electrorefining* the reversible emf is practically zero. Also the overvoltage is usually small, and the voltage drop is determined mainly by ohmic resistance. Here also a limiting current, due to concentration overvoltage as well as changes in reaction mechanism and deposition of other products may occur (Fig. 16-3 middle).

For a *galvanic cell* (Fig. 16-3 left) the overvoltage and the ohmic resistance work against the emf of the cell, and the obtained voltage decreases with increasing current density. Finally by short-circuiting of the electrodes the short-circuiting current is reached.

16-5 CURRENT AND ENERGY EFFICIENCY

Current efficiency is the ratio between amount of wanted product obtained and the amount which is to be expected from Faraday's law. Current efficiencies may be less than 100 percent for the following reasons:

1. Deposition of unwanted products, e.g., hydrogen in deposition of zinc
2. Electronic conductivity in the electrolyte
3. Chemical short-circuiting
4. Electrical short-circuiting inside or outside the electrolytic cell

Of these the first two have already been mentioned. Since the hydrogen overvoltage increases with increasing current density the ratio of zinc to hydrogen will also increase, i.e., the current efficiency increases with increasing current density.

Chemical short-circuiting will be illustrated by the electrolysis of an iron sulfate solution with *insoluble anode*. At the anode, in addition to O_2 evolution, Fe^{2+} ions will oxidize to form Fe^{3+}. The ferric ions diffuse to the cathode where they again are reduced to Fe^{2+}. Thus one Faraday has passed through the cell without any metal deposition. The part of the current which is transferred in this way depends on the rate of diffusion and convection in the electrolyte. For a given cell this occurs at a more or less constant rate, whereas the deposition of iron on the cathode increases with increasing current density. Thus again the current efficiency is increased by increased current density. One way to reduce chemical short-circuiting is to insert a semipermeable *diaphragm* between the anolyte and the catholyte. This cuts down the diffusion of the ferric ions and maintains a high concentration of ferrous ions in the catholyte. On the other hand the diaphragm causes a considerable increase in ohmic resistance.

Chemical short-circuiting is known to occur in many electrolytic processes. Thus in the electrolysis of fused NaOH the passage of 2 Faradays deposits 2Na on the cathode and 2OH on the anode. The latter reacts to give $H_2O + \frac{1}{2}O_2$. Of these water is quite soluble in the electrolyte and diffuses to the cathode where it reacts with one Na to give $NaOH + \frac{1}{2}H_2$. The overall reaction per 2 Faradays is then $NaOH = Na + \frac{1}{2}O_2 + \frac{1}{2}H_2$, corresponding to a current efficiency of 50 percent. In practice the efficiency is even less. Similar phenomena occur in the electrolytic production of aluminum (see Sec. 16-7).

The energy efficiency is usually defined as the product of the current efficiency and the ratio between theoretical (reversible) and actual voltage. It thus becomes equal to the ratio between the Gibbs energy of reaction, ΔG, and the actually applied electric energy, $-w'$. It should be pointed out, however, that for electrolytic processes where gases are evolved the entropy of reaction is usually positive, and hence the enthalpy of reaction is more positive than the Gibbs energy. According to the first law of thermodynamics (Sec. 2-2) we have for reactions at constant pressure and temperature

$$\Delta H = \Delta G + T\Delta S = q - w'$$

Therefore, if the supplied electric energy, $-w'$, were equal to ΔG, it could still be less than ΔH. The remaining energy $T\Delta S$ would then have to be drawn from the heat content of the electrolyte, which would cool, or it would have to be supplied from the surroundings. Thus in order to carry out an electrolytic process adiabatically and at constant temperature the electric energy must be equal to ΔH and not to ΔG of the reaction.

For the electrolysis of fused salts at elevated temperature energy is needed also to heat the reactants to the cell temperature. In addition there must be energy to cover heat losses. Of the total enthalpy requirement only ΔG needs to be supplied by electricity, the remainder may be supplied from fuel. In practice, however, in order to obtain a sufficiently high production rate, the applied voltage is usually much larger than the theoretical one, and it may even become a problem to avoid overheating of the cell.

In view of the facts given above it seems reasonable to redefine the energy efficiency as the ratio between *total enthalpy* required and actually *applied electric energy*. The difference between the two would then be the heat losses.

16-6 AQUEOUS PROCESSES

General Techniques

We distinguish between *winning* (production) and *refining* processes. For both types very much the same cell equipment is used. The cells are rectangular tanks with a cross section of about 1 m × 1 m and with a length which can be varied with the desired capacity, but which is usually between 5 and 10 m. The electrodes may have a cross section of about 0.9 m × 0.9 m, and are suspended vertically in the tank. Anodes and cathodes are placed alternately. In order to give a low resistance but without danger of metallic contact, the distances between anode and cathode may be 3 to 5 cm. An exception is represented by processes with diaphragms, where the distances may be about 10 cm. Usually all the anodes and all the cathodes in one tank are connected in *parallel*, making the voltage drop for the whole tank equal to the voltage drop between each set of electrodes. Since this voltage drop is usually much less than the available dc voltage several tanks are connected in series until they correspond to the voltage of the dc source. An exception is the so-called *series* system sometimes used for refining of copper. Here a number of electrodes of impure copper is placed in the cell, and the dc source is connected only to the two electrodes at the end of the tank, these acting as anode and cathode respectively. The current traverses all the electrodes, these acting as anode on one side and cathode on the other. This system has the advantage of a higher voltage drop per cell, and is suitable for small plants. Its disadvantages are many, however, and the system is not much used.

The current density for most aqueous plants is between 100 and 500 A/m², and is adjusted by means of the applied voltage as well as by the temperature and acidity of the electrolyte.

The electrolyte is usually made to flow from one tank to another. This is done partly to obtain the necessary stirring and partly to circulate the electrolyte between leaching and electrolysis or between electrolysis and purification of the electrolyte.

Aqueous Winning Processes

Electrolytic winning of copper from, e.g., leaching solutions, is a relatively simple process, copper being a rather noble metal, and impurities such as zinc or iron will remain in the electrolyte. Also, the electrolysis may be done from highly acid solutions and above room temperature without danger of hydrogen evolution. Copper is deposited on copper cathodes (starting sheets) and oxygen is liberated on insoluble (lead) anodes. For the electrolysis of a one molal $CuSO_4$ solution at pH = 0, the theoretical emf is -0.89 V. In practice about 2.5 V are used. Owing to the presence of ferrous-ferric ions in the solution the current efficiency is only about 80 to 90 percent. This makes the energy requirement about 2.5 kWh per kilogram of copper produced.

If the electrolytic plant works in combination with the leaching plant, the spent electrolyte, which still contains appreciable amounts of copper, is recirculated. Eventually the content of impurities builds up to a level where some electrolyte has to be withdrawn. The remaining copper may then be precipitated by cementation on scrap iron.

Electrolytic winning of zinc is more difficult. As already mentioned the reduction potential for zinc is more negative than for hydrogen. Zinc electrolysis is made possible by the high hydrogen overvoltage on zinc, which increases with increasing current density. The hydrogen overvoltage is lowered by the presence of certain impurities such as Co, Cu, and Sb, however, and it is important that the electrolyte be very pure. The solution from the leaching plant, therefore, has to be purified extensively as discussed in Sec. 15-4. The hydrogen overvoltage also increases with decreasing temperature, and the temperature of the electrolyte is therefore not allowed to rise much above 30°C. The only impurity which can be tolerated is manganese. Manganese is less noble than zinc and does not affect the cathode reaction. On the *anode*, however, it is partly oxidized to form solid MnO_2, which settles in the cell. This MnO_2 may be returned to the neutral leach where it assists in the oxidation of Fe^{2+} to Fe^{3+} (Sec. 15-4).

The reversible emf for a one molal Zn solution at pH = 0 is -1.99 V, and in practice about 3.5 V is used. The current efficiency is about 90 percent which makes the energy consumption per kilogram of zinc about 3.2 kWh.

The electrolysis is carried out with cathodes of hard rolled aluminum, the hydrogen overvoltage on Al also being quite high. Every 1 or 2 days the zinc product is stripped from the aluminum sheets. The anodes are made of lead or a lead–antimony or lead–silver alloy. Such alloys are used partly to make the anodes mechanically stronger and partly to lower the overvoltage for oxygen evolution, thus preventing the anode material from being dissolved.

Starting with an essentially neutral $ZnSO_4$ solution the electrolysis is carried to the point where about half the zinc content is deposited. The acid content of the electrolyte has then increased to more than 100 g H_2SO_4 per liter. This solution is returned to the leaching plant.

Manganese is even more difficult to deposit electrolytically than zinc, its standard reduction potential being 1.18 volts more negative than hydrogen. Since the hydrogen potential decreases with increasing pH the deposition of manganese is favored by a high pH. Since hydrogen ions are formed at the anode it is only possible to deposit manganese if the anolyte and the catholyte are separated by a diaphragm, usually canvas cloth. Fresh $MnSO_4$ solution with a pH of 8.2 to 8.4 is introduced at the cathode; higher pH would result in precipitation of manganese hydroxide. As the metal is being deposited the electrolyte seeps into the anode chamber, thus preventing the hydrogen ions from disturbing the cathode reaction. Nevertheless, the current efficiency for manganese is rather low, of the order of 40 percent. Also since rather high cell voltages, about 5 V, are used, the energy consumption is large, about 8.5 kWh/kg.

The acidified anolyte, which still contains appreciable amounts of manganese, is recirculated to the leaching plant.

Iron may be produced by electrolysis of a neutral $FeSO_4$ solution analogous to the Mn electrolysis. The diaphragm here has the additional function of preventing the ferric ions formed at the anode from entering the cathode chamber, where they would cause chemical short-circuiting. The energy consumption is about 3.5 kWh/kg, and the method has limited application.

Aqueous refining processes

Aqueous refining processes are used for copper, silver, and gold, as well as for nickel and, in certain cases, lead.

Copper refining is perhaps the most important aqueous refining process. It is used either if the copper contains harmful impurities such as antimony and arsenic, or if it contains enough silver and gold to justify the refining expenses. The anodes are cast from fire-refined copper or directly from blister copper. The starting sheets for the cathodes are usually electrodeposited copper which becomes part of the final product. The electrolyte is an acid solution of copper sulfate, and is heated to 50 to 60°C.

During electrolysis copper and less noble metals go into solution at the anode, whereas the more noble metals, silver and gold, remain unattacked. Also copper compounds with antimony, sulfur, and selenium remain undissolved, and together with the noble metals form the *anode mud*. Copper is deposited on the cathode whereas the less noble metals, nickel, iron, and arsenic, remain in solution. From time to time part of the electrolyte is withdrawn in order to remove these elements and prevent them from accumulating. Also because of the action of the air some copper dissolves chemically from the anode, thus increasing the copper content of the electrolyte. Part of the electrolyte therefore has to be electrolyzed with insoluble lead anodes in order to keep the copper content at a constant level.

For copper refining the theoretical potential is zero, but in order to overcome ohmic resistance in the electrolyte and possible overvoltages about 0.2 to 0.4 V is used. The current efficiency is about 95 percent, which gives 0.2 to 0.4 kWh per kilogram of refined copper. In addition about 1 kg of steam is needed to keep the electrolyte at the desired temperature.

Silver and *gold* are recovered from the anode mud. First the mud is treated with concentrated sulfuric acid to dissolve copper and other impurities. The remaining noble metals are cast into anodes and electrorefined. In the first refining stage silver is deposited on the cathode, less noble metals such as copper remain in the solution, whereas gold is left behind as the anode mud. This anode mud is once again cast into anodes and electrolyzed to give cathodic gold.

Lead may be electrorefined if it contains bismuth. Since lead forms insoluble compounds both with sulfuric and hydrochloric acids, an electrolyte of *silicofluoric acid* (H_2SiF_6) is used. The electrolysis is carried out very much as for copper, and bismuth remains undissolved in the anode mud from which it may be recovered.

Nickel refining is more difficult, since the main impurities copper, cobalt, and iron are rather close to nickel in the electromotive series. It might be possible to get a good separation under nearly equilibrium conditions, but this would require a very low rate of electrolysis. The methods which are used today are based on diaphragms which separate the cell into anode and cathode chambers. The purpose of the diaphragm is different from that in the manganese electrolysis, however. In the anode chamber nickel goes into solution together with copper, cobalt, and iron, whereas the more noble elements, such as the platinum metals, remain in the anode mud. The anolyte is withdrawn from the cell and purified by hydrometallurgical methods: iron is precipitated as hydroxide by oxidation with air, cobalt is precipitated by oxidation with chlorine, and copper is precipitated by cementation on nickel powder, in the manner described in Chap. 15. The purified solution is returned to the cell and introduced in the cathode chamber where nickel deposits on the cathode while the solution seeps through the canvas into the anode chamber. Because of the diaphragm the voltage employed is higher than for copper refining, 1.5 to 2 V, with correspondingly larger energy consumption.

The International Nickel Company[1] has developed a method whereby impure *nickel sulfide* is cast into anodes and electrolyzed directly. The overall reaction is

$$Ni_3S_2 = 3Ni + 2S$$

Elemental sulfur remains on the anode whereas nickel is deposited on the cathode. Also in this case a diaphragm is used, and the anolyte is removed and

[1] L. S. Renzoni et al.: *J. Met.*, **10**: 414 (1958).

purified in the same way as described above. The voltage employed is 3 to 4, and the anodic current efficiency is about 95 percent. The cathodic efficiency is higher, however, and in order to prevent the electrolyte from being depleted in nickel some nickel hydroxide or carbonate is added. A similar process has been tried for copper sulfide (white metal).

16-7 FUSED SALT PROCESSES

The most important fused salt electrolysis is the production of *aluminum* by the so-called *Hall-Héroult* process. The type of cell used is shown in Fig. 16-4. The cell pot, which is lined with carbon, acts as the cathode, and the current is taken out through heavy steel bars embedded in the bottom of the cell. The anode too is made of carbon, but because it is consumed during electrolysis it has to be continuously renewed. Two types of anodes are in use: *prebaked* and *self-baking*. Prebaked anodes consist of individual carbon blocks, and after one block has been nearly consumed a new one is put in its place. The self-baking electrode is illustrated in Fig. 16-4. The anode is made up of carbon paste held inside a steel frame. As it descends toward the hot zone the paste hardens, and as the anode is consumed from below new carbon paste is added. The electric current is introduced through steel studs which are embedded in the anode, and which are pulled out and moved up as the anode descends. As for aqueous

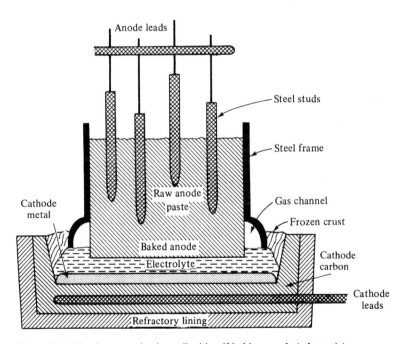

Figure 16-4 Aluminum production cell with self-baking anode (schematic).

processes a number of cells, or "pots," are connected in series to make the total voltage drop equal to that of the available dc source.

The electrolyte consists of molten cryolite, Na_3AlF_6, with 1 to 8 percent Al_2O_3. The melting point of cryolite is about 1010°C, but is lowered by the addition of Al_2O_3. Also other additions such as AlF_3 and CaF_2 may be made. According to existing information it is mostly likely that aluminum in the electrolyte is present as a complex oxyfluoride anion. During electrolysis aluminum is deposited on the cathode and oxygen is discharged on the anode where it reacts with carbon to give essentially CO_2. In the anode gas some CO is also found. As the electrolyte becomes depleted in Al_2O_3 fresh oxide is added. If the electrolyte should be completely depleted in oxide the resistance of the cell will increase drastically. As the total current through a pot-line in series is essentially fixed, this again means that the voltage drop across the cell increases drastically, from about 4 to about 40 volts. This is called the *anode effect*.

Impurities in the raw materials or in the anode which are more noble than aluminum, such as iron and silicon, will go into the cathode metal. It is therefore very important that the aluminum oxide as well as the anode paste are as free of impurities as possible. By careful control of these materials the aluminum produced may have a purity of 99.5 percent or better.

Sodium in the electrolyte is less noble than aluminum and will not be deposited on the cathode. Nevertheless, a certain sodium activity will be established, and sodium vapor may diffuse into the colder parts of the cathode, where it may "dissolve" in the carbon. This deposition of sodium tends to ruin the cathode carbon and may be one of the reasons for the limited lifetime of the cathode. Small amounts of aluminum carbide may also be formed at the cathode.

The current efficiency in aluminum production is between 80 and 95 percent. The loss of efficiency is believed to be caused by aluminum which dissolves in the electrolyte and is transferred to the anode where it is reoxidized. This reoxidation could possibly be electrochemical, but it is more likely caused by the anode gas, viz.,

$$3CO_2 + 2Al^* = Al_2O_3 + 3CO$$

where Al* denotes aluminum dissolved in the electrolyte. Indeed a correlation has been found between the current efficiency and the CO content in the escaping anode gas; the lower the current efficiency the higher is the CO content of the gas. The transport of aluminum to the anode decreases with increasing anode-to-cathode distance, giving an increased current efficiency. Also, for constant bath temperature, the current efficiency increases with increasing current density.

As mentioned in Sec. 16-2 (Example 16-3) the standard potential for the aluminum electrolysis with carbon anode is about 1.2 V. In addition there is a noticeable overvoltage for the evolution of CO_2 on the anode, about 0.4 V, whereas the cathodic overvoltage seems to be negligible. This gives a minimum voltage for the electrolysis of 1.6 V. In order to overcome ohmic resistance in the electrolyte and in the leads and electrodes, and to give a favorable production rate, cell voltages of about 4 are used. Combined with a current efficiency of, say, 88 percent this gives about 15 kWh per kilogram of aluminum. In comparison for the overall reaction

$$2Al_2O_3 \ (25°C) + 3C \ (25°C) = 4Al \ (950°C) + 3CO_2 \ (950°C)$$

the total theoretical energy requirement is 2420 kJ, or 6.22 kWh/kg Al. This means that the energy efficiency of the process is only about 40 percent, the rest being heat losses. It is clear, therefore, that much work has been done both to increase the current efficiency and to lower the necessary cell voltage. Thus whereas in 1930 about 25 kWh were needed, the best cells in 1980 use only 13 kWh/kg Al.

It should be pointed out, however, that not only the energy consumption but also the capital and labor expenses should be considered. Thus by decreasing the current density in a given cell the voltage drop and the energy consumption per kilogram of aluminum may be decreased. On the other hand the production rate per cell will decrease with a corresponding increase in capital expenses.

In addition to improvements of the traditional Hall–Héroult process much work has been done in recent years on alternative aluminum processes. Most interesting in this respect seems to be electrolysis of *aluminum-chloride* as developed by the Aluminum Company of America (Alcoa). In this process $AlCl_3$, dissolved in a suitable halide melt, is electrolyzed to give aluminum and chlorine. The chlorine gas is again used to produce the necessary chloride from aluminum oxide + carbon. The chloride process has the advantage that it can operate at a lower temperature than the Hall–Héroult process, and that the carbon anode is not consumed. The electric energy consumption is reported to be 8.9 kWh/kg Al for the electrolysis. On the other hand the process involves chlorination of oxide, which also may consume some energy.

Electrolytic refining of aluminum is done by the so-called "three-layer method." Here the electrolyte consists of a mixture of barium, sodium, and aluminum halides, which has a higher density than the pure aluminum cathode. The anode consists of aluminum scrap with enough copper added to make it *heavier* than the electrolyte. The electrolysis is carried out in cells which resemble the production cells, but with the exception that the anode is at the bottom and the cathode metal floats on top. The current from the cathode is taken out through graphite electrodes which dip into the cathode metal. Also, in order to prevent short-circuiting, the sides of the cell have to be of insulating material, e.g., refractory bricks. The cell temperature may be 700 to 900°C.

As for other refining processes the theoretical voltage is close to zero, but in order to generate enough heat in the relatively small cells, voltages around 5 are used. Also, in order to prevent contamination a larger anode-to-cathode distance (15 to 18 cm) is used. The current efficiency is better than for electrowinning, about 90 to 95 percent. This makes the energy consumption about the same as for the Hall–Héroult process, and it is evident that aluminum refining is economical only if a high purity metal is required.

Magnesium is mainly produced by electrolysis of molten $MgCl_2$–$CaCl_2$–NaCl mixtures at around 750°C. Since magnesium does not alloy with iron, iron may be used as material for the furnace pot, and also as the cathode. The anode, where chlorine is evolved, is made of graphite. The raw material may be either anhydrous magnesium chloride, made by methods described in Sec. 14-5, or a material with some oxide or crystal water may be used. In the latter case the chloride is charged around the carbon anode. As it heats up it reacts with chlorine and the anode, viz.,

$$2MgO + 2Cl_2 + C = 2MgCl_2 + CO_2$$

or

$$2H_2O + 2Cl_2 + C = 4HCl + CO_2$$

In that way there will be a certain consumption of the anode. If anhydrous chloride is used the anode consumption is small.

Magnesium electrolysis differs from aluminum electrolysis by the metal being lighter than the electrolyte. After its deposition on the iron cathode it rises to float on top of the electrolyte. Special precautions are needed in order to prevent it from coming into contact with the anode gas, such as a curtain which dips into the electrolyte and separates the anode and cathode compartments.

As previously mentioned the theoretical decomposition voltage for pure $MgCl_2$ is about 2.5. In practice 6 to 7 volts are used, which together with a current efficiency of 75 to 90 percent gives 18 to 20 kWh per kilogram of Mg produced.

Sodium and calcium may be produced similarly to magnesium. Sodium may be produced either by electrolysis of molten NaOH or molten NaCl. The latter gives a higher current efficiency. Since the melting point of NaCl is close to the boiling point of sodium, a NaCl–$CaCl_2$ eutectic electrolyte is used. As a result the cathode metal becomes a Na–Ca alloy. By cooling this alloy to just above the melting point of sodium, the melt segregates into solid calcium and essentially pure liquid sodium.

The rarer metals, **titanium, zirconium, and niobium** are not readily produced by electrolysis of fused salts. One reason is the low solubility of their halides in suitable salt mixtures, another their high melting points which means that the metals will be deposited in solid form. The main reason is, however, that the metals may form ions of different valency. Thus, titanium may occur as Ti^{2+}, Ti^{3+}, and Ti^{4+} ions. This gives rise to chemical short-circuiting, the current being used to oxidize the ions on the anode and reduce them on the cathode. Nevertheless, some work has been done with suitable diaphragms whereby the metal has been obtained.

For these metals electrolytic *refining* seems more interesting. If the anode were titanium scrap, this would go into solution and the titanium ions would remain in a low valence state. Thus chemical short-circuiting may be avoided.

Another possibility would be to use anodes of, e.g., titanium carbide formed by carbothermic reduction of the oxide. In that case the carbon would remain as anode mud. Such a process could possibly be a competitor to the present metallothermic reduction of the chlorides.

PROBLEMS

16-1 In the electrolytic refining of nickel the anode is a Cu–Ni alloy with 80 atom % Ni, whereas the cathode is pure nickel.

(a) Calculate the *reversible* emf at 25°C when the alloy is assumed to be an ideal solid solution.

(b) In *practice* both Ni and Cu dissolve from the anode. The anolyte is withdrawn and copper is removed by cementation as pure Cu on Ni sponge. Calculate the equilibrium ratio Cu/Ni in the purified electrolyte at 25°C, when the ionic activities are taken equal to their molalities.

(c) In practice the applied cell voltage is 2.0 and the cathodic current efficiency is 95 percent. Calculate the electric energy consumption in kilowatthours per kilogram of refined nickel.

16-2 An electrolytic cell operates at 700°C with a Pb–Mg alloy as the anode and pure Mg as the cathode. The electrolyte is a molten $MgCl_2$–NaCl mixture.

(a) Calculate the reversible emf when the Mg activity in the alloy, relative to pure liquid magnesium, is 0.04.

(b) Calculate the practical cell voltage at a current density of 1 A/cm^2 when the conductance of the electrolyte is 2.5 ohm^{-1} cm^{-1}, and the anode-to-cathode distance is 10 cm. Overvoltages are insignificant.

(c) Calculate the electric energy consumption in kilowatthours per kilogram of Mg, as well as the production rate in kilograms of Mg per square meter·hour for case (b) when the current efficiency is 90 percent.

16-3 In the electrolytic winning of Zn the cell voltage, V, and current efficiency, C.E., vary with the current density as given below:

A/cm^2	V	C.E.(%)
0.01	2.5	80
0.02	2.7	90
0.05	3.0	94
0.10	3.5	96

(a) Calculate the electric energy consumption in kilowatthours per kilogram of Zn for the four current densities.

(b) In practice also capital investments must be considered. Indicate which current density you would choose to give the lowest overall cost of electrolysis.

16-4 In the electrolytic winning of aluminum the overall reaction is $2Al_2O_3 + 3C = 4Al + 3CO_2$. The applied voltage is 5.0, and the current efficiency is 90 percent.

(a) Calculate the consumption of Al_2O_3, carbon, and electric energy all per kilogram of produced aluminum.

(b) The raw materials are introduced at 25°C and the products are withdrawn at 950°C. Make a heat balance for the process and calculate heat surplus or deficiency per kilogram of aluminum.

BIBLIOGRAPHY

Dobos, D.: "Electrochemical Data," Elsevier Sci. Publ. Comp., Amsterdam, 1975.

Gerard, G., and P. T. Stroup (ed.): "Extractive Metallurgy of Aluminum," AIME, New York, 1963.

Grjotheim, K., and B. J. Welch: "Aluminium Smelter Technology," Aluminium-Verlag, Düsseldorf, 1980.

Grjotheim, K., C. Krohn, M. Malinovský, K. Matiašovský, and J. Thonstad: "Aluminium Electrolysis," 2d ed., Aluminium-Verlag, Düsseldorf, 1982.

Latimer, W. M.: "Oxidation, Potentials," Prentice-Hall, Inc., Englewood Cliffs, N.J., 1952.

Mantell, C. L.: "Electrochemical Engineering," McGraw-Hill Book Co. Inc., New York, 1960.

Pourbaix, M.: "Atlas of Electrochemical Equilibria in Aqueous Solutions," Pergamon Press, London, 1966.

Queneau, P. (ed.): "Extractive Metallurgy of Copper, Nickel and Cobalt," AIME, New York, 1961.

Subbaroa, E. C. (ed.): "Solid Electrolytes and their Application," Plenum Press, New York, 1980.

Waddington, J.: "Refining of Aluminium" in "The Refining of Non-Ferrous Metals." Instn. of Min. Met., London, 1950.

ENTHALPIES OF FORMATION AT 25°C[1]

Most values refer to the formation of the various compounds from the stable modification of the elements, all at 25°C (298 K) and one atmosphere pressure. Where elements or compounds occur in different modifications, with different enthalpies, this is indicated in the tables. For the states of aggregation the following symbols are used: (s) = solid, (l) = liquid, (g) = gas. Special modifications are shown by name or accepted symbols. For the silicates and a few other substances the values refer to formation from the component oxides. This is indicated in the tables.

The enthalpy of formation at higher temperatures may be calculated from the equation

$$\Delta H_T^\circ = \Delta H_{298}^\circ + \Delta(H_T^\circ - H_{298}^\circ)$$

Here $\Delta(H_T^\circ - H_{298}^\circ)$ is the difference between the enthalpy increments, above 25°C, for the product and for the elements. These data are given in App. B.

The enthalpy of formation is, in principle, also a function of pressure. For practically all substances this dependency is very small, however, and the listed values may, without loss in accuracy, be used between zero and a few atmospheres, i.e., under all conditions normally encountered in metallurgical processes.

With few exceptions the values given in this part are converted to joules from data given by O. Kubaschewski et al.: "Metallurgical Thermochemistry," Pergamon Press, London 1979, and the reader is referred to that text for additional data.

Please note that the tables list $-\Delta H_{298}^\circ$, i.e., the enthalpy of formation with opposite sign. Thus ΔH_{298}° is positive when $-\Delta H_{298}^\circ$ is negative.

---

[1] This and the subsequent appendices are available as a separate booklet ("Thermochemical Data for Metallurgists" by the present author) from the Tapir Publ. Co., Trondheim, Norway.

Substance	Mol wt	$-\Delta H^\circ_{298}$ kJ/mol	Accuracy $\pm$ kJ
Ag(s)	107.9	0	
AgCl(s)	143.3	127.1	0.1
Ag_2O(s)	231.7	30.5	0.4
Ag_2S(s)	247.8	31.8	0.8
Al(s)	27.0	0	
AlF(g)	46.0	265	3
AlF_3(s)	84.0	1510	1.3
AlCl(g)	62.4	51	4
$AlCl_3$(s)	133.3	706	0.8
Al_2O_3(s)	102.0	1677	6
AlN(s)	41.0	318	2
Al_4C_3(s)	144.0	216	8
Andalusite	162.1	5.4†	2
Kyanite	162.1	7.9†	2
Sillimanite	162.1	2.5†	2
Mullite	426.0	−29 ‡	2
As(s)	74.9	0	
As_2O_3(s)	197.8	655	3
As_2O_5(s)	229.8	914	6
As_2S_3(s)	246.0	167	20
As_2S_5(s)	310.1	146	13
B(s)	10.8	0	
BN(s)	24.8	252	1.3
Ba(s)	137.3	0	
$BaCl_2$(s)	208.2	859	2
BaO(s)	153.3	554	19
BaO_2(s)	169.3	634	12
$BaSO_4$(s)	233.4	1473	17
$BaCO_3$(s)	197.3	1216	20
C (graphite)	12.0	0	
C (diamond)	12.0	−1.83	0.1
C (coke, etc.)	12.0	−13	6
CH_4(g)	16.0	74.9	0.4
CCl_4(l)	153.8	135	2
CCl_4(g)	153.8	103	2
$COCl_2$(g)	98.9	224	6
CO(g)	28.0	110.5	0.2
CO_2(g)	44.0	393.5	0.04
CS_2(l)	76.1	−89.6	1.0
CS_2(g)	76.1	119.7	0.8
COS(g)	60.1	138.4	0.8

† Al_2SiO_5 from oxides.
‡ $Al_6Si_2O_{13}$ from oxides.

Substance	Mol wt	$-\Delta H^\circ_{298}$ kJ/mol	Accuracy ± kJ
Ca(s)	40.1	0	
$CaF_2(s)$	78.1	1220	9
$CaCl_2(s)$	111.0	796	4
CaO(s)	56.1	634	1.7
CaS(s)	72.1	476	10
$CaSO_4(s)$	136.1	1434	4
$Ca_3P_2(s)$	182.2	506	25
$CaC_2(s)$	64.1	59	8
$CaCO_3(s)$	100.1	1207	3
CaSi(s)	68.2	151	8
$CaSi_2(s)$	96.3	151	13
$Ca_2Si(s)$	108.3	209	13
$CaSiO_3(s)$	116.2	89§	1.3
$Ca_2SiO_4(s)$	172.3	137§	6
$Ca_3SiO_5(s)$	228.3	113§	6
$CaAl_2(s)$	94.0	217	13
$CaAl_2O_4(s)$	158.0	15.1§	1.3
$Ca_3Al_2O_6(s)$	270.2	8.4§	3
Cd(s)	112.4	0	
$CdCl_2(s)$	183.3	391	1.3
CdO(s)	128.4	259	1.7
CdS(s)	144.5	149	2
$CdSO_4(s)$	208.5	929	2
Ce(s)	140.1	0	
$CeO_2(s)$	172.1	1090	1.7
Co(s)	58.9	0	
$CoCl_2(s)$	129.8	310	3
CoO(s)	74.9	239	2
$CoS_{0.89}$	87.5	95	4
CoS(s)	91.0	96	4
$Co_3S_4(s)$	305.0	359	25
$CoS_2(s)$	123.1	153	17
$CoSO_4(s)$	155.0	887	2
Cr(s)	52.0	0	
$CrCl_2(s)$	122.9	395	13
$CrCl_3(s)$	158.4	556	17
$Cr_2O_3(s)$	152.0	1130	11
$CrO_3(s)$	100.0	580	11
$Cr_4C(s)$	220.0	98	8
$Cr_7C_3(s)$	400.0	228	11
$Cr_3C_2(s)$	180.0	111	8
Cu(s)	63.5	0	
CuCl(s)	99.0	137	3

§ From oxides.

(continued over)

Substance	Mol wt	$-\Delta H^\circ_{298}$ kJ/mol	Accuracy $\pm$ kJ
$CuCl_2(s)$	134.4	218	8
$Cu_2O(s)$	143.1	167	3
$CuO(s)$	79.5	155	3
$Cu_2S(s)$	159.1	79.5	1.3
$CuS(s)$	95.6	52	4
$CuSO_4(s)$	159.6	771	1.3
$Fe(s)$	55.8	0	
$FeCl_2(s)$	126.8	342	0.8
$FeCl_3(s)$	162.2	400	0.8
$FeO(s)$	71.9	264¶	1.3
$Fe_3O_4(s)$	231.6	1117	4
$Fe_2O_3(s)$	159.7	821	3
$FeS(s)$	87.9	100	2
$FeS_2(s)$	120.0	172	8
$FeSO_4(s)$	151.9	920	13
$Fe_4N(s)$	237.4	15	4
$Fe_2N(s)$	125.7	4	8
$Fe_3P(s)$	198.5	164	8
$Fe_3C(s)$	179.6	-25	4
$FeCO_3(s)$	115.9	741	7
$FeSi(s)$	83.9	77	5
$Fe_2SiO_4(s)$	203.8	34††	6
$FeZn_2O_4(s)$	241.1	5§	4
$FeCr_2O_4(s)$	223.9	52§	2
$H_2(g)$	2.016	0	
$HF(g)$	20.0	273	1.5
$HCl(g)$	36.5	92.3	0.04
$H_2O(g)$	18.0	241.3	0.04
$H_2O(l)$	18.0	285.8	0.04
$H_2S(g)$	34.1	20.5	0.4
$Hg(l)$	200.6	0	
$HgCl(s)$	236.0	312	1.3
$HgCl_2(s)$	271.5	228	1.3
$HgO(s)$	216.6	91	0.8
$HgS(s)$	232.7	53	4
$K(s)$	39.1	0	
$KF(s)$	58.1	568.6	0.4
$KCl(s)$	74.6	436.7	0.8
$K_2O(s)$	94.2	363	3
$KOH(s)$	56.1	425	1.7
$K_2SO_4(s)$	174.3	1433	4
$K_2CO_3(s)$	138.2	1150	6
$K_2SiO_3(s)$	154.3	274§	13

§ From oxides.

¶ $Fe_{0.95}O$

†† From $0.1Fe + 2Fe_{0.95}O + SiO_2$

Substance	Mol Wt	$-\Delta H^{\circ}_{298}$ kJ/mol	Accuracy $\pm$ kJ
Mg(s)	24.3	0	
$MgCl_2$(s)	95.2	641.4	0.8
MgO(s)	40.3	601.2	0.8
MgS(s)	56.4	351	3
$MgSO_4$(s)	120.4	1285	13
Mg_2C_3(s)	84.7	-80	34
MgC_2(s)	48.3	-88	21
$MgCO_3$(s)	84.3	1112	15
Mg_2Si(s)	76.7	79	5
Mg_2SiO_4(s)	140.7	63§	2
$MgSiO_3$(s)	100.4	36§	3
Mn(s)	54.9	0	
$MnCl_2$(s)	125.8	482	2
MnO(s)	70.9	385	2
Mn_3O_4(s)	228.8	1387	4
Mn_2O_3(s)	157.9	957	5
MnO_2(s)	86.9	520	2
Mn_2O_7(s)	221.9	728	10
MnS(s)	87.0	213	2
MnS_2(s)	119.1	224	8
$MnSO_4$(s)	151.0	1065	4
Mn_3C(s)	176.8	15	13
$MnCO_3$(s)	115.0	894	5
$MnSiO_3$(s)	131.0	25§	1.7
Mn_2SiO_4(s)	202.0	49§	3
Mo(s)	95.9	0	
MoO_2(s)	127.9	588	1.7
MoO_3(s)	143.9	745	0.8
Mo_2S_3(s)	288.1	387	17
MoS_2(s)	160.1	275	7
Mo_2N(s)	205.9	70	2
Mo_2C(s)	203.9	-46	3
N_2(g)	28.0	0	
NH_3(g)	17.0	45.9	0.4
N_2O(g)	44.0	-82.0	0.4
NO(g)	30.0	-90.3	0.2
NO_2(g)	46.0	-33.1	0.8
Na(s)	23.0	0	
NaF(s)	42.0	575.4	0.8
NaCl(s)	58.5	412.6	0.4
Na_2O(s)	62.0	415	3
NaOH(s)	40.0	426	0.8
Na_2S(s)	78.1	374	13
Na_2SO_4(s)	142.1	1395	8

§ From oxides.

(*continued over*)

Substance	Mol Wt	$-\Delta H_{298}^{\circ}$ kJ/mol	Accuracy $\pm$ kJ
$Na_2CO_3(s)$	106.0	1131	1.7
$Na_2SiO_3(s)$	122.1	232§	15
$Na_2Si_2O_5(s,\alpha)$	182.2	231§	4
$Nb(s)$	92.9	0	
$NbO(s)$	108.9	420	13
$NbO_2(s)$	124.9	795	8
$Nb_2O_5(s)$	265.8	1900	5
$NbC(s)$	104.9	138	4
$Ni(s)$	58.7	0	
$NiCl(g)$	129.6	-230	38
$NiCl_2(s)$	129.6	305	2
$NiO(s)$	74.7	241	2
$Ni_3S_2(s)$	240.3	216	10
$NiS(s)$	90.8	94	4
$NiS_2(s)$	122.8	142	17
$NiSO_4(s)$	154.8	873	2
$Ni_3C(s)$	188.1	-38	6
$NiCO_3(s)$	118.7	680	13
$Ni(CO)_4(g)$	170.8	159‡‡	4
$P(s, white)$	31.0	0	
$P(s, red)$	31.0	17.3	0.6
$PCl_3(g)$	137.3	291	4
$PCl_3(l)$	137.3	321	4
$PCl_5(g)$	208.2	367	10
$PCl_5(s)$	208.2	444	13
$P_2O_5(s)$	141.9	1492	10
$Pb(s)$	207.2	0	
$PbCl_2(s)$	278.1	359.4	0.8
$PbO(s)$	223.2	219.4	0.8
$Pb_3O_4(s)$	685.6	719	6
$PbO_2(s)$	239.2	274	3
$PbS(s)$	239.3	98	2
$PbSO_4(s)$	303.3	920	1.7
$PbCO_3(s)$	267.2	700	5
$Pb_2SiO_4(s)$	506.5	17§	8
$PbSiO_3(s)$	283.3	18§	2
$S(s, rh)$	32.1	0	
$S(s, monocl.)$	32.1	-0.33	0.04
$S(g)$	32.1	-277.0	0.4
$S_2(g)$	64.1	-128.7	0.4

§ From oxides.
‡‡ From Ni + 4CO

Substance	Mol wt	$-\Delta H^{\circ}_{298}$ kJ/mol	Accuracy $\pm$ kJ
$SCl_4(l)$	173.9	60	13
$SO_2(g)$	64.1	296.8	0.2
$SO_3(g)$	80.1	396	3
$Sb(s)$	121.8	0	
$SbCl_3(s)$	228.1	382	1.6
$SbCl_5(l)$	299.0	394	10
$Sb_2O_3(s)$	291.5	709	3
$Sb_2S_3(s)$	339.7	205	21
$Sb_2(SO_4)_3(s)$	531.7	2402	20
$Si(s)$	28.1	0	
$SiF_4(g)$	104.1	1611	13
$SiCl_4(l)$	169.9	663	1.7
$SiO(g)$	44.1	98	10
$SiO_2(s)$§§	60.1	910	3
$SiO_2(s)$¶¶	60.1	906	4
$SiS_2(s)$	92.2	213	20
$SiS(g)$	60.2	-116	12
$SiC(s)$	40.1	69	6
$Sn(s, white)$	118.7	0	
$Sn(s, gray)$	118.7	1.9	0.2
$SnCl_2(s)$	189.6	331	6
$SnCl_4(l)$	260.5	529	15
$SnO_2(s)$	150.7	581	2
$SnS(s)$	150.8	108	3
$SnS_2(s)$	182.8	154	20
$Ti(s)$	47.9	0	
$TiCl_2(s)$	118.8	515	17
$TiCl_4(g)$	189.7	763	4
$TiCl_4(l)$	189.7	804	4
$TiO(s)$	63.9	543	20
$Ti_2O_3(s)$	143.8	1521	5
$Ti_3O_5(s)$	223.7	2459	3
$TiO_2(s)$	79.9	944	1.7
$TiC(s)$	59.9	184	6
$V(s)$	50.9	0	
$V_2O_3(s)$	149.9	1219	6
$V_2O_5(s)$	181.9	1551	6
$Zn(s)$	65.4	0	
$ZnCl_2(s)$	136.3	415.1	0.5

§§ α quartz
¶¶ β cristobalite stable > 1470°C

(*continued over*)

Substance	Mol wt	$-\Delta H^{\circ}_{298}$ kJ/mol	Accuracy $\pm$ kJ
ZnO(s)	81.4	350.6	0.4
ZnS(s)	97.4	205	3
ZnSO$_4$(s)	161.4	981	1.7
ZnCO$_3$(s)	125.4	818	1.3
Zn$_2$SiO$_4$(s)	222.8	33§	2
Zr(s)	91.2	0	
ZrO$_2$(s)	123.2	1101	2
ZrN(s)	105.2	368	3
ZrC(s)	103.2	185	6

§ From oxides.

ENTHALPY INCREMENTS ABOVE 25°C

This part gives the increase in enthalpy, $H_T - H_{298}$, when the substance is heated from 25°C to some higher temperature. All values apply to one atmosphere pressure. As mentioned in App. A most enthalpies vary only insignificantly with changing pressure, provided this does not involve phase changes, and the curves may, therefore, be used also at pressures different from one atmosphere. All boiling points change with pressure, however. In order to obtain the enthalpy increment for some vapor at low pressure the curve for the vapor may be extrapolated below its normal boiling point.

For some vapors, H_2O, CS_2, S_2, etc., the enthalpy increment is referred to the gas at 25°C, even though this is not stable at that temperature. In order to refer the enthalpy increment to the condensed form, the enthalpy of evaporation at 25°C, as given in part A, must be added. The enthalpy increment for equilibrium sulfur vapor is also given under "elements" and now referred to solid sulfur at 25°C. Since, on increased pressure, sulfur vapor undergoes polymerization, accompanied by a decrease in enthalpy, that particular curve applies only to atmospheric pressure.

The curves do not allow an accurate calculation of the enthalpy increment for molten alloys and slags, unless these correspond to some specific compound. Approximate values may be obtained by resolving the alloy or slag composition into combinations of nearby compounds, and adding the values for each compound in their relative amount. If these compounds are solid at a temperature when the alloy or the slag is liquid, their enthalpies of fusion must be added. For more accurate calculations enthalpies of solution are required, but these are regarded as outside the scope of the present tabulation.

This part is converted to joules from data given by:

Kelley, K. K.: *Bureau of Mines Bulletin*, **584**, 1960.
Wicks, C. E., and F. E. Block: *Bureau of Mines Bulletin*, **605**, 1963.
JANAF Thermochemical Tables, 1965–68.

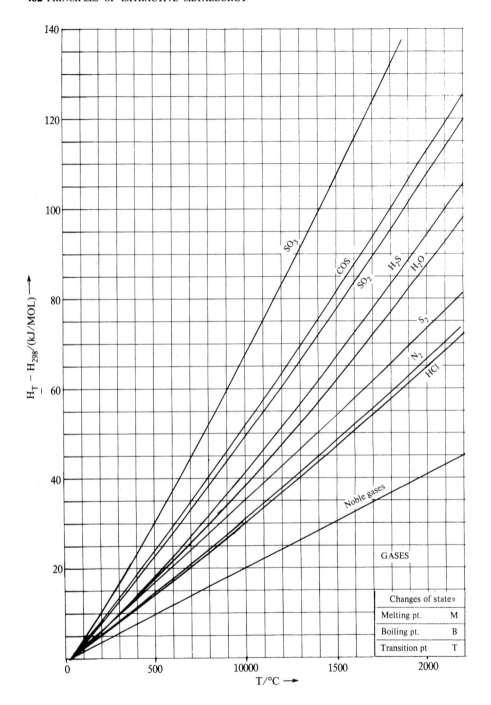

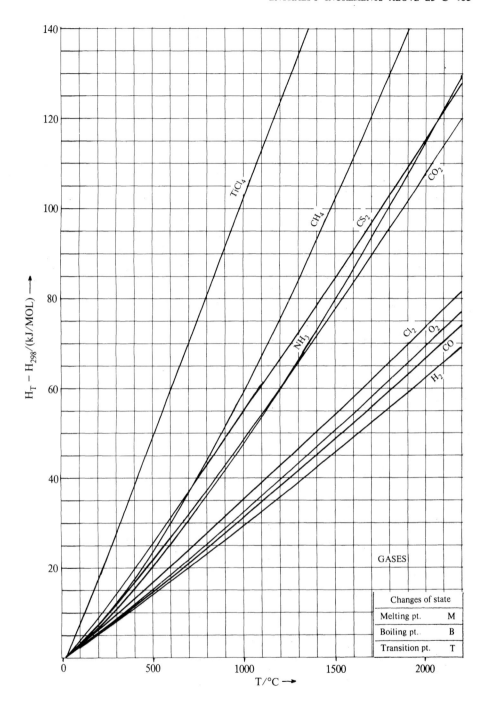

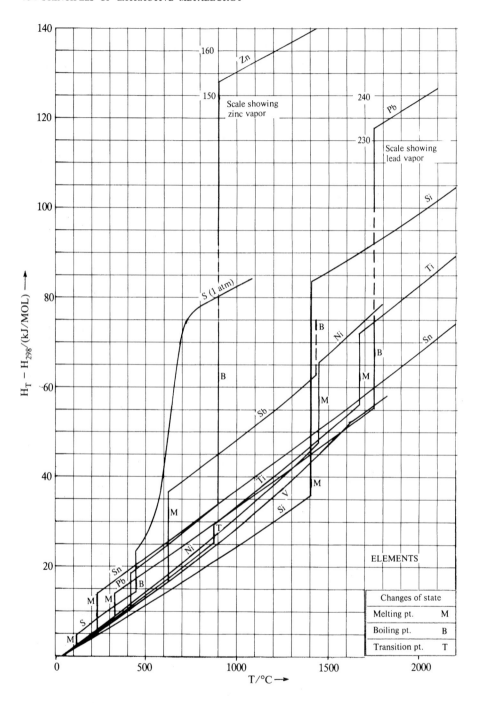

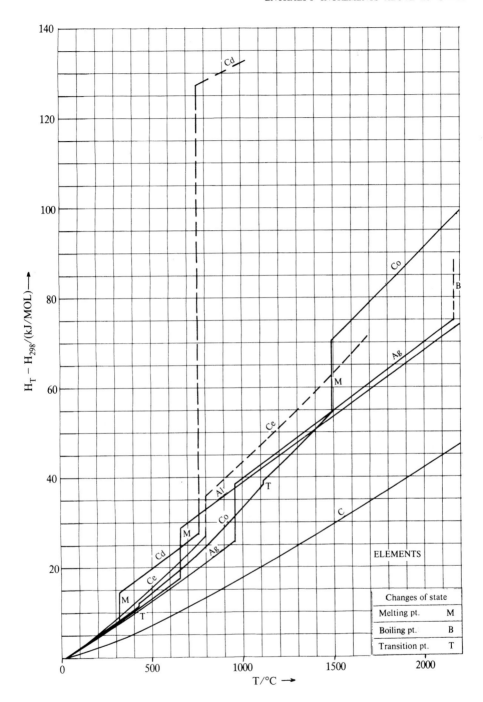

ELEMENTS

Changes of state	
Melting pt.	M
Boiling pt.	B
Transition pt.	T

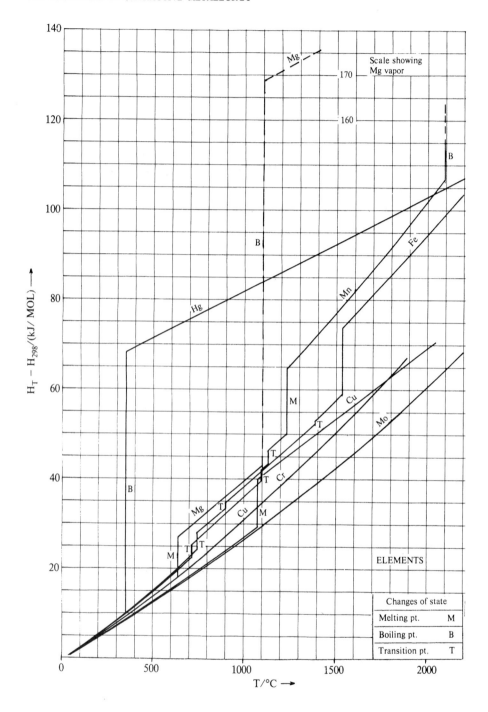

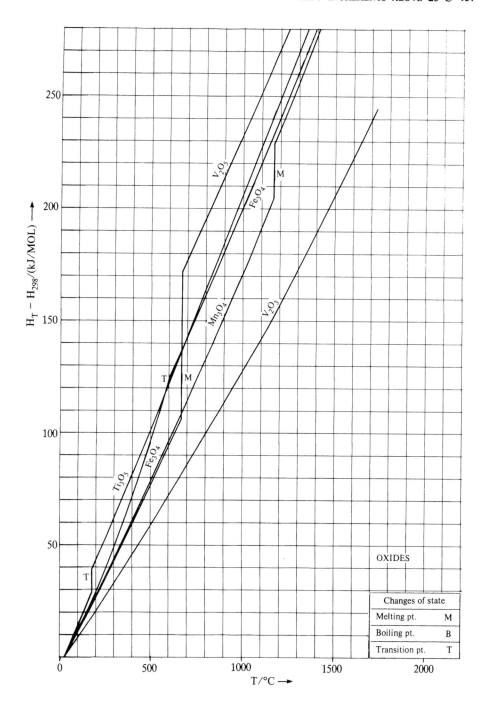

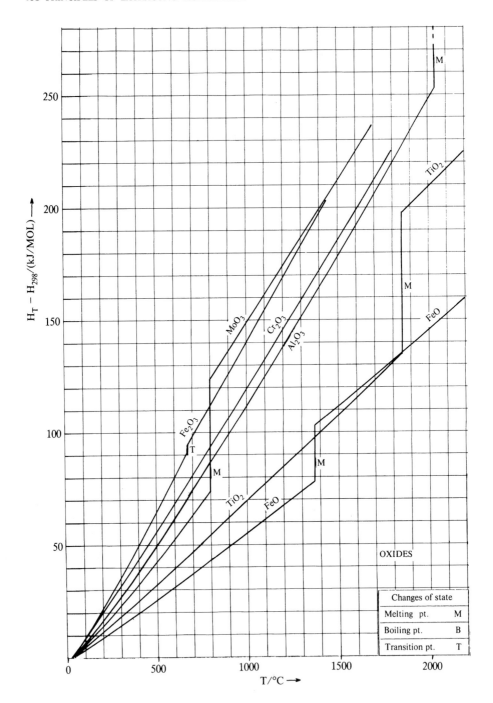

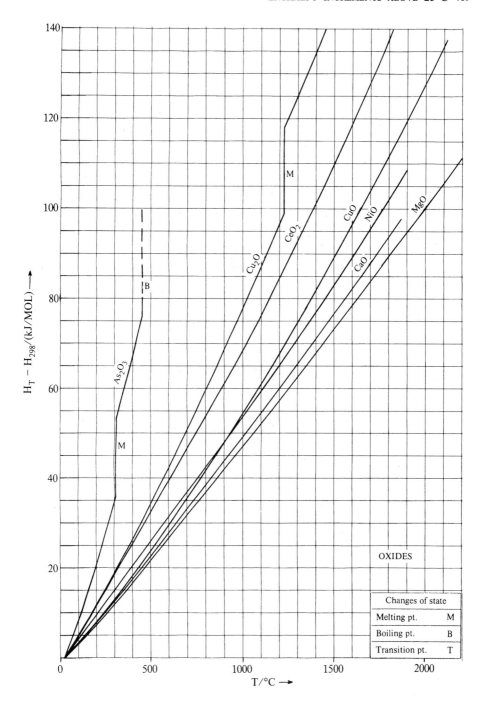

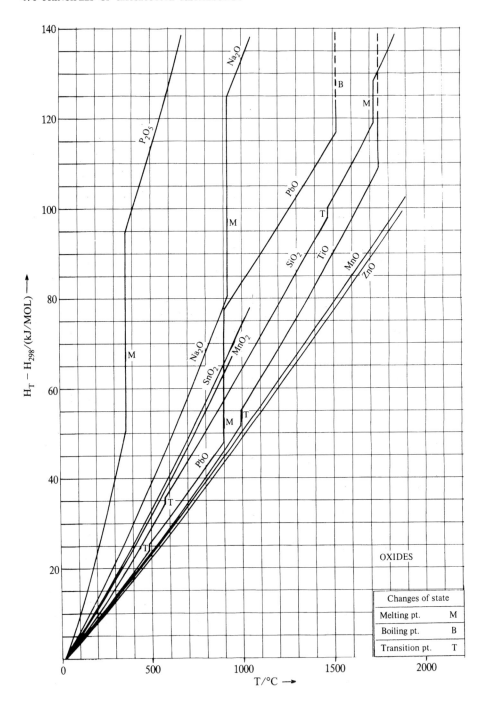

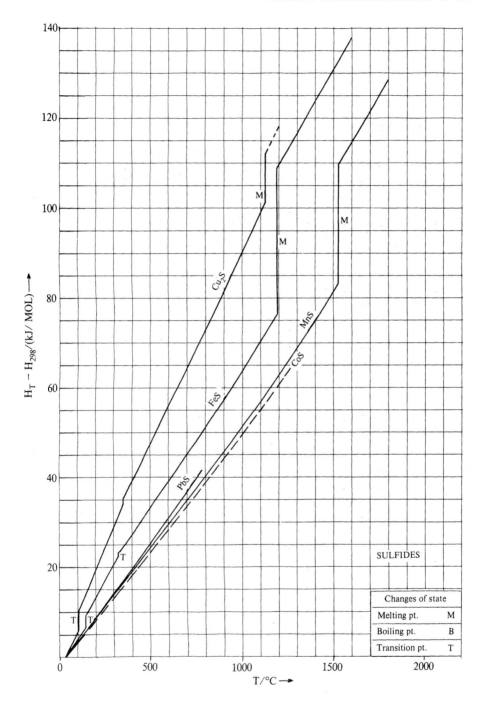

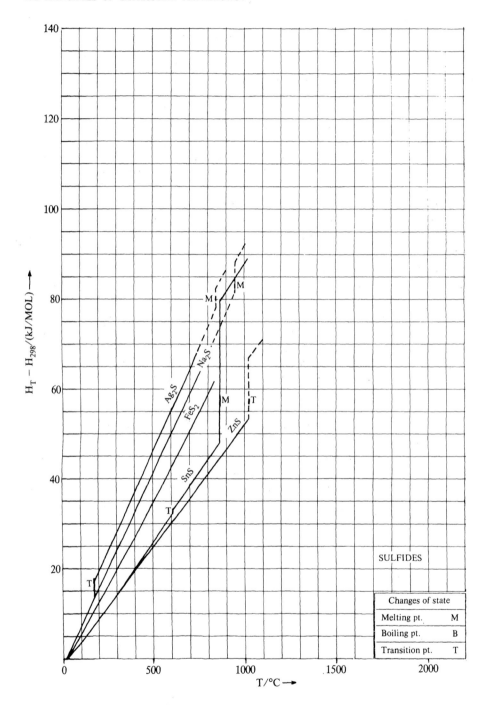

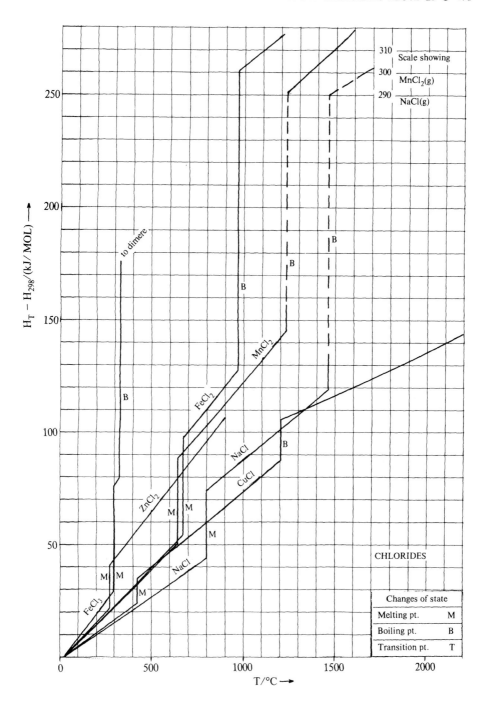

Scale showing
MnCl$_2$(g)
NaCl(g)

$H_T - H_{298}/(kJ/MOL) \longrightarrow$

to dimere

MnCl$_2$

FeCl$_2$

ZnCl$_2$

NaCl

CuCl

FeCl$_3$

NaCl

CHLORIDES

Changes of state	
Melting pt.	M
Boiling pt.	B
Transition pt.	T

$T/°C \longrightarrow$

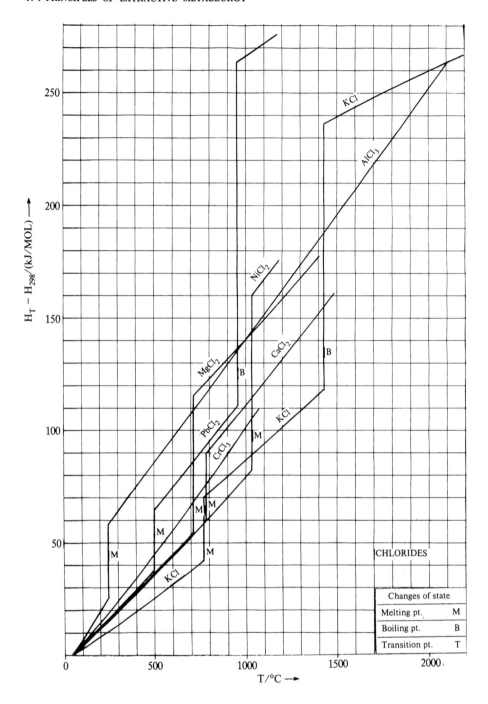

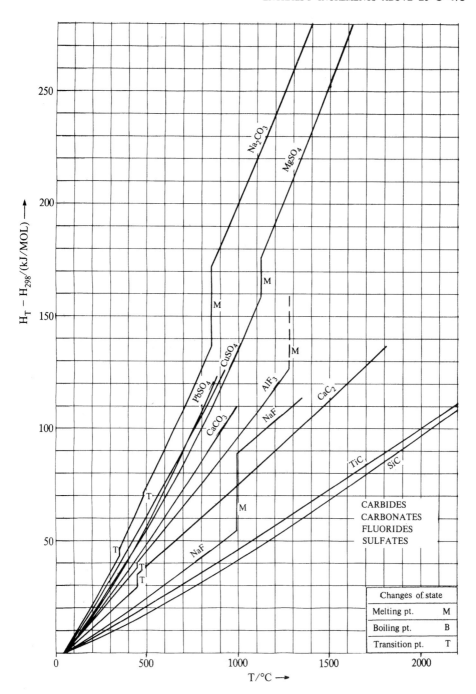

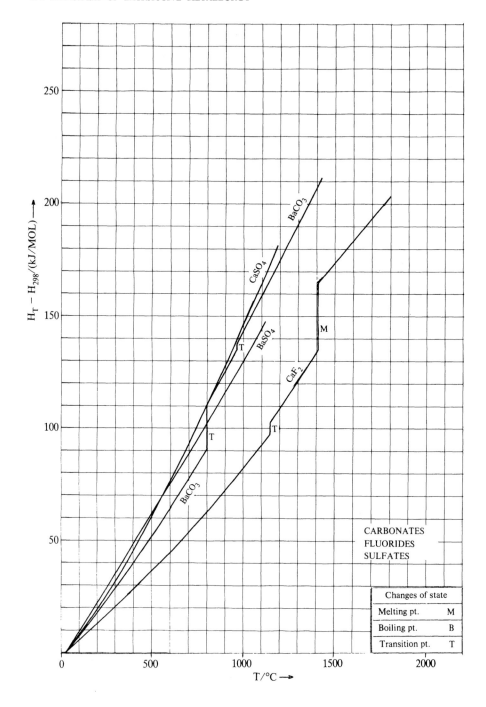

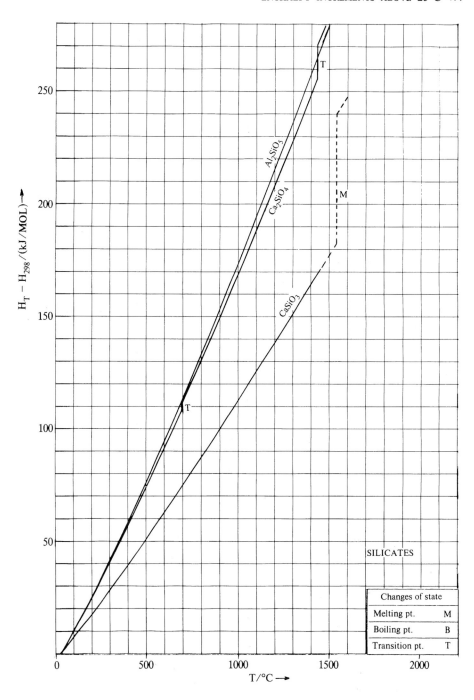

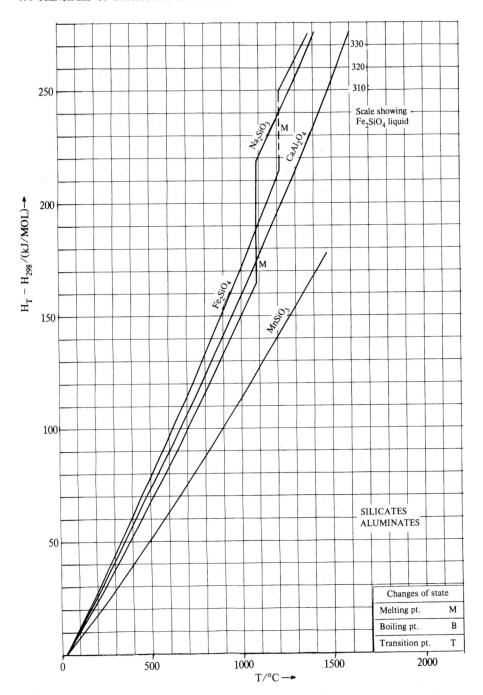

APPENDIX

C

STANDARD GIBBS ENERGIES OF FORMATION AND EVAPORATION

The standard Gibbs energy of formation, $\Delta G°$, is given for a number of compounds of metallurgical interest, oxides, sulfides, chlorides, etc. For each group the values refer to one mole of the common component, viz., $2Ag_2O$, $\frac{2}{5}V_2O_5$, SiO_2, etc. The standard states are the stable modifications of the condensed phases and gases at unit fugacity. For the sulfides the standard state is S_2 gas, even though this cannot exist at unit fugacity below 900°C. Under "sulfur compounds" special curves are given for the various polymers of sulfur vapor as well as for liquid and solid sulfur, all relative to S_2 gas.

In most cases the values refer to formation from the elements, for example $Ti + O_2 = TiO_2$, even though the compound and the element may not always coexist. For the iron oxides the values refer to the individual oxidation steps, $2Fe + O_2 = 2\text{"FeO"}$, $6\text{"FeO"} + O_2 = 2Fe_3O_4$, etc., and are strictly stoichiometric with respect to oxygen only.

For complex compounds such as sulfates, carbonates, silicates, etc., the values refer to formation from the component oxides. These values may be combined with corresponding values for these oxides to give the Gibbs energy of formation from the elements.

For some gaseous compounds, in particular chlorides, the vapor at unit fugacity consists of a mixture of various polymers, for example $CuCl(g)$ and $Cu_3Cl_3(g)$. When possible, the standard Gibbs energy of formation is given for each polymer. In cases where one polymer dominates under normal conditions, only values for this are given, even though, under extreme conditions the vapor may consist of other polymers. In such cases the curves should be used with care.

The values for evaporation are derived from the saturation vapor pressure, $p°$, by the equation: $\Delta G° = -RT \ln p°$, and no attempt has been made to separate possible polymers in the gas phase.

In order to save space it has sometimes been necessary to shift the curves relative to the scale. This is indicated on the curve. Thus $2Ag_2O - 150$ kJ means that the read value should be made 150 kJ more positive in order to give the true value. Likewise $2CaS + 400$ kJ means that the read value should be made 400 kJ more negative.

Most values are converted to joules from data drawn from the following compilations:

Kubaschewski, O. et al.: "Metallurgical Thermochemistry," Pergamon Press, London, 1967 and 1979.

JANAF Thermochemical Tables, 1965–68.

Wicks, C. E., and F. E. Block: "Thermodynamic Properties of 65 Elements, their Oxides, Halides, Carbides, and Nitrides," *Bureau of Mines Bulletin*, **605**, 1963.

Elliott, J. F., and M. Gleiser: "Thermochemistry for Steelmaking," Addison-Wesley Publ. Co. Inc., Reading, Mass., 1960.

Richardson, F. D., and J. H. E. Jeffes: The Thermodynamics of Substances of Interest in Iron and Steelmaking, *J. Iron Steel Inst.*, **160**: 261–270 (1948), **166**: 213–235 (1950), **171**: 165–175 (1952).

Larsen, H. R., and J. F. Elliott: The Standard Free Energy of Formation of Certain Sulfides of Transition Elements and Zinc, *Trans. AIME*, **239**: 1713–1720 (1967).

Villa, H.: Thermodynamic Data of the Metallic Chlorides, *J. Soc. Chem. Ind.* Suppl. Iss. No. 1: S9–S17 (1950).

Kellogg, H. H.: A Critical Review of Sulfation Equilibria, *Trans. AIME*, **230**: 1622–1634 (1964).

The reader is referred to the above literature for substances that are not included. Several of the curves are constructed from data drawn from several sources as well as from unpublished data. For that reason it has not been possible to give references to each curve. Also an attempt has been made to estimate the possible accuracy of the curves.

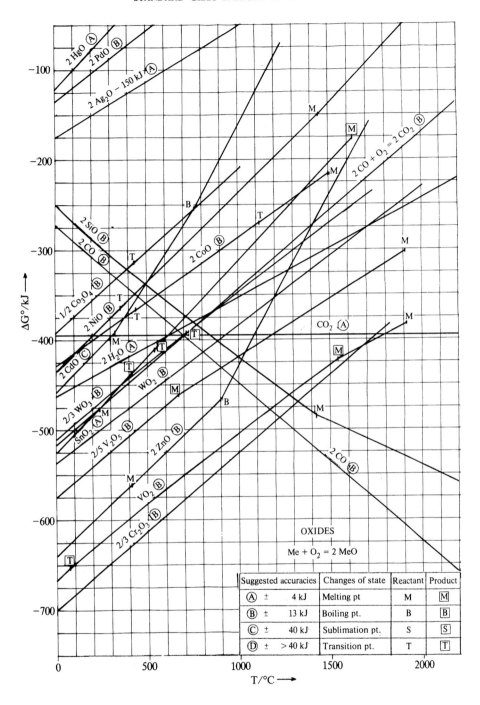

OXIDES

Me + O$_2$ = 2 MeO

Suggested accuracies		Changes of state	Reactant	Product
Ⓐ ±	4 kJ	Melting pt	M	☒
Ⓑ ±	13 kJ	Boiling pt.	B	☒
Ⓒ ±	40 kJ	Sublimation pt.	S	☒
Ⓓ ±	> 40 kJ	Transition pt.	T	☒

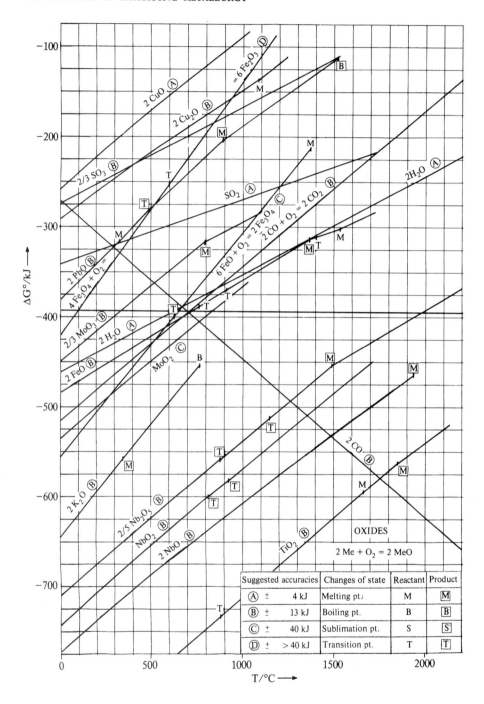

OXIDES

$2 Me + O_2 = 2 MeO$

Suggested accuracies		Changes of state	Reactant	Product
Ⓐ ±	4 kJ	Melting pt.	M	Ⓜ
Ⓑ ±	13 kJ	Boiling pt.	B	Ⓑ
Ⓒ ±	40 kJ	Sublimation pt.	S	Ⓢ
Ⓓ ±	> 40 kJ	Transition pt.	T	Ⓣ

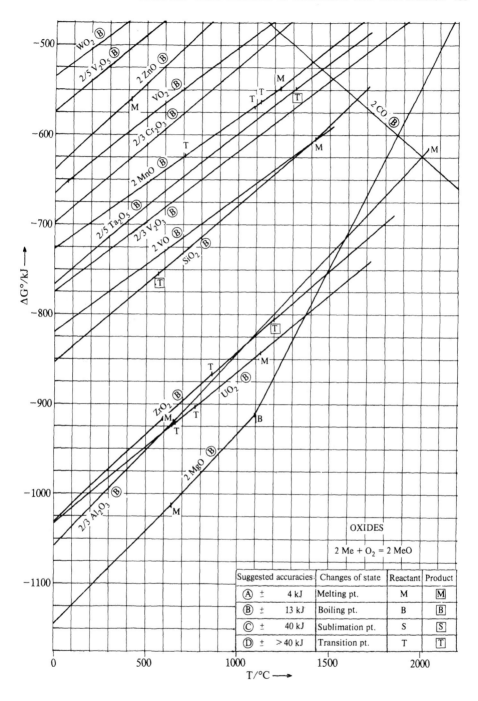

OXIDES

$$2 Me + O_2 = 2 MeO$$

Suggested accuracies:	Changes of state	Reactant	Product
Ⓐ ± 4 kJ	Melting pt.	M	M̄
Ⓑ ± 13 kJ	Boiling pt.	B	B̄
Ⓒ ± 40 kJ	Sublimation pt.	S	S̄
Ⓓ ± >40 kJ	Transition pt.	T	T̄

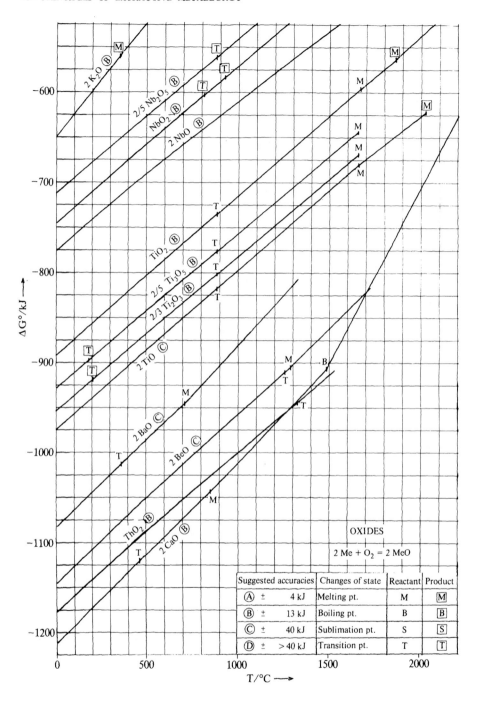

OXIDES

$$2\,Me + O_2 = 2\,MeO$$

Suggested accuracies		Changes of state	Reactant	Product
Ⓐ ±	4 kJ	Melting pt.	M	M̄
Ⓑ ±	13 kJ	Boiling pt.	B	B̄
Ⓒ ±	40 kJ	Sublimation pt.	S	S̄
Ⓓ ±	> 40 kJ	Transition pt.	T	T̄

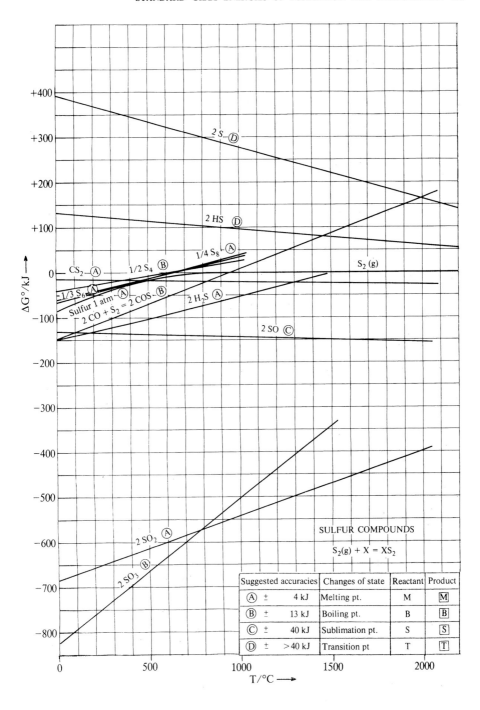

$\Delta G^\circ / kJ \longrightarrow$

$2 S - D$

$2 HS \quad D$

$1/4 S_8 - A$

$CS_2 - A$ $\quad 1/2 S_4 \quad B$ $\quad\quad S_2 (g)$

$1/3 S_6 \quad A$

Sulfur 1 atm - A

$2 CO + S_2 = 2 COS - B$ $\quad 2 H_2S \quad A$

$2 SO \quad C$

$2 SO_2 \quad A$

B

$2 SO_3$

SULFUR COMPOUNDS

$S_2(g) + X = XS_2$

Suggested accuracies		Changes of state	Reactant	Product
A $\pm$	4 kJ	Melting pt.	M	M
B $\pm$	13 kJ	Boiling pt.	B	B
C $\pm$	40 kJ	Sublimation pt.	S	S
D $\pm$	>40 kJ	Transition pt	T	T

$T/°C \longrightarrow$

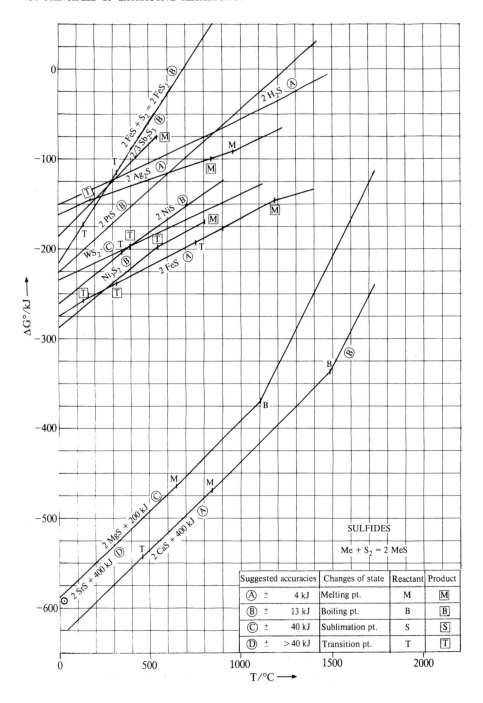

SULFIDES

$Me + S_2 = 2\,MeS$

Suggested accuracies		Changes of state	Reactant	Product
Ⓐ ±	4 kJ	Melting pt.	M	Ⓜ
Ⓑ ±	13 kJ	Boiling pt.	B	Ⓑ
Ⓒ ±	40 kJ	Sublimation pt.	S	Ⓢ
Ⓓ ±	> 40 kJ	Transition pt.	T	Ⓣ

$\Delta G^{\circ}/kJ \longrightarrow$

$T/^{\circ}C \longrightarrow$

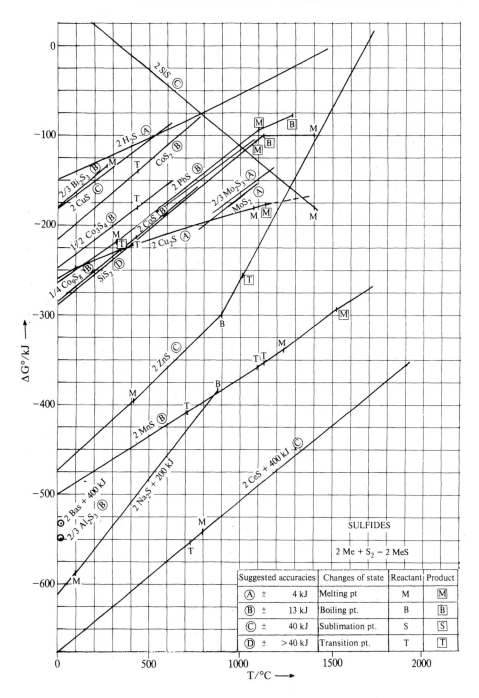

SULFIDES

$2 Me + S_2 = 2 MeS$

Suggested accuracies		Changes of state	Reactant	Product
Ⓐ ±	4 kJ	Melting pt	M	Ⓜ
Ⓑ ±	13 kJ	Boiling pt.	B	Ⓑ
Ⓒ ±	40 kJ	Sublimation pt.	S	Ⓢ
Ⓓ ±	>40 kJ	Transition pt.	T	Ⓣ

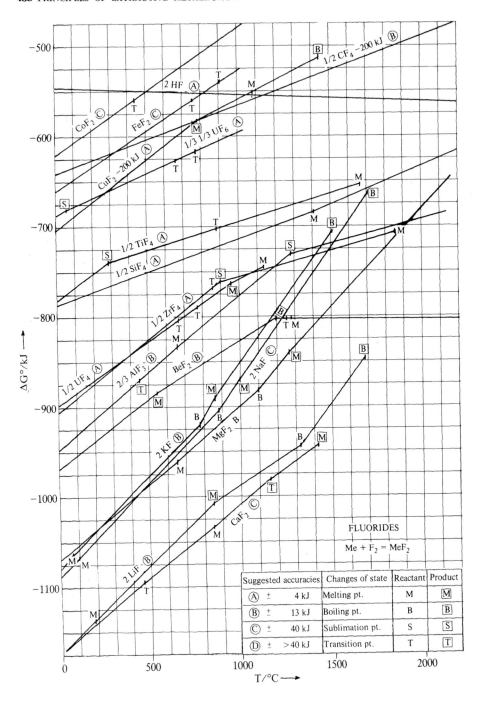

FLUORIDES

$$Me + F_2 = MeF_2$$

Suggested accuracies		Changes of state	Reactant	Product
Ⓐ ±	4 kJ	Melting pt.	M	Ⓜ
Ⓑ ±	13 kJ	Boiling pt.	B	Ⓑ
Ⓒ ±	40 kJ	Sublimation pt.	S	Ⓢ
Ⓓ ±	>40 kJ	Transition pt.	T	Ⓣ

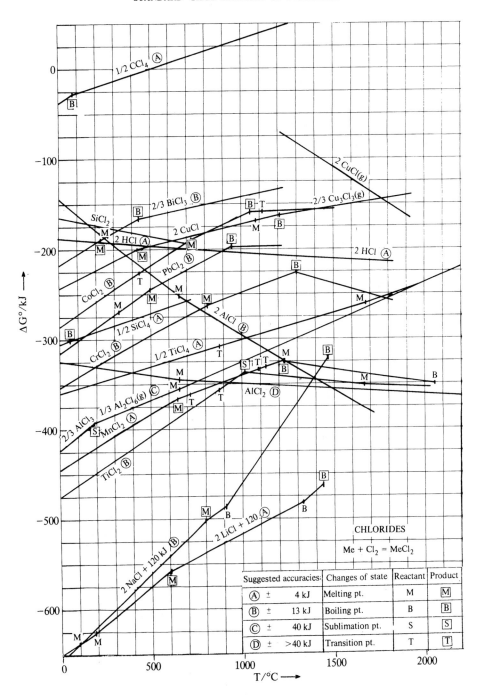

$\Delta G°/kJ \longrightarrow$

T/°C ⟶

CHLORIDES

$Me + Cl_2 = MeCl_2$

Suggested accuracies:	Changes of state	Reactant	Product
Ⓐ ± 4 kJ	Melting pt.	M	Ⓜ
Ⓑ ± 13 kJ	Boiling pt.	B	Ⓑ
Ⓒ ± 40 kJ	Sublimation pt.	S	Ⓢ
Ⓓ ± >40 kJ	Transition pt.	T	Ⓣ

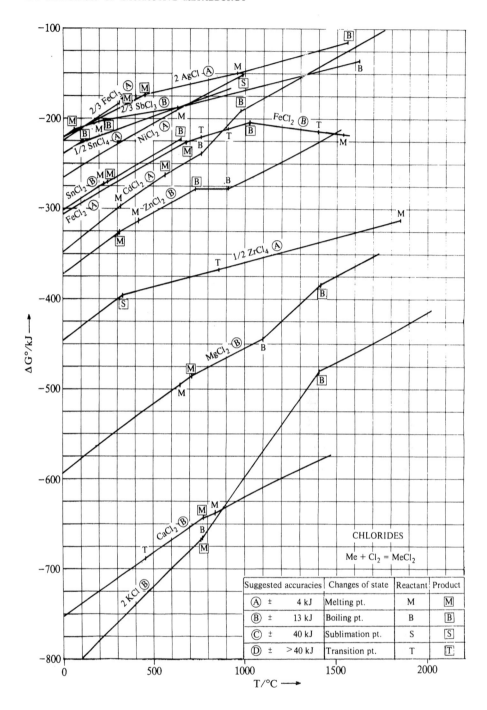

CHLORIDES

Me + Cl₂ = MeCl₂

Suggested accuracies		Changes of state	Reactant	Product
Ⓐ ±	4 kJ	Melting pt.	M	Ⓜ
Ⓑ ±	13 kJ	Boiling pt.	B	Ⓑ
Ⓒ ±	40 kJ	Sublimation pt.	S	Ⓢ
Ⓓ ±	>40 kJ	Transition pt.	T	Ⓣ

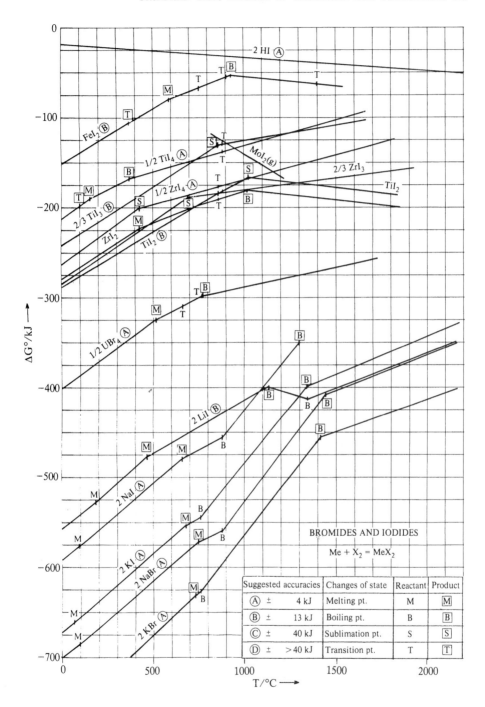

BROMIDES AND IODIDES

$Me + X_2 = MeX_2$

Suggested accuracies	Changes of state	Reactant	Product
Ⓐ ± 4 kJ	Melting pt.	M	Ⓜ
Ⓑ ± 13 kJ	Boiling pt.	B	Ⓑ
Ⓒ ± 40 kJ	Sublimation pt.	S	Ⓢ
Ⓓ ± >40 kJ	Transition pt.	T	Ⓣ

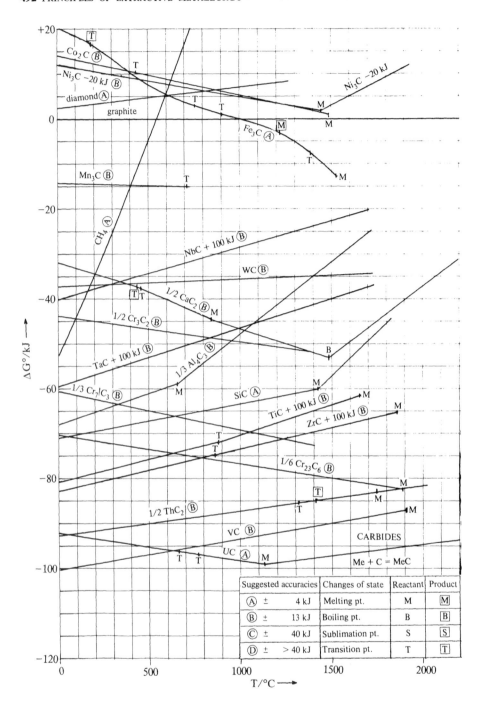

CARBIDES

Me + C = MeC

Suggested accuracies	Changes of state	Reactant	Product
Ⓐ ± 4 kJ	Melting pt.	M	Ⓜ
Ⓑ ± 13 kJ	Boiling pt.	B	Ⓑ
Ⓒ ± 40 kJ	Sublimation pt.	S	Ⓢ
Ⓓ ± > 40 kJ	Transition pt.	T	Ⓣ

$\Delta G°/kJ \longrightarrow$

$T/°C \longrightarrow$

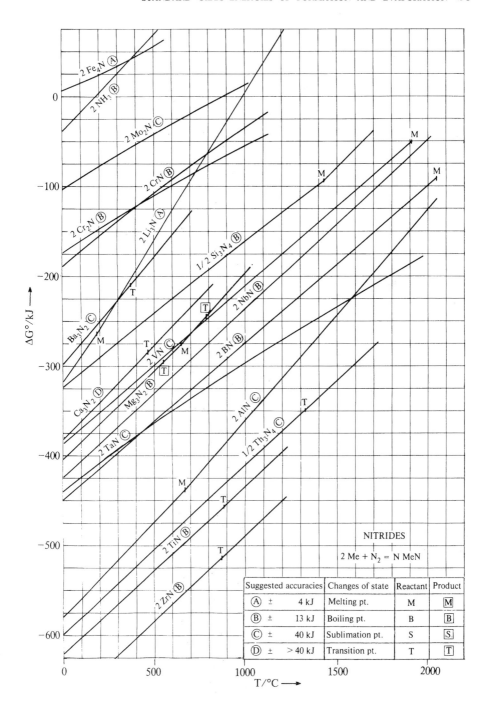

NITRIDES

$2\,Me + N_2 = N\,MeN$

Suggested accuracies		Changes of state	Reactant	Product
Ⓐ ±	4 kJ	Melting pt.	M	M̄
Ⓑ ±	13 kJ	Boiling pt.	B	B̄
Ⓒ ±	40 kJ	Sublimation pt.	S	S̄
Ⓓ ±	> 40 kJ	Transition pt.	T	T̄

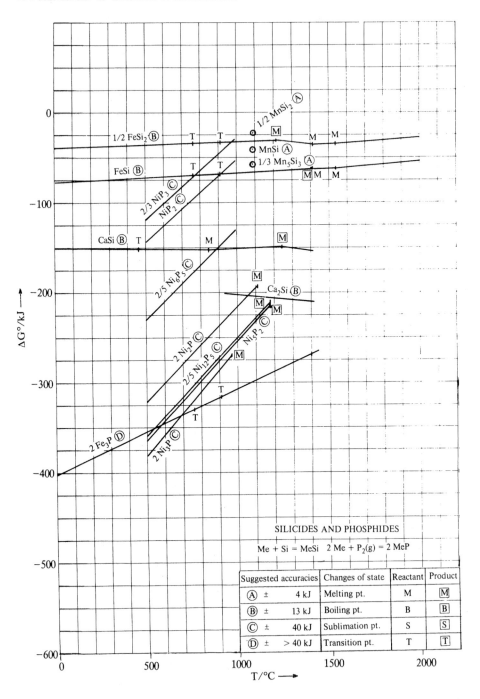

SILICIDES AND PHOSPHIDES

Me + Si = MeSi   2 Me + P₂(g) = 2 MeP

Suggested accuracies	Changes of state	Reactant	Product
Ⓐ ± 4 kJ	Melting pt.	M	Ⓜ
Ⓑ ± 13 kJ	Boiling pt.	B	Ⓑ
Ⓒ ± 40 kJ	Sublimation pt.	S	Ⓢ
Ⓓ ± > 40 kJ	Transition pt.	T	Ⓣ

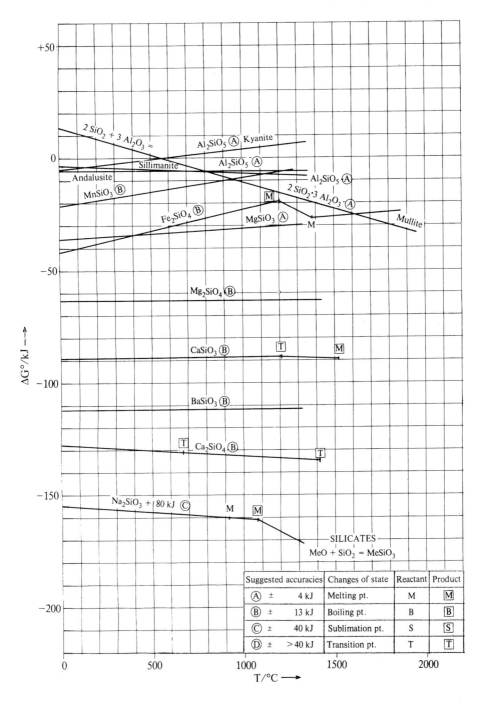

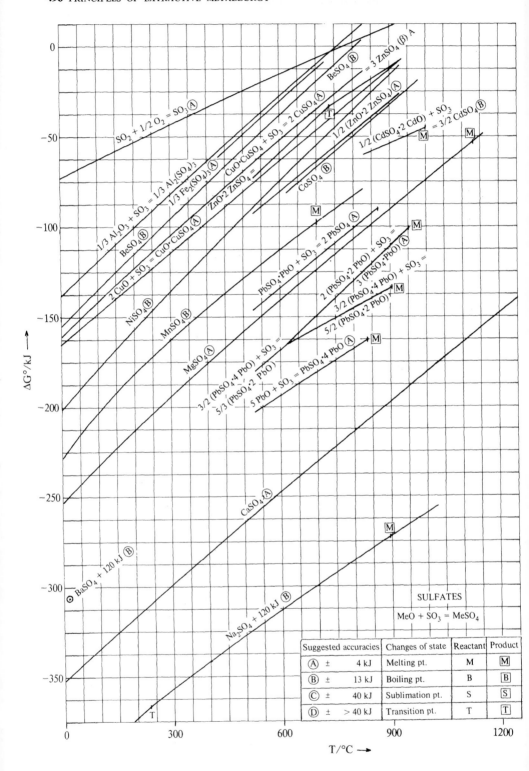

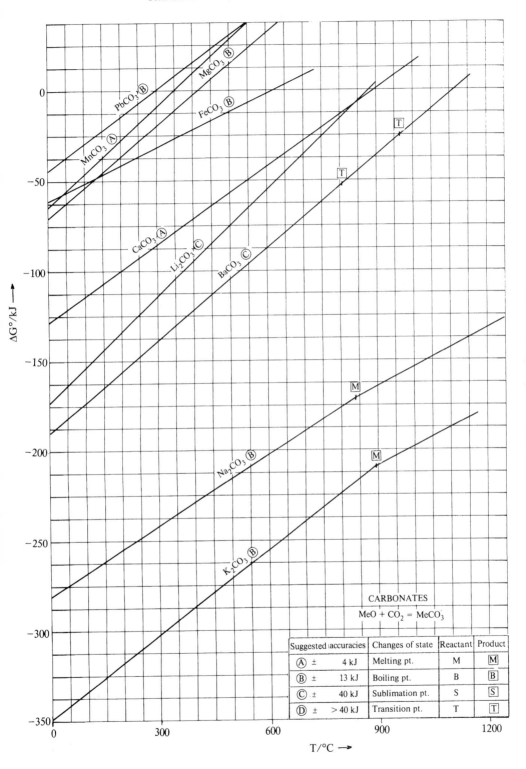

CARBONATES

$MeO + CO_2 = MeCO_3$

Suggested accuracies	Changes of state	Reactant	Product
Ⓐ ± 4 kJ	Melting pt.	M	M
Ⓑ ± 13 kJ	Boiling pt.	B	B
Ⓒ ± 40 kJ	Sublimation pt.	S	S
Ⓓ ± > 40 kJ	Transition pt.	T	T

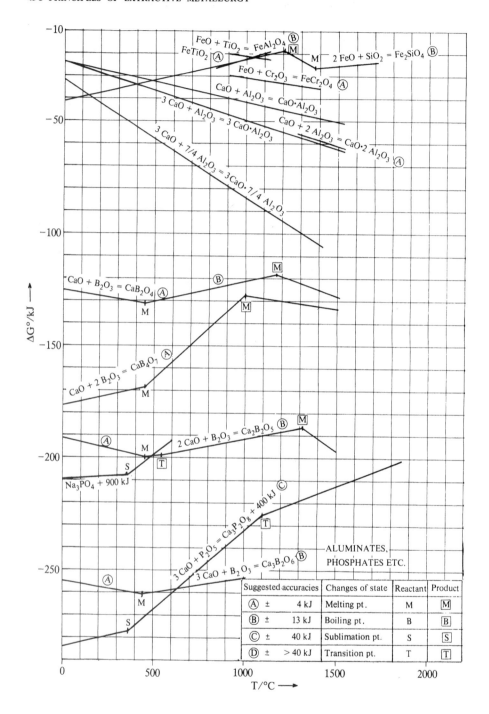

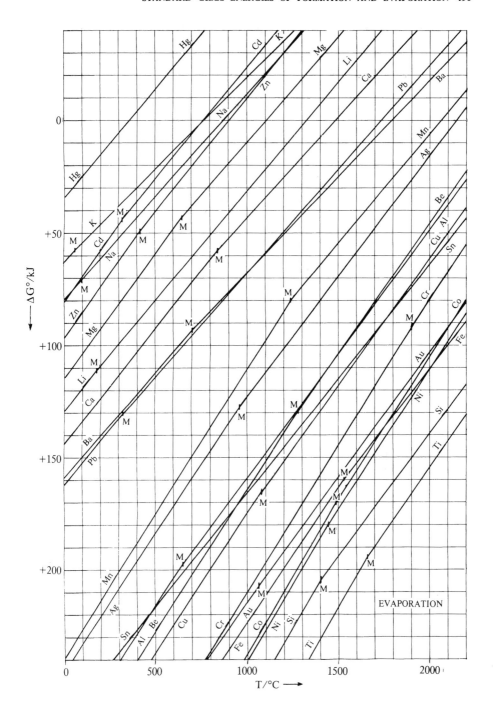

INDEX